# RADIO-FREQUENCY POWER IN PLASMAS
## EIGHTH TOPICAL CONFERENCE

# AIP
# CONFERENCE
# PROCEEDINGS 190

RITA G. LERNER
SERIES EDITOR

# RADIO-FREQUENCY
# POWER IN PLASMAS
## EIGHTH TOPICAL CONFERENCE
### IRVINE, CA 1989

EDITOR:

**ROGER McWILLIAMS**
DEPARTMENT OF PHYSICS
UNIVERSITY OF CALIFORNIA, IRVINE

**American Institute of Physics**                    **New York**

L.C. Catalog Card No. 89-045805
ISBN 0-88318-397-8
DOE CONF 8905120

Printed in the United States of America.

# CONTENTS

## CHAPTER 3
## ION CYCLOTRON RANGE OF FREQUENCIES

CHAPTER 4

GENERAL RF/PLASMA INTERACTIONS

# PREFACE

The Eighth Topical Conference on Radio Frequency Power in Plasmas was held in Irvine, California from 1–3 May 1989. The conference acted as an international forum for scientists with major interests in the interaction of radio frequency waves with plasmas. Following the trend of the last decade in plasma physics, the number of papers presented was up 10% compared to the last conference; a total of 125 papers, including 11 invited papers, were given.

The Organizing Committee, chaired by R. McWilliams, consisted of: Dr. Paul M. Bellan, Dr. Stefano Bernabei, Dr. Francesco De Marco, Dr. Claude Gormezano, Dr. Robert W. Harvey, Dr. Noah Hershkowitz, Dr. Stanley C. Luckhardt, Dr. D. Moreau, Dr. William M. Nevins, Dr. Jean-Marie Noterdaeme, Dr. Elio Sindoni, Dr. Gary Swanson, and Dr. Tetsuo Watari.

Mrs. Shirley Kaaz was the real organizer of this conference. The conference was fortunate to receive her services.

Roger McWilliams
Chairman

May 3, 1989

CHAPTER 1

ELECTRON CYCLOTRON RANGE OF FREQUENCIES

# ON ELECTRON CYCLOTRON RESONANCE HEATING
## AND CURRENT DRIVE
## IN THE W VII-AS STELLARATOR

V. Erckmann, WVII-AS Team *
Max-Planck-Institut für Plasmaphysik
Association Euratom, 8046 Garching, Fed.Rep.Germany
W.Kasparek, G.A.Müller, P.G.Schüller, M.Thumm
Institut für Plasmaforschung, Universität Stuttgart,
7000 Stuttgart 80, Fed.Rep.Germany

## ABSTRACT

The design of the Advanced Stellarator Wendelstein WVII-AS is based on optimization of the vacuum magnetic field configuration. Essential features of the underlying concept are the reduction of the Pfirsch-Schlüter currents and of the neoclassical heat transport losses. This optimization in general leads to a non-axisymmetric complex magnetic field configuration, which was realized by a set of modular twisted coils. Plasma operation started in October 1988. In the first experimental campaign the experiments concentrated on ECRH alone, combination with NBI is foreseen as the next step.

A 70 GHz ECRH system (0.8 MW for 3 s, 0.2 MW for 0.1 s) is installed for plasma build up, heating and current drive. A sophisticated microwave transmission and launching system gives full access to arbitrary poloidal and toroidal launching angles of the microwaves and consequently offers a wide experimental flexibility for heating and current drive investigations. First experimental results on plasma confinement, electron heating and current drive obtained with various kinds of microwave launching at 2.5 T and 1.25 T are discussed.

*W7AS-Team: E. Anabitarte[1], E. Ascasibar[1], S. Besshou[2], R. Brakel, R. Burhenn, G. Cattanei, A. Dodhy, D. Dorst, A. Elsner, K. Engelhardt, V. Erckmann, D. Evans, U. Gasparino, G. Grieger, P. Grigull, H. Hacker, H. Hailer[6], H.J. Hartfuss, H. Jäckel, R. Jaenicke, S. Jiang[3], J. Junker, M. Kick, H. Kroiss, G. Kuehner, I. Lakicevic, H. Maassberg, C. Mahn, R. Martin[1], G. Müller, H. Münch, A. Navarro[1], M. Ochando[1], W. Ohlendorf, M. Petrov[4], F. Rau, H. Renner, H. Ringler, J. Saffert, J. Sanchez[1], J. Sapper, A.V. Saposhnikov[5], F. Sardei, I.S. Sbitnikova[5], I. Schoenewolf, K. Schwoerer[6], F. Tabares[1], M. Tutter, A. Weller, H. Wobig, E. Würsching, M. Zippe.

1) Guest from CIEMAT, Madrid, 2) Guest from Kyoto University, 3) Guest from Southwestern Institute of Physics, Leshan, China, 4) Guest from Ioffe Institute, Leningrad, 5) Guest from General Physics Institute, Moscow, 6) Guest from IPF, University of Stuttgart

# 1. INTRODUCTION

Within the past few years, Electron Cyclotron Resonance Heating (ECRH) became a well established method of plasma heating for both, Stellarators and Tokamaks. Successful application of ECRH in the frequency range between 20 and 100 GHz up to the MW-microwave-power level was demonstrated in various devices /1,2,3,4,5,6/. In contrast to the Tokamaks, where ECRH is one candidate among others for additional heating, ECRH is of crucial importance for Stellarators, because of its capability to provide net-current-free plasma start up and heating. For the Wendelstein WVII-AS Stellarator a 1 MW (4 x 0.2 MW, 3 s and 1 x 0.2 MW, 0.1 s pulse-width) ECRH system was installed at a frequency of 70 GHz, which matches to the design magnetic field of 2.5 T (1st harmonic heating) of the W VII-AS and allows experiments at 2nd harmonic heating at a reduced magnetic field of 1.25 T. Based on the experimental results obtained in the previous W VII-A Stellarator, a sophisticated ECRH system was installed giving access to a wide variety of ECRH investigations such as on-/off-axis heating at ordinary and extraordinary wave polarization as well as current drive with arbitrary angles of incidence of the rf-beam /7/. This system was designed to make optimum use of the particular advantages of ECRH, such as narrow, almost complete (at sufficiently high temperatures) and well localized power deposition for electron temperature profile shaping and heat wave propagation studies. The current drive capability of EC-waves launched at oblique angles attracts increasing interest for Tokamaks and Stellarators as well. Systematic studies of ECCD were performed at the W VII-AS, where the small currents driven by EC-waves are not masked by large OH-currents as in Tokamaks. The small currents, which can be driven by ECCD are well suited to compensate possible pressure driven net currents (e.g. bootstrap current) in Stellarators, which is a crucial condition to maintain good confinement in low shear configurations /9/. A brief description of W VII-AS is given in Sec.2 and the ECRH transmission line and launcher is described in Sec.3. Experimental results on electron heating and confinement at 1st harmonic O-mode are discussed in Sec.4 along with first wave absorption mearurements. Some results on 2nd harmonic X-mode heating are presented in Sec.5, Sec.6 concentrates on preliminary results on ECCD at oblique wave launch, some conclusive remarks are given in Sec.7.

## 2. W VII-AS MACHINE PARAMETERS

The W VII-AS Stellarator has a high aspect ratio, low
shear magnetic field configuration with 5 field periods.
The effective major and minor radii are 2.0 and 0.18
meters, respectively, the nominal magnetic induction is
2.5 T. The rotational transform can be varied from 0.25
up to 0.65 by additive or subtractive superposition of a
simple toroidal field to the field generated by the
modular coil set alone. An outline of the coil system is
shown in Fig.1.

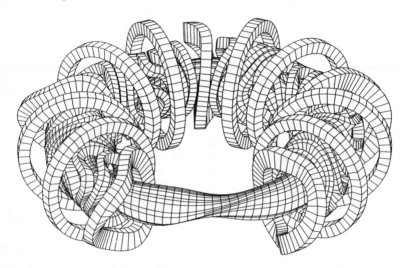

Fig.1: Design of the W VII-AS Stellarator. The set of
       twisted modular coils are shown as well as the
       additional toroidal field coils for variation of
       the rotational transform.

The shape of the flux surfaces changes along the toroidal
direction within one field period from almost elliptical
(indicating l=2 to be dominant) to almost triangular (l=3
dominant) with excursions from axisymmetry (l=1). The
elliptically shaped poloidal cross-sections show
tokamak-like magnetic field gradients and are therefore
used for ECRH to obtain maximum localisation of the power
deposition. Four out of five transmission lines are
combined at one poloidal plane.

## 3. ECRH SYSTEM AND RF-DIAGNOSTICS

The rf-power is generated by up to 5 commercially
available VARIAN gyrotrons, each of them delivering 200
kW output power at 70 GHz. Four gyrotrons have CW
capability and are located approximately 50 meters

away from the Stellarator in a separate hall. The
microwaves are transmitted to the Stellarator by means of
distinct circular oversized waveguides and are finally
combined at one poloidal plane by a quasioptical
launcher.

The TE 02 gyrotron output mode is converted to the low
loss TE 01 mode for long distance transmission. At the
torus end of the transmission lines, this mode is
converted by stages to the TE 11 mode and finally to
the HE 11 hybrid mode,
which has an almost
perfect linear
polarization (99 %).
The HE 11 mode is
radiated from an open
ended circular
corrugated waveguide
section inside the
vacuum chamber and
couples well to the
fundamental Gaussian
free space mode. The
beam launching is
achieved from the low
field side with a quasi
optical mirror system
inside the vacuum
chamber. As seen from
Fig.2, each rf-beam hits
a movable focussing
mirror and can be
directed into the plasma
at arbitrary poloidal
and toroidal angles with
respect to the
orientation of the
stellarator magnetic
field.

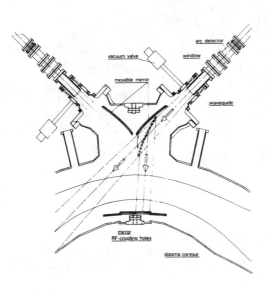

Fig.2: Layout of the quasi-optical EC-wave launcher. The
movable launching mirrors are seen in a plane
slightly above (below) the equatorial plane of
the vacuum vessel.

The curvature of the mirror surface is chosen to position
the beam waist with plane phase fronts in the plasma
centre (normal incidence). An additional segmented
graphite mirror is installed at the inner torus wall
opposite to the four launching mirrors, which reflects
the nonabsorbed fraction of the incident waves back to
the plasma in a well defined polarization /8/. An array
of 38 monomode waveguides is embedded in the central

section, which allows a direct measurement of the rf-beam quality with respect to beam divergence and cross-polarization (no plasma) as well as a measurement of the single-pass absorption and beam deflection in the presence of a plasma. A relative calibration of the sensitivity of the waveguide array at the high-field side mirror was achieved by sweeping one of the rf-beams across the array and adjusting the attenuation of each channel to the level of one arbitrarily chosen reference channel. The beam was found to be very close to the calculated Gaussian radiation pattern, the cross polarization is typically less than 3 %, the beam width at the plasma axis position is about 50 mm (1/e points).

### 4.EXPERIMENTAL RESULTS AT 1st HARMONIC O-MODE HEATING

The plasma parameters obtained so far with on-axis heating at an incident rf-power of 0.35 MW were an electron temperature of $T_e$ = 1.2 keV at an electron density of $n_e$ = 3.5 x 10$^{19}$ m-3 (about half the cut-off density)and $T_e$ = 1.7 keV at $n_e$ = 1.5 x 10$^{19}$ m-3 /9/.

The time development of the electron temperature measured by ECE diagnostics for a discharge, which was maintained with only 0.2 MW rf-power is shown in Fig.3. The density was kept constant at $n_{eo}$= 1.5 x 10$^{19}$ m-3 by feedback control in the later state of the discharge. The central electron temperatures measured by ECE and Thomson-scattering agree well, indicating, that suprathermal electrons play no significant role in this type of discharge.

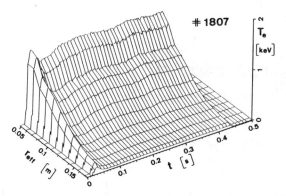

Fig.3: Electron temperature profile as a function of time for 1st harmonic O-mode on-axis heating with only 0.2 MW of incident rf-power. The plasma net current is controlled by the OH-transformer ($I_p$ < 0.2 kA).

The precise knowledge of the power deposition profile is

necessary to provide an electron heat transport analysis. The measurement of the rf-power transmitted through the plasma and consequently the single pass absorption was determined by means of the rf-waveguide array described

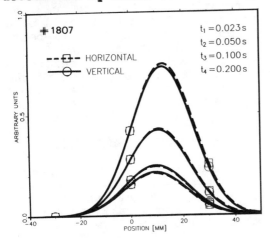

in Sec.3. An example is shown in Fig.4 for the same discharge as in Fig.3. The transmitted power is measured at 3 vertical and 3 horizontal positions across the rf-beam. The plot shows the rf-power profiles at 4 different times during the initial phase of the discharge.

Fig.4 : Horizontal (dashed line) and vertical (solid line) profiles of the rf-power transmitted through the plasma in the early phase of the discharge shown in Fig.3.

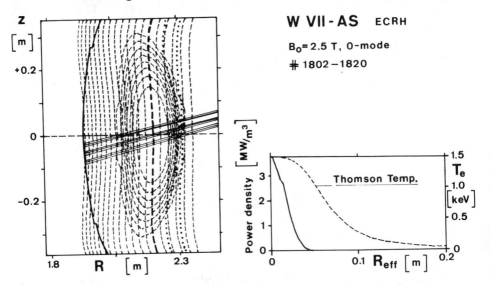

Fig.5 : Ray-tracing calculation (left) and the resulting power deposition profile (right) for the same discharge. The calculations are based on measured profiles of Te and ne (Thomson scattering), the single-pass absorption is 70 %.

At t = 0.023 s no plasma is present (1st ms after switch on of the beams) and the measurement agrees well with calibration. The transmitted rf-power decreases with time in the early phase of the discharge, where the electron temperature increases. The curves are Gaussian profiles determined by 5 data-points, which may be justified by hot calibration measurements without plasma, where the rf-beam was swept across the waveguide array resulting in an almost perfect Gaussian beam shape /7/. The single-pass absorption derived from this measurement increases with time and saturates at 75% -+ 5 % (t= 0.2 s) maximum value in the later phase of the discharge, where the temperature becomes stationary. No broadening of the beams by beam deflection is detected in agreement with ray tracing calculations shown in Fig.5 due to the low density. The electron density and temperature profiles were taken from Thomson scattering diagnostics.

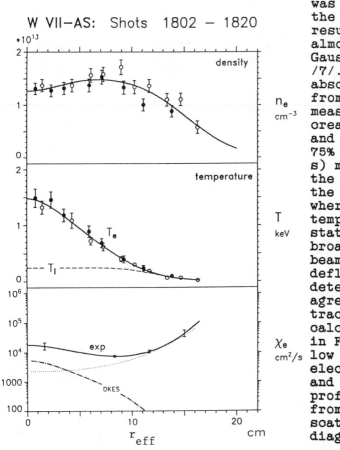

Fig.6 : Electron heat transport analysis based on profiles of ne (top) and Te (middle) measured by Thomson scattering diagnostics. The dashed-dotted line gives the result of neoclassical transport.

The agreement between measurement and ray-tracing result is excellent. This power deposition profile entered the electron heat transport analysis, where the electron heat diffusivity $\chi_e$ is calculated assuming steady state Te and ne profiles. The result is given in Fig.6, where all quantities are plotted as a function of the effective minor plasma radius.

## 5.EXPERIMENTAL RESULTS AT 2nd HARMONIC X-MODE HEATING

In general, total absorption of the rf-waves in a single pass was measured for electron temperatures exceeding 0.2 keV in agreement with the ray-tracing predictions. So this kind of heating is well suited for on-/off-axis heating and studies of the resulting temperature profile changes. Besides the well proven method for off-axis heating by tuning the magnetic field and consequently shifting the EC-resonance layer radially in- and outward, off-axis heating can be achieved also by adjusting the narrow rf-beams geometrically off axis. An example for measured profiles of Te and ne is shown in Fig.7 for on-axis and off-axis heating (perpendicular launch) with a heating power of 0.35 MW. A clear dependence of the electron temperature and density profiles on the position of the rf-power deposition is observed.

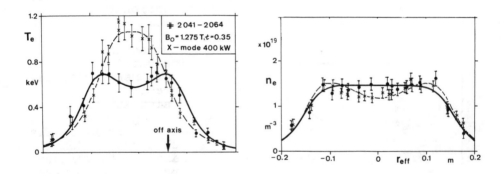

Fig.7 : Profiles of ne and Te (Thomson scattering) for on-axis and off-axis heating at 2nd harmonic X-mode (Bo = 1.25 T). Off-axis heating was performed by adjusting two rf-beams geometrically (arrow) at constant magnetic field.

## 6.ELECTRON CYCLOTRON CURRENT DRIVE

In W VII-AS small net currents in the range of 1-4 kA were observed for pure perpendicular EC-wave launch. The bootstrap current is a candidate responsible for the occurrance of such net currents, which change the vacuum magnetic field configuration

of the Stellarator in an uncontrolled manner. It is
important therefore, especially for almost shearless
magnetic field configurations as in W VII-AS, to
counteract such net currents to avoid a deterioration of
the global confinement by low order rational surfaces
appearing in the confinement region of the plasma and
giving rise to magnetic island formation /9/.

The net current can be balanced by application of a
proper loop voltage (OH transformer). EC-current drive
was found to be an alternative powerful tool for net
current control. Experimental results of ECCD are shown
in Fig.8. In these discharges the plasma net current was
kept close to zero (Ip < 0.2 kA) by feedback control
using the OH-transformer. The required loop voltage
change was measured as a function of the launching angle
of the EC-waves in co- and counter- direction with
respect to the direction of the bootstrap current.

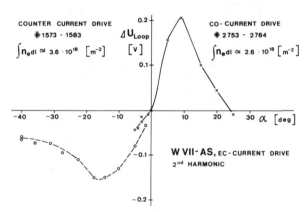

Fig.8 : Loop voltage change  required to keep
        the net plasma current at Ip < 0.2 kA,
        as a function of the launching angle of the
        rf-beams with respect to normal incidence.

The launching angle for maximum ECCD is in the  range  of
10 to 20 degrees with  respect to normal incidence in
agreement with theoretical current-drive modelling /10/.
In counter direction the EC-driven current is sufficient
to ballance the bootstrap current as can be seen in
Fig.9. The rf-power was switched to different levels
during the discharge, each intervall was chosen
sufficiently long to obtain steady state plasma
conditions. The two rf-beams were launched at an angle
of 10 degrees in counter direction. The measured net
current remains below 0.2 kA at all power levels, whereas
the total stored plasma energy derived from the

diamagnetic signal shows a clear dependence on the
incident rf-power.

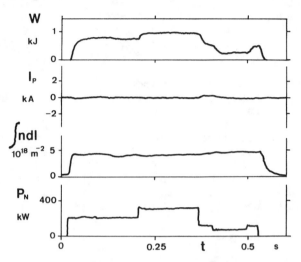

Fig.9 : Total stored plasma energy W from diamagnetic loop (top), plasma net current Ip,
line integrated density ∫ndl, and incident
rf-power (bottom) for counter ECCD at
10 degrees with respect to normal.

It should be noted, that all ECCD experiments were
performed with linearly polarized waves, whereas for
optimum launch an elliptically polarized wave is
required. Such experiments are under preparation.

## 7. CONCLUSIONS

During the first months, the W VII-AS operation
concentrated on exploring the accessible parameter range
rather than on investigating detailed phenomena. Plasma
start up and electron heating was successfully
demonstrated in the WVII-AS Stellarator at 2.5 T (1st
harmonic O-mode) and 1.25 T (2nd harmonic X-mode)
operation. Electron temperatures of 1.2 keV at ne = 3.5
x $10^{19}$ m-3 and 1.7 keV at 1.5 x $10^{19}$ m-3 were measured
at an injected rf-power of 0.35 MW. The single-pass
rf-absorption was measured and agrees well with ray
tracing modelling of discharges for both 1st harmonic
O-mode and 2nd harmonic X-mode heating. The electron
heat transport was calculated on the basis of the
measured power deposition profiles. EC-current drive
experiments were performed over a wide range of launching
angles at different rf-powers. By proper choice of the
launching angle the small net currents (1-4 kA) observed
at WVII-AS could be balanced.

# REFERENCES

/1/    J.Lohr et al., Phys.Rev.Letters 60, 2630 (1988)

/2/    V.Alikaev et al., Plasma Phys. and Contr. Fusion 29,
       1285 (1987); ibid 30, 381 (1988).

/3/    V.Erckmann et al., Plasma Phys. and Contr. Fusion
       28, 1277 (1986)

/4/    H.Zushi et al., Nuclear Fusion 28, 1801 (1988)

/5/    S.Okamura et al., Proc. 16th European Conf. on
       Contr. Fusion and Plasma Phys., Venice, Italy,
       1989, EPS Vol. 13B, Part II, p. 571

/6/    M.Murakami et al., Proc. 16th European Conf. on
       Contr. Fusion and Plasma Phys., Venice, Italy,
       1989, EPS Vol. 13B, Part II, p. 575

/7/    V.Erckmann et al., Fusion Technology, to be
       published

/8/    W.Kasparek et al., Proc. 15th Symp. on Fusion Techn.
       (SOFT), Utrecht (1988), B 19.

/9/    H.Renner et al., Proc. 16th European Conf. on
       Contr. Fusion and Plasma Phys., Venice, Italy,
       1989, Inv. Paper

/10/   U.Gasparino et al., Proc. 16th European Conf. on
       Contr. Fusion and Plasma Phys., Venice, Italy, 1989
       EPS Vol. 13B

# RECENT ECH AND LH CURRENT DRIVE STUDIES AT CULHAM

B Lloyd, M R O'Brien, N R Ainsworth, M W Alcock, S Arshad*, C Balkwill,
P R Collins, M Cox, A N Dellis, R O Dendy, T Edlington, G Fishpool•,
S J Fielding, T C Hender, J Hugill, P C Johnson, O J Kwon°,
C N Lashmore-Davies, S J Manhood, G F Matthews, A W Morris, R A Pitts,
A C Riviere, D C Robinson, P R Simpson, G Vayakis♣ G A Whitehurst

UKAEA Culham Laboratory/Euratom Fusion Association, Abingdon, Oxon,
OX14 3DB, England

* University College, University of Oxford
• Imperial College of Science & Technology, University of London
° Dept of Physics, University of California, San Diego
♣ Balliol College, University of Oxford

## ABSTRACT/INTRODUCTION

For typical reactor conditions lower hybrid (LH) waves are a strong candidate for bulk heating and for profile control via current drive in the outer regions of the plasma. However, limited accessibility, strong electron damping and the need to avoid damping on fusion-produced alpha particles probably exclude the use of such a scheme in the plasma core. Therefore, it is likely that lower hybrid heating/current drive in the outer regions of the plasma will be combined with either ECRF, ICRF or NBI in the core. Electron cyclotron (EC) waves are particularly attractive since they offer the possibility of localised absorption and in addition may be used to facilitate low voltage start-up of the tokamak[1]. Recent experimental and theoretical research on current drive at Culham has concentrated on some of the outstanding critical issues concerning the reactor applications of LH and EC waves. These include studies of the interaction of lower hybrid waves with energetic ions including alpha particles, the effect of LH waves on plasma stability, the efficiency of electron cyclotron current drive and the combined effects of LH and EC waves including applications to low-voltage start-up.

## INTERACTION OF LOWER HYBRID WAVES WITH ENERGETIC IONS

In plasmas with a significant hot ion population, ion damping may compete with the electron Landau damping (ELD) of LH waves desired for current drive. The LH waves may interact not only with alpha particles but also with energetic ions generated during high power ICRF or NBI as has been observed experimentally in JT-60 for example[2]. The extent to which LH waves interact with such ion tails is of particular interest with respect to forthcoming experiments in JET, the explicit purpose of which is profile control via LHCD in NBI and ICRF heated plasmas[3]. The degree of ion damping is very sensitive to the energy and spatial distribution of the hot ions. Therefore it is important to allow flexible specification of the tail parameters to enable a wide range of realistic scenarios to be addressed. Furthermore since the ray trajectory and location of ELD strongly influence whether the waves can interact with the energetic ions it is important when studying experiments such as JET to carry out calculations for realistic D-shaped equilibria. A ray tracing code has been developed which employs the warm plasma electromagnetic dispersion relation for calculation of the ray paths in circular or D-shaped equilibria. The latter are specified using a Fourier representation[4] which is routinely fitted to the experimental magnetic measurements on JET. Allowance is made for multiple reflection of the slow wave where necessary but the programme is terminated if mode conversion to the fast wave is encountered. Ultimately it is essential to compare the damping on energetic ions with that due to quasi-linear ELD but for the initial results presented here a Maxwellian electron distribution is assumed and the relative locations of linear ELD and ion damping are compared for single rays. The ion damping is evaluated numerically from the imaginary part of the susceptibility for arbitrary ion distribution functions assuming 'unmagnetised' ions since $k_\perp \rho_i \gg 1$. Physically the wave absorption may be attributed to perpendicular ion Landau damping which becomes large when $\omega \sim \underline{k}.v_i \sim k_\perp v_{\perp i}$. Temperature and density profiles are taken to be of the form $T = (T(o) - T_{edge})(1 - \rho^2(\psi)/a^2)^{\gamma_T} + T_{edge}$ and $n = (n(o) - n_{edge})(1 - \rho^2(\psi)/a^2)^{\gamma_n} + n_{edge}$ respectively where $2\rho(\psi)$ is the flux surface width in the equatorial plane.

## Damping by NBI Hot Ions

In JT-60 combined LH and NBI experiments[2], the acceleration of ions to energies above the beam energy (75keV H injected into H) was attributed to ion damping of the LH waves. We have attempted to model the damping using the following parameters:-

|  | JT-60 Parameters |
| --- | --- |

Shifted circular flux surfaces were assumed with the magnetic axis at R + 3.05m. The standard beam slowing-down distribution was assumed[5] including a tail above the injection energy with temperature $\sim T_e$. The density of hot beam ions was taken to be $n_{HOT} = 4.5(1 - \rho^2/a^2)^2 \times 10^{18}m^{-3}$.

| | |
| --- | --- |
| $n_{eo}$ | $3 \times 10^{19}m^{-3}$ |
| $n_{edge}$ | $1 \times 10^{18}m^{-3}$ |
| $\gamma_n$ | 1 |
| $T_{eo}$ | 4keV |
| $T_{io}$ | 4keV |
| $T_{edge}$ | 10eV |
| $\gamma_T$ | 1.5 |
| $B_{\phi o}$ | 4.8T |
| $I_p$ | 2.5MA |
| Initial $N_\parallel$ | 2.0 |
| $f_{LH}$ | 2 GHz |

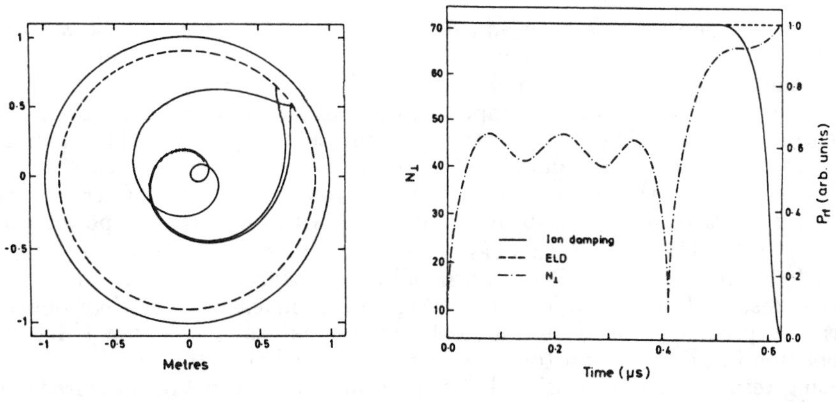

Figure 1 Damping by NBI hot ions in JT-60. The ray path and the evolution of $N_\perp$ and wave power are illustrated.

The wave was completely damped on ions with energies in the range $95 \lesssim E_\perp(keV) \lesssim 110$ i.e., above the injection energy (Figure 1). No linear ELD was observed since $N_\parallel$ did not rise above 2.8. When the calculation was repeated with no beam ions allowed above the injection energy no ion damping was observed before the ray was terminated (at the second reflection) since the peak $N_\perp$ achieved was 75, insufficient to damp on ions at the beam energy ($N_\perp \sim 79$ required). Any ion damping of LH waves might be expected to enhance the hot ion tail and thus lead to further ion damping earlier along the ray path. Therefore a self-consistent treatment involving both ray tracing and Fokker-Planck codes is being considered. For JET, in which combined NBI (140keV D beams) and LH experiments are planned, the code predicts no ion damping. This is because the higher wave frequency (3.7GHz) and the lower magnetic field (3.4T) in JET mean that $N_\perp$ cannot reach the values required for ion damping despite a higher electron temperature which allows the beam distribution to extend well above the injection energy.

### Damping on ICRF-Produced Ion Tails

In recent JET high fusion yield experiments employing $(He^3)$D minority ICRF a mean energy of the $He^3$ ions in the plasma centre of several MeV was inferred[6]. The damping of LH waves on the energetic ion tails in such plasmas has been studied using the following parameters:-

| | | JET Parameters (damping on ICRF tails) |
|---|---|---|
| The ion distribution is | $n_{eo}$ | $4.5 \times 10^{19} m^{-3}$ |
| is calculated on each flux | $n_{edge}$ | $6 \times 10^{18} m^{-3}$ |
| surface using the Stix model[7]. | $\gamma_n$ | 0.5 |
| The ICRF power deposition | $T_{eo}$ | 8.5keV |
| profile is taken to be of the | $T_{io}$ | 6keV |
| form $P \sim P_o \exp \{-(\rho(\psi) - \Delta)^2/w^2\}$. | $\gamma_T$ | 1 |
| For the chosen conditions | $T_{edge}$ | 350eV |
| there is total ELD of the | $B_{\phi o}$ | 3.3T |
| incident ray in the vicinity | $I_p$ | $\sim$3.3MA |
| of $\rho(\psi)/a \sim 0.6$ after one | Initial $N_\parallel$ | 2.245 |
| reflection from the plasma | $f_{LH}$ | 3.7GHz |
| boundary. In the code, linear | $n_{He^3}/n_e$ | 1% |

ELD is included for reference and does not deplete the incident power so that the ray continues until it is completely damped on the ions. For conditions typical of the JET experiment, namely $P_o = 3.3MW/m^3$, $w= 0.2a$ and $\Delta = 0$ (central deposition) the ray is totally damped on the ions in the region $0.3 \lesssim \rho(\psi)/a \lesssim 0.4$ after one reflection. Therefore electron damping is clearly expected to dominate in this case. In order for ion damping to precede ELD one must increase $\Delta$ to $0.3a$ (off-axis deposition). Alternatively one may maintain central ICRF deposition but increase $w$ to $0.5a$ (very broad deposition) with lower central power density, viz, $P_o = 2.2MW/m^3$. In this case (Figure 2) there is even some ion damping ($\sim 4\%$) in the first pass. The central tail temperature is $\sim$5MeV. Although the spatial distribution of the ion tail in Figure 2 is much broader than one would normally predict for JET, it should be noted that the calculated ICRF power deposition in JET is found to be much more peaked than the profiles of electron heating resulting from collisional slowing down of the ion tail measured after a sawtooth crash[6]. Possible causes of this discrepancy include redistribution of fast ions during the sawtooth crash, RF induced diffusion of the fast ions or finite orbit effects. Concerning the last of these effects it may be seen in Figure 2 that damping is complete before $N_\perp$ exceeds 36 and involves ions whose Larmor radius in the poloidal field is $\rho_{i\theta} \gtrsim 0.2a$.

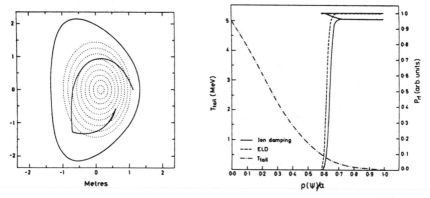

Figure 2 Damping on ICRF-produced ion tails in JET. The ion tail temperature and wave power are shown as functions of 'flux surface radius' for the ray illustrated.

**Damping on Fusion-Generated Alpha Particles**

Absorption of LH waves by alpha particles has been studied for JET-like high performance conditions using the so-called Rebut-Lallia profiles[8] in a D-T plasma. The standard alpha particle slowing-down distribution[9] was assumed. The following parameters were employed:-

|  |  | JET Parameters (damping on alpha particles) |
|---|---|---|
| It is seen in Figure 3 that | $n_{eo}$ | $1.7 \times 10^{20} m^{-3}$ |
| there are two regions of | $n_{edge}$ | $2 \times 10^{19} m^{-3}$ |
| ion damping separated by a | $\gamma_n$ | 4 |
| region of the ray where | $T_{eo}$ | 15keV |
| $N_\perp$ falls below $\sim$23 | $T_{io}$ | 15keV |
| which is the value of $N_\perp$ | $\gamma_T$ | 2.5 |
| for absorption by alpha | $T_{edge}$ | 1keV |
| particles at the birth energy | $B_{\phi o}$ | 3.24T |
| 3.5MeV. Although a high | $I_p$ | 6.1MA |
| energy tail above the birth | Initial $N_{\parallel}$ | 3.0 |
| energy is included it is of | $f_{LH}$ | 3.7GHz |
| limited extent since | $n_\alpha/n_e$ | 0.1% |

$T_e \ll 3.5 MeV$. There is also some damping on electrons. At lower values of launched $N_{\parallel}$ the rays skirt the plasma edge since accessibility is poor. Nevertheless damping on alpha particles can still occur. Although the alpha particle density assumed in Figure 3 is very low the spatial profile is taken to be the same as the electron density profile. If the alpha particle density is evaluated by balancing the alpha-particle production rate against the slowing down due to electron drag as in Reference 10, a much more peaked profile is predicted so that although the central alpha density is an order of magnitude higher than we have assumed, the density in the outer plasma regions is lower and the wave is damped on electrons. However, slight broadening of the profile by any of the mechanisms identified in connection with the ICRF-produced ion tails would lead to strong ion damping. In particular the finite drift orbit size may be sufficient to bring about this effect since $\rho_{i\theta} \sim 0.25a$ for the case considered.

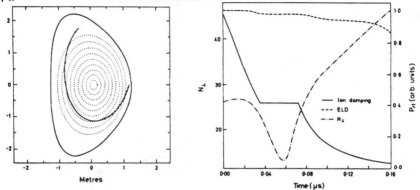

Figure 3 Damping on alpha particles in JET-like high performance conditions. The ray path and the evolution of $N_\perp$ and wave power are illustrated.

## EFFECT OF LHCD ON PLASMA STABILITY

The effect of LHCD on the stability of the m=2, n=1 tearing mode in JET has been studied using a $\Delta'$ analysis in cylindrical geometry. For typical JET conditions, ray tracing indicates a fairly narrow RF driven current profile $j_{rf}(r)$ with most of the current localised within a region $\sim 0.1 - 0.2a$ assuming that all of the rays are launched in the tokamak mid-plane[11].

In this case (assuming a Gaussian $j_{rf}$ with FWHM $\sim 9\%a$) it is found that an RF driven current $I_{rf} \sim 2.2\% I_{OH}$ is sufficient to fully stabilise ($\Delta' < 0$) the 2,1 tearing mode if $j_{rf}$ is centred exactly at q=2 (NB $q_a \sim 3.5$). The RF has a stabilising effect ($\Delta'$ reduced) only if $j_{rf}$ is centred in a narrow range ($\Delta r \sim \pm a/20$) around q=2. Inclusion of rays launched over the full poloidal extent of the grill leads to somewhat broader profiles of $j_{rf}$. In addition $j_{rf}(r)$ will be broadened by spatial diffusion of the fast electrons.

If we assume that $j_{rf}(r)$ is broadened by a factor of two as a result of these mechanisms then it is found that a minimum RF driven current $I_{rf} \sim 7\% I_{OH}$ is now required to fully stabilise the 2,1 tearing mode. However, stabilisation is less sensitive to the location of $j_{rf}$. $\Delta'$ is reduced provided $j_{rf}$ is centred within $\Delta r \approx \pm 0.1a$ of q=2. As in the previous case the reduction is a maximum if $j_{rf}$ is centred exactly at q=2.

## EFFICIENCY OF ELECTRON CYCLOTRON CURRENT DRIVE

Electron cyclotron current drive (ECCD) at the second harmonic has been studied in the CLEO tokamak at both 28GHz and 60GHz[12]. In both cases waves were launched from the low-field-side of the tokamak. At 28GHz no evidence of RF driven current could be detected. At 60GHz, using a $TE_{02}$ mode Vlasov antenna to launch waves at an angle $\sim 68°$ to $B_\phi$, RF driven currents of up to 5kA were observed at $\bar{n}_e \sim 4 \times 10^{18} m^{-3}$ for 185kW of injected power indicating an efficiency of $\eta \equiv \bar{n}_e I_{rf} R_o / P_{rf} = 0.001 (10^{20} m^{-3}, A, m, W^{-1})$. Experiments were performed by varying the toroidal direction of both the plasma current and the launched waves. The RF driven current at 60GHz was $\sim$ 30% of that predicted using a ray tracing code which includes a current drive model allowing for trapping effects[13] and the weakly relativistic resonance condition[14]. Trapping effects reduce $\eta$ by $\sim$ 20% when the cyclotron resonance is at the plasma centre. These linear calculations agreed well with the predictions of a bounce-averaged Fokker-Planck code[15] which showed that the energy of the resonant electrons was $\sim$ 3 − 4keV (NB $T_{eo}$ $\sim$ 1.2keV).

The collisional slowing down time for these electrons is of the same order or longer than the experimentally observed energy confinement time. It has been shown that this leads to a significant reduction in the predicted current drive efficiency[16]. Attempts to repeat the 60GHz second harmonic ECCD experiments in DITE[17] using a high-field-side $TE_{11}$ launch at 60° to $B_\phi$ were inconclusive. Because of the limited RF power available at the time and the enforced operating conditions of the tokamak it was predicted that only $\sim$ 5% of the plasma current would be driven by the waves. Such a low level is not detectable.

Experiments at the fundamental using a HFS launch have recently commenced. Preliminary investigations at $I_p$ $\sim$ 100$kA$, $\bar{n}_e$ $\sim$ 9 × 10$^{18}m^{-3}$, $B_{\phi o}$ $\sim$ 2.2$T$ with 310kW of power delivered to the plasma indicated a 20% difference in loop voltage $V_\ell$ when the toroidal direction of the launched waves was changed. However the observed $V_\ell$ difference was consistent with the expected change in plasma resistance resulting from the different heating efficiencies observed for the two wave directions. In each case the waves were directed at $\sim$ 45° to the major radius. For these conditions an RF driven current, allowing for trapping effects, of $\sim$ 10 − 20kA would have been expected.

## LHCD EXPERIMENTS IN DITE

LHCD experiments at 1.3GHz have recently commenced in DITE ($R/a$ $\sim$ 1.19/0.23$m$). The titanium grill antenna was originally used on the PETULA tokamak and is fed via a 60m transmission line by a high power klystron used previously on ASDEX. Prior to installation of the antenna several modifications were carried out including the addition of passive waveguides and poloidal shaping of the grill mouth to improve coupling of the waves to the DITE plasma. After baking of the grill and glow discharge cleaning in argon, full power operation (250kW) was established within a few days of operation.

Initial experiments have been conducted in hydrogen within the following parameter range; $I_p$ $\sim$ 100kA, $\bar{n}_e$ $\sim$ 0.5 − 1 × 10$^{19}m^{-3}$, $B_\phi$ $\sim$ 1.9 − 2.5$T$. Although $V_\ell$ can be decreased to a very low level ($\sim$ 0.3V) it has not yet been possible to reduce the loop voltage to zero. LHCD is usually accompanied by a small increase (decrease) in $\beta_\theta + \ell_i/2$ ($\beta_{dia}$). In the absence of gas puffing $\bar{n}_e$ is observed to fall by typically 20-30% and the total radiated power and soft X-ray emission decrease accordingly. In the plasma scrape-off layer(SOL) the electron density decreases by a factor ×2 − 3 and the total particle flux to the pump limiter, evaluated by interpolating between probes distributed over the limiter surfaces, falls by a factor ×3 − 4 (Fig. 4). The $H_\alpha$ emission decreases by a similar factor suggesting a significant improvement in particle confinement. There is a decrease in $T_e$ by typically $\sim$ 30% in the SOL. These observations are in complete contrast to those with ECRH where $\bar{n}_e$, $n_{edge}$, $T_{edge}$, $H_\alpha$ emission and the particle flux to the limiter all increase in the present DITE configuration. The probe measurements may be used to evaluate the convected power flux to the pump limiter assuming a constant power transmission factor for the sheath ($\delta = 10$). For OH and ECR heated plasmas this agrees closely with the difference between the total input power and the total radiated power. However during LHCD the convected power flux is found to be consistently low by an amount $\lesssim P_{LH}$ suggesting that a large fraction of the power deposition is local or via loss of energetic electrons (the single probes are biased within the range ±100V with respect to torus potential).

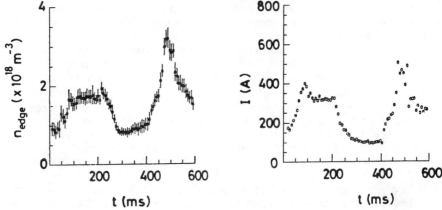

Figure 4 Electron density (7.5mm outside LCFS) and total particle flux to the pump limiter during LHCD. RF injection is from t=200ms to t = 400ms.

The switch-off of LH power usually leads to the onset of the Parail-Pogutse instability which has been stabilised in DITE by ECRH as in earlier experiments[18]. High power ECH (>300kW) during LH injection leads to a significant redistribution of reflected power over the active waveguides of the grill due to the increase in $n_{edge}$, but the global LH reflectivity does not change significantly.

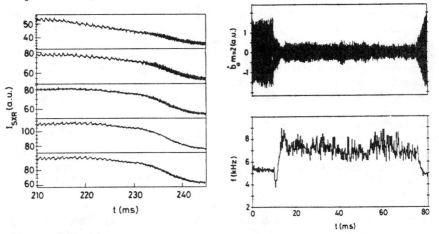

Figure 5 Sawtooth response during LHCD. The channels show line-integrated soft x-ray emission along vertical chords near the plasma core. Moving upwards in the figure corresponds to increasing major radius and successive channels are ∼ 15mm apart.

Figure 6 Feedback stabilisation of m=2 activity during LHCD. The feedback is activated between t=10ms and t=76ms. The lower figure shows the frequency of the m=2 activity.

During LHCD sawteeth are stabilised (Fig. 5) by $\sim$20 ms into the RF pulse. During this time there is normally a steady decrease in sawtooth amplitude and a slight increase in the period. The inversion radius does not appear to change. There is a slight peaking of the soft X-ray emission profile. Sawteeth are normally stabilised for the full duration of the LH pulse. When sawteeth have been observed to reappear during the RF pulse the inversion radius is the same as in the OH phase. Sawteeth stabilisation is usually accompanied by excitation of a large mode with m=2 and m=1 components. The m=1 oscillations, localised in the vicinity of the inversion radius, are clearly visible in Fig. 5. Simultaneous feedback stabilisation of the m=1 and m=2 oscillations has been achieved (Fig. 6) during LHCD using m=2 windings inside the DITE vacuum vessel[19]. This often leads to a further slight peaking of the soft X-ray emission profile and there are indications of a very small ($< 0.1$V) further decrease in $V_\ell$.

## Acknowledgements

The loan of LH equipment from IPP Garching and CEA Cadarache is gratefully acknowledged, as is the co-operation of Dr F Leuterer and Dr G Tonon. Similarly we are grateful to General Atomics for design of one of the ECRH antennae, and to Dr T Luce (GA) for his assistance. Fruitful discussions with Dr C Gormezano, Dr F Rimini, Dr M Lorentz-Gottardi, Dr G A Cottrell and Dr D F Start are gratefully acknowledged. Part of this work was carried out under a JET Task Agreement and an Article 14 contract for JET.

## References

[1] B Lloyd & T Edlington, Plas. Phys. & Contr. Fus., 909 (1986).
[2] K Ushigusa et al, Nuc. Fus. 29, 265 (1989).
[3] S Knowlton et al, Proc. 14th Eur. Conf. on Contr. Fus & Plas. Phys., Madrid, III, 827 (1987).
[4] L L Lao et al, Phys. Fluids 24, 1431 (1981).
[5] J G Cordey, Proc. 5th IAEA Conf. on Plas. Phys. & Contr. Nuc. Fus. Res., Tokyo, I, 623 (1974).
[6] L G Eriksson et al, Nuc. Fus. 29, 87 (1989).
[7] T H Stix, Nuc. Fus. 15, 737 (1975).
[8] P H Rebut et al, Proc. 11th IAEA Conf. on Plas. Phys. & Contr. Nuc. Fus. Res., Kyoto, I, 31 (1986).
[9] P T Bonoli & M Porkolab, Nuc. Fus., 27, 1341 (1987).
[10] K L Wong & M Ono, Nuc. Fus., 24, 615 (1984).
[11] F Rimini, private communication, (1989).
[12] B Lloyd et al, Nuc. Fus. 28, 1013 (1988).
[13] J G Cordey et al, Plas. Phys. & Contr. Fus., 24, 73 (1982).
[14] A Ferreira et al, Plas. Phys. & Contr. Fus., 26, 1565 (1984).
[15] M R O'Brien et al., Nuc. Fus. 26, 1625 (1986).
[16] R O Dendy & M R O'Brien, Nuc. Fus. 29, 480 (1989).
[17] M Ashraf et al, 12th IAEA Conf. on Plas. Phys. & Contr. Nuc. Fus. Rés., Nice, Paper IAEA-CN-50/E-1-3 (1988).
[18] S C Luckhardt et al, Proc. 3rd Joint Varenna-Grenoble Int. Symp. on Heating in Toroidal Plasmas, Grenoble, II, 529 (1982).
[19] A W Morris et al, Proc. 16th Eur. Conf. on Contr. Fus. & Plas. Phys., Venice, II, 541 (1989).

# A REVIEW OF RECENT RESULTS
# WITH ELECTRON CYCLOTRON HEATING IN TOKAMAKS*

R. Prater

General Atomics, San Diego, CA 92138

## ABSTRACT

Many experiments using Electron Cyclotron Heating (ECH) of plasmas in tokamaks have been reported over the past two years. At a power level of 4 MW, ECH has achieved electron temperatures as high as 10 keV in the T-10 tokamak, and the H–mode has been attained in divertor discharges in DIII–D and JFT-2M. Regarding global energy confinement in either L–mode or H–mode, ECH appears to be quite similar in efficiency to neutral injection, but in addition to bulk heating it has been useful for many purposes, including study of local electron heat diffusivity through pulse modulated heating; suppression of sawteeth, Edge Localized Modes, and other MHD activity; suppression of disruptions; preionization and startup; and current drive. In this paper, progress in these areas which has been reported since the IAEA meeting in 1986 will be summarized.

## INTRODUCTION

The technological advantages of Electron Cyclotron Heating have long been recognized. These advantages include insensitivity of wave propagation to the conditions of the plasma at the edge, provided the plasma density there is not too high ($\omega_{pe} < \omega$, where $\omega_{pe}$ is the local plasma frequency and $\omega$ is the applied frequency); ability to place the wave launchers distant from the plasma, since the wave propagates in vacuum; very high power density at the antenna (above 85 kW/cm$^2$ in current experiments at 60 GHz[1]); and excellent antenna optics even with small antennas, due to the short wavelengths involved. These advantages become even more important for the next generation of tokamaks, including CIT and ITER, for which a high intensity neutron flux will occur, since the characteristics of the ECH launchers can simplify the implementation of a neutron shield and the requirements of remote handling. In addition to the technological advantages, the strong single pass damping of EC waves motivates applications which require a high heating efficiency or which depend on a localized source of heat. The physics of ECH has been reviewed by Bornatici,[2] and the status of experiments has been reviewed by several authors.[3,4]

## ADVANCES IN BASIC PHYSICS OF ECH

Wave propagation and absorption are relatively well understood theoretically.[2] Three modes are of interest in plasmas of moderate temperature ($T_e < 10$ keV): the ordinary mode at the fundamental frequency and the extraordinary mode at the fundamental (which must be launched from the high field side of the tokamak in order to

---

* Work supported by U.S. DOE Contract DE-AC03-89ER51114.

avoid wave reflection at the right-hand cutoff) and at the second harmonic. Experiments which support the basic theory have been performed in tokamaks,[3] and recent experiments continue to support theory, at least in a qualitative manner.

In DIII–D, the experimentally determined efficiency of central heating of all three modes has been compared to theoretical calculations of first pass damping using the TORAY code,[5] as a function of density. In those experiments,[6] it was found that the density at which the heating efficiency began to drop was consistent with theory, and that heating at the highest densities was attained using the inside launch. The inside launch result was corroborated in work on DITE,[7] where increases in the poloidal beta were found up to the expected cutoff density.

The absorption of EC waves was measured directly in experiments on TFR.[8] Using a small microwave detector across the plasma from a wave launcher, the ratio of the transmitted power for various plasma conditions to that for vacuum conditions was compared with power transmission calculations using a ray tracing code. For cases with central heating, good agreement between the measured and calculated transmitted power was obtained, including cases in which the plasma temperature was increased by ECH at other toroidal locations.

These and previous experiments tend to corroborate the theory of wave propagation and absorption, and no clear violation of the theory as presently understood has been seen. Calculations in the form of ray tracing codes and Fokker-Planck codes appear to offer a predictive capability for future experiments.

## CONFINEMENT IN L–MODE

Studies of energy confinement under ECH have been reported in the T-10 tokamak[9-11] and in the DIII–D tokamak.[6,12,13] In these studies, the global energy confinement time is determined as a function of the plasma parameters such as density $n_e$, plasma current $I_p$, toroidal field $B_T$, applied heating power $P_{ECH}$, and so on. The resulting scalings can be compared with theoretical models and results from other tokamaks or other heating methods to determine the special properties of ECH. Transport analysis of steady-state equilibria with ECH can also be performed if sufficiently accurate profile information concerning $T_e(r)$, $T_i(r)$, $n_e(r)$, $Z_{eff}(r)$, $P_{rad}(r)$, and $P_{ECH}(r)$ is available.

A rather thorough analysis of energy confinement in T-10 was reported in Ref. 9. These experiments were performed with about 1 MW of ECH power, in the density range below the critical density where both the ohmic and the ECH confinement is proportional to density. At these densities, a "soft" saturation of $\tau_E$ is found with heating power as the auxiliary power is raised above the ohmic power, which may be contrasted to the hard saturation with neutral beam power found in the density range above the critical density in the usual L–mode of TFTR and ASDEX. The T-10 group argues that above the critical density, even a small auxiliary heating power causes the plasma to enter a state characterized by degraded electron heat confinement according to an expression $\tau_E = \tau_b + (\tau_E^{OH} - \tau_b)(P_{OH}/P_{tot})$, where $\tau_b$ is a measure of the confinement of the incremental energy, and $\tau_b \approx \frac{1}{4}\tau_E^{OH}$. In the low density regime, $\tau_E \propto P_{tot}^{-(0.4-0.5)}$, which is a little weaker dependence on power than exhibited in the L–mode scaling.[14] The increment to the stored plasma energy at a fixed heating power in T-10 does not depend on the magnitude of the plasma current, but the energy confinement time is a weak function of current: $\tau_E \propto I_p^{0.34}$.[11] This is in strong contrast to the L–mode, in which $\tau_E \propto I_p^{1.24}$.

It should be noted that the T-10 group uses the absorbed power determined from $dW_p/dt$ in the calculation of $\tau_E$, rather than the incident power in the usual convention. This practice disguises any effect of immediate decreases in confinement as observed, for example, on TFR.[8] The TFR experiments showed that while the transmitted power measurements indicated full wave absorption on a single pass, in many cases as little as a third of the power could be observed in $dW_p/dt$. In T-10 the absorbed power is usually 70%–75% of the incident power.

The experiments on DIII–D were performed in a density range including the critical density using several different modes of ECH and with the confinement in the L–mode regime.[6,12] These experiments, which were done with ECH power of about 1 MW, were limited by the operating range of density, which is limited by the usual density limit at the high density end and by locked modes at the low end. In order to increase the density by more than a factor of two, it was found necessary to increase $I_p$, which complicates the scaling studies. Nevertheless, it was determined that for second harmonic heating the confinement was consistent with the Kaye-Goldston scaling and for fundamental heating[12] it was about 1.7 times better than Kaye-Goldston. Discharges with NBI in the same power range showed similar behavior. These experiments showed little difference in global confinement and scaling between ECH and NBI.

The profile of the ECH power density was found to be a critical factor in maintaining good confinement in T-10.[15] For heating profiles broader than the initial ohmic heating profile, the plasma energy initially rises when the ECH is applied, but then decays on a time scale of the order of the resistive skin time, often to a level below the starting plasma energy. This effect is not seen for more centrally peaked heating profiles.

## TRANSPORT STUDIES

Transport has been studied directly in a number of tokamaks using the technique of ECH pulse modulation,[16] including TFR,[8,17] DITE,[18,19] DIII-D,[20] and TEXT.[21] In TFR,[17] the power of one gyrotron was modulated between 140 kW and 70 kW with a period of 4.5 msec, using central heat deposition. An array of SXR detectors was used to follow the propagation of the heat as it diffused radially. The electron thermal diffusivity $\chi_e$ found in this manner was 40% smaller than that found from analysis of the equilibrium profiles. The dominant uncertainty is the absorbed power, since only 20 kW of the modulated power of 70 kW was found in the plasma, and the authors conclude that an increase in $\chi_e$ occurs within 50 $\mu$sec of the application of the power. No scalings of $\chi_e$ with plasma parameters were reported.

Similar experiments were performed on DITE,[19] over a wide range of plasma density. Within experimental error of about 25%, the $\chi_e(r = a/3)$ determined from heat pulse propagation was consistent with that determined from the power balance at densities below $2 \times 10^{19}$ m$^{-3}$. Above that density, the diffusivity from pulse propagation remained constant and larger than the diffusivity from power balance which decreased with density. The results obtained are consistent with diffusive transport of heat in the electron channel, and measurement of the amplitude of the density fluctuations at the modulation frequency showed that off-diagonal elements of the transport matrix are small for the cases studied. The TEXT experiments[21] were performed at both high safety factor, $q = 6$, and at $q = 3$. At low $q$, $\chi_e$ from pulse propagation exceeded that from power balance, and it increased by 25% with the application of on-axis ECH in a manner similar to that found using analysis of the power balance.

These experiments show that pulse propagation studies of heat transport are a viable means of determining transport, but that more work is needed to develop the correspondence between measured values and global observations of confinement.

## PLASMAS WITH HIGH ELECTRON TEMPERATURE

High central electron temperatures, up to 10 keV, were observed in the T-10 tokamak with 4 MW of ECH.[15] These results were obtained using up to 4 MW of ECH power at a combination of two frequencies. Power from seven gyrotrons (about 2.5 MW generated power) at 81 GHz was combined with that from three (about 1.1 MW) at 75 GHz to achieve a total absorbed power of 2 MW. The line averaged density was $1.5 \times 10^{19}$ m$^{-3}$ and the toroidal field was 2.8 T. The resonance zones for the two frequencies were placed symmetrically about the center of the plasma, at a minor radius of 6.5 cm ($r/a = \pm 0.2$). The temperature was determined from electron cyclotron emission at the second harmonic, and pulse height analysis of SXR emission indicated electron temperatures of about 8 keV. The distribution function was found to be Maxwellian even though the ECH power density was very high (about 10 W/cm$^{-3}$, if all the power is absorbed within a minor radius of 8.5 cm). The waves propagate very nearly perpendicular to the magnetic field, and this reduces the tendency to generate nonMaxwellian tails through heating of electrons with large parallel velocity. Fokker-Planck calculations show that even at the very high power density and high temperature conditions of the T-10 experiments, the amplitude of the ECE emission tracks the average electron energy, and the SXR spectrum in the range below 25 keV should be consistent with a Maxwellian distribution,[22] as was observed in the experiments. The confinement time for energy fell from about 50 msec during ohmic heating to 6.8 msec during the high power ECH, compared to 8 msec for L–mode scaling.

High temperatures were also found in other tokamaks. In DIII–D, central temperatures of 5 keV were found with 1 MW of ECH, at a density of $0.7 \times 10^{19}$ m$^{-3}$.[6] In this case, $\tau_E = 85$ msec, or about double the value expected from L–mode scaling. This factor 2 in confinement comes from the scaling of $\tau_E$ with toroidal field which is found in the DIII–D data but not in the L–mode scaling; this scaling with $B_T$ is conjectured to be due to the effects of the dissipative trapped electron mode.[12] In TEXT at lower power,[21] strong increases in central electron temperature were found in non-sawtoothing discharges, and in TFR $T_e$ was found to have a strong peak well inside the sawtooth inversion surface.[17] These high temperature, low density plasmas offer ideal conditions for tests of noninductive current drive, and they show that even at very high power density the electron distribution may be maintained close to a Maxwellian.

## ECH H–MODE

A regime of improved confinement, called H–mode scaling, has been observed in several tokamaks. The H–mode was first discovered in divertor discharges in ASDEX when neutral injection was added at power levels above a threshold of about 2 MW. The result has since been reproduced in other tokamaks, including Doublet III, JFT-2M, JET, and DIII–D, all with high power neutral injection. The H–mode was also obtained using ion cyclotron heating (ICH) on ASDEX and on JFT-2M. The signature of the H–mode transition in all cases is an increase in the plasma energy, signifying an improvement in the global energy confinement time $\tau_E$; an increase in the plasma density and a decrease in the $D_\alpha$ emission, signifying an improvement in particle confinement time $\tau_p$; and a broadening of the pressure profile, with the development of a large gradient in electron temperature and/or density at the plasma edge.

The H–mode generated by ECH alonewas first found in the DIII–D tokamak, following experiments on JFT-2M in which the H–mode was obtained with combined ECH and neutral beam injection. In the JFT-2M case,[23] the H–mode was found through the heating of the plasma edge by ECH of plasma centrally preheated by NBI, with the plasma in the divertor configuration. The experiments were motivated by the observation that in NBI H–mode discharges in JFT-2M the edge electron temperature rises abruptly at the transition, and the intent was to induce the transition by raising the edge temperature directly. The experiment was done by comparing discharges with 319 kW of NBI with discharges having 101 kW of ECH plus 170 kW of NBI (271 kW total). The ECH resonance was located at $r/a = 0.87$. Under these conditions, a clear transition to H–mode occurred in the case with ECH plus NBI, but not in the case of NBI only. The H–mode did not occur under the same conditions when the resonance location was moved inward to $r/a = 0.32$ or outward to $r/a = 1.1$.

From this data it was concluded that heating near the outer edge of the plasma is more effective in JFT-2M than central heating in generating the H–mode. The transition takes place coincident with the crash phase of a sawtooth, which is associated with a transient heat pulse to the plasma edge, and the condition for the H–mode transition can be well described by a threshold in the electron temperature at the plasma edge.[24] In these discharges, the global energy confinement time of 40 msec changed very little (less than 10%) from the OH to the L–mode to the H–mode phases, but in the H–mode the particle confinement dramatically improved.

The first demonstration of the H–mode using ECH as the sole auxiliary heating method was in divertor discharges in the DIII–D tokamak.[12,25,26] In these experiments, it was found that for central heating using the second harmonic the application of 0.75 MW of ECH was sufficient to attain the H–mode, under conditions for which the threshold power for neutral injection was about a factor of 1.5 higher. The ECH data for the threshold power, which were obtained for both second harmonic (at $B_T = 1.1\,T$) and fundamental (2.1 T) heating, are consistent with a large body of data from several tokamaks using NBI heating which indicate that the H–mode threshold power is an increasing function of toroidal field with an approximately linear dependence.[13] Another similarity between the ECH and the NBI H–modes in DIII–D is that for both cases the plasma rotates in the poloidal direction following the transition, with a circumferential velocity near the separatrix of 25 km/sec (corresponding to an inward radial electric field of 200–300 V/cm).[13,27]

In DIII–D, both particle and energy confinement improve when the H–mode transition takes place. At the transition, the line-averaged density begins rising at a rate quite similar to that for neutral injection heating, indicating that the density rise normally associated with the H–mode is not dependent upon fueling by neutral beams, but is a consequence of improved particle confinement. Another consequence of improved particle confinement was improved confinement of impurities. The radiated power rose throughout the H–mode phase at a rate slightly faster than $n_e^2$, eventually equalling the input power and causing a reverse transition to the L–mode through power starvation. In order for the H–mode to be fully successful as a means of improving confinement in tokamaks, means must yet be found to control the accumulation of particles and impurities.

Energy confinement improved dramatically in DIII–D following the H–mode transition. The confinement time in the ECH H–mode with 0.9 MW of power was about the same as that in ohmic discharges, 130 msec, and 2 to 3 times greater than that observed during the L–mode phase of the discharge. A strong correlation was found between electromagnetic fluctuations observed in the divertor region and the global confinement of energy, suggesting that such turbulence may cause the deterioration of confinement found in the L–mode.[26]

Very recently, the H–mode was generated in JFT-2M by ECH alone. By raising the plasma density above a threshold density of $2 \times 10^{19}$ m$^{-3}$ which was found for JFT-2M for NBI heating, the H–mode transition was observed even at power as low as 120 kW.[28] In these experiments, the optimum location of the resonance was again found to be $r/a = 0.85$, even though this should require a higher power under the usual scaling law of $P_{th} \propto B_T$. The reason may be that the benefit of direct edge heating outweighs the effect of the higher threshold power for central heating.

## CONTROL OF MHD ACTIVITY

The localized nature of ECH has been used to affect or stabilize several MHD modes, including sawteeth, $m = 2$ activity leading to disruptions, and Edge Localized Modes (ELMs). This process has led to new regimes of operation and improved confinement.

Suppression of sawteeth, the relaxation oscillations found within and near the $q = 1$ surface, was first observed in T-10,[29] and then in other tokamaks including Doublet III, TFR, and DIII–D. In TFR,[30,31] it was found that placing the heating surface very near the sawtooth inversion layer was effective at suppressing the sawteeth. These experiments were corroborated in DIII–D,[32] where it was found that the sawtooth amplitude and period were very sensitive to the location of the heating. Qualitatively, the results fit models of Park[33] and Westerhof and Goedheer,[34] but with some discrepancies. At the best, periods up to 500 msec (limited by the ECH pulse length) were obtained free of sawteeth in DIII–D.[6]

Very strong effects on $m = 2$ fluctuations were observed in experiments on TFR [17] in which the heating location was placed a few cm outside the $q = 2$ surface, in discharges with edge safety factor $q_a$ near 3.4. Without ECH the discharges exhibited strong MHD activity, but 300 kW stabilized the mode. The TFR group attributes this stabilization to the heating of the plasma column, which causes the column to move radially by 1–2 cm. Since the limiter is on the inner wall, this increases $q(a)$ which stabilizes the mode. Calculations show that the requirements on heating location which must be satisfied to stabilize $m = 2$ activity in TFR through changes in the current density profile could not be met.

Disruptions due to excessive density have been avoided in T-10 through ECH,[10] thereby extending the operating range of the tokamak. The authors believe that disruptions at high density may be caused by shrinking of the plasma column due to cooling of the edge by heavy injection of gas. The shrinking column develops strong $m = 2$ oscillations which lead to disruption. By adding about 1 MW of ECH the density limit found during Ohmic heating could be exceeded by a factor of 2 without disruptions.

In H–mode discharges, the confinement is often limited by the effects of Edge Localized Modes on heat and particle transport. In experiments in DIII–D in which the H–mode was generated by heating by 3-4 MW of neutral injection, the application of 1 MW of ECH to the plasma outer edge near the midplane was found to stabilize the ELMs.[6] Concurrently with the disappearance of the ELMs, the global energy confinement time increased by 50%, and the electron temperature increased across the entire profile. The process by which the ECH affects the ELMs is under study.[13]

## PREIONIZATION

Present designs of the ITER/NET tokamak employ a thick-walled vacuum vessel, which limits the ohmic electric field to less than 0.3 V/m. ECH is seen as a means to obtain a reliable breakdown of the gas and a current ramp which makes efficient use of the magnetic flux available in the ohmic heating transformer. Present experiments support this application of ECH, but work remains to determine the scaling to future tokamaks.

Preionization has been studied using antennas on the low-field side of the vacuum vessel of the CLEO tokamak[35,36] at both the fundamental and second harmonic. Experiments tested the ordinary mode at the fundamental frequency and the extraordinary mode for both harmonics, as well as unpolarized power. All modes were effective in reducing the electric field needed for breakdown, to less than 0.2 V/m in the case of unpolarized power at the fundamental. The loop voltage was independent of rf power for $60 < P_{\mathrm{ECH}} < 135\,\mathrm{kW}$, and it fell to close to the level of $L d I_{\mathrm{p}}/dt$. Very weak dependence on the location of the resonance was found, provided it was within the plasma volume. Optimization of the startup procedure could probably provide further benefits in reduced loop voltage and flux savings.

## ECH CURRENT DRIVE

A demonstration of noninductive current drive using EC waves has been performed in the CLEO tokamak using second harmonic ECH at both 28 and 60 GHz.[37] At 60 GHz, driven currents up to 5 kA (30% of the total current) have been observed. This magnitude of current drive represents about 30% of that calculated with a ray tracing code which includes trapping effects and the weakly relativistic resonance condition. It was suggested by the authors that the discrepancy between the experiment and the theory may be attributable to the time scales involved, since in CLEO the confinement time for electrons is close to the electron-ion collision time, and the current drive mechanism relies on collisional relaxation of the heated electrons on the background ions.

The efficiency of electron cyclotron current drive (ECCD) tends to be low compared with that achieved in LHCD experiments because the waves interact primarily with electrons in the velocity range near the thermal velocity. One means of improving the current drive efficiency is to launch waves from the top or bottom of the torus so that they travel through the hot center of the plasma without approaching too close to the resonance, thereby damping on those high velocity electrons which generate a large Doppler shift. This absorption mechanism was first observed in experiments on PLT,[38] and subsequent experiments on WT-2,[39] DITE,[40] JFT-2M,[41] and WT-3[42] are in agreement with the theory of EC wave absorption on the tail of a distorted distribution function. This mechanism of frequency downshift may also be used to relieve the technical difficulties of generating ECH power at the frequencies needed for high field tokamaks like CIT.

Recently, calculations using a relativistic bounce-averaged Fokker-Planck code have shown that for outside launch the current drive efficiency rises as the power is increased.[42] This improvement of up to a factor 2 over the efficiency in the linear regime is due to a quasilinear increase in the tail electron population and to a decrease in the trapping of heated electrons due to relativistic effects.

## EXPERIMENTS IN THE NEAR FUTURE

During the next several years, new experiments on ECH of tokamak plasmas will start. The facilities planning ECH experiments include T-15, which will use 5 MW at 81 GHz; DIII-D and Tore Supra, each with 2 MW at 110 GHz; FTU, with 1 MW at 140 GHz; COMPASS with 2 MW at 60 GHz; and TEXT upgrade and RTP, with power at 60 GHz. The first experiments with extremely high peak power, several GW, are planned for the MTX tokamak in the summer of this year, using an induction-linac FEL as the source;[43] experiments will be at 140 GHz initially, and later at 250 GHz, with average power of 2 MW.

## REFERENCES

1 C.P. Moeller, R. Prater, A.C. Riviere, et al., *Proceedings of 6th Joint Workshop on Electron Cyclotron Emission and Electron Cyclotron Resonance Heating, Oxford, 1987*, Culham Report CLM-ECR (1987), p. 355.

2 M. Bornatici, et al., Nucl. Fusion **23**, 1153 (1983).

3 R. Prater, *Course and Workshop on Applications of RF Waves to Tokamak Plasmas* (International School of Plasma Physics, Italy, 1985) p. 354.

4 A.C. Riviere, *Applications of Radio-Frequency Power to Plasmas: Seventh Topical Conference, Kissimmee, FL 1987* (American Institute of Physics, New York, 1987), p. 1.

5 A.H. Kritz, H. Hsuan, R.C. Goldfinger, D.B. Batchelor, Proc. 3rd Joint Varenna-Grenoble International Symposium on Heating in Toroidal Plasmas, Grenoble (Euratom Report EUR7979EN, 1982) II, 707; this code was extended to noncircular tokamaks by G.R. Smith and others of Lawrence Livermore National Laboratory..

6 R. Prater, N.H. Brooks, K.H. Burrell, et al., Plasma Physics and Controlled Nuclear Fusion Research 1988 (Proc. 12th Int. Conf. Nice, 1988), E-1-2.

7 M. Ashraf, R. Barnsley, N. Delayanakis, et al., Plasma Physics and Controlled Nuclear Fusion Research 1988 (Proc. 12th Int. Conf. Nice, 1988), E-1-3.

8 FOM-ECRH Team and Equipe TFR, *Proceedings of 6th Joint Workshop on Electron Cyclotron Emission and Electron Cyclotron Resonance Heating, Oxford, 1987*, Culham Report CLM-ECR (1987), p. 257.

9 V.V. Alikaev, A.A. Bagdasarov, N.L. Vasin, et al., Sov. J. Plasma Phys. **13**, 1 (1987).

10 V.V. Alikaev, A.A. Bagdasarov, Yu. V. Balabanov, et al., Plasma Physics and Controlled Nuclear Fusion Research 1988 (Proc. 12th Int. Conf. Nice, 1988), E-1-1.

11 V.S. Strelkov, to be published in Proceedings of 16th European Conference on Controlled Fusion and Plasma Physics, Venice, Italy (1989).

12 B. Stallard, D. Content, R. Groebner, et al., "Heating and Confinement in H–mode and L–mode Plasmas in DIII–D using Outside Launch ECH," General Atomics Report GA-A19349 (1989); to be submitted to Nucl. Fusion.

13 J. Lohr, T. Edlington, V. Ilin, et al., "Recent Electron Cyclotron Heating Experiments with Low Field Launch of the Ordinary Mode on the DIII–D Tokamak," General Atomics Report GA-A19634 (1989).

14 S.M. Kaye and R.J. Goldston, Nucl. Fusion **25**, 65 (1985).

15 V. Alikaev, A. Bagdasarov, E. Berezovskii, et al., Plas. Phys. and Cont. Fusion **29**, 1285 (1987).

16 G.L. Jahns, S.K. Wong, R. Prater, et al., Nucl. Fusion **26**, 226 (1986).

17 TFR Group and FOM ECRH Team, Nucl. Fusion **28**, 1995 (1988).

18 N. Delayanakis, M. Cox, M.R. O'Brien, et al., *Proceedings of 6th Joint Workshop on Electron Cyclotron Emission and Electron Cyclotron Resonance Heating, Oxford, 1987*, Culham Report CLM-ECR (1987), p. 247.

19 M. Ashraf, C. Bishop, J.W. Connor, et al., Plasma Physics and Controlled Nuclear Fusion Research 1988 (Proc. 12th Int. Conf. Nice, 1988), A-V-2-1.

20 R. Stockdale, K.H. Burrell, W.M. Tang, Bull. Am. Phys. Soc. **33**, 1977 (1988).

21 D.C. Sing, M.E. Austin, R.V. Bravenec, et al., to be published in Proceedings of 16th European Conference on Controlled Fusion and Plasma Physics, Venice, Italy (1989).

22 R.W. Harvey, M.R. O'Brien, M.G. McCoy, and G.D. Kerbel, "ECE Spectra Calculated from ECRF Heated Distributions Obtained with a 3D Fokker Planck Code," General Atomics Report GA-A19667 (1989).

23 K. Hoshino, T. Yamamoto, N. Suzuki, et al., Nucl. Fusion **28**, 301 (1988).

24 K. Hoshino, T. Yamamoto, H. Kawashima, et al., Phys. Letters A **130**, 26 (1988).

25 J. Lohr, B.W. Stallard, R. Prater, et al., Phys. Rev. Letters **60**, 2630 (1988).

26  R. Prater, J.M. Lohr, T. Luce, et al., *Proceedings of 6th Joint Workshop on Electron Cyclotron Emission and Electron Cyclotron Resonance Heating, Oxford, 1987*, Culham Report CLM-ECR (1987), p. 195.

27  R.J. Groebner, P. Gohil, K.H. Burrell, et al., General Atomics Report GA-A19562 (1989); to be published in Proceedings of 16th European Conference on Controlled Fusion and Plasma Physics, Venice, Italy (1989).

28  K. Hoshino, T. Yamamoto, H. Kawashima, et al., "Observation of H–mode by Edge Heating Solely by Electron Cyclotron Heating in a Divertor Configuration of JFT-2M Tokamak," Japanese Atomic Energy Research Institute Report JAERI-M 89-038 (March 1989).

29  V.V. Alikaev, et al., Plasma Physics and Controlled Nuclear Fusion Research 1984 (Proc. 10th Int. Conf. London, 1984) I, 419.

30  H.P.L. deEsch, J.A. Hoekzema, W.J. Schrader, et al., Plasma Physics and Controlled Nuclear Fusion Research 1986 (Proc. 11th Int. Conf. Kyoto, 1986) F-III-4.

31  W.J. Goedheer, E. Westerhof, Nucl. Fusion **28**, 565 (1988).

32  R.T. Snider, D. Content, R. James, et al., Phys. Fluids B **1**, 404 (1989).

33  W. Park, D.A. Monticello, T.K. Chu, Phys. Fluids **30** 285 (1987).

34  E. Westerhof and W.J. Goedheer, in Controlled Fusion and Plasma Heating, 14–18 April 1986, Schliersee (European Physical Society, 1986) I 132.

35  B. Lloyd, T. Edlington, Plasma Physics and Controlled Fusion **28**, 909 (1986).

36  B. Lloyd, T. Edlington, M.W. Alcock, et al., in Controlled Fusion and Plasma Heating, 14–18 April 1986, Schliersee (European Physical Society, 1986).

37  B. Lloyd, T. Edlington, M.R. O'Brien, et al., Nucl. Fusion **28**, 1013 (1988).

38  E. Mazzucato, P. Efthimion, I. Fidone, Nucl. Fusion **25**, 1681 (1985).

39  A. Ando, et al., Nucl. Fusion **26**, 107 (1986).

40  A.C. Riviere, R. Barnsley, T. Edlington, et al., *Proceedings of 6th Joint Workshop on Electron Cyclotron Emission and Electron Cyclotron Resonance Heating, Oxford, 1987*, Culham Report CLM-ECR (1987), p. 237.

41  T. Yamamoto, K. Hoshino, H. Kawashima, et al., Phys. Rev. Letters **58**, 2220 (1987).

42  H. Tanaka, A. Ando, K. Ogura, et al., Phys. Rev. Letters **60**, 1033 (1988).

43  R.W. Harvey, M.G. McCoy, and G.D. Kerbel, Phys. Rev. Letters **62**, 426 (1989).

44  B.W. Stallard, G.R. Smith, R.A. James, et al., *Applications of Radio-Frequency Power to Plasmas: Seventh Topical Conference, Kissimmee, FL 1987* (American Institute of Physics, New York, 1987), p. 21.

# NUMERICAL SIMULATION OF ELECTRON CYCLOTRON EMISSION AND ABSORPTION IN TOKAMAK PLASMAS DURING LH CURRENT DRIVE EXPERIMENTS

A.Airoldi, A.Orefice and G.Ramponi

Istituto di Fisica del Plasma
Associazione EURATOM-ENEA-CNR
Via Bassini 15, 20133, Milano (Italy)

The feasibility of diagnosing the high energy electron tails driven by LH waves in Tokamak plasmas during LHCD experiments is investigated by computing the electron cyclotron emission (and absorption) spectra which could be measured both along lines of constant magnetic field and along the usual line of sight parallel to the horizontal midplane, making use of a suitable model for the electron distribution function.

The modelling and the calculation of the emission spectra are made by using a version of a toroidal ray tracing code /1,2/ which solves the radiation transport equation by evaluating the absorption coefficient given by the computation of the relativistic dielectric tensor and the emission coefficient obtained by a quasi-linear approach /3/. Besides the refractive effects, the effects of plasma diamagnetism, poloidal magnetic field, harmonic overlapping and finite acceptance angle of the antennae are also taken into account. The present version of the code is characterized by two basic features:

i) The plasma distribution function is a suitable analytical model of that provided by a lower hybrid ray tracing and self consistent Fokker-Planck absorption code /4,5/. We show in Fig.1 an example of such a distribution function, which is Maxwellian normally to the magnetic field while it exhibits two different flat tails in the positive and negative parallel direction, in the region of interaction with LH waves.

ii) In order to take into account high energy electron populations in the direction parallel to the magnetic field, a version of the relativistic dielectric tensor avoiding the "weakly" relativistic approximation for parallel energies is taken into account /6/. Although our code is conceived for working with D-shaped equilibria, the considered plasma cross section, for the time being, is taken to be circular, and the electron distribution function is mapped on 41 radial concentrical zones.

We present here the results of a simulation referring to a JET-like plasma, performed assuming the following parameters and profiles:

$$R_o = 3.06m \; ; \quad a = 1.07m \; ; \quad B(0) = 3.1T \; ; \quad q(a) = 3.2$$
$$n(r) = (n_o - n_a)(1 - (r/a)^2)^{\beta_n} + n_a$$
$$n_o = 6.77 \, 10^{13} cm^{-3}; \; n_a = 1.5 \, 10^{13} cm^{-3}; \; \beta_n = 0.44$$
$$T(r) = (T_o - T_a)(1 - (r/a)^2)^{\beta_T}$$
$$T_o = 6KeV; \; T_a = 0.7KeV; \; \beta_T = 2.$$

In the region r/a <0.5 , the distribution function is an unperturbed Maxwellian, while for r/a ≥ 0.5 it exhibits long flat tails both in the positive and negative parallel directions, up to energies $\simeq 200 KeV$ , and it has a Maxwellian behaviour with $T_\perp = 20 KeV$ in the perpendicular direction. The relative fractions of superthermal electrons with respect to the bulk density turn out to be functions of space as shown in Fig.2.

Aiming to evidence the sensitivity of the E.C. emission (and absorption) with respect to the main parameters of the superthermal population, we show in Fig.3 the emission spectra which should be measured (neglecting reflections) for the X-mode along a vertical line at R = 3.24 m.

Fig.4 shows the emission spectra for the O-mode obtained in the same conditions as those of Fig.3. All these graphs have been obtained by neglecting, for simplicity, the finiteness of the antenna beam and considering the central ray only.

In the vertical case, due to the fact that the magnetic field is constant along the ray trajectory (in absence of refraction and if the beam is sufficiently well collimated), the resonance condition $\omega - v\Omega/\gamma = 0$ $(n_\parallel = 0)$ ( where $\omega$ is the radiation frequency, $\Omega$ is the local electron cyclotron frequency, $v$ the harmonic number and $\gamma = (1 + (p/mc)^2)^{1/2}$ ), allows a one to one correspondence between the energy of the emitting (or absorbing) particles and the radiation frequency. Referring to the case of Fig 3a, the absorption and emission coefficients along the vertical trajectory are shown in Fig.5 for three different values of the frequency. In the case a), where the frequency is near the 3rd harmonic of the "local" electron cyclotron frequency, the radiation is emitted and absorbed by "bulk" electrons only $(3f_c/f = \gamma = 1.04)$. In the case b) the frequency value is such that $3f_c/f = \gamma = 1.165$, corresponding to a kinetic energy $E_k = 84 KeV$: the contribution to absorption and emission is due both to "bulk" and "tail" particles. In the case c) the frequency is "downshifted" $(3f_c/f = \gamma = 1.253)$ in such a way that only resonant superthermal electrons with $E_k \simeq 130 KeV$ are emitting and absorbing.

Fig.6 shows the emission spectra which could be measured, from the low field side, along a horizontal line of sight parallel to the equatorial plane. In this case, although the one to one correspondence between frequency and energy is lost because of the dependence on R of the magnetic field, the presence of superthermals electrons in the low field side region is evidenced by the features in the low frequency part ( 1st and 2nd harmonic region) of the spectra.

*(Work done under JET contract)*

REFERENCES

/1/A.Airoldi, A.Orefice, G.Ramponi, *Il Nuovo Cimento*, **6D**, 527(1985)

/2/G.Ramponi, A.Airoldi, D.V.Bartlett, M.Brusati. S.Nowak, A.Orefice, *Proc.5th Int. Workshop on ECE and ECH*, S.Diego (1985), *pg.*10

/3/A.Orefice , *Il Nuovo Cimento* **8D**, 318 (1986)

34

/4/A.Orefice: to be published
/5/M.Brusati,C.Gormezano,S.Knowlton,M.Lorentz Gottardi,F.Rimini: this
Conference
/6/A.Orefice *J.Plasma Phys.* **34**, 319 (1985)

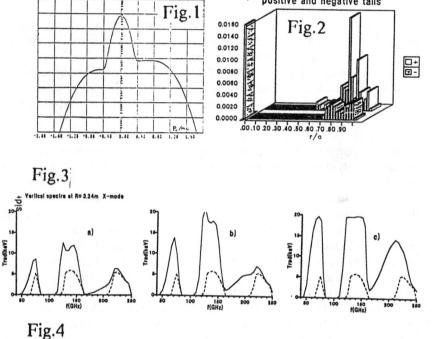

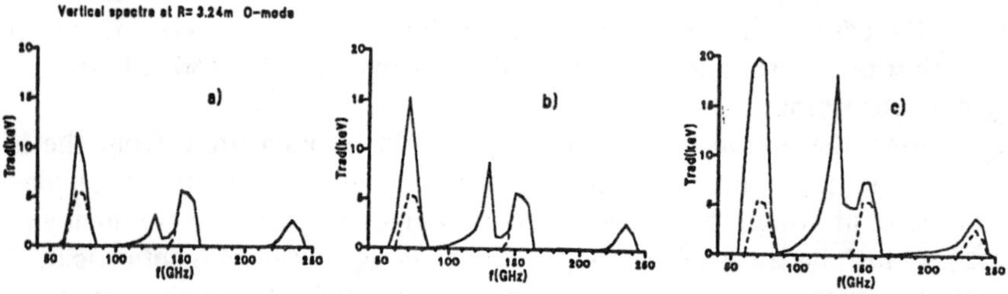

Fig.1:-Semilog plot of the electron distribution function vs. parallel momentum.

Fig.2:-Radial distribution of the relative fractions of electrons in positive and negative tails.

Fig.3:- Radiation spectra along a vertical chord for the X-mode. The continuous
line is obtained with superthermals with $T_\perp = 20\ KeV$ and relative densities
as shown in Fig.2; the dashed line is obtained without superthermals. In
case b) $T_\perp = 30\ Kev$. In case c), $T_\perp = 20\ KeV$ and superthermal fractions
increased by a factor 10.

Fig.4:-As Fig.3, for the O-mode.

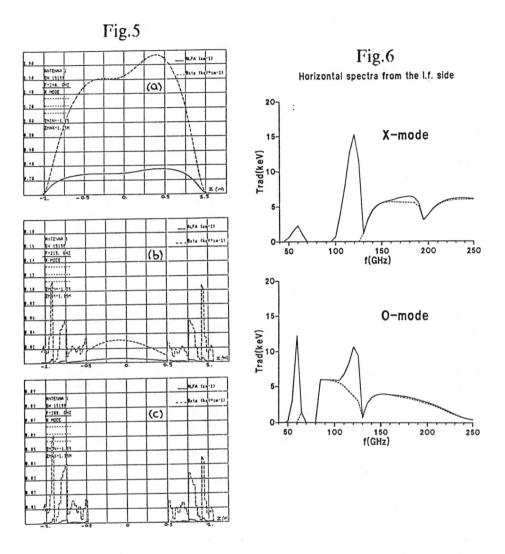

**Fig.5**

**Fig.6**

Horizontal spectra from the l.f. side

Fig.5:-Emission (dashed line) and absorption (continuous line) coefficients along the vertical trajectory, in the same conditions as Fig.3a). In the case (a) the frequency is f = 240 Ghz, in (b) f = 215 , in (c) f = 200 Ghz.

Fig.6:-Radiation spectra, with (continuous line) and without (dashed line) superthermals, "viewing" from the low field side along a line parallel to the equatorial plane.

# Preparation for Propagation and Absorption Experiments in MTX[*]

*J.A. Byers, R.H. Cohen, M.E. Fenstermacher, E.B. Hooper,*
*S. Meassick, T.D. Rognlien, G.R. Smith, and B.W. Stallard*
Lawrence Livermore National Laboratory, Livermore, CA 94550

## ABSTRACT

Preparatory calculations of microwave transmission through the MTX access duct, propagation of the waves through the plasma and the resulting power deposition profile on a calorimeter located on the tokamak inside wall have been performed. The microwave transmission calculations include the relative phase slippage of waveguide modes in the duct to determine the spatial structure of the wavefront at the duct exit. Ray-tracing calculations show substantial spreading of the beam in the poloidal direction at densities above $1.5 \times 10^{20}$ m$^{-3}$, well within the range of the experiments. Initial experiments with low or high toroidal field (cyclotron resonance outside the plasma) will investigate both diffraction and refraction effects, without absorption. Estimates of the fractional absorption of the beam in the initial experiments with the cyclotron resonance at the plasma axis have also been made.

## INTRODUCTION

The initial ECH absorption experiments in MTX will be done during the summer of 1989. The nominal major and minor radii in these experiments will be $R_o = 64$ cm and $a = 16.5$ cm, respectively, and the field on axis will be $B_o = 5$ T. A single pulse of 140 GHz microwave radiation from the ETA-II/ELF free electron laser (FEL) will be used per tokamak shot. The primary diagnostic for determining the fractional absorption of the FEL power by the plasma is a microwave calorimeter mounted on the inner wall of the vacuum vessel opposite the FEL access port. To calculate the fractional absorption, comparisons of the power absorbed by the calorimeter for shots without plasma, shots with plasma but with the cyclotron resonance outside the tokamak, and shots with the resonance in the plasma will be made. To interpret these experiments and to predict results for various plasma conditions, several theoretical tools have been developed which analyze the propagation and absorption of the beam through the MTX access port and the plasma. The tools and results described below include wave propagation with a Huygens integral code[1] to determine the effects of diffraction, non-linear absorption calculations at the resonance plane with an orbit code, and ray-tracing with the TORCH code[2] to determine the plasma refraction effects on the beam.

## MICROWAVE TRANSMISSION THROUGH THE MTX DUCT

The access duct to the MTX plasma through the toroidal field coils is a tall, narrow rectangular opening 4 cm wide by 25 cm high by 22 cm deep. To calculate the transmission of the microwave beam through this duct, the field impinging onto the entrance of the duct is represented as a set of duct waveguide modes, which then are transported to the end of the duct with the phase shifts appropriate for each mode. The transverse variations of both the incident amplitude $E_x(x, y)$ and phase $\phi(x, y)$ are obtained from the Huygens integral code MTH.[1] This code calculates the transport of the microwave beam from the FEL output through a system of several mirrors up to the duct. Two dimensional Fourier decompositions of an in–phase piece, $E_R = E_x(x, y) \cos \phi(x, y)$, and an out–of–phase piece, $E_I = E_x(x, y) \sin \phi(x, y)$, are required. The Fourier coefficients for even modes are given by

$$A_{Rmn} = \frac{\int_{-a}^{a} dx \int_{-b}^{b} dy \, E_R(x, y) \cos \frac{m\pi x}{2a} \cos \frac{n\pi y}{2b}}{ab}$$

---

[*] Work performed for the U. S. Department of Energy by Lawrence Livermore National Laboratory under contract W-7405-ENG-48.

and similarly for $A_{Imn}$ in terms of $E_I$. Each index $mn$ is associated with a waveguide mode with a transverse wavenumber given by $k_\perp^2 = \left(\frac{m\pi}{2a}\right)^2 + \left(\frac{n\pi}{2b}\right)^2$ and a corresponding axial wavenumber given by $k_z^2 = k^2 - k_\perp^2$.

The net phase at the end of the duct is the sum of the initial phase and the phase shift through the duct $\Delta_{mn}$. For each mode, the real and imaginary mode coefficients at the end of the duct can be expressed in terms of the real and imaginary mode coefficients at the beginning of the duct and the differential phase shift for each mode. The total real and imaginary fields at any point $x, y$ at $z = L$ are given by summing the Fourier series for $E_R$ and $E_I$. Then the total field amplitude and phase at $z = L$ are $E(x, y) = \sqrt{E_R^2(x, y) + E_I^2(x, y)}$ and $\phi(x, y) = \tan^{-1}\frac{E_I(x,y)}{E_R(x,y)}$. Using the MTH code this field is propagated another 18 cm to the plasma absorption region.

For the beam and duct parameters of the initial FEL experiments, the magnitude of the electric field exiting the duct on the horizontal midplane at the duct wall is 42% of the central field of the beam. This produces a mixture of launched modes which interfere as they propagate through the plasma. The horizontal field pattern at the plasma center is shown in Fig. 1. The electric field at the duct wall in the vertical midplane is only a few percent of the central field. The vertical field pattern in the plasma is nearly Gaussian.

The structure in the horizontal field pattern may complicate the interpretation of the initial experiments. Corrugated inserts along the duct walls may be used in later experiments to reduce the wall fields and smooth the field pattern to a nearly Gaussian shape. Tapering of the duct may also be used to spread the beam out toroidally in the plasma and thereby increase the nonlinear absorption.

## CALCULATION OF NONLINEAR ABSORPTION

The nonlinear absorption of the FEL is calculated using a computer code ORPAT which solves reduced relativistic equations of motion[3] for a Maxwellian ensemble of electrons streaming through the rf beam along the static magnetic field, $\mathbf{B}$. The electric field as seen by an electron is expanded into a Bessel function series where one term is resonant for each cyclotron harmonic number $\ell$ defined by the resonance condition $\omega - \ell\omega_c/\gamma - k_\parallel v_\parallel = 0$. Here $\gamma = (1 + p^2/m^2c^2)^{1/2}$ is the relativistic mass factor. Only one such harmonic is included for a typical calculation ($\ell = 1$ for MTX). The effect of the rf on the energy, magnetic moment, and gyrophase is calculated for each particle passing through the beam. The power absorbed is determined by taking the net change in the energy content of the ensemble. The reduced equations allow an economical evaluation of the absorption and the results have been checked with a code which solves the full relativistic equations of motion.

The nonlinear absorption is sensitive to the variation of the wavenumber $k_\parallel$ across the rf beam. This variation arises from the divergence of the beam and from the curvature of the toroidal field line, and causes the resonant energy to increase as the electrons stream across the beam. Thus, electrons trapped about the resonance are lifted to higher energy; for MTX, this process is the major contributor to the power absorption[4].

We use the variation of $k_\parallel$ and $E_\parallel$ calculated just in front of the absorption region by the MTH code. The ORPAT code then calculates the absorption on a series of flux surfaces (usually 5) as the microwave beam propagates through the resonance layer. The parallel electric field, $E_\parallel$, is attenuated according to the equation $dP/dR = -W_{rf}(s, R)$, where $P(s, R) = \epsilon_0 cNE_\parallel^2/2$ is the local Poynting flux in MKS units, $N$ is the refractive index, and $W_{rf}$ is the power absorbed per unit volume. Also, $s$ and $R$ are the directions along the magnetic field and the major radius, respectively. The value of $k_\parallel$ is assumed unaltered by the nonlinear interaction.

The calculated absorption for 140 GHz in MTX is shown in Fig. 2 at several densities. Without enhancing the spread in $k_\parallel$ by tapering the transmission duct, the absorption of a 2 GW pulse is predicted to be about 40% at $n_e = 1.8 \times 10^{20}$ m$^{-3}$ and $T_e = 1$ keV. The absorption is nearly proportional to $T_e$. The scaling of the total absorbed power with incident power is $P_{abs} = 0.62 + 0.28 \ln P_{inc}$ for 0.5 GW $< P_{inc} <$ 4 GW.

## RAY-TRACING CALCULATIONS

Ray-tracing for electron cyclotron wave propagation in MTX is done with the TORCH code[3] using the cold plasma approximation. For the purposes of determining the launch angles and locations of the rays, the microwave beam launched from the MTX port is modeled as an elliptical Gaussian beam.

The microwave transmission calculations show that in the horizontal direction the beam waist dimension, $w_0$, is about 1 cm. This means that the propagation lengths are of order 1–2 Rayleigh lengths, $Z_R \equiv \pi w_0^2 / \lambda \approx 15$ cm for $\lambda = 2.1$ mm, so that diffraction will be important. In addition, a mixture of modes are launched from the duct which interfere in the plasma. These effects are modeled crudely in the ray-tracing calculation by launching rays perpendicular to a wavefront shaped such that in the absence of plasma (refraction effects), the ray propagation to the center would yield a Gaussian approximation to the wavefront calculated by the Huygens integral code. The horizontal radius of curvature of this launched wavefront at the duct exit is 33.7 cm.

The Rayleigh length for the vertical direction, $Z_R \approx 4$ m, is much larger than the minor radius so that the plasma is in the near field of the duct for this direction. With $Z_R >> 2a$ and small fields on the duct walls, diffraction effects are negligible in this direction. The launched vertical wavefront at the duct exit, as calculated by the MTH code, is concave towards the plasma with a radius of curvature of 275 cm.

The ray launch locations are calculated such that each ray represents the same fraction of the total beam power. This assures that calculational time is not wasted on rays which do not represent significant beam power. The beam is subdivided into elliptical zones with minor and major semi-axes $a_i$ and $b_i$ respectively. Each annular zone $i$ is further sub-divided into azimuthal sectors. A ray is launched from the center of each of these wedge-shaped areas of the beam. The zone radii $a_i$ and $b_i$, and the azimuthal angles of the sector edges are determined by solving multiple integral equations such that the integral of the beam power in each wedge shaped region is the same.

The target plasma for the initial experiments has a central electron density and temperature, $n_e(0) = 1.8 \times 10^{20}$ m$^{-3}$ and $T_e(0) = 1.0$ keV respectively. The density profile is of the form $n_e(r) = n_e(0)[1 - (r/a)^2]^{\alpha_n}$ with $\alpha_n = 1.0$. A poloidal projection of the ray trajectories in this target plasma is shown in Fig. 3. Ray tracing calculations without absorption have been done for plasma densities in the range $1.0 \leq n_e(0) \leq 2.4 \times 10^{20}$ m$^{-3}$ and profiles in the range $0.5 \leq \alpha_n \leq 2.0$. The cut-off density for these experiments at 140 GHz is $n_c = 2.45 \times 10^{20}$ m$^{-3}$. For $n_e(0)$ less than the target density the fraction of the beam which misses the calorimeter is less than 10%. As the density is increased toward $n_c$, poloidal refraction of the beam center increases rapidly and a vertical cut of the power profile on the calorimeter begins to show a depression near the calorimeter center and maxima near the top and bottom.

The power profile on the calorimeter is only weakly sensitive to changes in the plasma density profile for central densities less than the target density. Calculations using the target density on axis and profiles in the range $0.5 \leq \alpha_n \leq 2.0$ show a slight increase in the bunching of the power at the extremes of the calorimeter for more peaked profiles. The power fraction missing the calorimeter did not exceed 20% in any of these cases.

## SUMMARY

Tools for calculating wave propagation, refraction and diffraction effects, and nonlinear absorption in the MTX experiments have been developed. Calculations of microwave transmission through the duct show that with the smooth walled duct the wave pattern in the plasma will have a complicated spatial structure. Nonlinear absorption calculations indicate that approximately 40% of a 2 GW beam pulse with this spatial structure will be absorbed with the resonance on axis. Calculations are in progress using corrugated duct walls to produce a smooth Gaussian wave pattern in the plasma and tapering in the duct which should increase the absorption by increasing the spread in $k_{\parallel}$ of the beam in the plasma.

Refraction of the beam by the plasma, as calculated by TORCH, produces a reduction in the power density on the inside of the torus compared with the outside. By moving the location of the resonance (by varying **B**) from outside to inside, an asymmetry in the nonlinear absorption, which would be absent for linear absorption, may be measurable. Experiments at fixed resonance location with various FEL powers will also be used to determine the nonlinear characteristics of the absorption.

## REFERENCES

[1] T. Samec, M. Makowski, and B. Stallard, Bull. Am. Phys. Soc. **32**, 1872 (1987).
[2] R. Myer, M. Porkolab, G.R. Smith, and A.H. Kritz, MIT PFC/JA-89-2 (1989).
[3] T.D. Rognlien, Phys. Fluids **26**, 1545 (1983).
[4] W.M. Nevins, T.D. Rognlien, and B.I. Cohen, Phys. Rev. Lett. **59**, 60 (1987).

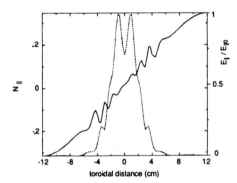

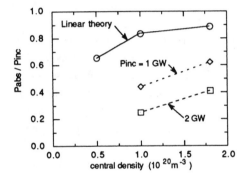

Figure 1. Parallel refractive index (solid) and parallel electric field (dashed) vs. toroidal distance within the beam.

Figure 2. Fractional absorption vs. central density for linear theory (o), and non-linear theory with 1 GW (◊) and 2 GW (□) pulses.

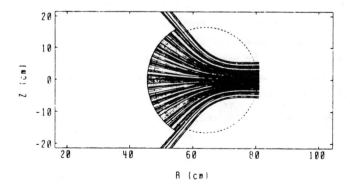

Figure 3. Poloidal projection of ray trajectories in the MTX target plasma

ELECTRON-CYCLOTRON CURRENT DRIVE IN THE TORE SUPRA TOKAMAK

I. FIDONE, G. GIRUZZI, R.L. MEYER[a], DRFC - CADARACHE, FRANCE

## Abstract

An experimental program on electron-cyclotron current drive in the TORE SUPRA tokamak at Cadarache is planned. The target plasma is a Maxwellian bulk with a fast tail generated by lower-hybrid waves. The aim of the program is twofold, i.e., flexible radial profile control of the RF driven current and experimental proof of the electron-cyclotron current drive by relativistic electrons. In this paper, we present theoretical predictions concerning the two problems.

## I. INTRODUCTION

TORE SUPRA is a large tokamak with superconducting coils ($N_b T_i$ cooled with superfluid $H_e$ at 1.8°K) with a = 0,7 m, R = 2.25 m and B (o) ≤ 4.5 Tesla. The experimental program of Tore Supra assigns an important role to non-inductive current drive. A first step towards this goal is the installation of 16 CW klystrons at 500 KW each at F = 3.7 GHz to drive toroidal current by lower-hybrid waves. A second step is to add 6 CW gyrotrons at 400 KW each and f = 110 GHz to interact with the lower-hybrid tail.

## II. ELECTRON-CYCLOTRON CURRENT DRIVE USING FAST ELECTRONS

### a) Current profile control

Recent theoretical studies have identified routes to second stability regimes in tokamak plasmas by raising q on axis to values significantly greater than unity[1]. To test experimentally these theoretical predictions, an efficient method of current profile control near the plasma axis is necessary. In low temperature plasmas, lower-hybrid travelling waves are very efficient for central current drive but not sufficiently flexible for radial profile control. We show that by coupling electron-cyclotron waves with the lower-hybrid sustained tail[2], q(r) is appropriately changed at constant total current. This problem is studied by a 3-D bounce-averaged Fokker-Planck code (i.e., two-dimensional in momentum space and one-dimensional in real space). Specifically, the final current profile is obtained by making the following steps. We first compute the electron velocity distribution in the presence of lower-hybrid waves using the quasilinear equation

$$\frac{\partial f}{\partial t} = \langle \frac{\partial}{\partial p_{\|}} D_{LH} \frac{\partial f}{\partial p_{\|}} \rangle + \langle (\frac{\partial f}{\partial t})_{coll} \rangle \qquad (1)$$

where $(\partial f/\partial t)_{coll}$ is the Coulomb collision operator and the bracket denotes a bounce-average over the electron orbit. Next we apply the EC wave
described by the operator

$$(\frac{\partial f}{\partial t})_{cy} = 4\pi e^2 \int_{-\infty}^{\infty} dN_{||} \frac{\gamma}{p_\perp} L D_{cy} \delta(\gamma - \omega_c/\omega - N_{||} p_{||}/mc) \text{ I f, where } \gamma L = (\omega_c/\omega)$$

$(\partial/\partial p_\perp) + (N p_{||}/mc) (\partial/\partial p_{||})$, $D_{cy}$ is the electron-cyclotron diffusion coefficient and $\gamma$ is the relativistic factor. The mechanism for which EC waves
modify the LH driven current locally, is the reduction of the collision
frequency of a selected group of electrons in the LH tail. This reduces
the electron drag and enhances the parallel pushing of the LH waves. We
consider the following plasma and wave parameters :

$n_e(r) = 0.5 \times 10^{14} cm^{-3} \times (1-r^2/a^2)$, Te $(r) = 3$ kev $\times (1-r^2/a^2)^{3/2}$,
$D_{LH} = const = 0.8$ for $c/4 \le V_\emptyset \le c/2$ and $D_{LH} = 0$ outside. For the EC waves
we consider both extraordinary (B = 4.5 T) and ordinary (B = 3.8 T) for
top launching near normal to B (10° - 15°). For the two modes, the poloidal projection of the ray trajectories and the corresponding damping are
shown in Figs.1(a) and 1(b) respectively. The current density profile for
the 0-modes without (dashed) and with EC waves is shown in Fig.2 for I =
0.464 MA and a) $P_{EC} = 2$ MW ($P_{TOT} = 3.8$ MW) and b) $P_{EC} = 4$MW ($P_{TOT} = 5.1$).
It appears that the progressive substitution betwen the two waves results
in a degradation of the overall efficiency. This is due to the fact that
the EC wave is coupled with the low energy part of the tail.

b)    Current drive by relativistic electrons

Current drive by electron-cyclotron waves is presently considered as a potential candidate for central current generation in a high temperature plasma and the implementation of the method is envisaged for
steady-state operation of a tokamak reactor. In a hot plasma with $T_e(0) >$
25 Kev the current carrying electrons are in the relativistic range of
energy and the current drive efficiency attains its maximum value. In low-
temperature plasmas, current drive by relativistic electron is possible if
the electron-cyclotron wave is coupled with the far end of the lower-
hybrid tail. According to the present understanding of the tail formation,
the upper limit of the lower-hydrid phase-velocity spectrum is determined
by the condition $n_{||} \ge n_c$, where $n_{||}$ is the lower-hybrid parallel retarda-
tion index. Therefore, the lower-hybrid sustained tail terminates at the
parallel momentum $p_{|| c} = mc (n_c^2 - 1)^{-1/2}$, where c is the speed of light.
However, a relatively slowly decaying electron population exists

for $p_{\parallel} > p_{\parallel c}$ with a momentum distribution f determined by Coulomb collisions only. This part of the lower-hybrid tail can be the target for electron-cyclotron wave absorption which, if the electrons are well confined, generates an additional current with the same efficiency as in a high temperature Maxwellian plasma. Using the impulse-response method[3], for $N_{\parallel} = 0$ we obtain

$$I_{cy}/W_{cy} = (15.6/\Lambda) \frac{2Q(\omega_c/\omega)}{n_e(0)} \langle V_{\parallel} \rangle \ (A/W),$$

where

$$Q(\gamma) = (\gamma + 1)^{-1/2} \gamma^{-2} (\gamma - 1)^{-5/2} [2\gamma^2(\gamma + 2)\ln\gamma - (4\gamma - 1)(\gamma^2 - 1)],$$

and

$$\langle V_{\parallel} \rangle = \int_{V_{\parallel}c}^{V+} dV_{\parallel} V_{\parallel} P(V_{\parallel})/P, \quad P = \int_{V_{\parallel}c}^{V+} dV_{\parallel} P(V_{\parallel}),$$

$$P(V_{\parallel}) = -(\pi m^3 c^3/16\omega) \ \omega_p^2 \ (\omega_c/\omega) |E_x - iE_x + (N_{\perp}V_{\parallel}\omega/\omega_c)E_z|^2 (V_{\perp} \frac{\partial f}{\partial V_{\perp}})_{V_{\perp} = V_{\perp o}},$$

$\overline{V} = \overline{p}/mc$, $V_+ = [(\omega_c/\omega)^2 - 1]^{1/2}$, $V_{\perp o}^2 = (\omega_c/\omega)^2 - 1 - V_{\parallel}^2$. In the range of interest $(p_{\parallel c} > 0.5 \ mc)$, $\omega_c/\omega > 1.3$ and $Q = 0.2$, thus $\eta = n_e R I_{ey}/\omega_{ey} = 0.4 \langle V_{\parallel} \rangle$, where $n_e$ is in $10^{20}/m^3$, and R in meters.

III. CONCLUSION

We have shown that the combined system of EC and LH waves is appropriate for both radial control of the rf current and the experimental proof of the EC current drive by relativistic electrons.

REFERENCES

1/ M.J GERVER, J. KESNER, and J.J. RAMOS, Phys. Fluids 31, 2674 (1988).

2/ I. FIDONE, G. GIRUZZI, V. KRIVENSKI, E. MAZZUCATO, and L.F. ZIEBELL, Nuc. Fusion 27,579 (1987)).

3/ I. FIDONE, G.GRANATA, and J.JOHNER, Phys. Fluids 31,2300 (1988)

a) Laboratoire de Physique des Milieux Ionisés, U.A CNRS 835 Université de Nancy I, FRANCE.

43

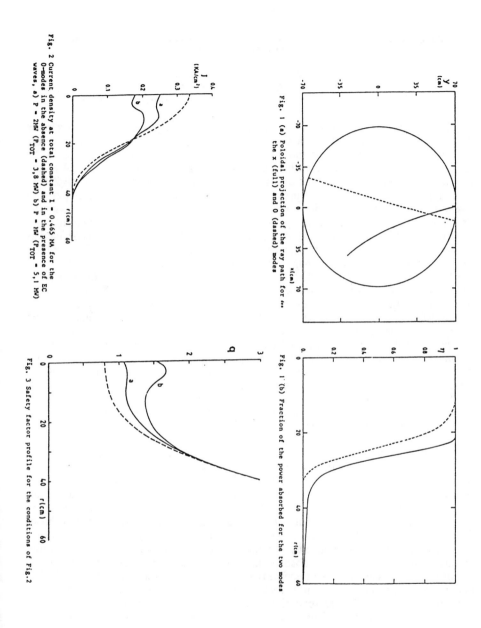

Fig. 1 (a) Poloidal projection of the ray path for the x (full) and 0 (dashed) modes

Fig. 1 (b) Fraction of the power absorbed for the two modes

Fig. 2 Current density at total constant I = 0.465 MA for the 0—modes in the absence (dashed) and in the presence of EC waves, a) P = 2MW ($P_{TOT}$ = 3,8 MW) b) P = MW ($P_{TOT}$ = 5,1 MW)

Fig. 3 Safety factor profile for the conditions of Fig.2

# ELECTRON CYCLOTRON CURRENT DRIVE AND PROFILE CONTROL IN NET

G. Giruzzi, T.J. Schep, E. Westerhof
*Association EURATOM-FOM, FOM Instituut voor Plasmafysica*
*"Rijnhuizen", Nieuwegein, The Netherlands*

**1. Introduction.** A potentially attractive scenario for steady-state operations in the Next European Torus relies on the use of lower-hybrid (LH) waves for non-inductive current drive in the plasma periphery and of electron cyclotron (EC) waves in the central region [1]. Current drive by EC waves can also be used for control of the current density profile around the q=2 surface to prevent the growth of the m=2, n=1 tearing mode and, thereby, to prevent major disruptions [2]. In this paper, we investigate both these applications of current drive by EC waves theoretically with the aim of determining for each of the applications the best options for the EC current drive system and of evaluating the expected current drive efficiency. The latter is defined as $\eta = I_{CD}R_0\bar{n}_e/P_{EC}$ (AW$^{-1}$ $10^{20}$ m$^{-2}$), where $I_{CD}$ is the total driven current, $P_{EC}$ the total dissipated wave power, $R_0$ the major radius and $\bar{n}_e$ the average density.
The basic NET parameters are taken to be: $R_0$ = 5.25 m, minor radius in the equatorial plane a = 1.4 m, elongation $\kappa$ = 2.2, $B_T$ = 5.5 T, $I_p \approx$ 11 MA, $\bar{n}_e$ = 0.75 $\times$ $10^{20}$ m$^{-3}$, and $T_e$ = 15 keV. The profiles chosen for this study are: $n_e = n_e(0)(1-\rho^6)$, $T_e = T_e(0)(1-\rho^2)$, where $\rho^2 = (\psi - \psi_0)/(\psi_b - \psi_0)$, $\psi$ is the poloidal flux function, and $\psi_0$, $\psi_b$ its values at the plasma axis and boundary, respectively. The EC waves trajectories in the NET magnetic configuration, obtained from an equilibrium code, are evaluated by means of the toroidal ray-tracing code TORAY [3]. Furthermore, the ray-tracing is coupled to a 3-D bounce-averaged quasilinear Fokker-Planck code [4] which determines the modifications of the electron distribution function due to the absorption of EC wave beams of finite angular spread, the driven current and evaluates the wave damping selfconsistently. The Coulomb collission operator is approximated in the code by its relativistic high velocity limit.

**2. Central current drive for steady state operation.** Current drive by EC waves (ECCD) requires oblique propagation (i.e., parallel refractive index $N_{||} \neq 0$) and wave absorption at frequency $\omega$ significantly different from the EC frequency $\omega_c$, because of the relativistic and Doppler detuning mechanisms. Two scenarios are possible: wave absorption at downshifted frequency ($\omega < \omega_c$) or at upshifted frequency ($\omega > \omega_c$). The use of waves at downshifted frequencies requires wave injection from the high-field side of the resonance, which in practice means injection from the top or bottom of the torus, and allows the use of X-mode waves which for oblique propagation show the highest absorption. On the other hand, the use of waves at upshifted frequencies allows injection from the outside in the equatorial plane, but requires the use of O-mode polarization. Extensive ray-tracing calculations have been performed for each of these scenarios showing that in both scenarios good central power deposition can be obtained and that the deposition profile is easily changed by changing the injection direction or the frequency of the waves. For a number of selected cases, we then performed Fokker-Planck quasi-linear calculations to calculate the current drive efficiencies. In these caculations a wave beam with a cross-section of 100 cm$^2$ and with a divergence of 5° has been used. The results of these calculations are given in Table I. The first three columns give the results for the injection of 80 GHz X-mode waves from the top of the torus. The injection angles $\theta$, the angle between the vertical and the direction of wave injection, and $\phi$, the angle between the major radius and the direction of wave injection, are chosen to obtain central deposition or slightly off-axis deposition to obtain an idea of the trapped particle effects on the current drive efficiency. The other columns give the results for the equatorial injection of O-mode waves at different frequencies and injection angles. The current density profiles corresponding to the cases of the last three columns are given in Fig. 1. From these results we can draw the following conclusions. The use of downshifted frequencies, although in principle very attractive because of the high absorption of the X-mode at oblique propagation, yields current drive efficiencies which are well below the best obtained using upshifted frequencies. Moreover, the efficiency drops by about a factor of three for off-axis deposition as a consequence of trapped particle effects [4, 5]. For the O-mode using an upshifted frequency, the efficiency deterioration due to trapping effects is less than a factor of 2 when the location of wave power deposition is shifted from the centre to $\rho \approx 0.5$. Using upshifted frequencies, it further appears, that the position of the driven current can be varied easily by changing only the toroidal injection angle, and that the power dependence of the efficiency is very weak, as expected for $\omega > \omega_c$ [6]. The efficiency can be significantly improved by increasing the wave frequency, since this generally enhances the resonant energy. It must be noted, however, that in the case of 220 GHz and central power deposition some 10% of the wave power will be absorbed due to overlap with the second harmonic and this power will (at best [7]) not contribute to the driven current. This effect is not accounted for in the numbers given in Table I and limits the maximum usable frequency to around 240 GHz and the maximum obtainable efficiency to around $\eta \approx 0.3$.

## TABLE I

Results of Fokker-Planck quasi-linear calculations for central current drive. A beam with a cross-section of 100 cm$^2$ and with a divergence of 5° is used.

| frequency (GHz) | 80 | 80 | 80 | 220 | 220 | 180 [a)] | 180 [b)] | 180 [c)] |
|---|---|---|---|---|---|---|---|---|
| mode | X | X | X | O | O | O | O | O |
| injection | top | top | top | equatorial | eq. | eq. | eq. | eq. |
| $\phi$ | –90° | –90° | –90° | 145° | 145° | 155° | 150° | 145° |
| $\theta$ | 155° | 145° | 145° | 90° | 90° | 90° | 90° | 90° |
| $P_{EC}$ (MW) | 11 | 45 | 90 | 13 | 6.5 | 10 | 75 | 75 |
| $I_{CD}$ (MA) | 0.34 | 0.51 | 0.76 | 1.12 | 0.54 | 0.53 | 2.45 | 2.25 |
| $\eta$ | 0.12 | 0.045 | 0.035 | 0.34 | 0.33 | 0.21 | 0.13 | 0.12 |
| $\rho_{abs}$ | 0.1 | 0.4 | 0.4 | 0.1 | 0.1 | 0.1 | 0.3 | 0.5 |

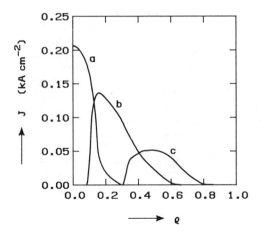

**Fig.1:** Profiles of the driven current density for O-mode beams of frequency 180 GHz, half-width $\Delta\phi = \Delta\theta = 5°$, injected in the equatorial plane, and:
(a) $\phi = 155°$, W = 10 MW; (b) $\phi = 150°$, W = 75 MW; (c) $\phi = 145°$, W = 75 MW.

**3. The use of ECCD for current profile control.** Here, we study the use of ECCD for current profile control around the q=2 surface in order to stabilize the m=2, n=1 tearing mode. A sufficient condition for stabilization of the m=2 tearing mode is the complete flattening of the current density profile just inside the q = 2 surface. Assuming a Gaussian profile for the driven current $j_{CD} \sim \exp(-\beta(\rho-\rho_{CD})^2)$, where $\rho$ is the normalized minor radius and $\beta$ the peakedness, this condition was shown to lead to the following requirement on the amplitude $I_{CD}$ and localization of the driven current [8]:

$$\left|\frac{I_{CD}}{I_p}\right| \approx \frac{10}{\beta}, \text{ and } \rho_{CD}-\rho(q=2) \approx -H(-I_{CD}/I_p)\sqrt{2/\beta} \pm \sqrt{1/\beta}, \quad (1)$$

where H is the Heaviside function. In general, the requirement given by Eq. (1), is an overestimate of the current required for stabilization by about a factor of 2 [8]. To optimize the efficiency of the current profile control both the current drive efficiency $\eta$ and the peakedness $\beta$ must be maximized. Based on the results of the previous section, the O-mode at upshifted frequencies is chosen to obtain the optimum $\eta$. The waves are launched in the equatorial plane, because this minimizes the spread in the position of power depostion due to the finite width of beam, while it fully uses the small natural line width of the absorption process for EC waves to obtain highest peakedness.

Extensive ray-tracing calculations have been performed, in which a first estimate for the current drive efficiency $\eta$ is obtained from the Fisch formula generalized to include trapped particle effects [4]. The wave parameters are chosen such that the power is deposited around the q=2 surface which for the magnetic equilibrium of NET used here, is located at $\rho = 0.85$. To analyze the influence of the temperature at q=2, various temperature profiles $T_e = T_e(o) (1 - \rho^2)^\alpha$ are used with fixed $T_e(o) = 30$ keV. For $\alpha = 1, 3/2, 2$ we have $T_e(q=2) = 8, 4, 2$ keV, respectively. A Gaussian beam with a divergence of $2.5^o$ and a cross-section of 100 $cm^2$ is injected from R = 7 m. The results are presented in Fig. 2 for the case with $T_e(q=2) = 8$ keV. In Fig. 2a the peakedness $\beta_{beam}$ of the deposition profile of the total beam and the peakedness $\beta_{ray}$ of the deposition along the central ray of the beam are given as a function of the toroidal injection angle $\phi$. In Fig. 2b the wave frequency normalized to the cyclotron frequency on q=2, $f/f_c(q=2)$ with $f_c(q=2) = 138$ GHz, and the estimate for the current drive efficiency $\eta$ are given. A maximum for both $\beta_{beam}$ and $\eta$ is found for $\phi \approx 210^o$ where $\beta_{beam} \approx 250$ and $\eta \approx 0.07$ [$AW^{-1} 10^{20}$ $m^{-2}$]. Similar results are obtained for the cases with a temperature of 4 and 2 keV at the q=2 surface.

A more accurate estimate of the current drive efficiency is obtained from Fokker-Planck quasi-linear calculations. The results are given in Table II for each of the three different temperatures. The peakedness of the driven current density profile is also given. Substituting the obtained values for the peakedness into Eq. (1), and using the fact that Eq. (1) is an overestimate of the required current for stabilization of m=2 tearing mode by about a factor of 2 [8], one finds that the currents that must be driven to stabilize the m=2 mode are 215 kA in the case with $T_e(q=2) = 8$ keV, 155 kA in the case with $T_e(q=2) = 4$ keV, and 135 kA when $T_e(q=2) = 2$ keV. Thus we can conclude that, in both the 8 keV and the 4 keV case, the current driven by 10 MW is sufficient for stabilization. In the 2 keV case, the maximum efficiency is no longer obtained at $\phi = 210^o$, but at $\phi \approx 200^o$. In the last column of Table I, the results for a calculation for the injection of 10 MW with $\phi = 200^o$ and a smaller divergence of only $1^o$ are presented, which shows that also in this case 10 MW of EC power suffices for stabilization. At the high frequencies used here, such a small divergence is not unrealistic.

**TABLE II**

Results of Fokker-Planck quasi-linear calculations for ECCD around the q=2 surface.
A beam with a cross-section of 100 cm$^2$ and with a divergence of 2.5$^0$ is used.

| $T_e(q=2)$ | 8 keV | 4 keV | 2 keV | 2 keV $^*$ |
|---|---|---|---|---|
| $\phi$ | 210$^0$ | 210$^0$ | 210$^0$ | 200$^0$ |
| $\dfrac{f}{f_c(q=2)}$ | 1.108 | 1.080 | 1.056 | 1.033 |
| $P_{EC}$ | 10 MW | 10 MW | 10 MW | 10 MW |
| $I_{CD}$ | 308 kA | 147 kA | 38 kA | 83 kA |
| $\eta$ [AW$^{-1}$ 10$^{20}$ m$^{-2}$] | 0.12 | 0.06 | 0.016 | 0.032 |
| $\beta_{beam}$ | 250 | 350 | 400 | 700 |

* In this case the beam divergence is reduced to 1$^0$ in order to optimize the localization.

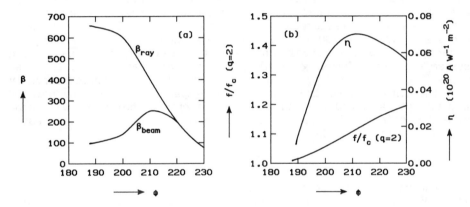

**Fig. 2** The results of ray-tracing calculations. A beam of O-mode waves is injected in the equatorial plane from the low-field side at an angle $\phi$ with respect to the major radius. (a) the peakedness of the power deposition for the total beam and for the central ray; (b) the normalized wave frequency and the current drive efficiency estimated from the generalized Fisch formula

**Acknowledgements.** We thank the NET team for providing the NET magnetic equilibrium. This work was performed under the Euratom-FOM association agreement, with financial support from NWO and Euratom (NET contract no. 88-151).

**REFERENCES**
[1] WEGROWE, J.G., et al., Fusion Techn. **14** (1988) 165.
[2] WADDELL, B.V., et al., Phys. Rev. Letters **41** (1978) 1386; TURNER, M.F., WESSON, J.A., Nucl. Fusion **22** (1982) 1069; HOLMES, J.A., et al., Nucl. Fusion **19** (1979) 1333.
[3] BATCHELOR, D.B., GOLDFINGER, R.C., Rep. ORNL/TM-6844; KRITZ, A.H., et al., proc. of 3$^{rd}$ Int. Symp. on Heating in Toroidal Plasmas, Grenoble, 22-26 March 1982, Vol. II p. 707.
[4] GIRUZZI, G., Phys. Fluids **31** (1988) 3305.
[5] CHAN, V.S., et al., Nucl. Fusion **22** (1982) 787; START, D.F.H., et al., Plasma Phys. **25** (1983) 447; COHEN, R.H., Phys. Fluids **30** (1987) 2442; YOSHIOKA, K., ANTONSEN, T.M., Nucl. Fusion **26** (1986) 839.
[6] HARVEY, R.W., et al., in *Applications of RF Power to Plasmas*, (AIP, New York, 1987) p. 49.
[7] SMITH, G.R., et al., Phys. Fluids **30** (1987) 3633.
[8] WESTERHOF, E., Nuclear Fusion **27** (1987) 1929.

48

QUASI-OPTICAL 140 GHz ECRH SYSTEM ON THE ADVANCED W VII-AS STELLARATOR

W. Henle, W. Kasparek, H. Kumric, G.A. Müller, P.G. Schüller, M. Thumm
Institut für Plasmaforschung, Univ. Stuttgart, D-7000 Stuttgart 80, FRG

V. Erckmann
Max-Planck-Institut für Plasmaphysik, EURATOM Ass., D-8046 Garching, FRG

ABSTRACT

A design is presented of a 140 GHz 2nd harmonic electron cyclo-tron resonance heating (ECRH) system to be used on the advanced stel-larator W VII-AS at IPP Garching. The primary objectives of these ECRH experiments will be (1) to extend the density range for ECRH and for combined heating (NBI and/or ICRH), (2) to investigate the local elec-tron heat transport in high density plasmas by the heat-wave techni-que, (3) to develop new millimeter-wave transmission line concepts and antenna systems for reactor-compatible multi-megawatt heating of future large-size stellarators. The planned 140 GHz ECRH facility will use a 200 kW/100 to 200 ms TE03-mode gyrotron from KfK Karlsruhe com-bined with a quasi-optical fundamental Gaussian-mode transmission system employing focusing metallic mirrors as phase correcting ele-ments. The unpolarized gyrotron output mode will be converted directly into the linearly polarized Gaussian free-space beam with the help of a quasi-optical coupler. A universal quasi-optical polarizer will pro-vide the optimum polarization state in order to get the best accessi-bility conditions to the plasma.

INTRODUCTION

During the first period of 70 GHz ECRH-experiments on the modu-lar W VII-AS stellarator at IPP Garching using outside (low-field side) launch of the waves, electron heating at fundamental resonance (B = 2.5 T, O-mode) and at second harmonic frequency (B = 1.25 T, X-mode) has been studied [1]. Plasma generation from the neutral gas and electron heating with quasi-stationary plasma behaviour at microwave pulse lengths of up to 700 ms with injected power levels up to 0.6 MW (3 gyrotrons) were attained. Owing to the flexibility of the quasi-optical HE11 microwave beam injection antenna [2], different energy deposition profiles could be realized. For central energy absorption peaked $T_e$ profiles (with $P_{rf}$ = 380 kW and $n_{eo}$ = $n_{cutoff}$/2: $T_{eo}$ = 1.2 keV, $n_{eo}$ = $3 \cdot 10^{19} m^{-3}$ at B = 2.5 T and $T_{eo}$ = 1.0 keV, $n_{eo}$ = $1.5 \cdot 10^{19} m^{-3}$ at B = 1.25 T) were measured, whereas flattened $T_e$ profiles were observed with off-axis heating. Experiments with EC-current drive have been successful in compensating small plasma net-currents (e.g. boot-strap current) in order to reduce its influence on the rotational transform profile and thus to maintain the good confinement properties of the nearly shearless stellarator [1].

However, these investigations on purely ECRH-sustained plasmas are restricted to densities below $6.2 \cdot 10^{19} m^{-3}$, the cut-off density for the O-mode at the fundamental resonance. To extend the density range for ECRH and for combined heating together with 1.5 MW NBI and

3 MW ICRH the application of 140 GHz 2nd harmonic X-mode ECRH is required. This 140 GHz ECRH can be run together with the 70 GHz ECRH.

## PERSPECTIVES OF 140 GHz ECRH ON W VII-AS

The primary objectives of the planned 2nd harmonic 140 GHz ECRH experiments on W VII-AS (at B = 2.5 T) are:

(1) Extension of the density range ($n_e \leq 1.2 \cdot 10^{20} m^{-3}$) for ECRH and for combined heating (together with NBI and/or ICRH) with controlled power deposition and local current drive. Study of MHD stability for $\langle\beta\rangle > 1\%$. Ray-tracing calculations [3] show for $n_{eo} = 1.1 \cdot 10^{20} m^{-3}$ and $T_{eo} = 1$ keV (with parabolic profiles) 93% single-pass absorption within a well-localized region in the plasma center (see Fig. 1).

(2) Investigation of the local electron heat transport approaching the critical $\beta$ for MHD stability by the heat-wave technique [4].

(3) Studies on impurity transport. Modification of impurity influx by power deposition and ionization at the plasma edge. Variation of the magnetic configuration by inhomogeneous current densities (islandization).

(4) Development of new millimeter-wave mode-conversion and transmission-line concepts and antenna systems for reactor-compatible multi-megawatt heating of future large-size fusion devices, in particular of the planned next stellarator W VII-X.

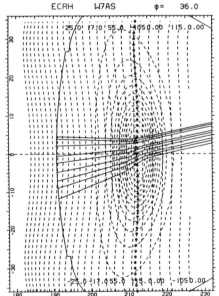
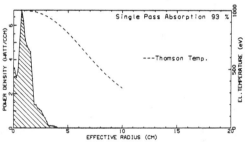

Fig. 1. Ray paths (left side) and power deposition profile (right side) for a 140 GHz 2nd harmonic X-mode ECRH on W VII-AS. Parabolic plasma temperature and density profiles are assumed with $T_{eo} = 1$ keV and $n_{eo} = 1.1 \cdot 10^{20}$ m$^{-3}$.

There are two major concepts for low-loss millimeter-wave transmission with approx. 1 MW power per transmission line [5]:

(a) Closed, highly oversized tubular $HE_{11}$-mode waveguide (corrugated or dielectrically lined).

(b) Quasi-optical $TEM_{00}$ transmission through a Gaussian beam waveguide using focusing metallic reflectors as phase correcting elements.

In the proposed experiments which are planned for 1990 a quasi-optical mirror waveguide will be developed and tested with respect to high transmission efficiency and mode purity. The mode-control features of such transmission systems already can be studied with available gyrotron unit powers of around 200 kW whereas the full power carrying capability can be tested only using long-pulse 1 MW sources.

## QUASI-OPTICAL 140 GHz ECRH SYSTEM

A schematic drawing of the quasi-optical transmission line, the location of the gyrotron and the stellarator is shown in Fig. 2. The geometrical parameters of the ellipsoidal mirrors are summarized in Table I. This 140 GHz ECRH facility uses a 200 kW/100 to 200 ms $TE_{03}$-mode gyrotron from KfK Karlsruhe [6]. The unpolarized gyrotron output mode will be converted directly into a linearly polarized Gaussian-like free-space beam with the help of an improved quasi-optical mode-transducing antenna [7] mounted on the top of the tube and combined with a matching optics (mirrors M1 and M2). Such a converter is basically an overmoded waveguide slot radiator with tapered edges (reduced diffraction losses) combined with a cylindrical reflector of optimized cross-section. The achievable mode conversion efficiencies of advanced quasi-optical mode transducers are almost 95%.

Table I    Parameters of the ellipsoidal reflectors

| Mirror | M1 | M2 | R3 | R4 | R5 |
|---|---|---|---|---|---|
| length a(m) | 0.3 | 0.55 | 1.11 | 0.55 | 0.25 |
| width b(m) | 0.2 | 0.47 | 0.47 | 0.47 | 0.1 |
| radius of curvature | | | | | |
| $r_a$(m) | -0.76 | 2.44 | 70.37 | 7.06 | 3.58 |
| $r_b$(m) | -0.44 | 2.36 | 12.57 | 5.80 | 0.55 |

The quasi-optical beam waveguide consisting of at least 3 reflectors transports the beam towards the vessel entrance port and simultaneously performs proper focusing to the plasma center in the X-mode from the low-field side (outside launch) in the equatorial plane. The beam waveguide is used in H-plane orientation to minimize ohmic losses at the mirrors (< 0.2% per polished Al reflector) [8]. Optimization of the mirrors with respect to minimization of diffraction losses (< 0.1% per mirror) results in typical sizes as given in Table I. The beam diameter varies between 0.34 and 0.47 m along the transmission line. At the entrance port (barrier window) a beam waist of 0.06 m is chosen (2% coupling losses).

A universal quasi-optical polarizer [5] provides the optimum polarization state in order to get the best accessibility conditions to the plasma (linear polarization for perpendicular wave launching and arbitrary elliptical polarization in the case of oblique wave launching). This device is composed of a smooth roof-top mirror facing a flat, corrugated surface reflector having fixed groove depth (nominal $\lambda/8$). By rotating the whole instrument around the axis of the incoming (and outgoing) beam, the output polarization will rotate

twice as much. Rotation of the corrugated reflector around a perpendicular axis allows elliptical polarization to be arbitrarily adjusted.

A high precision with respect to machining quality (0.05 mm to-
lerance typically) of the various reflectors and their adjustment is
required. The supporting elements for the mirrors shall provide long-
term mechanical stability. The overall losses of the quasi-optical
transmission line are estimated to be approx. 5% (losses of the mode
transducing antenna are not included). For personal safety (microwave
leakage) and for gas purging of the beam-waveguide to reduce the atmospheric absorption of almost 1.5%, the whole beam path is surrounded
by a shielding duct.

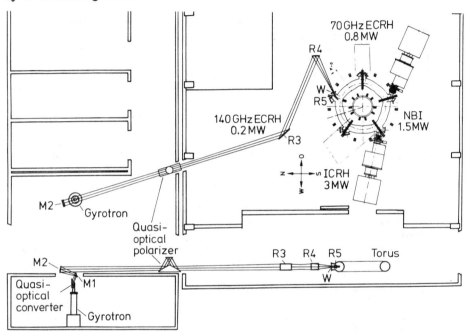

Fig. 2.   Outline drawing (ground-plane and vertical projection) of the
quasi-optical 140 GHz ECRH system.

### REFERENCES

1. H. Renner et al., Proc. 16th European Conf. Contr. Fusion and
   Plasma Physics, Venice, 1989, Invited Paper.
2. W. Kasparek et al., Proc. 15th Symp. Fusion Technology (SOFT),
   Utrecht, 1988, Contr. Paper B19.
3. M. Tutter (W VII-AS Team), private communication.
4. H.J. Hartfuss et al., Nuclear Fusion 26, 678 (1986).
5. M. Thumm et al., Conf. Digest 13th Int. Conf. Infrared and Millimeter Waves, Honolulu, 1988, p. 111.
6. G. Dammertz et al., Int. J. Electronics 64, 29 (1988).
7. S.N. Vlasov et al., Izv. VUZ Radiofisika 31, 1482 (1988).
8. W. Kasparek, M. Thumm, Conf. Digest 11th Int. Conf. Infrared and
   Millimeter Waves, Pisa, 1986, p. 212.

# H-MODE BY EDGE ECH IN JFT-2M

K.HOSHINO, T.YAMAMOTO, H.KAWASHIMA, H.TAMAI, T.OGAWA, K.ODAJIMA,
N.SUZUKI, Y.UESUGI, M.MORI, H.AIKAWA, S.KASAI, T.KAWAKAMI,
T.MATSUDA, Y.MIURA, H.OGAWA, H.OHTSUKA, T.SHOJI, T.YAMAUCHI,
T.KONDO, I.NAKAZAWA[+], C.R.NEUFELD[++] and H.MAEDA

Japan Atomic Energy Research Institute, Tokai, Ibaraki, Japan
+ Mitsubishi Electric Co., Tokyo, Japan
++Hydro-Quebec Research Institute, Quebec, Canada

## ABSTRACT

H-mode transition occured by the edge heating solely by ECH.
The launched wave has frequency of 60GHz and extraordinary mode
polarization. The threshold power for the H-mode transition is as
low as 120 kW, which is the least threshold power observed in the
JFT-2M tokamak. The second harmonic ECR layer is located at (0.7-
0.9)a (minor radius). The H transition by edge heating occurs not
only in the lower-single-null divertor configuration but also
in the D-shape limiter configuration. These results show that no
additional power in the core plasma region is required to obtain
the H-mode, and that the increase of the peripheral electron
temperature is closely related to the trigger of the H-mode.

## INTRODUCTION

Since the first discovery of the H-mode,which has an
improved particle/energy confinement during the additional heating phase, in the ASDEX tokamak[1], the physics of the H-mode has
become one of the major problems in the research of the tokamak
confinement.The H-mode was found in the plasma under the various
additional heating methods, such as neutral beam injection heat-
ing(NBH) and radio frequency heating the frequency of which is in
the range of the ion cyclotron frequency(ICRFH). Recently, the H-
mode was found to occur even by the electron cyclotron heating(EC
H) of the plasma edge preheated by the NBH in JFT-2M tokamak[2,3]
, and in the core ECH of Doublet III tokamak[4]. In the Doublet
III tokamak, H-mode has not been found by the edge heating.

In the previous experiment in the JFT-2M tokamak[2,3], the
ECH power was restricted to 100 kW. This time, the ECH power is
raised to maximum 230 kW, and in the higher density plasma than
in the previous experiment, it was found that H-mode is produced
by the edge heating solely by ECH[5]. Here we present the experimental results when the plasma density is changed, the first
result of the generation of the limiter-H-mode by edge ECH, and
generation and distruction of the H-mode by controlling the edge
electron temperature.

The JFT-2M tokamak is a non-circular D-shaped tokamak which
has plasma major radius of 1.31m and minor radii of 0.35m x 0.53m
in its full size operation. Maximum strength of the toroidal
field is 1.5 T.

# ECH H-MODE IN THE SINGLE NULL DIVERTOR CONFIGURATION

Figure 1(a) shows a typical time evolution of the H-mode induced by edge(2nd harmonic ECR layer is located at normalized radius of r0/a=0.85) electron heating solely by ECH in the single null plasma configuration [5]. There observed an increase of the edge(0.88a) electron temperature over 100eV and clear suppression of the $D_\alpha$ signal occurs. There is an increase of the stored energy of 4kJ measured by diamagnetism.

Figure 1(b) shows the dependence of the incremental stored energy on the position of the ECR layer. Measured density and electron temperature profiles are also shown. The calculated deposition power in single path $\eta$ decreases toward the plasma edge. H-mode is produced when the ECR layer has good access($n_e<$ $2.15\times10^{19}m^{-3}$). Thus edge electron heating without power deposition at the plasma core due to the cutoff,produces H-mode.

Edge ECH effect is negligible at the low plasma density, presumably due to the small power deposition at the plasma edge. Figure 2 compares the threshold power for the H-mode transition by the neutral beam(NBH) and by the combination of ECH+NBH [2]. The launched ECH power for the combination case is shown by the broken line. At $\bar{n}_e$=1.6x10$^{19}$m$^{-3}$, the threshold power in the combined heating is larger(600 kW) than in NBH only case(450KW). Therefore edge ECH power of more than 150kW out of launched power 190kW is lost at such low density. On the other hand, at the higher density the threshold power in the combination case is lower than the NBH only case, presumably due to the increase of the ECH damping at the plasma edge. Thus high density($3\times10^{19}m^{-3}$) is required for the H transition by edge ECH.

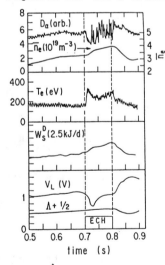
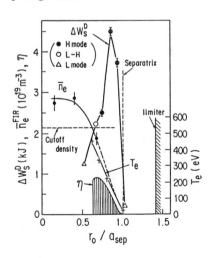

Fig.1(a) Time evolution of the H-mode.ECH power $P_{EC}$=194kW.$B_{t0}$= 1.23T.$I_p$=201kA.

Fig.1(b) Incremental stored energy $\Delta W_s^D$.$P_{EC}$=(210-230)kW.$B_{t0}$=(1.16- 1.27)T. $\eta$ is the calculated single path absorption rate.

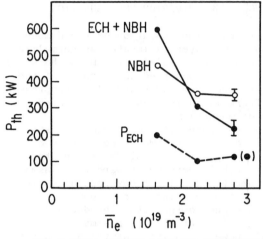

Fig.2

Density dependence
of the threshold
power P th for the
combination ECH+NBH
(ECH power is shown
by the broken line),
and for NBH only.
$I_p$ =(200-255)kA.
$B_{t0}$=(1.23-1.25)T.

ECH H-MODE IN THE LIMITER CONFIGURATION

The edge ECH is found to produce the H-mode even in the
limiter configuration as shown in Fig.3(a). The plasma position
is slightly shifted outside. The H-mode is produced by the ECH
(193kW) superposed on ICH(473kW,17.5MHz) as shown in Fig.3(b). In
this case, the H-mode could not be produced by ICH maximum power
of 561 kW. The ECH induced H-mode terminates in 50ms, which may
be due to the appearance of the cutoff at the ECR layer due to
the linear increase in density during the H-mode. After the H to
L transition, the density begins to decrease, and the H-mode is
set on again. There is a marked increase of the temperature at
0.67a measured by ECE. But the increase of the core electron
temperature at 0.15a (from ECE 78.4GHz) is negligible. Therefore
the ECH effect to the plasma parameters is to raise the peripheral electron temperature as is already reported in ref.3.
Therefore it is shown that H-mode is induced even in the
limiter configuration by edge electron heating.

DESTRUCTION OF THE H-MODE BY PERIPHERAL COOLING

It was found in ref.6 that launching the 27MHz wave(ion
Bernstein wave frequency) into the H-mode terminates the H-mode.
It is observed that the launched wave in the ohmic heated plasma
cools the peripheral electron temperature but not that of the
plasma core. The inhancement of the radiation loss by the
increased heavy impurity content brings the cooling. There observed a decrease in edge electron temperature(at pedestal) when the
27MHz wave is injected into the H-mode,and then H-mode terminates

From these experimental results, we may conclude that edge
(pedestal region) electron heating brings the H-mode, and edge
electron cooling destroys the H-mode.

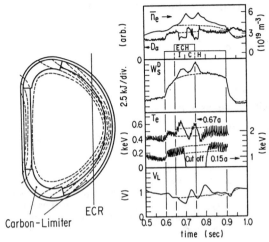

Fig.3(a) Limiter configuration. ECR layer is located at 0.71a. $I_p$ =232kA. $B_{t0}$ =1.27T. Fig.3(b) Limiter H-mode. Broken line is for the case without ECH.

## DISCUSSION

It is shown experimentally that increase of the edge electron temperature is closely related to trigger the H-mode both in divertor configuration and in limiter configuration by these experiments. By the careful inspection of the behaviour of the edge electron temperature during the H-mode in ref.3,7,8, we can define the threshold electron temperature. But the threshold electron temperature is dependent on the plasma current and weakly on density(ref.8). Rather, the edge gradient is more suited to characterize the H-mode( ref.9). To identify the H-mode mechanism, we cannot exclude the effect of the ion temperature. The ion temperature may increase due to the energy transfer from heated electrons. But the electron-ion collision time at the typical pedestal parameter of $T_e$ =300eV, $n_e$ =(1-2 )x10$^{19}$ m$^{-3}$, with $Z_{eff}$ 2 is 2-4ms, which seems too small to explain the time interval of the marginal state(typically 30-70ms) which is observed at the threshold ECH power level in ref.3. Rather, these time scale may well in the range of current diffusion time which is typically 20-50ms with $T_e$ =(0.2-0.4)keV, $Z_{eff}$ =2, and the scale length of 0.1m. Finally, we note that ECH is a sure method of the local modification of the boundary plasma parameter such as electron temperature to improve the tokamak confinement, without the impurity problems.

## REFERENCES
1.F.Wagner et al.,Phys.Rev.Lett.49,1408(1982).
2.K.Hoshino et al.,Nucl.Fusion 28,301(1988).
3.K.Hoshino et al.,Phys.Lett.A130,26(1988).
4.J.Lohr et al.,Phys.Rev.Lett.60,2630(1988).
5.K.Hoshino et al.,Japan Atomic Energy Research Institute Report
   JAERI-M 89-038(1989).
6.H.Tamai et al., JAERI-M 89-036(1989), and this conference.
7.K.Hoshino et al.,J.Phys.Soc.Japan 56,1750(1987).
8.K.Hoshino et al.,Phys.Lett.A124,299(1987).
9.K.Hoshino et al.,J.Phys.Soc.Japan 58,1248(1989).

# Increasing Efficiency of Electron Cyclotron Current Drive

V. I. Krivenski[*]

Institute for Fusion Studies
The University of Texas at Austin
Austin, Texas 78712

## Abstract

The deformation of the electron distribution function during electron- cyclotron current drive is studied by means of a 2D bounce-averaged Fokker-Planck code as a function of the parallel refractive index and power of the wave beam, for toroidal geometry and down-shifted frequency. It is shown that two regimes exist: in the first current drive efficiency decreases for increasing power, in the second the efficiency is nearly indipendent from the power level. The characteristics of an optimized spectrum for electron-cyclotron current drive are discussed.

## 1. Introduction

Electron-cyclotron (EC) current drive by an extraordinary wave with a frequency smaller than the electron-cyclotron frequency has several attractive features:[1,2] a) it requires a relatively low frequency source, b) has a higher density cut-off, c) a large range of values of the parallel refractive index allow the wave propagation in the plasma, d) the current drive efficiency is a slow function of the magnetic field.

Current drive by EC waves at down-shifted freaquencies is based only on the mechanism of asymmetric modification of the plasma resistivity,[3] since, in general, the momentum transfer from the wave to the electrons is in the opposite direction to the electron drift and the electrons, while they are increasing their perpendicular energy, loose a part of their parallel energy. This effect can be particularly deleterious in the presence of trapped electrons since for increasing level of power the electrons are pushed towards the trapping region and the efficiency with which they are contributing to the current generation decreases sharply.[4,5]

Here a study of this problem is presented for the Tore Supra tokamak parameters, with

$$T_e(0) = 3 \text{ keV}, \quad n_e(0) = 5 \times 10^{13} \text{ cm}^{-3}, \quad B_0 = 45 \text{ kG}, \quad f \approx 110 \text{ GHz}.$$

For simplicity, we will discuss the modification of the distribution function only in one point in the space, corresponding to the innermost point of the $r = 20$ cm ($r/a \approx 0.3$) flux surface. The local results we present are a fair approximation of the global electron behaviour for top injection and vertical propagation of the wave packet.[6]

## 2. Quasilinear Diffusion in Toroidal Geometry

The effect of the applied wave power on the electrons is to make them diffuse in the resonant part of the momentum space along a path parallel to the direction defined by the vector

$$\mathbf{d} = \frac{s\,\omega_c}{\omega}\,\mathbf{e}_\perp + \left(\gamma - \frac{\omega_c}{\omega}\right)\frac{p_\perp}{p_\parallel}\,\mathbf{e}_\parallel$$

For a down-shifted frequency, $\omega < \omega_c$, and the term in paranthesis is negative in the region of momentum space where the maximum of the wave absorption is localized; therefore in that region the parallel component of the diffusion is opposite to $p_\parallel$. Here we have considered the local direction, at

---

a) Permanent address: Asociación EURATOM/CIEMAT para Fusion, CIEMAT, 28040 Madrid.

a given point in space, however the effective diffusion direction is the average of **d** along the electron orbit. If we consider the absorption region to be localized in the internal part of the magnetic surface, then there exists a pitch-angle $\theta_0$ , such that for $\theta > \theta_0$ ( or $\theta < \pi - \theta_0$ ) the parallel component of the average diffusion direction , $< \mathbf{d} >_{||}$ , has the same sign as $p_{||}$ . The current drive efficiency increases for the electrons diffusing in this region of the momentum space, while it decreases for the others.

## 3. Numerical Results

To illustrate this effect we have numerically solved the bounce-averaged Fokker-Planck equation for the distribution function.

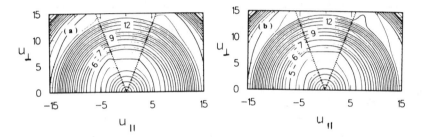

FIG. 1.

In Fig. 1(a) - 1(b) the level curves of the distribution function for an angle of propagation to the normal, $\psi = 70^0$ , and a half-width of the wave beam , $\Delta\psi = 15^0$ , are shown for wave power $P_0 =$ 1, 3 MW respectively. We use momentum normalized to the thermal one and the level lines for a Maxwellian are equally spaced, with labels corresponding to the normalized momentum. The current produced is mainly carried by electrons near $u_{||} \approx -5$. The deformation of the distribution function is large, but it is almost exactly compensated by the deformation near $u_{||} \approx 5$ which is generated by pitch-angle scattering through the trapped particle region ( the region between the dashed lines ) . The modification of the current drive efficiency, defined by $\eta = J / 2 \pi R_0 P$ , is given in Table I .

Table I. Current density, absorbed power, and current drive efficiency for $\psi = 70^0$ , $\Delta\psi = 15^0$ , f = 110 GHz and increasing applied power.

| $P_0$ | 10 kW | 1 MW | 2 MW | 3MW |
|---|---|---|---|---|
| J ( A / cm $^2$) | $4.81\times10^{-4}$ | $3.35\times10^{-2}$ | $4.55\times10^{-2}$ | $5.03\times10^{-2}$ |
| P (W / cm $^3$) | $7.74\times10^{-6}$ | $9.97\times10^{-4}$ | $2.10\times10^{-3}$ | $3.20\times10^{-3}$ |
| $\eta$ (A / kW ) | 44 | 24 | 15 | 11 |

We observe a degradation of the efficiency, which increases with the power. This trend can be modified if we succeed to make more electrons interact with the part of the spectrum in which the average parallel diffusion has the same direction as the electron drift.

58

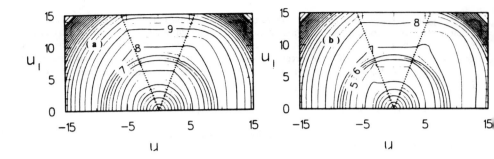

FIG. 2.

This is shown in Fig. 2(a) - 2(b) where now $\psi = 30^0$ , $\Delta\psi = 8^0$ , and f = 120 GHz .

Table II. As in Table I but for $\psi = 30^0$ , $\Delta\psi = 8^0$ , and f = 120 GHz .

| $P_0$ | 10 kW | 1 MW | 2 MW | 3MW |
|---|---|---|---|---|
| J ( A / cm $^2$) | 4.57×10$^{-4}$ | 5.13×10$^{-2}$ | 1.04×10$^{-1}$ | 1.54×10$^{-1}$ |
| P (W / cm $^3$) | 7.42×10$^{-6}$ | 8.64×10$^{-4}$ | 1.85×10$^{-3}$ | 2.97×10$^{-3}$ |
| η (A / kW ) | 44 | 42 | 40 | 37 |

Table II shows that the larger modification of the distribution function near $u_\parallel \approx - 5$ now slows down the degradation of the current drive efficiency . In Fig.3 the comparison between the two parallel functions, for $P_0 = 3$ MW , is presented.

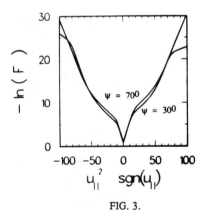

FIG. 3.

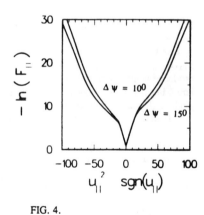

FIG. 4.

The width of the absorption region in momentum space is proportional to the wave beam parallel width and decreases for increasing $u_\perp$ . Since the region in momentum space which can increase the current drive efficiency is located at large $u_\perp$ , a smaller spectrum width results in a sharp decrease of the current drive efficiency. This effect is shown in Table III and in Fig.4 .

Table III. As in Table I but for $\Delta\psi = 10^0$ .

| $P_0$ | 10 kW | 3MW |
|---|---|---|
| $J \, ( A / cm^2)$ | $4.73\times10^{-4}$ | $1.92\times10^{-2}$ |
| $P \, (W / cm^3)$ | $7.56\times10^{-6}$ | $2.84\times10^{-3}$ |
| $\eta \, (A / kW )$ | 44 | 5 |

We can fully exploit this effect by trying to extend the resonant region. In Fig.5 we present the parallel function and in Fig.6 the level lines of the distribution function for a wave spectrum in which three frequencies are superposed, f = 120, 110,100 GHz, the other parameters are the same as in Fig. 2(b). The corresponding efficiency is given in Table IV .

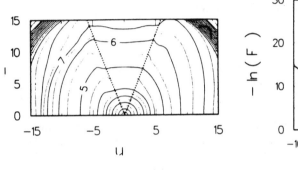

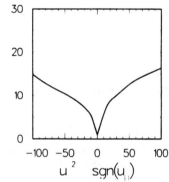

FIG. 5.                    FIG. 6.

Table IV. As in Table II, but for f = 120, 110, 100 GHz .

| $P_0$ | 10 kW | 3MW |
|---|---|---|
| $J \, ( A / cm^2)$ | $3.02\times10^{-4}$ | $1.11\times10^{-1}$ |
| $P \, (W / cm^3)$ | $5.13\times10^{-6}$ | $1.92\times10^{-3}$ |
| $\eta \, (A / kW )$ | 42 | 41 |

### References

1   I. Fidone, G. Giruzzi, E. Mazzucato, Phys. Fluids **28**, 1244 (1985).
2   I. Fidone, G. Giruzzi, V. Krivenski, L. F. Ziebell, E. Mazzucato, Phys. Fluids **29**, 803 (1986).
3   N. J. Fisch, Phys. Rev. Lett. **41**, 873 (1978).
4   G. Giruzzi, Nucl. Fusion 27, 1934 (1987);  Phys. Fluids **31**, 3305 (1988).
5   R. W. Harvey, M. G. McCoy, G. D. Kerbel, Phys. Rev. Lett. **62**, 426 (1989).
6   I. Fidone, G. Giruzzi, V. Krivenski, E. Mazzucato, L. F. Ziebell, Nucl. Fusion **27**, 579 (1987).

# CURRENT DRIVE USING A RESONANT DECAY INSTABILITY

KYOKO MATSUDA

General Atomics, San Diego, CA 92138-5608

## ABSTRACT

The ordinary mode near the second harmonic of electron cyclotron frequency obliquely launched to a tokamak with central density higher than the cutoff density may excite a Langmuir wave by a parametric decay instability. It is demonstrated that a steady traveling Langmuir wave is formed if pump power is sufficiently strong and continuous. The Langmuir wave may drive current at an overall efficiency several times larger than that of the linear electron cyclotron wave.

## I. INTRODUCTION

Recently, steady current drive for a tokamak by electron cyclotron heating has been intensively studied using ray tracing codes and Fokker-Planck codes. However, the efficiency is found to be rather modest due to the fact that electron cyclotron waves are absorbed mostly by thermal electrons especially in a high temperature plasma. It is the motivation of this work to find a new mechanism in which a launched wave of this frequency range transfers its parallel momentum to superthermal electrons. The mechanism should excite a weakly damped Langmuir wave in the plasma by an externally launched electromagnetic wave. The beat wave current drive scheme proposed by B.I. Cohen[1] is based on a similar idea. It uses two electromagnetic waves of different frequencies to excite a Langmuir wave. However, for this mechanism to work, the plasma in the beat field must retain a density and a temperature so as to satisfy the Manley-Rowe relation, or the power must be strong enough to execute a modulational instability. A mechanism which has more flexible plasma conditions and works for relatively low power is needed.

We have numerically explored several possibilities to excite a weakly damped Langmuir wave by decay instabilities of an electromagnetic wave by the kinetic theory. We have considered the situations in which an electromagnetic wave launched obliquely at the edge of the plasma becomes an almost circularly polarized wave propagating parallel to the magnetic field near the cutoff. In this case, a resonant Langmuir wave and also another daughter wave propagate parallel to the magnetic field. We may ignore perpendicular convective loss and expect an instability to grow if the dimension of the pump field along the magnetic field is large enough compared with the $e$-holding length for the instability. Near the cutoff, possible resonant decay instabilities have been explored for both the O–mode and the X–mode. It is found that a left-handed wave with large index of refraction can excite a Langmuir wave near the cutoff with $\omega_{ce}/\omega_0 \sim 0.45$. Such a pump wave can be launched as an O–mode from the outside of the plasma.

It is found that the phase velocity of the excited Langmuir wave is about five times the electron thermal velocity. If the net decay rate is close to unity, we may expect a factor five improvement over the linear electron cyclotron current drive efficiency. The estimation of net decay rate requires a more extensive study including the effects of inhomogenities and the nonlinear saturation of the Langmuir wave. This current drive scheme is well localized and suited to current profile control.

## II. A KINETIC THEORY

The general formalism for a nonlinear dispersion relation for such a system has been given by Liu and Tripathi.[2] Applying the formalism for the case of propagation parallel to a steady magnetic field, the coupled equations can be obtained for a left wave pump $E_0^+ \exp[-i(\omega_0 t - k_0 z)]$ as

$$
\epsilon_s E_s = i \frac{\omega_{pe}^2}{4 k_s v_e} \frac{eE_0^+}{m_e \omega_0 v_e} \left\{ \frac{k_0}{k_{-1}} \frac{\partial}{\partial \omega_s} \frac{Z(u_s) - Z(u_{-1})}{u_s - u_{-1}} \right.
$$

$$
\left. + \frac{k_{-1}}{k_0} \frac{\omega_0}{\omega_{-1}} \frac{\partial}{\partial \omega_s} \frac{Z(u_s) - Z(u_{01})}{u_s - u_{01}} \right\} E_{-1}
$$

$$
+ i \frac{\omega_{pe}^2}{4 k_s v_e} \frac{e(E_0^+)^*}{m_e \omega_0 v_e} \left\{ \frac{k_0}{k_1} \frac{\partial}{\partial \omega_s} \frac{Z(u_s) - Z(u_1)}{u_s - u_1} \right.
$$

$$
\left. + \frac{k_1}{k_0} \frac{\omega_0}{\omega_1} \frac{\partial}{\partial \omega_s} \frac{Z(u_s) - Z(u_{01})}{u_s - u_{01}} \right\} E_1 \quad , \tag{1a}
$$

$$
\epsilon_+ E_{-1} = i \frac{\omega_{pe}^2}{\omega_{-1} k_{-1} v_e} \frac{e(E_0^+)^*}{m_e \omega_0 v_e} \left\{ \frac{k_0}{k_s} + u_s \frac{u_s - u_0}{u_s - u_1} Z(u_s) \right.
$$

$$
+ u_{-1} \frac{u_1 - u_0}{u_1 - u_s} Z(u_1) + \frac{1}{2} \frac{k_{-1} k_0 \omega_0}{k_s \omega_{-1}} \frac{\partial}{\partial k_{-1}} \frac{Z(u_s) - Z(u_{-1})}{u_s - u_{-1}}
$$

$$
\left. + \frac{1}{2} \frac{k_1 \omega_0}{k_0} \frac{\partial}{\partial \omega_{-1}} \frac{Z(u_{-1}) - Z(u_{01})}{u_{-1} - u_{01}} \right\} E_s \quad , \tag{1b}
$$

$$
\epsilon_- E_1 = i \frac{\omega_{pe}^2}{\omega_1 k_1 v_e} \frac{eE_0^+}{m_e \omega_0 v_e} \left\{ \frac{k_0}{k_s} + u_s \frac{u_s - u_0}{u_s - u_1} Z(u_s) \right.
$$

$$
+ u_1 \frac{u_1 - u_0}{u_1 - u_s} Z(u_1) + \frac{1}{2} \frac{k_1 k_0 \omega_0}{k_s \omega_1} \frac{\partial}{\partial k_1} \frac{Z(u_s) - Z(u_1)}{u_s - u_1}
$$

$$
\left. + \frac{1}{2} \frac{k_1 \omega_0}{k_0} \frac{\partial}{\partial \omega_1} \frac{Z(u_1) - Z(u_{01})}{u_1 - u_{01}} \right\} E_s \quad , \tag{1c}
$$

where

$$
\omega_\ell = \omega_s + \ell \omega_0 \quad , \qquad k_\ell = k_s + \ell k_0 \quad , \qquad u_s = \frac{\omega_s}{\sqrt{2} k_s v_e} \quad ,
$$

$$
u_{\mp 1} = \frac{\omega_{-1} \pm \Omega_e}{\sqrt{2} k_{\mp 1} v_e} \quad , \qquad u_{01} = \frac{\omega_0 - \Omega_e}{\sqrt{2} k_0 v_e} \quad , \qquad u_0 = \frac{\omega_0}{\sqrt{2} k_0 v_e} \quad ,
$$

$$\epsilon_s(\omega, k) = 1 + \sum_\alpha \frac{4\pi n_\alpha q_\alpha^2}{m_\alpha k^2 v_\alpha^2} \left[1 + u_\alpha Z(u_\alpha)\right] \quad , \qquad u_\alpha = \frac{\omega}{\sqrt{2}\, k v_\alpha} \quad ,$$

$$\epsilon_\pm(\omega, k) = 1 + \frac{\omega_{pe}^2 \omega}{\omega^2 - c^2 k^2} \frac{Z(u_{\pm 1})}{\sqrt{2}\, k v_e} \quad , \qquad v_e = \left(\frac{T_e}{m_e}\right)^{1/2} \quad .$$

Three decay waves are $E_s \exp[-i(\omega_s t - k_s z)]$ and $E_\pm \exp[-i(\omega_\pm t - k_\pm z)]$. Considering the case where both waves $(\omega_s, k_s)$ and $(\omega_{-1}, k_{-1})$ are resonant modes of the plasma and linear damping of $(\omega_{-1}, k_{-1})$ is negligibly small, the growth rate of the instability is given by

$$\gamma = \left[\frac{\partial(Re\,\epsilon_+)}{\partial\omega}\right]^{-1} Im\left(\frac{M_{s+} M_{+s}}{\epsilon_s}\right) \quad , \tag{2}$$

where $M_{s+}$ is the coefficient for $E_{-1}$ in Eq. (1a) and $M_{+s}$ is one for $E_s$ in Eq. (1b). The O–mode launched very obliquely to a tokamak becomes a left-handed wave at a cutoff density. Ray tracing studies confirm that rays are propagating almost parallel to the confining magnetic field for some distance. For this pump field, there are possible resonant couplings; a coupling to a Langmuir wave with a left-handed wave near the fundamental frequency as a lower side-band wave, and a coupling to a Langmuir wave with a higher harmonic left wave as an upper side-band wave. Each competing couplings has two different directions of propagation. Left-handed modes of this frequency range are not damped until their wave vectors gain some perpendicular component. Thus, their nonlinear growth rates are given by Eq. (2). The couplings to backward Langmuir waves are found to be stable in a relatively low power regime due to their negligible Landau damping rates. The forward Langmuir wave coupled with the backward left-handed wave near the fundamental frequency may grow with realistic power.

A left handed wave with $n_0 \sim 0.8$ and $\omega_{ce}/\omega_0 \sim 0.45$ is a good choice for the pump where $n_0$ is the parallel index of refraction of the pump wave. The instability is expected for a wide range of temperature as shown in Fig. 1. The details of the coupling waves considered in Fig. 1 are listed in Table I. If the density of target plasma is higher than the expected, the instability occurs at a location slightly outer side since the mechanism does not strongly depend on the magnetic field. Thus, the required condition for a target plasma is relatively broad.

From Table I and Fig. 1, it is noticed that the nonlinear growth rate is highest when the phase velocity of Langmuir wave is about $5.2\, v_e$. Thus, the current drive efficiency of this Langmuir wave itself should be quite high. The other daughter wave, the lower frequency left-handed wave, is propagating backward where the external magnetic field is stronger than the fundamental resonance and absorbed after gaining some perpendicular component for the wave vector. Current driven by the secondary left wave is in the same direction as one driven by the Langmuir wave although much smaller. Comparing with linear electron cyclotron current drive (LECCD) where an average resonance velocity is usually about $2\, v_e$ for $T_e \sim 5$ keV and even smaller for a higher temperature, the net current drive efficiency of this process can be expressed using Fisch and Boozer's formula[3] as

$$\eta = \frac{4}{3}\left(\frac{5.2}{2}\right)^2 \frac{\omega_s}{\omega_0} \frac{P_{\text{DECAY}}}{P_{\text{INPUT}}} (1 - \alpha)^{-1} \eta_{\text{LECCD}} \quad , \tag{4}$$

where $4/3$ is the ratio obtained by Fisch and Boozer for parallel heating versus perpendicular heating, and $\alpha$ is dimensionless reduction factor due to the banana trapping.

Estimation of the decay rate, $P_{\text{DECAY}}/P_{\text{INPUT}}$, requires a much more extensive study such as a particle simulation study including the nonlinear effects of the Langmuir wave itself and inhomogeneities of the plasma and a static magnetic field.

This is a report of work sponsored by the U.S. Department of Energy under Contract No. DE-AC03-89ER53277.

## REFERENCES

1. B.I. Cohen, R.H. Cohen, B.G. Logan, W.M. Nevins, G.R. Smith, A.V. Kluge, and A.H. Kritz, UCRL-97288, submitted to Nucl. Fusion, 1988.
2. C.S. Liu and V.K. Tripathi, Phys. Rep. **130**, 143 (1986).
3. N.J. Fisch and A.H. Boozer, Phys. Rev. Lett. **45**, 720 (1980).

Table I
Frequencies and Wavenumbers of Coupling Resonant Waves
for Left Wave Pumps at Various Parameters

| T (keV) | $n_0^2$ | $\omega_{ce}/\omega_0$ | $\omega_{pe}/\omega_0$ | $\omega_s/\omega_0$ | $\omega_{-1}/\omega_0$ | $k_0 v_e/\omega_0$ | $k_s v_e/\omega_0$ | $k_{-1} v_e/\omega_0$ | $\omega_s/k_s v_e$ |
|---|---|---|---|---|---|---|---|---|---|
| 10 | 0.78 | 0.44 | 0.5629 | 0.61180 | 0.38817 | $1.233 \times 10^{-1}$ | $1.298 \times 10^{-1}$ | $6.528 \times 10^{-3}$ | 4.71 |
| 10 | 0.80 | 0.44 | 0.5367 | 0.60427 | 0.39577 | $1.249 \times 10^{-1}$ | $1.447 \times 10^{-1}$ | $1.986 \times 10^{-2}$ | 4.18 |
| 7 | 0.78 | 0.44 | 0.5629 | 0.59995 | 0.40010 | $1.032 \times 10^{-1}$ | $1.144 \times 10^{-1}$ | $1.121 \times 10^{-2}$ | 5.24 |
| 7 | 0.80 | 0.45 | 0.5385 | 0.58638 | 0.41362 | $1.045 \times 10^{-1}$ | $1.255 \times 10^{-1}$ | $2.095 \times 10^{-2}$ | 4.67 |
| 5 | 0.78 | 0.44 | 0.5629 | 0.59031 | 0.40968 | $8.732 \times 10^{-2}$ | $9.936 \times 10^{-2}$ | $1.213 \times 10^{-2}$ | 5.94 |
| 5 | 0.80 | 0.45 | 0.5385 | 0.57324 | 0.42677 | $8.835 \times 10^{-2}$ | $1.083 \times 10^{-1}$ | $1.999 \times 10^{-2}$ | 5.29 |
| 3 | 0.85 | 0.46 | 0.4680 | 0.50129 | 0.49876 | $7.056 \times 10^{-2}$ | $9.866 \times 10^{-2}$ | $2.810 \times 10^{-2}$ | 5.08 |
| 3 | 0.85 | 0.55 | 0.4822 | 0.51386 | 0.48613 | $7.056 \times 10^{-2}$ | $9.786 \times 10^{-2}$ | $2.730 \times 10^{-2}$ | 5.25 |
| 1 | 0.90 | 0.70 | 0.4123 | 0.42780 | 0.57223 | $4.193 \times 10^{-2}$ | $6.407 \times 10^{-2}$ | $2.214 \times 10^{-2}$ | 6.68 |

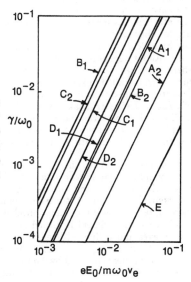

FIG 1. Growth rates for various $T_e$; lines A, B, C, D, and E are $T_e =$ 10, 7, 5, 3, and 1 keV, respectively. Line $A_1$ for $n_0^2 = 0.78$ and $\omega_{ce}/\omega_0 = 0.44$, abbreviated as (0.78, 0.44), $A_2$ for (0.80, 0.44), $B_1$ for (0.78, 0.44), $B_2$ for (0.80, 0.45), $C_1$ for (0.78, 0.44), $C_2$ for (0.80, 0.45), $D_1$ for (0.85, 0.46), $D_2$ for (0.85, 0.55), and E for (0.90, 0.70).

# Electron Cyclotron Current Drive by Radial Transport of Particles in a Tokamak

Sanghyun Park and Robert J. Taylor
Tokamak Fusion Laboratory
University of California, Los Angeles, California 90024-1597

## ABSTRACT

Plasma currents have been generated by injecting microwaves at the electron cyclotron resonance frequencies to the Continuous Current Tokamak in the absence of any other power input. The radial profiles of plasma current and particle densities show that the current is driven by the radial flow of electrons acted upon by the self–consistent vertical magnetic fields in a way a dynamo acts. Also the energy spectrum of the current carrying electrons shows that the current is carried by the bulk of the plasma electrons and no superthermal electron contribution was detected.

## INTRODUCTION

Non–inductive current drive in a tokamak geometry with radio frequency (rf) power at the electron cyclotron resonance frequency has been extensively studied both theoretically[1,2] and experimentally with EC in Levitron[3] device, Tosca[4], Cleo[5], WT-2, and WT-3[6] tokamaks. In the Levitron experiment, the target plasma was produced by another EC power at the lower power level and in the latter four tokamak experiments the target plasma was ohmically heated or complemented by lower hybrid waves. We report the observation of plasma current in a tokamak driven by a sole power of EC waves in quasi–steady state fashion and its dependence on radial flow of particles, which is controlled by externally imposed toroidal and poloidal magnetic fields, rf power, and the neutral particle density.

## EXPERIMENTAL

The Continuous Current Tokamak (CCT) has been in operation since 1986 at UCLA[7] and it is suitable for the basic research since it is capable of maintaining steady state toroidal magnetic fields of up to 0.9 kG with the major (minor) radius of $R = 150(a = 40)$ centimeters. The minor radius is defined by water – cooled copper liners covering the whole plasma surface. The low energy content of the plasma makes it possible to use metallic probes inside the plasma over the entire minor cross section. Fig. 1 shows plane cutaway view of the CCT device with all the attachments irrelevant to EC current drive not indicated. The device also has an excellent accessibility for the power input and diagnostics of the plasma. There are ten 10 cm ports along the high field side midplane allowing a direct launching of the microwaves from the high field side without depending on reflector to steer the beam. Along the midplane at the low field side, there are three 12.5 cm ports exclusively for the microwave launching. The waveguides there have been installed in such a way that the microwave beams make 60° (co), 90° (out), and –60° (counter), respectively at the plasma boundary. The rf sources are magnetrons with a frequency of $\omega/2\pi = 2.45$ GHz, peak power of $P = 1.5$ kW each, and the duty cycle is 50 % modulated by 60 Hz. This rf power is transported to the CCT via circular waveguide with $TE_{11}$ mode. power adapter The other source has frequency of 9.4 GHz, power levels of up to 30 kW, but the pulse length is only 1 mS with 0.5 Hz repetition rate.

Preionization:    When the chamber is filled with neutral gas at the pressure where the number density approximately corresponds to the critical density ($\omega_p/\omega = 1$), the microwave is injected to the chamber with the toroidal magnetic field of 600 Gauss ($\omega_c/\omega = .8$ at $r=0$) to preionize the neutrals. plasma A tungsten filament at the radial location of the liner is heated to a glowing temperature providing initial seed electrons even without any biasing. Once discharge is initiated, the filament is turned off without affecting the discharge. is are accelerated by the microwaves and collide with neutrals resulting in ionization in less than one millisecond with plasma density of $n_e = (1 - 2) \times 10^{10} cm^{-3}$ and temperature $T_e = (20 - 30)eV$. A set of two Langmuir probes separated by 2 mm with two active surfaces facing toroidally opposite direction is independently biased in steps of voltages. This two– sided Langmuir probe measures plasma current density, electron density and temperature, space potential, and the energy spectrum of the current carrying electrons. is

Polarization and Launch Schemes: When the microwave is launched from the high field side at the midplane with the wave polarization of ordinary(O) or extraordinary(X) mode, the phase measurements of the wave indicate that the single pass absorption is very low for the O-mode but the reflections off the wall scrambles the wave polarization and the current drive efficiency is comparable to that of X-mode within a factor of two. The single pass absorption of the X-mode is almost complete when the wave reaches an upper hybrid layer and mode converted to electron Bernstein waves. The current drive efficiency is about the same for the low field side launches with the waves propagating perpendicular to or at an angle of $\pm 60°$ with the toroidal field. The interferometric measurements of the wave inside the plasma show the spatial attenuation of the wave amplitude but the effect of refraction and reflection could not be quantified with a reasonable precision. Typically, temperature and density profiles are singly peaked at different radial locations and the current channel coincides with peak of the electron pressure ($P_e = n_e k T_e$).

## RESULTS

The current drive efficiency is most strongly dependent on the poloidal magnetic fields. The Fig. 2(a) shows the total plasma current driven by EC waves as a function of the vertical field. When this field is not optimized for the plasma confinement, the peak plasma density is degraded and the number of electrons available to carry the current is reduced accordingly. This vertical field dependence of the plasma current can be empirically fitted to $I_p \sim x \exp(-x^2)$ for $-\infty < x < \infty$ where $x$ is magnitude of normalized vertical field. In Fig. 2(b), the plasma current driven is shown as a function of the position of EC resonance layer. When the layer is too close to the high field side limiter, the initial neutral breakdown process is hampered by loss of electrons which would have been making ionizing collisions with neutrals otherwise. As the layer approaches the minor axis, symmetric outflow of particles results in antisymmetric plasma current distribution, which add up to a null current. It has been suggested that this is due to the trapped particle effect[8]. In our experiment, the diminishing plasma current as the resonance layer is moved outwards seems to be due to the fact that the wave absorption decreases as the upper hybrid resonance layer shifts along with the EC resonance layer and the geometry of the vertical field in the region of the current channel, in addition to the trapped particle effect. Further increase in toroidal field was limited by power supply to the coils. As has been shown in Fig. 1, there is no substantial difference in plasma current by the four different launch schemes. This also indicates that wave momentum transfer is *not* the mechanism of EC current drive. Each wave absorption process is different but randomization of polarization and multiple pass of the waves caused by reflection off the liner lead to complete wave absorption within $\pm 90°$ range of toroidal angle. With all the external magnetics optimized, the plasma current and its $L/R$ decay time after the microwave pulse is turned off was measured against the neutral fill pressure, which is shown in Fig. 3. The neutral pressure is normalized to the critical density of the X-band frequency of the wave. The Plasma current as well as the $L/R$ time increases as the pressure is lowered until the normalized pressure is one, and begins to decline sharply thereafter. In the S-band operation, the radial profiles show that the electron saturation current and plasma current are roughly inversely proportional to the neutral pressure with the radial range of current channel remaining unchanged. In the low density side, the low neutral density requires longer time for plasma density buildup due to longer mean free path of the electrons accelerated by microwaves. And in the high density side, recombination process is responsible for reduced current.

Radial Profiles: According to the theories[1,2] of current carrying superthermal electrons, the bulk of the electron distribution has distribution function symmetric in velocity space with the tail of the population making up an asymmetry and thus carrying the plasma current. In our experiments, however, the energy spectrum of the current carrying electrons was measured by subtracting the electron current drawn by one probe from the other in a range of bias potential values and the energy distribution of the electrons constituting the plasma current is peaked at the energy comparable to the bulk plasma temperature. This observation and the vertical field dependence of the plasma current suggest that the plasma current is driven by radial flow of electrons acted upon by the vertical field in a way similar to the dynamo action in an electric motor. So the asymmetry in toroidal current is due to antisymmetry in the vertical field modified by the poloidal field generated by the plasma current itself. When the microwave–plasma system reaches equilibrium, the steady flow of particles from the density peak to the plasma–wall boundary is outward in minor radial direction and the particle flux is almost symmetric. The radial profile of the plasma current density in Fig. 4(a) shows roughly antisymmetric distribution and the total plasma current is null. In an effort to break the symmetry in radial particle flow, a spindle cusp field was applied so that its field lines are curved inward producing a good curvature at the high field side with the vertical field remaining essentially unchanged. The radial current distribution in Fig. 4(b) shows a current that is toroidally unidirectional. Also noticed here is the MHD fluctuations on the low field side where the plasma is forced to move by the cusp field.

## DISCUSSION

The plasma current in tokamaks driven by the particle density gradient has been studied theoretically by Bickerton[9], and Sigmar[10]. In many EC heating experiments on tokamak plasmas, no increase in the plasma current was reported except for Tosca[5], where the EC power of ten times the ohmic heating (OH) power was injected to a small volume of plasma. If the rf power deposition is at the peak of the particle pressure at the minor axis and the rf power is not high enough to alter the radial profile of the particle distribution, the increase in density gradient will be only a small perturbation and its effect on total plasma current may well be masked by OH driven current and it may be further obscured if the change in current is radially antisymmetric. Similar effects of the vertical field on the plasma current have been observed in other CCT current drive experiments[11] with lower hybrid waves and electron beam. Also the fact that the EC current drive efficiency is comparable for various modes of microwave injection suggests that the current drive is due to the radial particle transport of the bulk electrons acted upon by the Lorentz force from the self–consistent vertical field. This point of view is further supported by direct measurements of the particle profile evolution. peaks

This work was supported by USDOE under Contract No. DE–FG03–86ER53225

## REFERENCES

1 N.J. Fisch, Rev. Mod. Phys. 59, 175 (1987) and references therein
2 D.F.H. Start, M.R. O'Brien, and P.M.V. Grace, Plasma Phys. 25, 1431 (1983)
3 D.F.H. Start et al., Phys. Rev. Lett. 48, 624 (1982)
4 M.W. Alcock et al., Proc. Nineth Int. Conf. Plasma Phys. Cont. Fusion Res., Vol. II, IAEA, Vienna, 1983, p. 51.
5 D.C. Robinson et al., in Proc. of the Eleventh Int. Conf. Plasma Phys. Cont. Fusion Res., IAEA, Vienna, 1986, Paper No.IAEA–CN–47/F–III-2
6 A. Ando et al., Phys.Rev.Lett. 56, 2180 (1986), and H. Tanaka et al., Phys.Rev. Lett. 60, 1033 (1988) Tokamaks,
7 R.J. Taylor et al., Bull.Am.Phys.Soc. 30,1623 (1985)
8 V.S.Chan et al., Nucl.Fusion 22, 787 (1982)
9 R.J. Bickerton, J.W. Conner, and J.B. Taylor, Nature Phys.Sci. 229, 110 (1971)
10 D.J. Sigmar, Nucl. Fusion 13, 17 (1973)
11 R.J. Taylor et al., Bull.Am.Phys.Soc. 32, 1747 (1987)

## FIGURE CAPTIONS

Fig. 1 Cutaway plane view of the experimental arrangement.
Fig. 2 Plasma current as a function of vertical field(a) and position of EC resonance layer(b) for four different launch schemes.
Fig. 3 Plasma current and $L/R$ decay time as a function of the neutral density normalized to the critical density.
Fig. 4 Radial profiles of interferometric traces, electron saturation current, and plasma current density with the spindle cusp field turned off(a) and on(b).

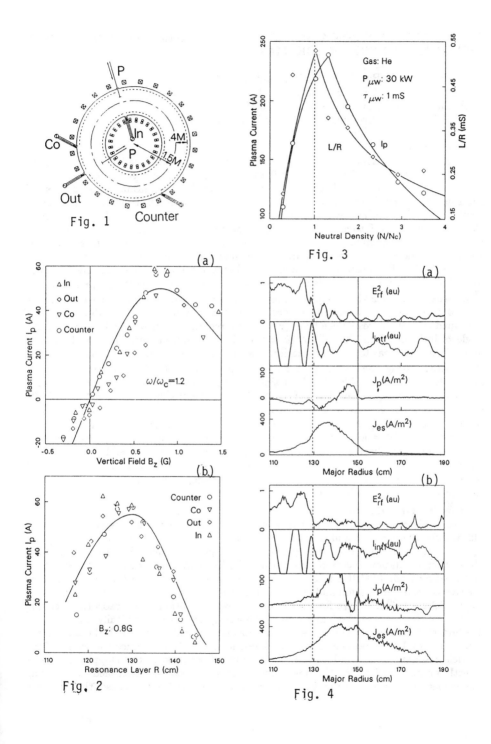

Fig. 1

Fig. 3

Fig. 2

Fig. 4

# ELECTRON CYCLOTRON ASSISTED STARTUP AND HEATING SCENARIOS IN THE COMPACT IGNITION TOKAMAK*

M. Porkolab, P. T. Bonoli, R. Englade and R. C. Myer
PLASMA FUSION CENTER, MIT, Cambridge, MA 02139 USA
G. R. Smith, LLNL, Livermore, CA 94550 USA
A. H. Kritz, Hunter College, CUNY, New York, NY 10021 USA

## I. Introduction

The Compact Ignition Tokamak (CIT) operating scenario calls for ramping the toroidal magnetic field from $B_T = 8.0$ to 10.0 Tesla in a few seconds, followed by a burn cycle and a ramp-down cycle. Simultaneously, the plasma must be heated from an initial low beta equilibrium ($\bar{\beta} \simeq 0.44\%$ at 7.0 to 8.0 Tesla) to a final burn equilibrium ($\bar{\beta} = 2.8\%$) having 10.0 Tesla on the magnetic axis [1]. Here we propose ECRF heating of CIT to ignition utilizing a constant source frequency but with a time dependent, variable angle of injection. Thus, initially $N_\parallel$ is large enough so that the Doppler broadened resonance of particles on the magnetic axis with f = 280 GHz and $B_T = 7.0$ - 8.0 Tesla can provide adequate absorption [2]. As the resonance layer is moved toward the magnetic axis the microwave beam is swept toward perpendicular to $\vec{B}$ in order to reduce the Doppler width and avoid heating the plasma edge. At $B_T = 10.0$ Tesla for normal wave incidence strong absorption occurs immediately on the high field side of the resonance layer (relativistic regime) [3]. However, by reducing the angle of incidence from 90° to 60° ($\theta = 0$ to 30°) absorption near the q = 1 surface, and concomitant sawtooth stabilization may be attained even in a burning plasma. We envisage using the ordinary mode (O-mode, $\vec{E}_{RF} \parallel \vec{B}$) of polarization which is accessible from the outside (low-field side) of the torus provided the density is such that $\omega_{pe} \leq \omega \approx \omega_{ce}$. Considering f = 280 GHz for central heating at $B(0) = 10.0T$, the maximum cutoff density is at $n_{crit} \approx 9.7 \times 10^{20} m^{-3}$ which is above the maximum central density ($n_{max} \simeq 7 \times 10^{20} m^{-3}$) in CIT. Electron heating in CIT is a viable option since equilibration of temperature between electrons and ions ($\tau_{EQ}$) is expected to be significantly shorter than typical energy confinement times, $\tau_E$. For example, at $n_e \approx 2.0 \times 10^{20} m^{-3}$, $T_e = 5$ keV, $Z_{eff} \approx 1.5$, we estimate $\tau_{EQ} \approx 30$ msec, while at $n_e(0) \approx 8 \times 10^{20} m^{-3}$, $T_e \approx T_i \approx 20$ keV, $\tau_{EQ} \approx 60$ msec, both significantly shorter than the expected energy confinement time.

## II. Ray Tracing Results

To study single pass absorption of waves in equilibria representative of the CIT plasma, the temperature and density profiles are taken to be $T_e(\psi) = T_e(0)[1 - \psi]$ and $n_e(\psi) = n_e(0)[1 - \psi]$ where $\psi$ is the normalized poloidal flux. The ray tracing and absorption simulation was performed using the Toroidal Ray Tracing, Current Drive and Heating Code (TORCH) developed by Smith and Kritz[4]. In Fig. 1 (a) we show the case of wave penetration and absorption for $B_T = 8.5$ T, $\theta = 20°$, $T_e(0) = 10$ keV and $n_e(0) = 3.2 \times 10^{20} m^{-3}$. We find 100% single pass absorption with absorption peaking near $r \simeq 0$. Notice that the half width of the absorption layer is typically $\Delta r \simeq 10$ cm. We find that relativistic effects (which are included in this code) shift the absorption toward the cyclotron resonance layer by amounts $\Delta r \gtrsim 10$ cm. Since the width of the particle resonance and the location of maximum absorption in the Doppler regime is directly proportional to $N_\parallel$, the power deposition profile can be controlled by changing the incident wave propagation angle. We find that for the 7.5 Tesla case, the optimum angle is approximately $\theta = 30°$ to the normal [5]. In order to keep the absorption close to the magnetic axis as the beta and magnetic fields are increasing during the ramp-up, the angle must be swept toward normal incidence at $B_T = 10.0$ Tesla. This ensures wave penetration to the center at full field and beta (relativistic regime, $N_\parallel < (T_e/m_e c^2)^{\frac{1}{2}}$) [3] and central heating results for $\theta \gtrsim 10°$ (see Fig. 1(b) for $\theta = 0$). By changing $\theta$ to $30°$ away from the normal, absorption at $r/a \simeq 0.5$, near the $q = 1$ surface results which may be useful for sawtooth stabilization (see Fig. 1(c)).

The efficiency of coupling to the O-mode at the edge of the plasma as a function of the angle of incidence has been calculated, and the result is

$$\frac{P_O}{P_T} = \frac{1}{4}\left(1 + \frac{\sin^2\theta}{\eta}\right)^2 + \frac{\cos^2\theta}{\beta\eta^2}, \tag{1}$$

where $\beta = \omega_{ce}^2/\omega^2$ and $\eta = (\sin^4\theta + 4\cos^2\theta/\beta)^{\frac{1}{2}}$. For nonresonant heating ($B_T \geq 7$ T, $\theta \leq 30°$), Eq. 1 predicts that at least 68% of the power injected will couple to the O-mode at the edge.

We have also examined the importance of scattering of EC rays by low frequency density fluctuations and find that for $< \delta n_e/n_e > \gtrsim 0.1$, scattering is not important. Monte Carlo calculations will be incorporated into the ray tracing code in the near future to further quantify these predictions.

### III. Transport Code Simulations

A version of the combined equilibrium and transport code BALDUR1-1/2D originally developed by G. Bateman [6] has been used to simulate some important aspects of the ECH heating scenario for CIT. We have assumed that the electron heat flux can be written as $q_e = -M\kappa_e \nabla T_e - (M-1)\alpha_T \kappa_e T_e \nabla V/V(a)$ (conduction and inward heat pinch), with $\kappa_e = [\text{CI}(\rho)\text{V}^2(\text{a})\text{A}^{-3/2}/T_e(\rho)|\nabla V|^2][1 + \gamma_0(1 - P_{OH}/P_{TOT})^2 < \beta_p >]$ [7,8]. Here $\rho$ is a flux surface label, $I_\rho$ is the current within $\rho$, A is the total cross-sectional area, $V = V(\rho)$ is the volume within $\rho$, $\rho = a$ designates the plasma boundary, and $\alpha_T$ describes a "canonical" profile shape $T_c \propto \exp(-\alpha_T V/V(a))$ which the transport model seeks to enforce. The constant C is chosen to fit low density Ohmic experiments, and $\gamma_0 \simeq 9$ reproduces L-mode experimental results with auxiliary heating, provided that ion heat transport of sufficient magnitude to reproduce ion temperature measurements is also present. This is illustrated in Fig. (2) showing the power dependence of electron stored energy and central ion temperature for 2.2 MA TFTR deuterium discharges with neutral beam injection and $\bar{n} \simeq 4.4 \times 10^{19} \text{m}^{-3}$ at steady state [9,10]. Similar fits for 1.4 MA data are obtained by reducing the $\chi_i/\chi_e$ values by 35%. The increase of the effective ion heat transport with input power appears consistent with an ion temperature gradient instability. The simulation points were obtained by neglecting the heat pinch [M = 1] and assuming an anomalous ion thermal conductivity with the same spatial dependence as the electron conductivity. It may be seen that while at low powers $\chi_i/\chi_e \gtrsim 0.5$, at high powers (P $\lesssim 10$ MW) $\chi_i/\chi_e \simeq 1$ - 2 is required to fit the data. We have taken $\alpha_T = 3.33$ for all runs and have generally assumed that M = 3. H-mode transport is assumed to be reproduced by taking $\gamma_o \simeq 4.5$. ECH absorption per unit volume by electrons is represented by the form $P(\rho) = P_0\exp[-(x - x_0)^2/225]$, where $x(\rho)$ is the half-width in cm of a flux surface in the meridian plane, $x_0$ designates the location of maximum absorption, and $P_0$ is proportional to the total power launched in the O-mode.

In our previous report [5], we assumed $\chi_i = 0.5 \chi_e$, and considered both L-mode ($\gamma_o = 9$) and H-mode ($\gamma_o = 4.5$) confinement. To achieve ignition, L-mode confinement was acceptable as long as no sawteeth were present, and, depending upon density profiles, a confinement time of $\tau_E \simeq 0.60$ sec was predicted at ignition for $P_{RF} \simeq 17 - 25$ MW, $n_e(0) \simeq 6.6 \times 10^{20} \text{m}^{-3}$, $Z_{eff} = 1.5$, $\langle \beta \rangle \simeq 2.8\%$. However, under these conditions, $P_\alpha \simeq$ 70 MW, and, therefore, $\chi_i/\chi_e = 0.5$ is too optimistic.

Therefore, in the present paper we examine the importance of varying the anomalous conductivity ratio $\chi_i/\chi_e$ for high density CIT discharges with parabolic density profiles. The 3 sec long ECH pulse is deposited halfway out from the magnetic axis near the q = 1 surface and profile consistency is assumed [8]. Assuming L-mode electron confinement and no sawteeth, 45 MW of auxiliary power is required to achieve ignition ($\tau_E = 0.5$ sec, $\langle \beta \rangle = 4.4\%$) for $\chi_i/\chi_e = 1.0$, while 75 MW ($\tau_E = 0.43$ sec, $\langle \beta \rangle$ = 5.8%) is needed for $\chi_i/\chi_e = 1.5$. Ignition does not occur for $\chi_i/\chi_e = 2.0$. It may be difficult to handle such an equilibrium and $\alpha$ power ($\sim$ 160 MW)and neutron flux ($\sim$ 640 MW) in CIT. Therefore, we must consider H-mode confinement, no sawteeth, and $\chi_i/\chi_e = 1.0$. In this case, 17 MW is sufficient to achieve ignition with $\tau_E = 0.73$ sec, $\langle \beta \rangle = 1.65\%$, (see Fig. 3(a)). Assuming $\chi_i/\chi_e = 2.0$, H-mode confinement ($\gamma_o = 4.5$), 36 MW of ECH power is needed for ignition with $\tau_E = 0.57$ sec, $\langle \beta \rangle = 2.7\%$ and $P_\alpha =$ 80 MW (see Fig. 3(b)).

To summarize the results of Ref. 5 and this paper, it is concluded that low to moderate values of $\chi_i/\chi_e$($\gtrsim 1$) are desirable for ignition at reasonable RF and alpha powers, with confinement times of the order of $\tau_E \gtrsim 0.6$ sec. Peaked density profiles are most beneficial in this regard since in this case the $\eta_i$ modes are expected to be stable, $\chi_i \simeq \chi_i^{NC}$, and ignition near the center will easily commence. Finally, sawtooth control by ECH deposited near the q = 1 surface may be of considerable importance.

*Work supported by the U. S. Department of Energy.

**Figure Captions**
Fig. 1. ECH ray trajectories. a) B = 8.5 T, $T_e$ = 10 keV, $n_e(0)$ = 3.2 $\times 10^{20} \text{m}^{-3}$, $\theta = 20°$; b) B = 10 T, $T_e$ = 20 keV, $n_e(0)$ = 8 $\times 10^{20} \text{m}^{-3}$, $\theta = 0°$; c) B = 10 T, $T_e$ = 20 keV, $n_e(0)$ = 6 $\times 10^{20} \text{m}^{-3}$, $\theta = 30°$. Each solid circle represents 20% power absorption.
Fig. 2. Comparison of experimental (open squares) and simulation (circled numbers) results for TFTR, 2.2 MA deuterium discharges with neutral beam injection and $\bar{n} \simeq$ $4.4 \times 10^{19} \text{m}^{-3}$. a) Electron stored energy, including Thompson scattering recalibration; b) Central ion temperature.
Fig. 3. CIT discharge evolution for $n_e(0) = 6.6 \times 10^{20} \text{m}^{-3}$, ECH power deposited at r/a = 0.5. H-mode confinement, no sawteeth. a) $P_{RF} = 17$ MW, $\chi_i/\chi_e = 1.0$; b) $P_{RF}$ = 36 MW, $\chi_i/\chi_e = 2.0$. $T_e$ and $T_i$ are the central, peak values.

**References**
[1] R. Parker, et al., paper IAEA-CN-50/J-I-1, presented at the 12th Int. Conf. on Plasma Physics and Contr. Nuclear Fusion Research, Nice, France, 1988.
[2] R. Myer, M. Porkolab, G. R. Smith, A. H. Kritz, MIT PFC/JA-89-2 (1989).
[3] M. Bornatici, et al, Nuclear Fusion **23**, (1983) 1153.
[4] G. R. Smith, et al, Bull. Am. Phys. Soc. **31**, (1986) 1516.
[5] M. Porkolab, et al, Paper presented at the 16th European Conf. on Contr. Fusion and Plasma Physics, Venice, Italy, 3/13-17, 1989, Vol. 13B , p. 1155.
[6] G. Bateman, Spring College on Plasma Physics, Trieste, Italy (1985).
[7] B. Coppi, Fizica Plazmy **11**, (1985) 83. Also: MIT Report PTP-85-16 (1985).
[8] R. Englade, MIT Report PTP-87-12 (1987). To be published in Nuc. Fusion.
[9] R. Goldston, et al, Proc. 11th Int. Conf. of Plasma Phys. and Contr. Fusion, Kyoto, Japan [IAEA, Vienna] Vol. 1, p. 75 (1987).
[10] R. Hawryluk, et al, Ref. 9, Vol. 1, p. 51 (1987).

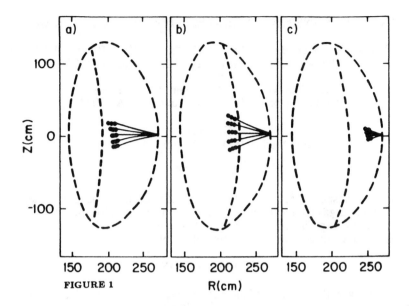

FIGURE 1

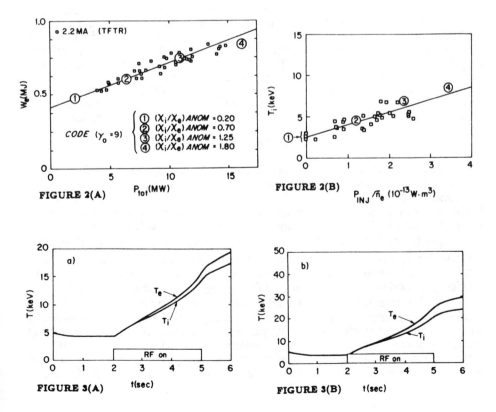

FIGURE 2(A)

FIGURE 2(B)

FIGURE 3(A)

FIGURE 3(B)

# PARTICLE TRANSPORT DURING ECRH ON THE TEXT TOKAMAK

B. Richards, K.W. Gentle, D.C. Sing, P.E. Phillips, W.L. Rowan, A.J.Wootton
Fusion Research Center, University of Texas, Austin, TX 78712

D.L. Brower, N.C. Luhmann, W.A. Peebles
University of California, Los Angeles, CA 90024

G. Cima
Consiglo Nazionale Delle Ricerche, Milano, Italy

## ABSTRACT

Particle transport coefficients during a TEXT discharge have been determined using a modulated-gas feed technique. Comparisons made before and during electron cyclotron heating experiments have shown that the largest effect on transport is an increase in particle diffusion at the periphery of the plasma.

## INTRODUCTION

Beginning with the some of the earliest experiments using electron cyclotron resonance heating (ECRH) on tokamaks[1], it has been observed that application of ECRH generally reduces the density. This effect has been seen in low-power ($P_{ech} \leq P_{oh}$) experiments on the TEXT tokamak, where, for constant gas feed, a decrease in the central density is observed for clean low density ($\overline{n_e} \leq 2 \times 10^{19}$ m$^{-3}$) discharges upon application of rf power. In the case of higher density discharges ($\overline{n_e} \leq 5 \times 10^{19}$ m$^{-3}$), where little heating is observed, a decreased is also observed, but somewhat displaced from the center of the discharge. The decrease is also seen in the case of off-axis heating. The pumpout effect is seen in spite of an increase of the edge particle source, as seen by an increase of $H_\alpha$. We have measured the diffusion coefficient and convective velocity for the central-resonance low-density case and have found that the major effect is an increase in the particle diffusion in the periphery of the plasma.

## EXPERIMENT

TEXT is a circular-cross-section tokamak of 1 m major radius, 0.26 m minor radius, and full poloidal TiC-coated graphite limiter[2]. Operating parameters for the experiment reported here were: 2.1 T toroidal field, 200 kA plasma current, and a $2 \times 10^{19}$ m$^{-3}$ average density. The central electron temperature is approximately 800 eV for reasonably clean hydrogen discharges. These are typical low-$\beta$ sawtoothing discharges, $\beta_p \leq 0.2$. The ECRH is provided by one 60 GHz gyrotron which supplies 200 kW for 80 ms and is coupled to the plasma with an O-mode outside-launch antenna. Total coupling efficiency is estimated to be approximately 70%. Heating efficiency is good for central resonance at this density; $T_{eo}$ increases by 25%[3].

For this case, the line-averaged density typically decreased by 10% while $H_\alpha$ increased by a similar amount. However, the density could be maintained constant by an appropriate modest adjustment of the gas feed. Such effects could be the result of changes in plasma transport, or they might reflect primarily changes in edge conditions: recycling, limiter pumping, etc. To determine if genuine changes in particle transport were induced by ECRH, gas-feed modulation experiments[4] were performed to measure the transport coefficients. A modulation frequency of 31 Hz was sufficient to provide two full periods of quasi-equilibrium modulation during the ECRH pulse. Typical traces are shown in Fig.1. The amplitude and phase of the density modulation were constructed from multi-chord FIR interferometry. The instrument could be moved to scan the region from 20 cm inside the major radius to the limiter outside. By combining results from several discharges, chord integrals of the flux-surface averaged density perturbation over the range of minor radius from 0 to 18 cm could be obtained.

Transport coefficients were extracted by comparing the measured modulation amplitude and phase with line-integrated solutions of the diffusion-convection equation:

$$\frac{1}{r}\frac{d}{dr}\left(r\,D\,\frac{d\tilde{n}}{dr}\right) + \frac{1}{r}\frac{d}{dr}(r\,V\,\tilde{n}) = \tilde{P} - i\omega\tilde{n} \tag{1}$$

where $\tilde{P}$ is the plasma source modulation and $\tilde{n}$ is the complex modulated density. We have used models for the plasma source obtained from $H_\alpha$ measurements and from calculations for comparable ASDEX sources. To avoid errors which would be associated with quantitative source measurements, both the model calculation and the experimental data are normalized so that the amplitude is unity and the phase zero for the central chord. For $D(r)$ and $V(r)$, we have used a model consisting of an interior region, an exterior region, and a linear connecting region. We varied the levels and boundaries of the inner and outer regions to obtain the best fits to measured amplitude and phase. A typical fit to the data during ECRH is shown in Fig. 2, which shows the amplitude and phase of the perturbation as observed at 6, 9, 12, 15, and 18 cm together with the curves from the model calculation of best fit. In addition, the Fig. 2c shows the equilibrium (chord-integrated) density profile implied by the transport coefficients deduced from the modulation experiment together with the experimental points. The fits to the ohmic phase are even better because the longer time interval permits better extraction of the modulation amplitude and phase.

The transport coefficients which provide the fits for the ohmic and ECRH plasmas are plotted in Fig. 3. Although the ECRH resonance is at the center and the heating is strongly peaked within q=1, the particle transport coefficients are unchanged in the central region. The principal change is an increase of more than 25% in the peripheral particle diffusion coefficient. (There is also a slight decrease in peripheral convective velocity in this case, but that is not consistently seen for all cases analyzed; the increase in D is always found.) The increase in diffusion occurs in the entire outer portion (r≥15 cm) of the plasma; it is not purely an edge effect. In fact, since the analysis uses data only out to 18 cm, purely edge effects have little influence on the result. The strongest evidence that the increase in D is uniform in the outer region is that fact that the increase in D deduced here is consistent in magnitude with the reduction in global particle confinement time inferred from $H_\alpha$, which is largely determined by D in the edge source region.

Furthermore, the amplitude of the $H_\alpha$ modulation remains constant from OH to ECRH during the discharge, but the absolute magnitudes of the density perturbations decrease during ECRH, suggesting the same increase in D near the edge during ECRH. (Since only the profile of relative perturbation amplitudes is used in the analysis above, this effect is independent of those results.)

We have also tried to analyze the data using a "model-independent" technique in which we first invert the modulation data and do a straightforward fit to the diffusion-convection equation. We first compiled a "composite" data set by scanning the plasma with 12 shots of data, each shot covering a 12 cm width with 9 vertical chords. Shots were picked from a larger data-base on the basis of having comparable $H_\alpha$ and horizontal position traces and were within 5% of $2\times10^{13}$ cm$^{-3}$ line average density. The resulting data set had the equivalent of about 30 chords of data.

The resulting data set showed an in-out symmetry in the steady state profiles; however, they were asymmetric in the modulation profiles to the extent that the inversion technique did not produce meaningful results. On the assumption that the majority of the asymmetry comes from the plasma source region (e.g., behind the limiter) rather that from large asymmetries in the transport coefficients in the plasma interior, we symmetrized the data about the center of the steady-state profile. Work is underway to model this asymmetry more fully. The data are smoothed using a spline approximation routine. An Abel inversion technique[5] is applied separately to the sin $\omega$t and cos $\omega$t components of the oscillation. This inversion technique fits the line-integrated density data using a set of nested disks, with no pre-conceived form for D(r) or V(r). The analysis assumes only that ñ(r) is constant on toroidal surfaces, which could be, but are not necessarily, flux surfaces. No transport code or steady state profile information was used to calculate fluxes or profiles for this inversion.

The diffusion-convection equation can be integrated once to obtain:

$$D\frac{d\tilde{n}}{dr} + V\tilde{n} = \frac{1}{r}\int_0^r r\,(\tilde{P} - i\omega\tilde{n})\,dr \qquad (2)$$

Equating the real and imaginary parts allows us to solve for D(r) and V(r) directly. Results of this inversion are shown in Fig. 4. As before, we find that we have diffusion coefficients and convective velocities which are small near the center of the plasma and start increasing about half-way out to the plasma edge. Also, the conclusions regarding the transport changes during ECH are in agreement. We see little change in either transport parameter during heating in the center of the plasma, but we see a definite increase in the diffusion coefficient toward the outer half of the plasma. We have not as yet performed a detailed error analysis for this inversion technique. We have found that the values of D and V close to the plasma edge can vary by 50%, depending on the exact shape of $\tilde{P}$(r). However, the effects of ECRH on the diffusion coefficient are consistent for all reasonable source profiles.

## CONCLUSION

The application of central ECRH power induces an increase in the particle diffusion coefficient in the outer region of the plasma, but no significant effect in the interior. This is certainly consistent with the decrease of density generally associated with ECRH. The specific mechanism, however, remains an enigma, for the outer region seems the region least affected, either directly or indirectly, by the ECRH. The local parameters are very slightly changed. Similar density decreases are observed at higher densities, for which central heating no longer occurs, and with off-axis heating. Additional experiments of this type will be required to confirm if the same increase in particle diffusion is responsible in these cases as well.

This research is supported by the US Department of Energy, Contract Nos. DE-FG05-88ER53267 and DE-FG03-86ER53225.

1. R.M. Gilgenbach, et al. Phys. Rev. Lett. **44**, 647 (1980).
2. R.V. Bravenec, et al. Plasma Physics and Controlled Fusion **27**,1335 (1985).
3. D.C. Sing, et. al., this meeting.
4. K.W. Gentle, B. Richards, and F. WaelBroeck, Plasma Physics and Controlled Fusion **29**, 1077 (1987)
5. N. Gottardi, J. Appl. Phys. **50**, 2647 (1979)

## Typical Discharge

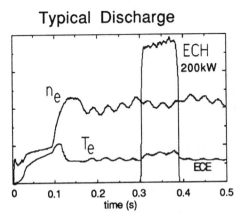

Fig. 1. Time traces of the modulated gas feed experiment. The central interferometer chord, central electron temperature from ECE, and ECRH power are shown.

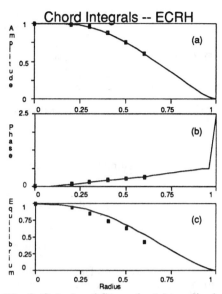

Chord Integrals -- ECRH

Fig. 2. Calculated fits to the (a) amplitude and (b) phase of the density modulation and the (c) equilibrium profile using the transport coefficients which best fit the modulation data. Radius is normalized to the minor radius.

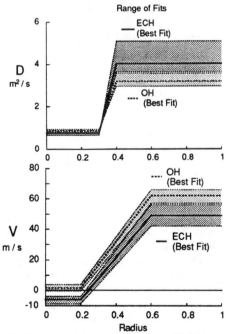

Fig. 3. The model transport coefficients, $D(r)$ and $V(r)$, which best fit the data of Fig. 2.

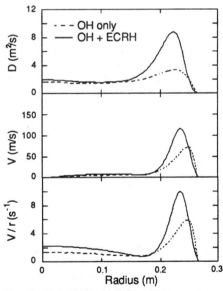

Fig. 4. "Model-independent" fits to the transport coefficients and to $V(r)/r$.

75

# THERMAL TRANSPORT DURING ELECTRON CYCLOTRON HEATING IN THE TEXT[*] TOKAMAK

D. C. Sing, R. V. Bravenec, M.S. Foster, K.W. Gentle, P.E. Phillips, A. Ouroua, B. Richards, B.A. Smith[§], J.C. Wiley, A.J. Wootton.
The Fusion Research Center, The University of Texas at Austin, Austin TX 78712

M. E. Austin, R. Gandy.
Auburn University, Auburn AL

J.Y. Chen, Z.M. Zhang.
Institute of Plasma Physics, Academia Sinica, Hefei, PRC

G. Cima.
Consiglio Nazionale Delle Ricerche, Milano, Italy

Kevin W. Wenzel. Plasma Fusion Center
Massachussetts Institute of Technology, Cambridge, MA

Electron Cyclotron Heating (ECH) experiments have been performed on the TEXT tokamak ($R_0 = 1$ m, a = .26 m) using a Varian 60 Ghz, 200 kW gyrotron. Pulses as long as 89 ms have been delivered to the plasma. The $TE_{11}$ mode is launched along the midplane with the O-mode polarization at $10°$ from perpendicular to $B_T$. Typical gyrotron power is 200 kW, which combined with the transmission and mode conversion efficiencies results in 157 kW launched into the tokamak. Experiments have been performed in two regimes; a low-q ($I_P \sim 200$ kA, $B_T \sim 1.9$-2.4 T, q(a) $\sim$ 3, $\bar{n}_e \sim 1.5$-2.5 x $10^{19}$ m$^{-3}$) sawtoothing discharge, and a high-q ($I_P \sim 110$-140 kA, $B_T \sim 1.9$-2.4 T, q(a)$\sim$5-6, $\bar{n}_e \sim 1.5$-2.0 x $10^{19}$ m$^{-3}$) non-sawtoothing discharge.

Figure 1 shows the changes in $T_e(0)$ (as measured by ECE) for typical low-q ($I_P = 120$ kA) and high-q ($I_P = 200$ kA) discharges with central ECH heating applied. The density drop usually associated with ECH in tokamaks is compensated by additional gas puffing to keep the line averaged density nearly constant at 1.6 x $10^{19}$ m$^{-3}$. The loop voltage drops 0.2 V (from 1.6 V to 1.4 V) in the low-q discharge and 0.4 V for the high-q discharge. For both cases $P_{oh}$ decreases by $\sim$40 kW. $T_e(0)$ for the low-q case shows a moderate increase ($\sim$30%), and the temperature sawteeth increase in amplitude by a factor of 3. For the high-q discharge $\Delta T_e(0)$ is much larger ($\sim$80%). Sawteeth occasionally occur after the ECH pulse turns off.

The $T_e$ profiles for central ECH heating as measured by Thomson scattering are shown in Fig. 2a for low-q (with $\bar{n}_e = 2.5$ x $10^{19}$m$^{-3}$ and in Fig. 2b for high-q ($\bar{n}_e = 1.8$ x $10^{19}$ m$^{-3}$) discharges. In low-q discharges $T_e(0)$ increases but the shape of the profile remains similar to the pre-ECH case. The application of ECH to the high-q discharge changes the electron profile shape considerably, as well as increasing the central electron temperature as much as 80%. The increase due to ECH drops off rapidly from the plasma center; outside of $\rho \equiv$ (r/a)$\sim$0.2 the increase of $T_e$ is much smaller. The variation in $\Delta T_e(0)$ with resonance position depends on $I_P$; for low-q discharges $\Delta T_e(0)$ remains nearly constant when the resonance is within the mixing radius ($\sim$7 cm), and disappears when the resonance is moved outside of that radius. For the high-q discharge large $\Delta T_e$'s are obtained only when the resonance is within 3 cm of the magnetic axis.

## CHANGES IN ELECTRON THERMAL TRANSPORT

The measured profiles of $T_e$, $n_e$, $p_{rad}$, and the central $T_i(0)$ and the changes in $V_{loop}$ and $I_P$ are used to determine the changes in $\chi_e$ by power balance analysis. The ECH power density was calculated using the TORAY ray tracing code[1] coupled with the measured gyrotron power output and known transmission and mode conversion losses. A lower limit to $p_{ech}$ is calculated assuming single pass absorption (typically 75%) of a Gaussian beam by the plasma. The upper limit to $p_{ech}$ is calculated assuming the remaining 20 to 30% is absorbed by the plasma according to the absorption profile shape calculated assuming uniform illumination of the plasma.

Figure 3 shows the $\chi_e$ profile determined by power balance analysis of the low-q discharge with central ECH heating. The heat transport inside the mixing radius ($\rho \sim 0.3$) is dominated by sawteeth and $\chi_e$ is not shown for $\rho < 0.3$. In the confinement region ($\rho \sim 0.5$) $\chi_e$ is increased by 35%. Measurement of $\chi_{e,HP}$ by the sawtooth heat pulse propagation method is possible for the low-q discharge and provides an independent measure of changes in confinement with ECH. Figure 4 shows the variation in $\chi_{e,HP}$ versus density for the low-q regime. As is usually observed, $\chi_{e,HP} > \chi_e$ from power balance. On average, $\chi_{e,HP}$ increases by 25% with on-axis ECH, which is comparable to the change in $\chi_e$ measured by power balance analysis, given the uncertainties in the $\chi_e$ measurements.

Figure. 5 shows $\chi_e$ as measured by the power balance method for the non-sawtoothing high-q discharge. Also shown are the $\chi_e$ profiles used with the CHAPO transport code to simulate the ohmic and ECH discharges.[2] The power balance $\chi_e$ increases with ECH, but the error bars are large, primarily due to large uncertainties in computing $dT_e/dr$. The CHAPO calculated $T_e$ profile and the measured $T_e$ profile are shown in Figure 6. A 30% increase in $\chi_e$ from the ohmic case is required to match the $T_e(0)$ changes, which is consistent with $\chi_e$ scaling as $\sqrt{T_e}$. The agreement between the calculated and measured $T_e$ profiles is reasonable, although the data (see Figure 2b) shows an inflection point in the $T_e$ profile between $\rho = 0.1$ and $0.2$ not reproduced by the simulation.

## ECH POWER MODULATION EXPERIMENTS

Experiments were undertaken in which the gyrotron was switched on and off at a 125 Hz rate. The tokamak was run in the standard high-q condition ($I_p = 120$ kA) with the ECH resonance on the magnetic axis. The horizontally viewing SXR array was used to view the outwardly moving heat wave. Phase delays with respect to the central X-ray channel were obtained using the analysis of Jahns et. al.[3] A linear relationship was found between the phase delay $\varphi$ and the radial view r of the SXR detectors, and the relation $\chi_{e,mod} = 3/2\omega(\Delta r/\Delta\varphi)^2$ is used to calculate the heat diffusivity. Figure 7a shows a plot of the observed phase delays as a function of the SXR detector viewing radius. The heat diffusivity $\chi_{e,mod}$ deduced from this analysis is large, typically 15 $m^2$/sec, which is much larger than the $\chi_e$ obtained from power balance considerations (2-3 $m^2$/sec). A possible explanation for the large value of $\chi_e$ is that the ECH antenna

78

pattern is relatively large, and that at a given radial position direct ECH heating (with zero phase delay) partially offsets the phase shift due to the diffusively transported heat. Figure 7b shows the fraction of launched ECH power (single pass absorption only) absorbed within a given radial position for conditions typical of the modulated ECH discharge. The absorbed power curve does not appreciably flatten until the 12 cm radius, which includes much of the region over which the phase delays are included for the calculation of the modulated heat diffusivity.

## CONCLUSIONS

Moderate degradation of electron energy confinement is observed with ECH heating in the TEXT tokamak. This is observed for sawtoothing and non-sawtoothing discharges. Sharp $T_e$ profiles are produced in high-q discharges by extremely localized central ECH power deposition coupled with the lack of sawtooth driven transport.

## REFERENCES

1. A.H. Kritz, et al., in "Heating in Toroidal Plasmas", vol.2, 707 (1982).
2. J. C. Wiley, "CHAPO, A One Dimensional Transport Code", FRC report 328, The University of Texas Fusion Research Center, Austin TX (1988) .
3. G.L. Jahns, et. al., Nuclear Fusion **26**, 226 (1986).

*Operated by The University of Texas at Austin under DOE grant No. DE-FG05-88ER53267. § Supported by a USDOE Magnetic Fusion Energy Fellowship administered by the Oak Ridge Associated Universities.

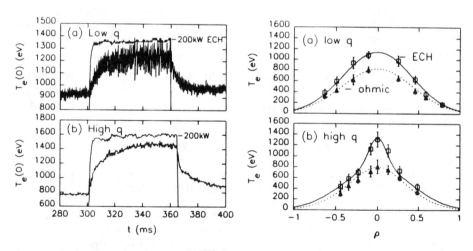

Fig. 1. $T_e(0)$ vs. time.(a) Low-q. (b) High-q     Fig. 2. $T_e$ profile. (a) Low-q (b) High-q

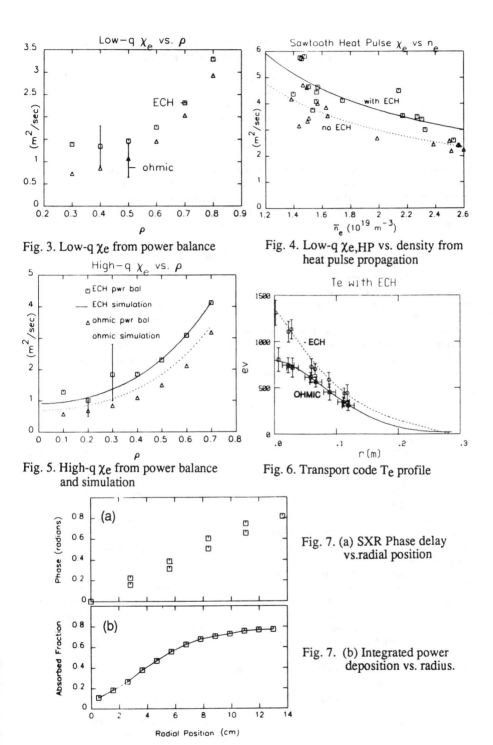

Fig. 3. Low-q χe from power balance

Fig. 4. Low-q χe,HP vs. density from heat pulse propagation

Fig. 5. High-q χe from power balance and simulation

Fig. 6. Transport code Te profile

Fig. 7. (a) SXR Phase delay vs.radial position

Fig. 7. (b) Integrated power deposition vs. radius.

# Optimum Launching of Electron-Cyclotron Power for Localized Current Drive in a Hot Tokamak

*Gary R. Smith*
Lawrence Livermore National Laboratory
University of California
Livermore, CA 94550

## ABSTRACT

Optimum launch parameters are determined for localized electron-cyclotron current drive near the magnetic axis and the $q = 2$ surface by solving several minimization problems. For central current drive, equatorial and bottom launch are compared. Localized current drive near $q = 2$ is studied for equatorial launch and for an alternative outside launch geometry that may be better for suppressing tearing modes and controlling disruptions.

## INTRODUCTION

One of the best uses of electron-cyclotron power in a tokamak is for localized current drive, because superior localizability tends to compensate for lower efficiency compared to other current-drive schemes. Localization near the magnetic axis and the $q = 1$ and $q = 2$ surfaces is of interest. To determine launch scenarios that yield an optimum combination of efficiency and localizability, one must search through a parameter space that includes the poloidal launch location, the launch direction, and the ratio $\Omega_{axis}/\omega$ of on-axis cyclotron frequency to wave frequency. This paper determines optimum launch parameters automatically by solving several minimization problems. Application is made to the ITER tokamak.

## PLASMA MODEL

We use flux surfaces computed at Livermore for the ITER Basic Engineering Device during the technology phase. The specific case "j023" corresponds to the start of burn. At the nominal major radius of $R_0 = 5.5$ m, the magnetic field is $B_0 = 5.3$ T. The magnetic axis is at $R_{axis} = 5.76$ m, where the field is $B_{axis} = 5.08$ T. The safety factor $q$ increases monotonically from $q_{axis} = 1.14$ at the magnetic axis to $q_{95} = 3.64$ at $\psi = 0.95$. The poloidal flux $\psi$ is normalized so that $\psi = 0$ at the axis and $\psi = 1$ at the separatrix. The $q = 2$ surface is located at $\psi = \psi_2 = 0.725$.

The density and temperature profiles are not consistent with the pressure in the flux-surface calculation but have the widely used forms $n(\psi) = \hat{n}(1 - \psi)^{\alpha_n}$ and $T_e(\psi) = \hat{T}_e(1 - \psi)^{\alpha_T}$ with $T_i = T_e$ and $\alpha_n = 0.5$ and $\alpha_T = 1$. The peak (on-axis) density and temperature are $\hat{n} = 10^{20}$ m$^{-3}$ and $\hat{T}_e = 35$ keV.

Ray tracing and absorption is based on the weakly relativistic dielectric tensor of Shkarofsky.[1] Current-drive efficiency is calculated with the inclusion of trapped-particle effects[2] but with the omission of momentum transfer from hot to bulk electrons.[3] The latter effect enhances current-drive efficiency by roughly 50% near the magnetic axis but much less on outer flux surfaces. Current-drive efficiency is reduced from that found in a pure plasma by our choice of $Z_{eff} = 2.2$. In studying all physical situations, we employ single-ray calculations to understand the absorption and current-drive physics, then we complete the study with multi-ray calculations to learn about the sensitivities to finite size and divergence of realistic microwave beams.

## FORMULATION OF OPTIMIZATION CALCULATIONS

Optimization calculations are performed with the aid of IMSL routine N0ONF, which is based on work by Schittkowski.[4] This routine attempts to minimize an objective function $F(x)$, subject to constraints and lower and upper bounds on the variables $x$. In this work we use no constraints but specify bounds to limit the parameter space searched by the solver. Depending on the current-drive problem of interest we make various choices of $F$ and $x$.

To maximize the current driven near the magnetic axis, we choose $F(x) = -I_d(\psi_c)$ with $\psi_c = 0.1$, where $I_d(\psi)$ is the current driven on all flux surfaces between the axis ($\psi = 0$) and $\psi$. We model power launched along the equatorial plane ($Z = 0$) by a single ray, and we choose $x$ to consist of the microwave frequency $f$ and the angle $\phi_\ell$ between the launch direction and a line joining the launch point ($X = 750$ cm, $Y = Z = 0$) with the machine center ($X = Y = Z = 0$); symbolically, $x = \{f, \phi_\ell\}$. For power launched below the plasma ("bottom launch"), more variables are required to specify the launch location and direction. Using a cylindrical coordinate system ($R, \Phi, Z$), we fix the radial and azimuthal components of the launch location at $R_\ell = 830$ cm and $\Phi_\ell = 9.25°$ and vary the vertical component of the launch location $Z_\ell$ and all components ($R_t, \Phi_t, Z_t$) of a target location that determines the launch direction. Thus, we use $x = \{f, Z_\ell, R_t, \Phi_t, Z_t\}$.

Current drive just inside of the $q = 2$ surface is capable, at least theoretically, of suppressing tearing modes, which may help to reduce the number or severity of disruptions. The current required drops dramatically as localization improves,[5] so we choose to maximize $dI_d/d\psi$ just inside $\psi = \psi_2$. For power launched along the equatorial plane, we use $x = \{f, \phi_\ell\}$ as before. An alternative launch geometry fixes the launch location below the equatorial plane at ($X = 750$ cm, $Y = 0$, $Z = -100$ cm) and varies two launch angles $\theta_\ell$ and $\phi_\ell$, which are related to the wavevector components at the launch location by $\tan \theta_\ell = -k_Z/(k_X^2 + k_Y^2)^{1/2}$ and $\tan \phi_\ell = k_Y/k_X$. Thus, we use $x = \{f, \theta_\ell, \phi_\ell\}$.

## OPTIMUM LAUNCH PARAMETERS FOR ITER

Power launched along the equatorial plane drives maximum current near the magnetic axis for $f = 198$ GHz and $\phi_\ell = 39°$. In terms of dimensionless parameters that play important roles in the absorption and current-drive physics, this result can be expressed as $\Omega_{axis}/\omega = 0.72$ and, at the absorption peak, $N_\parallel = 0.82$. These optimum parameters are similar to those found in Ref. 6. However, that work found a substantially higher current-drive efficiency than we do at the optimum launch parameters, for reasons that are not presently understood. Taking momentum transfer into account (which Ref. 6 did not), we find a total driven current per incident watt of $I_d/P_i = 0.05$ A/W, or $\eta \equiv \langle n \rangle I_d R_{axis}/P_i = 0.20 \times 10^{20}$ A/Wm$^2$.

Bottom launch is optimum for central current drive at $f = 204$ GHz and with launch parameters $Z_\ell = -324$ cm, $R_t = 557$ cm, $\Phi_t = -48°$, and $Z_t = 100$ cm. A projection of the ray trajectory into the poloidal plane is shown in Fig. 1. There is little refraction for this ray. The apparent curvature seen in Fig. 1 is mostly an effect of projection onto the non-Cartesian $RZ$ plane. The wave frequency corresponds to $\Omega_{axis}/\omega = 0.7$ and, at the absorption peak, $N_\parallel = 0.82$. Although the bottom-launch geometry deposits power on electrons with higher mean energy than does launch in the equatorial plane, current-drive efficiency is only slightly better: $\eta = 0.21 \times 10^{20}$ A/Wm$^2$. The single-ray bottom-launch calculation shows that over 90% of the driven current is concentrated within the flux surface $\psi = 0.05$, which compares with $\psi = 0.03$ for equatorial launch.

Power launched along the equatorial plane with the optimal $f = 137$ GHz and $\phi_\ell = 25°$ drives a localized current with peak $dI_d/d\psi$ at $\psi = 0.675$. The total driven current, ignoring the small momentum-transfer effect on this outer flux surface, is 0.022 A/W, which corresponds to $\eta = 0.09 \times 10^{20}$ A/Wm$^2$. To characterize the localization of the driven current, we quote the width $\Delta\psi = 0.15$ that contains the middle 80% of the current.

The alternative launch geometry for localized current drive near $q = 2$ has optimum ray trajectories shown in Fig. 2. To account crudely for finite beam divergence, we model a Gaussian beam emerging from a corrugated waveguide with i.d. of 9 cm by means of three rays. All rays have the same $\phi_\ell$, but the outer rays have $\theta_\ell$ that differ by $\pm 1°$ from the central ray. For an optimal $f = 148$ GHz, the maximum $dI_d/d\psi$ at $\psi = 0.7$ occurs if the central ray is launched

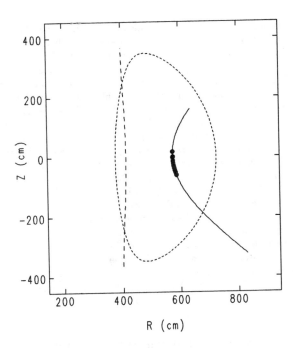

Fig. 1. Poloidal projection of a ray trajectory launched far below the equatorial plane and passing up through the plasma center. The short-dash curve is the $\psi = 0.95$ flux contour of an equilibrium calculated for the ITER tokamak. The long-dash curve is the fundamental cyclotron resonance. The filled circles indicate the location of wave absorption.

with $\phi_\ell = 45.5°$ and $\theta_\ell = 23°$. At the absorption peak, $N_\parallel \approx 0.5$, slightly more than the 0.4 found for equatorial launch. The total driven current is 0.030 A/W, which corresponds to $\eta = 0.12 \times 10^{20}$ A/Wm$^2$. The middle 80% of the current occurs in a width $\Delta\psi = 0.03$. Thus, the efficiency and especially the localization are significantly better for the alternative launch geometry than for equatorial launch.

## CONCLUSIONS

Central current drive, optimized for either equatorial or bottom launch, is found to have an efficiency of $\eta \approx 0.2 \times 10^{20}$ A/Wm$^2$. Localization is somewhat better with equatorial launch.

Current drive just inside the $q = 2$ surface is more efficient and much more localized if the power is launched so that the power is absorbed while the microwave beam is nearly parallel to the flux surface. A specific launch geometry with these characteristics is found here for ITER. Further studies should address the degradation of efficiency and localization that would occur if $T_e$ were lower at $q = 2$ or if scattering by drift-wave fluctuations caused the beam to broaden significantly.

For all of the current-drive schemes discussed here, we must assess the prospects for tracking moving flux surfaces using the simplest possible technology.

## ACKNOWLEDGMENTS

Many people aided the author in the performance of this work. R. E. Bulmer provided

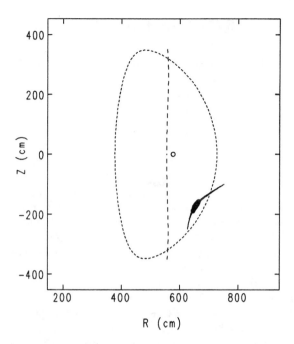

Fig. 2. Poloidal projection of three ray trajectories launched below the equatorial plane at angles designed to drive current localized just inside the $q = 2$ surface. The open circle is the magnetic axis.

the equilibrium magnetic-field results. A. V. Kluge, A. H. Kritz, and I. P. Shkarofsky made major contributions to the ray-tracing code. R. H. Cohen made his calculations of current-drive efficiency available.

This work was performed under the auspices of the U.S. Department of Energy by Lawrence Livermore National Laboratory under contract number W-7405-ENG-48.

## REFERENCES

[1] I. P. Shkarofsky, *J. Plasma Phys.* **35**, 319 (1986); I. P. Shkarofsky and A. K. Ghosh, Technical Report 232-1, MPB Technologies Inc. (1984).

[2] R. H. Cohen, *Phys. Fluids* **30**, 2442 (1987) and **31**, 421 (1988).

[3] R. W. Harvey, K. D. Marx, and M. G. McCoy, *Nucl. Fusion* **21**, 153 (1981); C. F. F. Karney and N. J. Fisch, *Phys. Fluids* **28**, 116 (1985).

[4] K. Schittkowski, *Annals of Operations Research* **5**, 485 (1986).

[5] E. Westerhof, *Nucl. Fusion* **27**, 1929 (1987).

[6] G. Giruzzi, T. J. Schep, and E. Westerhof, "Current Drive by Electron Cyclotron Waves in NET," presented at Meeting of the European Physical Society (Venice, March 1989).

# ECH on the MTX[*]

B. Stallard, J. Byers, B. Hooper, S. Meassick, B. Rice,
T. Rognlien, and J. Verboncoeur
Lawrence Livermore National Laboratory, Livermore, CA. 94550
M. A. Makowski
TRW, Redondo Beach, Ca. 90278

## Abstract

The Microwave Tokamak Experiment (MTX) at LLNL is investigating the heating of high density Tokamak plasmas using an intense pulse FEL. Our first experiments, now beginning, will study the absorption and plasma heating of single FEL pulses (20 ns pulse length and peak power up to 2 GW) at a frequency of 140 GHz. A later phase of experiments also at 140 GHz (FY 90) will study FEL heating at 5 kHz rate for a pulse train up to 50 pulses (35 ns pulse length and peak power up to 4 GW). Future operations are planned at 250 GHz with an average power of 2 MW for a pulse train of 0.5 s. The microwave output of the FEL is transported quasi-optically to the tokamak through a window-less, evacuated pipe of 20 in diameter, using a six mirror system. Computational modelling of the non-linear absorption for the MTX geometry predicts single-pass absorption of 40% at a density and temperature of $1.8 \times 10^{20} m^{-3}$ and 1 keV, respectively. To measure plasma microwave absorption and backscatter, diagnostics are available to measure forward and reflected power (parallel wire grid beam-splitter and mirror directional couplers) and power transmitted through the plasma (segmented calorimeter and waveguide detector). Other fast diagnostics include ECE, Thompson scattering, soft x-rays, and fast magnetic probes.

## Introduction

Intense pulse FEL electron heating of a tokamak plasma will soon begin (Summer, 1989) in the MTX experiment at LLNL. In addition to technology demonstrations for the FEL and microwave transport system, important physics issues to be addressed in the initial experiments are the reduction in absorption (from linear theory) by non-linear effects, the significance of parametric instabilities which may cause backscatter, and the possibility of beam filamentation. The initial phase of experiments at 140 GHz will investigate the absorption of single FEL pulses with 20 ns pulse duration and up to 2 GW peak power. The short pulse electromagnets of the ELF wiggler limit heating to a single FEL pulse per plasma discharge [1]. A second phase of experiments (FY 90), using the IMP wiggler with DC magnets [2], will heat at higher power (35 ns pulses and peak power up to 4 GW) and at a 5 kHz rate for a pulse train of 50 pulses. These

[*]This work was performed under the auspices of the U.S. Department of Energy by Lawrence Livermore National Laboratory under Contract W-7405-Eng-48.

experiments will focus on multi-pulse heating, radial transport of FEL absorbed power, and ECH of pellet-fuelled plasmas.

## ECH Transmission

The transport of FEL output power to the tokamak is accomplished by six mirror quasi-optical transmission, as shown in Fig. 1. The transmission system is window-less and mirrors are enclosed within a 50 cm diameter evacuated pipe. To avoid the need of an achromatic jog in the transport pipe for the e-beam which drives the FEL, two additional mirrors ($J_0$ and $J_1$) were added to the original design previously described [3]. Mirrors $J_0$, $M_2$, and $M_4$ are focussing optics, and the remaining mirrors are flats. The dominant output mode of the wiggler is $TE_{01}$ in WR 229 rectangular waveguide (5.82 cm x 2.91 cm). Using the MTH code [4] the overall transmission efficiency at 140 GHz is 89%. Most of the loss is clipping of side-lobe power of the mode at the first three mirrors. For the second phase of experiments using the IMP wiggler the waveguide mode will be $TE_{11}^\circ$ in 3.25 cm circular waveguide. To transport this mode mirrors $J_0$ and $M_2$ must be modified. The final optic ($M_4$) focusses the microwave beam into an elliptical cross-section (about 6:1 ellipticity) for transmision through the narrow port of MTX (4 cm horizontal x 30 cm vertical x 22 cm duct length.)

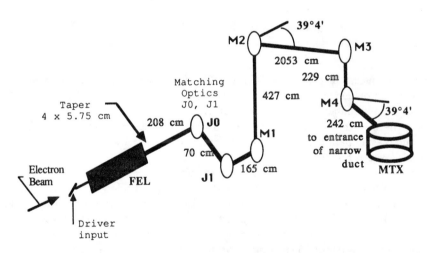

Figure 1: MTX microwave transport. Design for 140 GHz (not to scale).

## Non-linear Absorption

Because of the intense electric fields of the FEL beam ($E_{max} \sim$ 200 to 300 keV/cm), the absorption is non-linear and reduced from the linear theory value [5]. The expected absorption was modelled by the following multi-step process: 1) the calculated beam profile at the port entrance (MTH code) is decomposed by Fourier analysis into the appropriate set of waveguide modes excited by the incident beam, 2) the waveguide modes are propagated to the end of the 22 cm long duct, taking into account the differential phase shift between modes, 3) the MTH code calculates the electric field at the plasma using diffraction theory and the mode amplitudes and phases at the duct exit, and 4) attenuation of the beam and heating of electrons as they pass through the beam

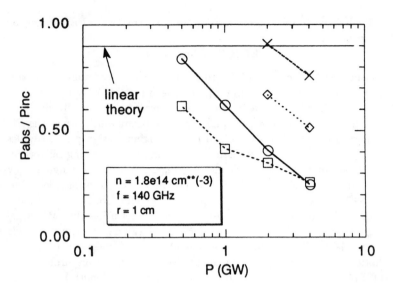

Figure 2: Fraction of incident power absorbed for various cases. (o) Smooth duct; dashed curves are corrugated: (□) no taper, (◇) 2:1 taper, (x) 4:1 taper.

are computed using the orbit code ORPAT. The differential phase shift between modes after propagation along the duct produces significant distortion in the electric field profile at the plasma and also increases wall electric fields to $E \sim 0.5\ E_{max}$ at the duct exit. For the smooth wall duct the calculated single-pass absorption is $\sim 40\%$ for 2 GW power and plasma parameters $n_{e0} = 1.8 \times 10^{20} m^{-3}$ and $T_{e0} = 1$ keV. The variation with power is shown in Fig. 2.

To increase the absorption efficiency and reduce wall electric fields, we are exploring the use of parabolic down-tapers with smooth or corrugated side walls (corrugations on vertical walls, perpendicular to E, with corrugation depth and period $\lambda/4$ and $\sim \lambda/3$, respectively) [6,7]. The electric field profile incident upon the entrance port is near Gaussian and strongly couples to the lowest order hybrid mode $HE_{\parallel}$ of the corrugated waveguide. A 4:1 taper of 25 cm length can fit within the entrance port.

The increase in absorption through use of the taper results from the reduced electric field and the greater beam divergence ($k_{\parallel}$ spread) at the plasma caused by diffraction from the small beam waist near the duct exit [5]. Fig. 2 compares several down-tapers in the corrugated duct with the un-tapered, smooth wall duct. Absorption with the 4:1 taper is more than double the absorption at 2 GW and equal to the linear absorption value.

## Absorption and Fast Response Diagnostics

To measure single pulse absorption several diagnostics have been developed to measure forward and reflected power at the input port and the power transmitted through the plasma. From these measurements the single-pass absorption can be inferred. We measure forward and reflected power by 1) a parallel wire grid beam splitter oriented at 45 deg to the beam axis and 2) single mode waveguides near the beam center on $M_4$, viewing the forward and reflected directions. The wire grid is a diffraction grating where the electric field and grid wires lie in the plane of incidence. For wire diameter 0.005 in and wire spacing $\lambda/d = 0.080$, the coupling coefficient is 0.023 [8]. Mirror optics focus the coupled power onto fast response photon drag detectors. On the inside wall we measure power transmitted through the plasma by 1) a small microwave horn located near the beam center and 2) a segmented calorimeter consisting of 2 cm x 9 cm (toroidal direction) silicon carbide tiles, backed by thermisters, and segmented in the poloidal direction. The calorimeter can measure a large fraction of the total transmitted power and the effects of plasma refraction.

In addition to direct measurements of absorption, the localization and magnitude of heating can be assessed using various diagnostics with fast time response ($\leq 1 \mu$sec). These include soft x-rays, Thomson scattering, an ECE polychrometer, and fast magnetic probes. The time scale for equilibrium after injection of an FEL pulse is estimated to be several $\mu$sec (Alfven times and toroidal equilibrium of initially trapped electrons). Because the fractional energy increase for a single pulse is small ($\Delta W/W \sim 1\%$), measurement of the energy increase for a single pulse, although difficult, may be possible using fast magnetic probes. For time scales $< 100 \mu$sec, the vacuum wall of the MTX vessel is flux conserving. The estimated magnitudes of field changes for the initial experiments are poloidal (toroidal) field $\delta B_\theta \sim$ few gauss ($\delta B_T \sim (\frac{1}{10}) \delta B_\theta$). If successful, these measurements will be useful to study the plasma transient response and also to measure single pulse heating.

## References

1. T. Orzechowski et al., Phys. Rev. Lett. 57 (1986) 2172.
2. R. A. Jong et al., Rev. Sci. Instrum., 60 (1989) 186.
3. B. W. Stallard et al., Seventh Top. Conf. on Appl. of Radio-Frequency Power to Plasmas, Kissimmee, Fla.,(1987) 21.
4. T. Samec, Bull. Amer. Phys. Soc., 32 (1987) 1872.
5. W. M. Nevins et al., Phys. Rev. Lett., 59 (1987) 60.
6. J. Doane, Internat. J. of Infrared and Millimeter Waves, 5, (1984) 737; J. Doane, Internat. J. of Infrared and Millimeter Waves, 8, (1987) 13.
7. J. Doane; mode coupling coefficients in rectangular corrugated waveguide, private communication.
8. J. Lorbeck, Lawrence Livermore National Laboratory Report, UCID-21644, April, 1989.

ELECTRON CYCLOTRON CURRENT DRIVE
AND HEATING EXPERIMENTS ON THE WT-3 TOKAMAK

Y. Terumichi, H. Tanaka, K. Ogura, S. Ide,
M. Iida, T. Itoh, M. Iwamasa, K. Hanada, H. Sakakibara,
T. Minami, M. Yoshida, T. Maekawa and S. Tanaka
Department of Physics, Faculty of Science, Kyoto University,
Kyoto 606 Japan

## ABSTRACT

The second harmonic electron cyclotron current drive ($2\Omega e$ ECCD) and heating ($2\Omega e$ ECH) studies on WT-3 have been carried out with 56 GHz EC power up to 200 kW which is injected in the x-mode into a target Ohmic heated (OH) plasma from the low field side. $2\Omega e$ EC driven current up to 70 kA can be sustained without OH power during the EC pulse. The safety factor $q_a$ = 5, the electron density $\bar{n}_e = 2 \times 10^{12}$ cm$^{-3}$ and the central electron temperature Te(0) $\simeq$ 500 eV. The current drive efficiency is 0.06 ($10^{19}$ A/Wm$^2$ ). In the $2\Omega e$ ECH experiments, it is observed that the period of the sawtooth oscillations becomes long and the amplitude saturates when the EC power is nearly comparable to the OH power and the $2\Omega e$ ECR layer is located near (just outside) the q = 1 surface.

## INTRODUCTION

There has been much interest in ECCD and in the profile control by the local ECH in the tokamaks. In this paper, we report the $2\Omega e$ ECCD[1] and the control of the sawtooth oscillations by $2\Omega e$ ECH. Experiments have been carried out on the WT-3 tokamak which has the major radius, $R_0$, of 65 cm, the minor radius, a, of 20 cm and the maximum toroidal field, $B_T$ (0), of 1.75 T. A 56 GHz, 200 kW, 100 ms pulse gyrotron operated in TE$_{02}$ mode was used as a microwave power source. A Vlasov antenna with a parabolic reflector was used to convert the TE$_{02}$ mode to a linearly polarized one and to launch the x-mode microwave radiation from the low field side into the plasma at the incident angle of 120° to the toroidal field.[2] To study an electron distribution function, soft and hard x-ray (HX) detector arrays, HX pulse height analysers, 4-channel radiometers and 5-channel HCN interferometer were used.

## $2\Omega e$ ECCD EXPERIMENTS

In our experimental situation, the low density target plasma was produced by normal tokamak discharge. The high energy tail electrons existed in the target plasma. The EC power selectively heated up the slide-away tail electrons. The plasma current were carried by these electrons. The typical waveforms are shown in Fig. 1. The dotted curves show the waveforms of the initial

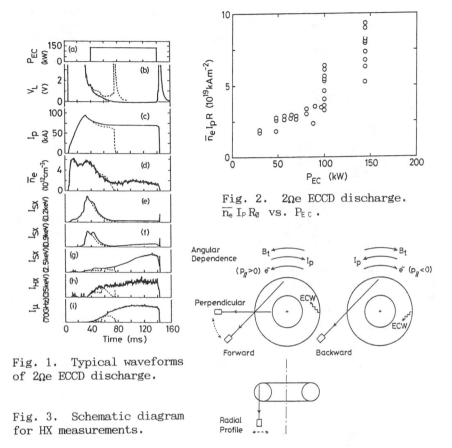

Fig. 2. $2\Omega e$ ECCD discharge. $\overline{n}_e\,I_P\,R_\emptyset$ vs. $P_{EC}$.

Fig. 1. Typical waveforms of $2\Omega e$ ECCD discharge.

Fig. 3. Schematic diagram for HX measurements.

target plasma and the solid curves show the waveforms with the ECH pulse. The plasma current, $I_P$, of 70 kA was sustained without OH power during the 144 kW EC pulse. The ECCD plasma had the line averaged electron density, $\overline{n}_e$, of $2 \times 10^{12}$ cm$^{-3}$ and the central electron temperature $Te(0)$ of about 500 eV which was measured with Thomson scattering. $\overline{n}_e\,I_P\,R_\emptyset$ for various ECCD discharges were plotted versus the injected EC power in Fig. 2. The $2\Omega e$ ECCD efficiency, $\eta_{CD}$, is about 0.06 ($10^{19}$ A/Wm$^2$).

A trial to get the tail electron velocity distribution function was pursued on good reproducible ECCD discharges ($I_P = 25$ kA, $\overline{n}_e = 2 \times 10^{12}$ cm$^{-3}$, $Te(0) = 200$ eV). The HX spectra emitted from the plasma were measured with a multi-channel pulse height analyser in various direction to the magnetic field $B_T$ and in various radial positions. The photons were counted up in every 10 ms interval from just before the EC pulse to the end of it. The configuration is schematically shown in Fig. 3. From the measurement of the radial profile of the HX spectra, the perpendicular temperature is nearly the same in the half of the minor radius and the intensity is rather peaked. On the basis of this fact, we assume that the shape of the tail electron velocity distribution function is uniform in space. The observed photon counts normalized to the effective code length were plotted versus detection angle to $B_T$. We assume the distribution function as follows ;

forward    $n_F(r)\exp-[mc^2\{1+(p_{||}/mc)^2\}^{1/2}/T_{||F}+mc^2\{1+(p_\perp/mc)^2\}^{1/2}/T_\perp]$

backward    $n_B(r)\exp-[mc^2\{1+(p_{||}/mc)^2\}^{1/2}/T_{||B}+mc^2\{1+(p_\perp/mc)^2\}^{1/2}/T_\perp]$

and calculate the the HX spectra at each detection angle to $B_T$ including the effect of the radial profile. The normalized photon counts just before the end of EC pulse are plotted versus detection angle in Fig. 4. The solid curves are best fitted ones which are obtained when $Z_{eff}$ is uniform in space and is unity, and $T_\perp$ = 75 keV, $T_{||F}$ = 100 keV, $T_{||B}$ = 50 keV and $n_B(0)/n_F(0)$ = 0.1 were chosen. The radial profile of photon counts $N_{HX}(r)$ is proportional to $Z_{eff}(r)$ $\times$ $n_{tail}(r) \times n_{bulk}(r)$.

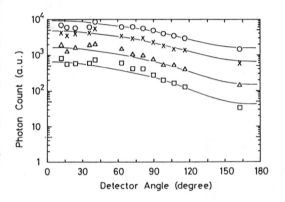

Fig. 4. Normalized photon counts as a function of detection angle to $B_T$.

By using the observed radial profiles of the photon counts and the bulk electron density, the tail electron density profile are estimated to be $n_{tail}(r) \propto (1 - r^2/a^2)^{3.4}$. If the tail electrons carried the total plasma current $I_P$, of 25 kA, $n_F(0)$ was estimated to be $8.6 \times 10^{10}$ cm$^{-3}$. The EC power deposition profile was calculated with the ray tracing code by using the induced distribution function for tail electrons and the Maxwellian distribution for bulk electrons. The one pass absorbed power was about 9 % of the injected power for bulk electrons and 18 % for tail electrons (16 % for the forward electrons and 2 % for the backward electrons).

## CONTROL OF THE SAWTOOTH OSCILLATION BY 2Ωe ECH

When 2Ωe ECH was applied to the normal tokamak discharge with m = 2 MHD oscillations and/or the sawtooth oscillations, the period and the amplitude of the sawtooth oscillations increased but there was no remarkable change in the MHD oscillations. When the EC power comparable to the OH power was injected into the OH plasma with $q_a$ = 2.4, the amplitude of some sawtooth oscillations became saturated. The saturation of the amplitude occurred in the outer region at first and expanded toward the center. The saturated state continued for about 1 ms and crashed abruptly. The typical waveforms are shown in Fig. 5. The characteristic of these sawteeth is that there is no precursor and the crash is very fast (in about 40 $\mu$s). In Fig. 6, the period of the sawtooth oscillations is plotted as a function of the position of the $2\Omega$e ECR layer. The period remarkably increased when the $2\Omega$e ECR layer located just outside of the inversion radius. When the EC power was less than the OH power, the period increased with the EC power but the amplitude of sawteeth did not saturate.

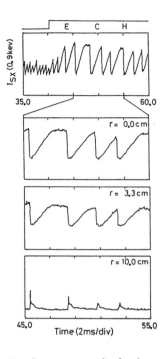

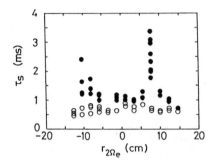

Fig. 6. Period of sawteeth vs. location of $2\Omega$e ECR layer.

Figl. 5. Sawteeth during $2\Omega$e ECH.

## SUMMARY

$2\Omega$e EC driven current up to 70 kA can be sustained without OH power by 144 kW, 100 ms EC pulse. The ECCD plasma has $q_a$ of about 5, $\overline{n_e}$ of $2 \times 10^{12}$ cm$^{-3}$ and Te(0) of about 500 eV. $\eta_{CD}$ is about 0.06 ($10^{19}$ A/Wm$^2$). We tried to estimate the electron velocity distribution function of the $2\Omega$e ECCD plasma by HX spectra. The period and the amplitude of the sawtooth increase by $2\Omega$e ECH when the resonance layer is located just outside of q = 1 surface. The saturation of the amplitude is observed when the EC power is comparable to the OH power.

## ACKNOWLEDGEMENT

This work is supported by a Grant-inAid for Scientific Research from the Ministry of Education in Japan.

## REFERENCES

1. H. Tanaka et al., Phys. Rev. Lett. 60, 1033 (1988).
2. Y. Terumichi et al., Int. J. of Electronics, 65, 691 (1988).

# CHAPTER 2

# LOWER HYBRID RANGE OF FREQUENCIES

# First Results on Lower Hybrid Current Drive
## at 2.45 GHz in  A S D E X

F. Leuterer, F.X. Söldner, R. Büchse, A. Carlson, A. Eberhagen,

H. Fahrbach, O. Gehre, F. Haßenpflug, W. Herrmann, G. Janeschitz,

M. Kornherr, T. Luce*), K. McCormick, F. Monaco, M. Münich,

H. Murmann, M. Pelicano, K. Steuer, M. Zouhar, and ASDEX-team

Max-Planck-Institut für Plasmaphysik

EURATOM Association, D-8046 Garching, F.R.G.


R. Bartiromo, R. DeAngelis, V. Pericoli, F. Santini, A. Tuccillo,

Assoziatione EURATOM-ENEA sulla Fusione, C.R.E. Frascati

C.P. 65-00044-Frascati, Roma, Italy


S. Bernabei, C. Forrest

Princeton University, Plasma Physics Laboratory,

Princeton, New Jersey, USA

## Abstract

A new lower hybrid system with 2.45 GHz/3 MW/1 sec has started operation on ASDEX. Current drive effects have been identified up to a density of $\bar{n}_e$ = 4.7 · $10^{13}$ cm$^{-3}$. Full curent drive at $I_p$ = 420 kV was achieved up to a density of $\bar{n}_e$ = 2.1 · $10^{13}$ cm$^{-3}$. The efficiency was maximum at $\bar{n}_e$ = 1.35 · $10^{13}$ cm$^{-3}$ and reached $\eta$ = 1.46 ($10^{13}$ cm$^{-3}$ · A · m/W). The electron temperature is peaking and reached peak values up to 6 keV, while the electron density profile flattens. Sawteeth have been stabilized up to a density of $\bar{n}_e$ = 3.4 · $10^{13}$ cm$^{-3}$. The global confinement time decreases with increasing rf-power. The scaling can be described by an offset linear relation. At low density global confinement is better during the LH-phase than in the OH-phase at the same total power input.

---

*) General Atomic, San Diego, USA

## I. Introduction

Lower hybrid experiments have been performed in ASDEX for many years at a frequency of 1.3 GHz. Heating, current drive, transformer recharging, sawtooth stabilization, and profile modification had been studied, however, limited in density to $\bar{n}_e \leq 2 \cdot 10^{13}$ cm$^{-3}$ because of diminuishing accessibility and onset of ion interaction /1-5/. For this reason a new system with f = 2.45 GHz, P = 3 MW, and T = 1 sec has been constructed in cooperation between ENEA-Frascati, PPPL-Princeton and IPP-Garching, combining the klystrons which were previously used at the FT- and PLT-tokamaks. This system started operation in December 1989 and in the meantime an incident power of 2 MW ($\hat{=}$ 3.8 kW/cm$^2$) was reached without any particular conditioning.

## II. The lower hybrid system

The transmitter consists of 2 groups of 3 klystrons each, which are fed by a common dc-power supply. Although being driven by one common master-oscillator, they can be operated independently. Their outputs are amplitude- and phase controlled. In a power splitter unit the power of these 6 klystrons is divided into 6 x 8 outputs, whose phase can be arbitrarily set, and which feed a 2 x 24 waveguides grill. Thus we are able to generate current drive spectra, opposite current drive spectra and symmetric sectra at variabe $N_{\shortparallel}$ (1 < $N_{\shortparallel}$ < 4.4, $\Delta N_{\shortparallel}$ = 0.4 full half width). Due to an excellent decoupling the forward waves are insensitive to arbitrary reflections.

The grill consists of 2 arrays of 24 waveguides with inner dimensions of 10 · 109 mm and 4 mm walls and is surrounded by graphite tiles protruding 3 mm beyond the actual grill surface. 2 front window blocks of PLT-design are located about 25 cm from the plasma surface. The front ends are made of stainless steel and their inner surfaces are coated with a rough gold layer to prevent multipactors. The upper and lower waveguide arrays are connected to the two groups of klystrons and can thus be operated independently with different powers and different spectra.

## III. Coupling

At low $\bar{N}_{\shortparallel}$, up to 2.2, the global power reflection coefficient $\langle R \rangle$ is about 10 to 25 %, depending on density and power. At higher $\bar{N}_{\shortparallel}$ it increases. In the individual waveguides the reflection is, however, quite different. Launching a symmetric wave spectrum leads to a symmetric distribution of the individual reflection coefficients, while asymmetric spectra lead to a correspondingly asymmetric distribution. In the latter case the distribution reverses when switching from a normal current drive spectrum ($\Delta\psi = +90°$) to a spectrum for opposite current drive ($\Delta\psi = -90°$). A variation of the relative phase between the upper and lower grills had only a negligeable influence on each reflection pattern.

## IV. Current Drive Experiments

Current drive experiments have been performed in the parameter range $\bar{n}_e = 1.3 - 4.7 \cdot 10^{13}$ cm$^{-3}$, $I_p = 300 - 420$ kA, $B_t = 1.86 - 2.8$ T. The experiments are run with a feedback controlled constant plasma current /2/. Depending upon the rf-driven current, only a fraction of the total current has then to be driven inductively. Therefore a lower loop voltage and a lower ohmic heating power are required during the LH-phase. The phase dependence of the drop in loop voltage shows a broad maximum for a grill phasing around $\Delta\psi = 75°$ to $90°$, corresponding to $\bar{N}_{\shortparallel}$ between 1.8 and 2.2. For lower and higher $\bar{N}_{\shortparallel}$ the drop in $U_{loop}$ is much less. However, the phase dependence is not as pronounced as expected theoretically by considering only the accessible part of the launched Brambilla spectrum. with the assumption of homogeneous profiles. In Fig. 1 we show measured characteristics of $\Delta U/U_{OH}$ as a function of rf-power for different densities. They show that $\Delta U/U_{OH}$ is a nonlinear function of the rf-power as already found in our previous experiment at 1.3 GHz /1/. This is explained by the current drive theory which describes the dependence of the current drive efficiency on the dc-electric field /6,7/. However, a detailed comparison taking into account the increase of electron temperature and $Z_{eff}$ along these characteristics has not yet been done.

We also recognize from Fig. 1 that a remarkable reduction in the loop voltage can still be found up to a density of $4.7 \times 10^{13}$ cm$^{-3}$, indicating that a density limit is not yet reached. Indeed, nonthermal ECE- and HX-radiation indicate still the generation of suprathermal electrons, although at this density interaction with ions is already starting. Thus up to these densities the waves are mainly absorbed by fast electrons which is consistent with results previously obtained in the FT-tokamak /8/.

It was also found, as indicated by other experiments /9/, that the drop in loop voltage does not depend much on the total plasma current. This suggests a currentdrive efficiency improving with the current. From shots where the loop voltage was nearly zero, i.e. $\Delta U/U_{OH} \approx 1$, we can derive the current drive efficiency, $\eta = \bar{n}_e \cdot I_p \cdot R/P_{LH}$ ($10^{13}$ cm$^{-3}$ A·m/W). In Fig. 2 the results are compared with those obtained in the previous experiment at 1.3 GHz where we used an eight waveguide coupler with $\Delta N_{\shortparallel} \approx 1.0$. With the higher frequency and the narrower spectrum zero loop voltage was now achieved at densities up to $2.1 \times 10^{13}$ cm$^{-3}$. Comparing the points obtained with $B_t = 2.17$ T and $B_t = 2.8$ T we see, clearer than from the phase dependence, the importance of the better accessibility. Comparing our efficiencies with those reported from the PLT experiment, /10/, (16 wave-guide-coupler, $B_t = 3.1$ T) we find about the same $\eta$'s for the $B_t = 2.17$ T points, indicating that what we gain with the narrower spectrum is lost in the worse accessibility. However the points at $B_t = 2.8$ T where accessibility is comparable to that in the PLT experiment, further confirm their result that the efficiency improves with a narrower spectrum. There are three pairs of points for different plasma current which all show an improved efficiency at the higher current.

In one series of shots taken at $\bar{n}_e = 2.1 \cdot 10^{13}$ cm$^{-3}$ we launched the waves in the opposite direction in order to generate an rf-current opposite to the inductive driven current. In this case, however, the loop voltage did hardly change in the rf-power range up to 1 MW. We also have changed the spectrum width $\Delta N_{\shortparallel}$ by feeding only a reduced number of waveguides and found less loop voltage drop for the wider spectra. However, this effect was not as pronounced as expected.

Strong parametric decay signals have also been observed in these experiments. They are described in a companion paper at this conference /11/.

## V. **Heating and confinement**

With LH-current drive we also observe heating of the plasma. In Fig. 3 we show as a function of the total input power, $P_{tot} = P_{LH} + P_{OH}$, the total energy content, $W_p = (W_{equ} + W_{dia}/2) \cdot 2/3$, as derived from magnetic measurements, together with the energy content $W_e$ in the thermal electrons determined from the YAG Thomson scattering profile. In general the energy content increases with $P_{tot}$. However, at low density ($\bar{n}_e = 1.35 \cdot 10^{13}$ cm$^{-3}$) we find a range were $P_{tot}$ can be less than the value of $P_{OH}$ in the OH-phase. Yet the total energy content is higher, by about 50 % in the case of LH-application as compared to Ohmic heating alone at the same total power input, resulting in a higher global confinement time. But if we compare the thermal electron energy content $W_e$, no such difference is seen between the OH and LH plasmas. Such a behaviour was already found in the previous 1.3 GHz experiment, however at much lower densities ($\bar{n}_e < 0.7 \cdot 10^{13}$ cm$^{-3}$), /12/. At higher densities, $\bar{n}_e \gtrsim 2 \cdot 10^{13}$ cm$^{-3}$, the same total power input leads to the same total energy content, both in the LH- and the pure OH-phase.

In Fig. 4 we study the low density case in more detail, plotting the increments of the energy content as derived from the different measurements. During LH-current drive we find $\Delta W_{equ} > \Delta W_{dia}$ owing to the preferential energy transfer from the waves to suprathermal electrons in parallel direction. We also find $\Delta W_{dia} > \Delta W_e$ since $W_e$ accounts only for the thermal electrons whereas $W_{dia}$ includes also the pitch angle scattered fast electrons which are seen also in the HX-ray and nonthermal ECE-emissions. Ion heating is not observed at these low densities according to charge exchange measurements. Thus it is obvious that the increase in total stored energy at low density, as determined by magnetic measurements, is only due to the generated suprathermal electrons. The total stored energy Wp and the thermal energy $W_e$ increase about linearly with power in the whole range $0.3 \lesssim P_{tot} \lesssim 1.5$ MW. This scaling extends also to powers below the value for pure ohmic current drive, which are accessible only through combined operation of OH- and LH-current drive. From the slope of the energy increase with power, an incremental confinement time, $\tau_{E,inc} = dWp/dP_{tot}$, of about 20 msec is obtained which is comparable to the value obtained during neutral injection or ion cyclotron heating in similar discharges. Increasing the density above $2 \cdot 10^{13}$ cm$^{-3}$ we find $\Delta W_{dia} \approx \Delta W_e$ and less difference between $\Delta W_{equ}$ and $\Delta W_{dia}$, indicating a decreasing pressure anisotropy. In these conditions the incremental confinement time also decreases.

However, by increasing the plasma density at the relatively low toroidal magnetic field of ASDEX we expect that an increasing fraction of the launched power cannot access to the plasma core. This should ultimately lead to a reduced power absorption by the plasma. We can determine the absorbed power from the rate of change of the total plasma energy (i.e. kinetic and magnetic) at the switch-on and the switch-off of the LH-pulse. The resulting absorption coefficient $\alpha$ does not show any appreciable dependence on the launched power, both at low and high density. In a limited density range ($\bar{n}_e \approx 2.1 \cdot 10^{13}$ cm$^{-3}$) we have changed $\bar{N}_{\shortparallel}$ and the width of the spectrum without appreciably changing the value of $\alpha$. For $\bar{N}_{\shortparallel} = 1.8$ and the same density we have also changed the plasma current in order to vary the q-profile, which should influence the wave trajectory in the plasma. But this did also not affect the absorption coefficient $\alpha$.

We have found, however, that $\alpha$ shows a remarkable decrease with density as expected from accessibility considerations and as found already in the 1.3 GHz-experiment /12/. In Fig. 5 we show the results of a density scan and we compare them with the fraction of power satisfying the accessibility condition at a plasma density equal to the line averaged density and a magnetic field equal to the toroidal field on axis. From this we conclude that the decrease in $\alpha$ is related to the accessibility, although at lower magnetic field we also found absorption in cases where no power should reach the plasma core.

Taking the measured absorption coefficient into account the increase in the diamagnetic energy as a function of the total absorbed power, $P_{tot,abs} = \alpha P_{LH} + P_{OH}$, turns out to be comparable for densities ranging from $1.35 \cdot 10^{13}$ to $3.4 \cdot 10^{13}$ cm$^{-3}$, as shown in Fig. 6. The incremental confinement time in this case is about 27 msec, roughly independent of density and in close agreement with results from neutral injection and ion cyclotron heating. A preliminary investigation of the influence of plasma current on confinement during LH current drive shows that for equal total power the energy increase in the plasma is larger at lower current in contrast with the prediction of the standard L-mode scaling.

## VI. Profile Modifications

Plasma heating by absorption of the LH waves through suprathermal electrons leads to a peaking of the electron temperature profile. From the

Thomson scattering measurements of the radial electron temperature profile central values of up to $T_{eo} \approx 6$ keV are determined similar to the observations in PLT /13/. The peaking of the electron temperature profile increases with increasing power as found already earlier /14/. The profile factor $Q_{Te} = T_{eo} / \langle T_e \rangle$ is plotted in Fig. 7a versus total power. For $P_{tot} \approx 1.5$ MW a value of $Q_{Te}$ near to 5 is obtained. The electron density profile broadens at the same time. The profile factor $Q_{n_e} = n_{eo}/\langle n_e \rangle$ drops with power as shown in Fig. 7b. The variation of the density profile with LH becomes smaller at higher density.

At the low values of $\bar{N}_{\parallel}$ used for current drive the current density profile does not seem to change according to magnetic measurments. At low density, $\bar{n}_e = 1.35 \cdot 10^{13}$ cm$^{-3}$ this is confirmed by direct measurements with a Lithium beam. Only at higher values of $\bar{N}_{\parallel}$, the magnetic measurements indicate a decreasing internal inductance. However, we also found that for current drive near $U_{loop} \approx 0$ we often, but not always, excite m=2 modes which often lead to a disruption.

## VII. Sawtooth Stabilization

The sawtooth period increases with increasing LH power. Above a threshold power sawteeth are completely stabilized. This has been found up to a density of $3.4 \cdot 10^{13}$ cm$^{-3}$. The threshold power rises with plasma current. The fraction of plasma current which has to be driven by the LH drops with increasing density. This fraction of LH driven current can be expressed by the drop in Ohmic power input at the transition from OH- to LH-phases, $\Delta P_{OH}(LH)/P_{OH}(OH)$. In Fig. 8 this drop at the stability margin of sawteeth is plotted versus density in the range of sawtooth stabilization experiments with the actual 2.45 GHz system and with the previous 1.3 GHz system. In both cases the fraction of LH driven current required for stabilization drops with density. But a larger fraction of LH-current drive is necessary with the higher frequency.

Based on the explanation of our 1.3 GHz results on sawtooth stabilization by broadening of the current density profile /3,4/, the difference between the two experiments might be attributed to the different width $\Delta N_{\parallel}$ of the wavespectra ($\Delta N_{\parallel} = 0.4$ as compared to $\Delta N_{\parallel} = 1$ at 1.3 GHz) at the about the same $\bar{N}_{\parallel}$, which should affect the power deposition profile. However in the present experiment the mechanism for sawtooth stabilization might be quite different. Not only that we have no direct indications for a current profile broadening, but m=1 modes are still observed from SX ray measurements after the disappearance of the sawteeth. This suggests that a q=1 surface still exists in the sawtoothfree phase, a situation which was also reported from the PLT-experiment /13/.

In the forthcoming experiments we shall study whether the profile of j(r) can be locally modified by controlling the $N_{\shortparallel}$-spectrum. There are already first indications at higher $N_{\shortparallel}$. Furthermore LH-current drive will be combined with neutral beam injection and ion cyclotron heating.

## Figure captions

Fig. 1: Loop voltage drop as a function of LH-Power.
$B_t$ = 2.8 T; $I_p$ = 420 kA; $\bar{N}_{\shortparallel}$ = 1.8; $D_2$

Fig. 2: Experimental current drive efficiency as a function of density obtained in experiments at 2.45 GHz and at 1.3 GHz.

Fig. 3: Total energy $W_p$ and thermal electron energy $W_e$ versus total input power in lower hybrid current drive discharges.
Parameters as in Fig. 1.

Fig. 4: Increase in energy content as determined from diamagnetic, equilibrium and kinetic measurements at $\bar{n}_e$ = 1.35 $\cdot$ $10^{13}$ cm$^{-3}$.
The arrow indicates the power input in the pure OH-phase.
Parameters as in Fig. 1.

Fig. 5: Experimental absorption coefficient compared with fraction of accessible power to the centre. Parameters as in Fig. 1.

Fig. 6: Increase in energy content as measured with the diamagnetic loop as a function of total absorbed power. Parameters as in Fig. 1.

Fig. 7: Density and temperature profile factors as function of total power. Parameters as in Fig. 1.

Fig. 8: Drop in Ohmic power input at the sawtooth stability threshold for experiments at 1.3 GHz and 2.45 GHz. $B_t$ = 2.17 T.

# References

/1/    F. Leuterer, F. Söldner, D. Eckhartt et.al.
Plasma Phys. and Contr. Fus. $\underline{27}$, 1399, (1985)

/2/    F. Leuterer, D. Eckhartt, F. Söldner et.al.
Phys. Rev. Lett. $\underline{55}$, 491, (1987)

/3/    F. Söldner, K. McCormick, D. Eckhartt et.al.
Phys. Rev. Lett. $\underline{57}$, 1137, (1986)

/4/    K. McCormick, F. Söldner, D. Eckhartt et.al.
Phys. Rev. Lett. $\underline{58}$, 491, (1987)

/5/    F. Leuterer et.al.
13th Europ.Conf.Contr.Fus. and Plasma Phys., Schliersee 1986
Europhysics Conf.Abstracts, Vol.10c, part II, page 409

/6/    C. Karney, N. Fisch,
Phys. Fluids $\underline{29}$, 180, (1986)

/7/    K. Yoshioka et.al.
Phys. Fluids $\underline{31}$, 1224, (1988)

/8/    A. Alladio et.al.
Mucl. Fus. $\underline{24}$, 725, (1984)

/9/    C. Gormezano et.al.
13th Europ.Conf.Contr.Fus. and Plasma Phys., Schliersee 1986,
Europhysics Conf.Abstracts, Vol 10C, part II, page 311

/10/   J. Stevens et.al.
Nucl. Fus. $\underline{28}$, 217, (1988)

/11/   V. Pericoli et.al.
this conference

/12/   F. Söldner et.al.
12th Europ.Conf.Contr.Fus. and Plasma Phys., Budapest 1985
Europhysics Conf.Abstracts, Vol. 9F, part II, page 244

/13/   T.K. Chu et.al.
Nucl. Fus. $\underline{26}$, 666, (1986)

/14/   F. Söldner et.al.
14th Europ.Conf.Contr.Fus. and Plasma Phys., Madrid 1987
Europhysics Conf. Abstr., Vol. 11D, part III, page 831.

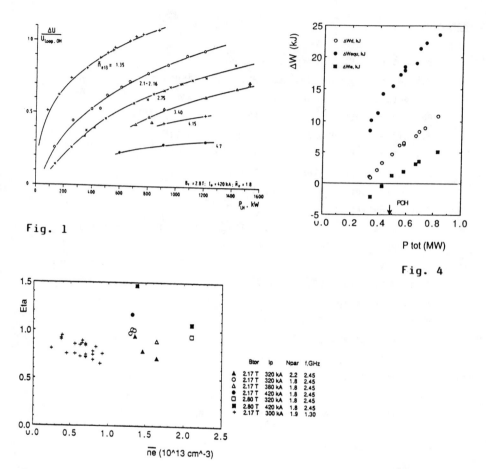

Fig. 1

Fig. 4

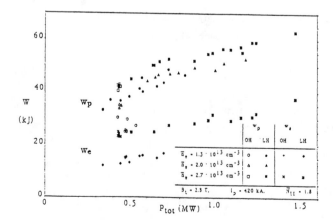

Fig. 2

Fig. 3

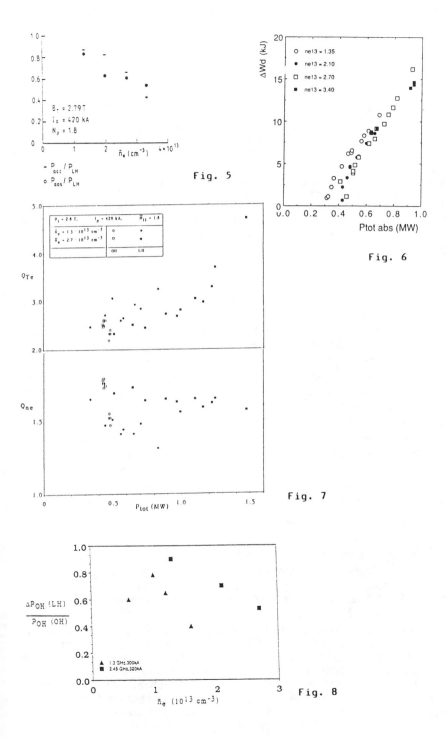

Fig. 5

Fig. 6

Fig. 7

Fig. 8

# RF CURRENT DRIVE AND HEATING IN JT-60

K. Uehara, H. Kimura and JT-60 Team

Naka Fusion Research Establishment, Japan Atomic Energy Research Institute,
801-1, Naka-machi, Naka-gun, Ibaraki-ken, 311-01 Japan

## ABSTRACT

RF current drive and heating experiment with Lower Hybrid Range of Frequencies (LHRF) and Ion Cyclotron Range of Frequencies (ICRF) in JT-60 are presented. In LHRF, high efficient current drive and profile control with various $N_{//}$ are demonstrated by the multi-junction launcher in successful. In ICRF, optimization of the second harmonic heating with various methods, and beam acceleration and heating by third harmonics are presented.

## I. LHRF RESULTS

### 1. INTRODOUCTION

JT-60 is the only machine among the four large tokamaks (TFTR, JET, T-15 and JT-60)[1] that the Lower Hybrid Wave stressed to the first priority for the rf heating. In JT-60, 24 MW LHRF at 2 GHz band are installed,in which we have already performed 2 MA for 2.5 sec steady current drive and high efficient current drive of $1.7\text{-}2.8 \times 10^{19}$ $AW^{-1}m^{-2}$ with the conventional 4 x 8 phased array waveguides launcher.[2] We have improved some controlling system and one of the 4 x 8 waveguide launchers is changed into 4 x 24 multi-junction type launcher. For the sake of above improvements, we can get relatively quick performance of launcher aging. We can further enlarge the driving efficiency of LHCD up to $3.4 \times 10^{19}$ A $W^{-1}$ $m^{-2}$ by using the multi-junction launcher and can get successful profile control with various $N_{//}$. In section 2 and 3, system description and LHRF heating results are given,respectively. Discussion and conclusion are presented in section 4.

### 2. SYSTEM DESCRIPTION

In JT-60,a new divertor coil was installed to produce a lower X point configuration in early 1988.[3] Experiments described in this part were performed in the hydrogen plasma in the range of the average plasma density $\bar{n}_e = (0.8\text{-}3.0) \times 10^{19}$ $m^{-3}$, plasma current $I_p = 1$ -1.5 MA, toroidal magnetic field $B_t$ =4 -4.5 T and the effective q value $q_{eff} = 3$ -4, respectively.

RF heating system[4] has performed some improvements of the control system in order to obtain efficient launcher conditioning and good operational maintenance. Especially,the notching circuit of the LHRF reflection power are equipped. When the reflection power to the klystron is higher than the setting value,then rf power is cut off and we cannot retry until the shot is over, however, the operational efficiency becomes worse when we perform the high power experiments. So,we improve the system so as to retry again in a sequence. By these improvement of the control system we can contribute to shorten the conditioning time and the efficient operation can be realized.

The multi-junction launcher with the sharp wave spectrum and good directivity to improve the current drive efficiency is newly installed in JT-60. Previously we have two launchers of wide width waveguide for the plasma heating and one launcher of narrow waveguide width for the current drive. We have changed one of the heating launchers into the current drive launcher. The new current drive launcher is multi-junction type[5] with 4 x 24 phased array consisted by 8 modules. Each waveguide of the conventional launcher is divided into three sections at the top of the launcher to form one module. The geometrical phase shifters with taper type are equipped at the middle position of the launcher. The top of the waveguide piece is carbon coated to get a low secondary emission and the copper is plating inside the waveguide. The most difficult points during manufacturing the launcher are reduction of distortion due to the welding and the difficulty of copper plating inside waveguides.[6] The phase difference between adjacent waveguide is set 70 degree. Directivity is improve by 50 % compared with the conventional one.

Improvement of the control system and new launcher lead to quicken the aging time ,that is, we can get about 2 MW injection by performing 10 hours in vacuum conditioning,16 hours in TDC conditioning and 30 shots plasma injection,whereas we need about 100 shots vacuum and plasma injection in the former operation. The reflection is reduced by 20 - 25 % compared with the conventional one.[6]

## 3. CURRENT DRIVE EXPERIMENT

### 3.1 Current drive efficiency

Figure 1 shows the obtained driving current of $\bar{n}_e R I_{RF}$ vs the LH power,where $\bar{n}_e$ is the line average density,R is the major radius and $I_{RF}$ is the driving current. We can see that JT-60 LHCD results exceeds more than eight times than other tokamaks and that the continuous progress of the currnt drive is obtained with increase of the rf power and ten times progress has been performed since the initiation of LHCD experiments in the world. Figure 2 shows the $\eta_{CD}$ vs $\bar{n}_e$ ,where the experimental current drive efficiency is defined as

$$\eta_{CD} = \frac{\bar{n}_e I_{RF} R}{P_{LH}} \tag{1}$$

We can see the increasing of $\eta_{CD}$ with $\bar{n}_e$ up to 2.7 x $10^{19}$ m$^{-3}$ for plasma current $I_p$ = 1 MA and the higher plasma current tends to have higher efficiency for the same $\bar{n}_e$. The highest efficiency of 3.4 x $10^{19}$AW$^{-1}$m$^{-2}$ is obtained at $\bar{n}_e$ = 1.5 x $10^{19}$ m$^{-3}$ with $N_{//}peak$= 1.3 and $<T_e>$ = 2 keV,where $N_{//}peak$ means the peak value of rf spectrum in the new multi-junction launcher. The efficiency is compared with various $N_{//}$ by varying the phase difference between adjacent modules of the multi-junction launcher. The efficiency clearly increases than the conventional one and changes with $N_{//}$,in which we can see that $\eta_{CD}$ increases with decrease of $N_{//}peak$ and decreases beyond $N_{//}peak$ = 1.3. The optimum $\eta_{CD}$ is obtained at $N_{//}peak$ = 1.3 and the value of $\eta_{CD}$ shows a similar behaviour to the Fisch prediction qualitatively.[7] The hard X ray emission (E> 200 keV) normalized by the LH power shows the same behaviour as that of $\eta_{CD}$ with $N_{//}peak$, that is,the signal from higher energy electrons behaves in corresponding to the variation of $\eta_{CD}$. The temperature dependence of $\eta_{CD}$ is shown in Fig.3,in which we can see that the higher temperature may lead the higher efficient current drive.

It should be noted that the conventional theoretical prediction may give the maximum value,because the theoretical efficiency is usually defined as

$$\eta_{CD} = \frac{\bar{n}_e \int j \, 2\pi r dr}{\int P_d \, 2\pi r dr} \tag{2}$$

where $\int P_d \, 2\pi r dr$ is the absorbed power by plasmas to hold the current[7] which differs to $P_{rf}$ in eq.(1) and is rather smaller than this depending on the extent of power absorption.[8] Strictly speaking, we cannot observe $P_d$ by the experiment. Experimental results in Figs 2 and 3 suggest that the extent of power absorption may be a function of $<T_e>$ and $\bar{n}_e$. The effect of $Z_{eff}$ may be also considered, since $Z_{eff}$ becomes small with increase of the density.

### 3.2 Volt-Sec saving experiment

Volt-sec saving by the LHCD was performed during the plasma current ramping up from 0.7 MA to 1.5 MA with $B_t = 4.5$ T and $\bar{n}_e = 1 \times 10^{19}$ m$^{-3}$ by varying the ramp-up speed of $\dot{I}_p$. The loop voltage of 1.9 V without LHCD is reduced to 0.9 V during 2 sec of LHCD for $\dot{I}_p = 0.4$ MA/s discharge. The saved volt-sec is 0.9 V x 2 sec =1.8 V·sec and the saving volt-sec is proportional to the injected LH energy, $P_{LH}\Delta t$, and is higher for larger current ramping up rate.

### 3.3 Profile control with $N_{//}$

Using the advantage of sharp $N_{//}$ and high directivity for wide $N_{//}$ range, we demonstrate the profile control experiments with various $N_{//}$ in successful. Time derivative of internal inductance $l_i$ against $N_{//}$ shows the higher decreasing rate for the larger $N_{//}$, which indicates that the higher $N_{//}$ wave may flatten the current profile more. Correspondingly, the spatial distribution of the hard X ray signal vs $N_{//}$ shows the same behaviour. We equipped four channel X ray diagnostics in the radial direction as shown in Fig.4 (a). The Abel transformed radial profile using emission signals measured at $r/a=0.23$(ch1), $r/a=0.57$(ch.2) and 0.86(ch.3) are shown in Fig.4 (b) and the dependence on $N_{//}^{peak}$ is shown in Fig.4 (c), in which we can see that the relatively larger number of higher energy electrons are localized at the center with small $N_{//}$ and small number of the tails are in the outer with large $N_{//}$.

We also observe the increase of the coherent m=2 and m=3 oscillation with large $N_{//}$ which is accompanying the decrease of li and the suppression of the sawtooth oscillation of NB heated plasma with small $N_{//}$ of 1.3, which is characterized by the delay of the starting time of the sawtooth oscillation. The suppression period $\tau_{st}$ of sawtooth oscillation increases with $P_{LH}$ and $\tau_{st} = 1.8$ sec is obtained for $P_{LH} = 2$MW. The decay time of the stored energy after the LHCD cut off shows the twice as much that of NB or LH alone.

## 4. DISCUSSION AND CONCLUSIONS

Dependence of the electron temperature and the density must be further refined to fit the theoretical understanding of LHCD. The effects which are not consider in the quasi-linear theory may be included such as the multi-path, the density fluctuation, the forbbiden condition of mode conversion, the non-linear effect and so on.[9] Many experimental results obtained in JT-60 can be expected to refine the theory of LHCD.

The success of the profile control by the LHCD can open new frontier in tokamaks for the various possibility such as the controlling the plasma disruption, the improvement of plasma confinement and so on. It is stressed that the higher energy electron tail caused in LHCD is independent on the bulk plasma and it is confirmed that the LHCD can really affect the plasma profile by varying $N_{//}$, whereas it may be very difficult to vary the current profile by affecting bulk electrons as is shown by the scheme of profile consistency.[10]

In conclusions, multi-junction launcher successfully brought the further improvement of the driving efficiency and the profile control is demonstrated with various $N_{//}$. Many experimental results on the LHCD in JT-60 can give the informations on the further verification on the LHCD including the quasi-linear theory.

## II. ICRF RESULTS

### 1. INTRODUCTION

An experimental study of the second harmonic and even much higher (up to 4th) harmonic ICRF heating on a large tokamak is being carried out in JT-60. Up to now, we have investigated most intensively the second harmonic heating with ohmic and NBI-heated plasmas. Phase control in the toroidal direction has been found to play an important role in optimizing the second harmonic heating[11]. Significant enhancement of the plasma stored energy associated with strong beam acceleration has been observed in combination with high power NBI heating[12]. Combination with pellet fuelling has also been examined[13]. In Section 2, system description is presented. In Section 3, optimization of the second harmonic heating is discussed from the point of view of phase control, species effects and dependence on plasma current. Most recently, we have observed significant beam acceleration and effective heating via third harmonic resonance in combination with NBI. These results are described in Section 4. Conclusions are given in Section 5.

### 2. SYSTEM DESCRIPTION[4]

The total generator output is 6MW in the frequency range of 108-131MHz, which is delivered by eight lines of amplifier chains. The frequency is set at 131MHz for the present experiment. The phased 2x2 loop antenna array is used. The maximum injected power so far is 3MW. The corresponding power density at the antenna is $1.6kW/cm^2$. New functions, reflection power limiter and frequency feed-back control, have been introduced. The former is useful to continue power injection without cut-off even in the case of bad matching due to rapid change of the antenna loading. The latter is effective to maintain good matching against practical change of the antenna loading, although long line effect is not applied.

### 3. OPTIMIZATION OF SECOND HARMONIC HEATING

3.1 Phase Control

Up to now, two phasing modes, (0,0) mode and ($\pi$,0) mode, have been mainly investigated. The former in the parenthesis is the toroidal phase difference and the latter is the poloidal one, respectively. (0,0) mode is characterized by a large coupling resistance but moderate heating efficiency. ($\pi$,0) mode has a smaller coupling resistance but excellent heating efficiency. Figure 5 shows incremental energy confinement time $\tau_E^{inc}(\equiv\Delta W/\Delta P$, $\Delta W$ is incremental stored energy due to additional power $\Delta P$) as a function of the line averaged electron density $\bar{n}_e$ for various heating conditions. Circles and squares denote (0,0) mode and ($\pi$,0) mode, respectively. Apparently, $\tau_E^{inc}$ of ($\pi$,0) mode is much larger than that of (0,0) mode. $\tau_E^{inc}$ of (0,0) mode tends to decrease with increasing electron density. Significant scattering of $\tau_E^{inc}$ of (0,0) mode is not only due to random error but also due to appearance of two distinct modes[14]. $\tau_E^{inc}$ of (0,0) mode is kept at ~50ms even in the high density regime with pellet fuelling as shown in Fig.5.

## 3.2 H Minority Second Harmonic Heating in He Discharge

Heating efficiency of the second harmonic heating has been further improved when $(\pi,0)$ mode was applied to the hydrogen minority second harmonic regime in the helium discharge. In this experimental run, mixture gas of 90% He and 10% H was used. The operational range for ICRF experiment has been extended significantly with helium discharge, i.e., the highest $\bar{n}_e$ and Ip and the lowest $q_{eff}$ achieved so far are $8.3 \times 10^{19}$ m$^{-3}$, 2.8MA and 2.2, respectively. Part of the data corresponding to this scheme are also indicated in Fig.5. $\tau_E^{inc}$ of 100~120ms was obtained in the wide range of the electron density. Giant sawteeth were observed during the heating even in the high density regime $(\bar{n}_e \sim 7 \times 10^{19}$ m$^{-3})$ . Period of the giant sawteeth seems to be independent of $\bar{n}_e$. Typical waveforms of the hydrogen minority second harmonic heating is shown in Fig.6.

## 3.3 Dependence on Plasma Current

We have observed that $\tau_E^{inc}$ of (0,0) mode increased with $I_p$ unlike the NBI heating in JT-60. $\tau_E^{inc}$ reached about 100 ms at $I_p$=2MA, whereas typical value of $\tau_E^{inc}$ at $I_p$=1.5MA is 50ms. Therefore, $\tau_E^{inc}$ of 100ms obtained at $I_p$=2MA means considerable good confinement. However, the good confinement shots showed some strange behaviours in their time evolutions. Typical example is shown in Fig.7. A minor disruption (M.I.D.) took place twice in the course of the ICRF pulse. The plasma stored energy increased dramatically just after the second M.I.D. Both electron and ion temperatures increased in the plasma core, but $\bar{n}_e$ at r≈ 0.5a did not change appreciably after the second M.I.D. It seems that M.I.D. produces some favourable conditions for the heating of (0,0) mode.

## 4. THIRD AND FOURTH HARMONIC BEAM ACCELERATION

We have examined whether third and fourth harmonic beam accelerations occur in the central region, varying the toroidal magnetic field, $B_T$. The beam acceleration is measured in the incremental tail ion temperature, $\Delta T_i^{tail}$ [15], which is the difference of the slope of the ion energy spectra above the injection energy between NBI only and NBI+ICRF. $\Delta T_i^{tail}$ was measured by a charge exchange neutral analyzer, whose line of sight intersected with specific beam lines of NBI in the plasma core, so that we could obtain the ion energy spectra in the plasma core[16]. From the data of $\Delta T_i^{tail}$, we have confirmed that the third and fourth harmonic beam accelerations actually occur in the plasma core. Degree of the beam acceleration becomes weak with increasing order of harmonics.

Heating effects on the bulk plasma by the third harmonic beam acceleration are found to be as strong as that of the second harmonics. Figure 8 illustrates time evolutions of the plasma stored energy, the central electron temperature, the charge exchange neutral flux at 92 keV and $\bar{n}_e$ at r≈ 0.5a in the case of the third harmonic beam acceleration. Enhancement of the central electron temperature and sawteeth period was seen with increasing population of the energetic ions. $\tau_E^{inc}$ of ICRF of this shot is 80ms, which is comparable to the one of the second harmonic heating. Heating effects by the fourth harmonics is not so strong up to now.

## 5. CONCLUSIONS

Recipes for improving the second harmonic heating has been elucidated. Phase control, pellet injection, helium discharge, higher plasma current as well as beam acceleration improve $\tau_E^{inc}$ of the second harmonic heating. Beam acceleration with third and fourth harmonics has been observed for the first time. Heating effects by the third harmonics are as strong as those by the second harmonics.

## III. RF PLAN FOR JT-60 UPGRADE

We are planning the up-grade programme for JT-60 (named JT-60U).[17] RF plan for JT-60U is in the following. In LHRF, two units of RF lines are jointed to form one rf injection with horizontal direction and one unit keeps with oblique injection. The horizontal launcher consists of 4 x 4 module multi-junction with four 18 waveguide columns at the center and eight 12 waveguides columns at the top, bottom and side.[18] We also expect further power up of klystrons with some improvements. In ICRF, present antenna will be replaced by two new antennae, which are also 2 x 2 loop array and have larger width ( ~ 90 cm) to ensure large coupling for $(\pi,0)$ mode and H-mode. The generator output will be increased up to 10 MW by replacing the present tetrode 8973 with X-2242.

## ACKOWLEDGEMENTS

The continuing support of Drs. M. Yoshikawa and M. Tanaka is greatly appreciated.

## REFERENCES

1. A.H. Spano (compiler) , Nucl. Fusion **15** 909 (1975)
2. T. Imai et al., Nucl.Fusion **28** 1341 (1988)
3. JT-60 Team presented by H. Kishimoto et al., Plasma Phys. and Contr. Fusion A-I-4
4. T. Nagashima and K. Uehara et al., Fusion Eng.& Design **5** 101 (1987)
5. T.K. Nguyen et al., Fusion Tech. **2**, 1381 (1882) and G. Gormezano et al., Nucl.Fusion **25** 419 (1985)
6. Y. Ikeda et al, "First operation of multi-junction launcher on JT-60" this conference
7. N. Fisch, Phys. Rev. Letters **41**,873 (1978)
8. G. Tonon, Plasma Phys.Contr.Fusion **26** 45 (1984)
9. K Uehara, M. Nemoto et al., Nucl. Fusion **29** May 1989 (in press)
10 F. Wagner., et al., Phys. Rev. Lett. **56** 2187 (1986)
11. H. Kimura et al., in Contr. Fusion and Plasma Phys. (Proc. 14th Europ. Conf. Madrid, 1987) EPS, vol.11D, Pt.3 p.857
12. T. Fujii et al., in Plasma Phys. and Contr. Nucl. Fusion Research 1988 (Proc. 12th Int. Conf. Nice, 1988) IAEA, Paper IAEA-CN-50/E-2-4
13. JT-60 Team, Japan Atomic Energy Research Institute Report JAERI-M 89-033 (1989) p.185
14. ibid., p.181
15. M. Yamagiwa et al., Plasma Phys. Controlled Fusion **30** 943 (1988)
16. H. Kimura et al., Japan Atomic Energy Research Institute Report, JAERI-M 88-123 (1988)
17 . M. Kikuchi et al., 15th SOFT,Utrecht
18. M. Seki et al., "Design of new launcher on JT-60 Up-grade" this conference

112

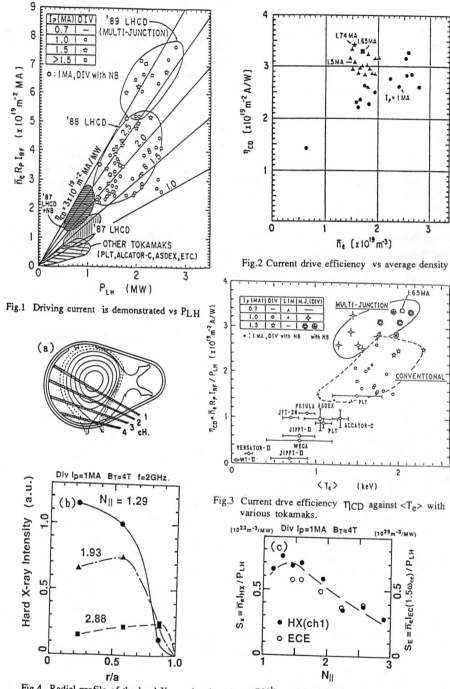

Fig.1 Driving current is demonstrated vs $P_{LH}$

Fig.2 Current drive efficiency vs average density

Fig.3 Current drve efficiency $\eta_{CD}$ against $<T_e>$ with various tokamaks.

Fig.4 Radial profile of the hard X ray signal against $N^{peak}_{\parallel}$ and hard X ray diagnostics with four channel

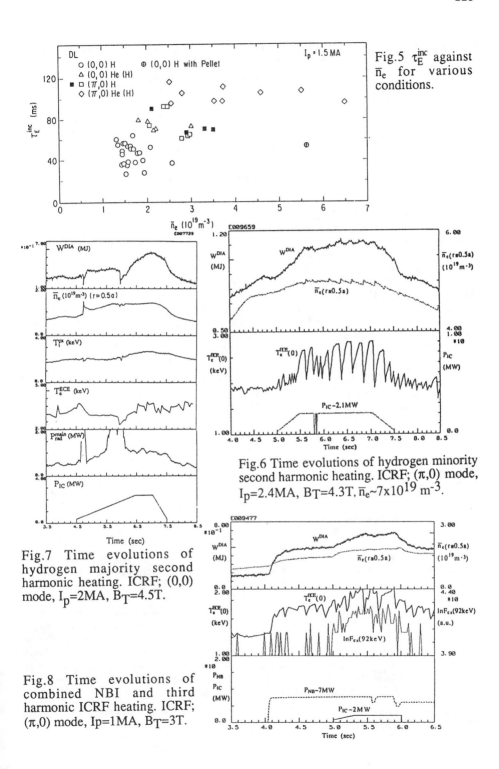

Fig.5 $\tau_E^{inc}$ against $\bar{n}_e$ for various conditions.

Fig.6 Time evolutions of hydrogen minority second harmonic heating. ICRF; $(\pi,0)$ mode, $I_p=2.4MA$, $B_T=4.3T$, $\bar{n}_e \sim 7 \times 10^{19}$ m$^{-3}$.

Fig.7 Time evolutions of hydrogen majority second harmonic heating. ICRF; $(0,0)$ mode, $I_p=2MA$, $B_T=4.5T$.

Fig.8 Time evolutions of combined NBI and third harmonic ICRF heating. ICRF; $(\pi,0)$ mode, $I_p=1MA$, $B_T=3T$.

# MODELING OF LH CURRENT DRIVE IN SELF-CONSISTENT ELONGATED TOKAMAK EQUILIBRIA[*]

D. T. Blackfield(TRW), R. S. Devoto, M. E. Fenstermacher
Lawrence Livermore National Laboratory, Livermore, CA 94550
P. T. Bonoli, M. Porkolab
Plasma Fusion Center, MIT, Cambridge, MA 02139
J. Yugo (TRW)
Oak Ridge National Laboratory, Oak Ridge, TN 37831

## INTRODUCTION

Calculations of non-inductive current drive typically have been used with model MHD equilibria which are independentaly generated from an assumed toroidal current profile or from a fit to an experiment. Such a method can lead to serious errors since (1) the driven current can dramatically alter the equilibrium and (2) changes in the equilibrium B-fields can dramatically alter the current drive. The latter effect is quite pronounced in LH current drive where the ray trajectories are sensitive to the local values of the magnetic shear and the density gradient. In order to overcome these problems, we have modified a LH simulation code[1] to accomodate elongated plasmas with numerically generated equilibria. The new LH module has been added to the ACCOME code[2] which solves for current drive by neutral beams, electric fields, and bootstrap effects in a self-consistent 2-D equilibrium. We briefly describe the model in the next section and then present results of a study of LH current drive in ITER.

## DESCRIPTION OF MODEL

The LH model[1] includes ray tracing using a dispersion relation with electromagnetic and thermal effects and the solution of a relativistically correct 1-D Fokker-Planck equation with quasi-linear diffusion due to the absorbed LH waves on each radial ($\psi$) surface. The distribution function in $v_\perp$ is modelled by a Maxwellian with an effective temperature, $T_\perp \geq T_e$. In the present simulation, the ray equations are integrated in Cartesian geometry using spatial derivatives obtained with the aid of a bicubic spline fit to the numerical solution of the Grad-Shafranov (G-S) equation. Computations along typical rays indicate that the dispersion relation is generally satisfied to within $10^{-5}$.

Before the computation of the LH and the other currents, an initial MHD equilibrium must be available. We find it desirable to start from an equilbrium with certain desired properties, namely a specified toroidal current, $I_p$, and current profile which yields a specified safety factor on axis, $q_0$. We assume $j_\phi \propto (1 - \psi)^{\alpha_j}$, where $\psi$ is the poloidal magnetic flux, with a specified $\beta_p$, and allow the equilibrium code to adjust $\alpha_j$ until it obtains the desired $q_0$. During this process, the currents in the external coils are also adjusted to obtain the desired plasma size and shape. With the aid of this initial equilibrium, we compute the ray trajectories, the deposition of neutral beam atoms, and all the components of the current. The computed total current is then used to construct a new right-hand-side for the G-S equation. We compute new values of $ff' = -\mu_0(f^2 p' + f < j_\parallel B >)/ < B^2 >$, where $f(\psi) \equiv RB_\phi$, $p(\psi)$ is the pressure from the thermal species, the fast ions and the fast alpha particles, $\mathbf{B}$ is the magnetic field with $B_\phi$ its toroidal component, $R$ is the major radius coordinate, $j_\parallel$ is the current parallel to $\mathbf{B}$, and $<>$ denotes a flux surface average. Neither $f$ nor $< B^2 >$ change rapidly between iterations, so values from the previous equilibrium along with $p'$ and the computed $< j_\parallel B >$ are used on the right-hand side of the equation for $ff'$. The toroidal current density is then found from $j_\phi = -Rp' - ff'/\mu_0 R$ which is used in solving the G-S equation.

[*]Work performed under the auspices of the U.S. Department of Energy by Lawrence Livermore Laboratory under Contract W-7405-ENG-48

| $\alpha_n(\alpha_T) =$ | 0.5(0.5) | 0.5(1.0) | 1.0(0.5) | 1.0(1.0) |
|---|---|---|---|---|
| $\hat{n}_\parallel = 1.7$ | 0.40(1) | — | 0.48(1) | — |
| 1.8 | 0.40 | 0.28(2) | 0.48 | 0.33 |
| 1.9 | 0.45 | 0.32 | 0.53 | 0.35 |
| 2.0 | 0.43 | 0.34 | 0.62 | 0.40 |
| 2.1 | 0.47 | 0.33 | 0.64 | 0.40 |
| 2.2 | 0.46 | 0.34 | 0.65 | 0.38 |
| 2.3 | 0.53 | 0.36 | 0.66 | 0.39 |
| 2.4 | 0.52 | 0.35 | 0.70 | 0.47 |
| 2.5 | 0.49 | 0.35 | 0.84 | 0.49 |

Table 1: Lower hybrid figure of merit, $\gamma_{lh}$ for different profile exponents, $\alpha_n, \alpha_T$, and central refractive index, $\hat{n}_\parallel$ with fixed $\Delta n_\parallel$. Numbers in parentheses indicate the number of reflections of the rays at the plasma boundary

## APPLICATION TO ITER

In the technology phase, the proposed ITER design specifies a major radius, $R_0 = 5.5$ m, a minor radius, $a = 1.8$ m, a toroidal field, $B_0 = 5.3$ T, a plasma current, $I_p = 18$ MA, and $Z_{eff} = 2.2$. We take an elongation at the separatrix of 2.28 and a triangularity of 0.35 with density and temperature profiles of the form $x(0)(1 - \psi)^{\alpha_x}$ where $x = n, T$. We fix $n_e(0)$ and $T_e(0)$ to obtain a volume-average electron density of approximately $7.3 \times 10^{19}$ m$^{-3}$ and a density-weighted average temperature of 22 keV. We first examine the effect of varying the spectrum and the shape of the density and temperature profiles on the figure of merit $\gamma = \langle n_{e20} \rangle_v R_0 I / P$ in a fixed equilibrium. The volume-average electron density is $\langle n_{e20} \rangle_v$ in $10^{20}$ m$^{-3}$, and $I$ and $P$ are the current and absorbed power. The LH frequency is 5.5 GHz with a Gaussian spectrum of half width $\Delta n_\parallel = 0.05$ centered about a central $\hat{n}_\parallel$ ranging from 1.7—2.5. The LH waves are launched from the midplane on the low field side.

The results of this study are shown in Table 1. The numbers in parentheses in Table 1 indicate the number of reflections which the central ray undergoes before it is completely absorbed. At low $n_\parallel$ the LH rays reach the limit on accessibility before they are totally absorbed, refract back to lower density and temperature and then reflect from the plasma edge. A subsequent upshift in $n_\parallel$ aids the absorption, although additional reflections may occur before absorption is complete. An example of ray trajectories for the case with $\hat{n}_\parallel = 1.8$ and two reflections is shown in Fig. 1. An example with strong absorption at a higher frequency, $f = 8$ $GHz$ is shown in Fig. 2. Except for the frequency change, this case has the same parameters as the 5.5 GHz case shown in Fig. 1. The much improved absorption arises from the scaling of the ratio of wave group velocity components perpendicular and parallel to the field, $v_{g\perp}/v_{g\parallel} \propto f$. The rays at higher frequency penetrate sooner to the region of strong damping.

The general trend in Table 1 is for $\gamma_{lh}$ to increase as $n_\parallel$ increases. However, the LH current drive profile peaks towards the outer edge as $n_\parallel$ increases as can be inferred from the scaling formula for the temperature at which strong electron Landau damping occurs, $T_e(keV) \approx 50/n_\parallel$. In the cases with strong single pass absorption, there is little shift in $n_\parallel$ as the LH waves penetrate the plasma. The highest values of $\gamma$ occur when $\alpha_n = 1.0$ and $\alpha_T = 0.5$ i.e. peaked density and flat temperature profiles. The ITER design guidelines call for $\alpha_n = 0.5$ and $\alpha_T = 1.0$. We see from Table 1 that these profiles have the lowest figure of merit, approximately half of the values obtained with a peaked density and flat temperature.

In the next study, also for a fixed equilibrium, we have investigated the effect of increasing the LH power for $f = 5.5$ GHz and central $n_\parallel = 2.2$ for the ITER baseline, $\alpha_n = 0.5$ and $\alpha_T = 1.0$. As the RF power level is increased from 25 to 100 MW, a greater flattening of

| Case | P19D | I19D | P19C | I19C | P119D | I119D |
|---|---|---|---|---|---|---|
| Number of iterations | 1 | 5 | 1 | 5 | 1 | 5 |
| $f_{RF}$(GHz) | 5.5 | 5.5 | 5.5 | 5.5 | 8.0 | 8.0 |
| $P_{lh}$(MW) | 24.7 | 24.7 | 49.5 | 49.5 | 24.7 | 24.7 |
| $< n_e > (10^{19} m^{-3})$ | 7.34 | 6.72 | 7.34 | 6.92 | 7.34 | 6.71 |
| $< T_e >$ (keV) | 22.3 | 20.6 | 22.3 | 21.1 | 22.3 | 20.6 |
| $\gamma_{lh}$ | 0.35 | 0.32 | 0.33 | 0.33 | 0.34 | 0.31 |
| $\gamma_{nb}$ | 0.55 | 0.51 | 0.55 | 0.65 | 0.55 | 0.51 |
| $I_{lh}$(MA) | 2.0 | 2.0 | 3.7 | 4.0 | 0 | 0 |
| $I_{nb}$(MA) | 11.5 | 11.4 | 9.8 | 8.1 | 11.6 | 11.5 |
| $I_{bs}$(MA) | 4.5 | 4.6 | 4.5 | 6.0 | 4.5 | 4.6 |

Table 2: Comparison between fixed profiles and self-consistent profiles for launched RF power with $f = 5.5$ and 8.0 GHz, $\hat{n}_{\parallel} = 2.2$ and $\Delta n_{\parallel} = 0.05$.

the plateau in the electron distribution function occurs which decreases the electron Landau damping and allows the LH waves to penetrate further into the plasma. However, the increase in penetration is only slight–full absorption with 25 MW occurs by $T_e = 14$ keV while 100 MW penetrates to $T_e = 19$. The figure of merit also decreases slightly with higher power, reflecting the higher density at which the absorption occurs.

We next show the effect of varying the Gaussian half-width, $\Delta n_{\parallel}$, of the initial RF power spectrum over the range 0.05—0.14. As the Gaussian-width increases, the velocity-width of the plateau of the distribution function increases but the slope of this plateau also increases. This produces a broader RF current distribution which peaks towards the outside as $\Delta n_{\parallel}$ increases. The increasing slope in the distribution lowers the current drive efficiency of the RF and the figure of merit decreases from 0.35 to 0.26 as $\Delta n_{\parallel}$ increases from 0.05 to 0.14.

All of the above cases have been computed for a fixed current profile, i.e. without iterating to find self-consistent current and magnetic fields. We show some examples of results obtained when ACCOME is used to obtain self-consistent parameters in Table 2. In these cases, the central densities and temperatures are fixed but, since the profiles depend on the spatial distribution of $\psi$, their profiles in real space are altered during the iteration. Both the electron density and temperature profiles become somewhat more peaked as ACCOME iterates to find a consistent solution with a corresponding decrease in the average electron density and temperature. Results for the current and safety-factor profiles for a case at 5.5 GHz are shown in Figs. 3–4. The peak in $j_{\parallel}(\psi)$ due to the LH waves is reflected in a plateau in the $q(\psi)$. Hopefully, the plateau and the loss of magnetic shear will not cause loss of ideal MHD stability. On the other hand, if such a plateau could be generated at a rational surface, such waves could be used to stabilize resistive modes. When the RF power is raised to 50 MW, much more of a dip occurs in the $q(\psi)$ profile due to the LH. At higher power, the bootstrap current is increased due to the flattening of the current profile when more of the driven current occurs at large radii. Shown in Figs. 5–6 are current profiles for LH at 8 GHz with 25 and 50 MW of power and $\hat{n}_{\parallel} = 1.8$. Comparison of Figs. 3 and 5 shows that considerably deeper penetration is possible at the higher frequency and lower $n_{\parallel}$. The penetration can be further enhanced by doubling the power, as shown in Fig. 6. Although the LH efficiencies are similar between these two frequencies, 8.0 GHz demonstrates the capability of driving current closer to the plasma center because of the scaling of $v_{\perp g}$ with frequency.

[1] P. T. Bonoli and R. C. Englade, Phys. Fluids **29**, 2937 (1986).

[2] R. S. Devoto, D. T. Blackfield, M. E. Fenstermacher, P. T. Bonoli, and M. Porkolab, in *Sixteenth European Conference on Controlled Fusion and Plasma Physics, Venezia*, p. 1295 (1989).

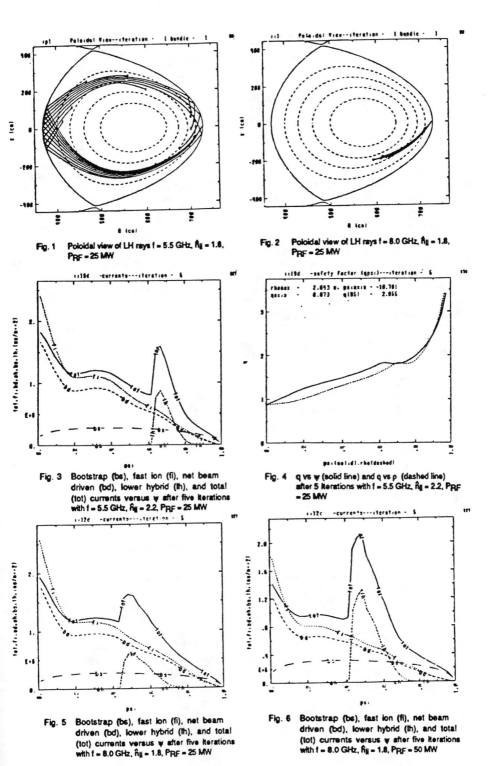

Fig. 1   Poloidal view of LH rays f = 5.5 GHz, ñ₋ = 1.8, P_RF = 25 MW

Fig. 2   Poloidal view of LH rays f = 8.0 GHz, ñ₋ = 1.8, P_RF = 25 MW

Fig. 3   Bootstrap (bs), fast ion (fi), net beam driven (bd), lower hybrid (lh), and total (tot) currents versus ψ after five iterations with f = 5.5 GHz, ñ₋ = 2.2, P_RF = 25 MW

Fig. 4   q vs ψ (solid line) and q vs ρ (dashed line) after 5 iterations with f = 5.5 GHz, ñ₋ = 2.2, P_RF = 25 MW

Fig. 5   Bootstrap (bs), fast ion (fi), net beam driven (bd), lower hybrid (lh), and total (tot) currents versus ψ after five iterations with f = 8.0 GHz, ñ₋ = 1.8, P_RF = 25 MW

Fig. 6   Bootstrap (bs), fast ion (fi), net beam driven (bd), lower hybrid (lh), and total (tot) currents versus ψ after five iterations with f = 8.0 GHz, ñ₋ = 1.8, P_RF = 50 MW

# Achievement of Second Stability by Means of Lower Hybrid Current Drive*

P. T. Bonoli and M. Porkolab
PLASMA FUSION CENTER, MIT, Cambridge, MA 02139
D. T. Blackfield
TRW/LLNL, Redondo Beach, CA 90278
R. S. Devoto and M. E. Fenstermacher
LLNL, Livermore, CA 94550

## Abstract
It is shown that profiles of the safety factor, necessary to access a high beta operating regime in shaped, low aspect ratio tokamaks, can be achieved via off-axis lower hybrid current drive (LHCD). In order to accurately access this RF current generation, a simulation model for LHCD has been extended to noncircular equilibria and combined with an MHD equilibrium solver. Model results will be presented for Versator Upgrade parameters.

## I. Introduction

It has been shown[1,2] that for low aspect ratio ($R_o/a = 3$) shaped, tokamak equilibria with safety factor $2 \gtrsim q(\psi) \gtrsim 7$, a sequence of MHD equilibria exist which allow a stable path to a high beta operating regime. Access to this region of second stability would greatly improve the operating space for D-T fusion reactors and would make steady state reactor operation more feasible by an increase in the bootstrap current. The required $q(\psi)$ profiles can be achieved via off-axis lower hybrid current drive (LHCD). In order to accurately access this RF current generation, a simulation model for LHCD[3] has been modified for noncircular equilibria and added to a code which self-consistently computes free or fixed boundary MHD equilibria[4] and current generated by neutral beams[5], ohmic electric fields and bootstrap effects[6]. Model results are presented for Versator Upgrade parameters[7].

## II. Model Calculation

Free boundary MHD equilibria are obtained by solving the Grad-Shafranov equation[8] using a source term which includes ohmic ($J_{OH}$), lower hybrid ($J_{RF}$), bootstrap ($J_{BS}$), and neutral beam ($J_{NB}$) driven currents:

$$\Delta^*\psi = -\mu_o R^2 \frac{d}{d\psi}p(\psi) - F(\psi)\frac{dF(\psi)}{d\psi}, \tag{1a}$$

$$\Delta^*\psi = R\frac{\partial}{\partial R}\left(\frac{1}{R}\frac{\partial\psi}{\partial R}\right) + \frac{\partial^2\psi}{\partial z^2}, \tag{1b}$$

$$F\frac{dF}{d\psi} = -\mu_o\left[F^2\frac{d}{d\psi}p(\psi) + F\langle J_\parallel B\rangle\right]/\langle B^2\rangle. \tag{1c}$$

Here, $\psi$ is the poloidal flux function, $F = RB_t$ is the toroidal function, $B_t$ is the toroidal magnetic field, $p(\psi)$ is a specified pressure profile, $\langle\rangle$ denotes a flux surface average, $R = (x^2 + y^2)^{1/2}$ is the major radial position (measured in the equatorial plane of the tokamak), and $J_\parallel = J_{OH} + J_{RF} + J_{BS} + J_{NB}$ is the total current density (along $\underline{B}$). In the present work only the ohmic and LHRF current densities are considered (i.e., $J_{NB} = J_{BS} = 0$). The iteration procedure used to solve Eq. (1) is started by first assuming that $J_{RF} = 0$ and the plasma current is purely ohmic, i.e., $J_{OH} = E_\parallel\eta_\parallel$ and $E_\parallel$ is adjusted so that the total plasma current $I_p = \int J_\parallel(\psi)dA(\psi)$ is equal to a specified value. Equation (1) is then solved to obtain a first approximation to the MHD equilibrium and the LHRF current is re-calculated based on this new MHD equilibrium. $E_\parallel$ is again adjusted to keep $I_p$ constant and the Grad-Shafranov equation is solved a second time using the new source term. This process is repeated until the MHD equilibrium and $J_{RF}$ no longer change. Typically, five to ten iterations between the MHD solver and the LHCD code are required in order to obtain convergence.

The LHCD package has been described in detail in Ref. 3. The calculation has been extended to noncircular geometry by integrating the ray equations in cartesian geometry using a LH disperson relation $[\epsilon(\underline{x}, \underline{k}, \omega) = 0]$ which includes electromagnetic and thermal effects. The plasma quantities, $n_e(\psi)$, $T_e(\psi)$, $T_i(\psi)$, $\underline{B}(R, \psi)$ and their spatial derivatives are given in cartesian coordinates by a bi-cubic spline interpolation of the equilibrium results of the MHD solver.

A parallel velocity Fokker Planck calculation[3] is carried out on each $\psi$ surface in the plasma (where $\psi$ labels a magnetic surface). The calculation is relativistically correct (i.e., $p_\parallel = \gamma m_e v_\parallel$), includes the effect of finite electron tail confinement $(\tau_L)$, an effective perpendicular electron temperature due to pitch angle scattering $(T_\perp \geq T_e)$, but ignores the effect of the parallel DC electric field $(E_\parallel)$. The quasilinear diffusion coefficient due to the RF waves, $D_{RF}(p_\parallel, \psi)$ is consistent with the local wave damping. In the calculations presented here, we assume $T_\perp = 5 \times T_e$, $\tau_L = \tau_o \gamma^3$, and $\tau_o \sim \tau_E \sim 8.1 \times 10^{-3}$ sec (for the parameters given below).

### III. Model Results

The parameters used to study LHRF current profile control are typical of the proposed Versator Upgrade device[6] where $a = 0.3$ m, $R_o = 0.9$ m, $B_{to} = 1.0$ T, $I_p = 150$ kA, $\kappa = 1.4$, $\delta = 0.3$, $Z_{eff} = 1.5$, hydrogen gas, $n_e(0) = 3 \times 10^{19} \text{m}^{-3}$, and $T_{eo} = T_{io} = (1.5 - 2.0)$ keV. The plasma profiles were chosen to be $n_e(\psi) = n_{eo}(1 - \psi_n)^{\gamma_n}$, $T_a(\psi) = T_{ao}(1 - \psi_n)^{\gamma_{ta}} (a = e, i)$, $\gamma_n = 1.0$, $\gamma_{te} = \gamma_{ti} = 1.25$, $\psi_n = (\psi - \psi_a)/(\psi_o - \psi_a)$, $\psi_a = \psi(a)$, and $\psi_o \equiv \psi(0)$. The pressure profile is determined from $p(\psi) = n_e(\psi)[T_e(\psi) + T_i(\psi)]$. The RF parameters used were $f_o = 2.45$ GHz, $P_{LH} \leq 500$ kW, and a superposition of the RF power spectra shown in Fig. 1 for relative waveguide phasings of $\Delta\phi = \pi/2$ and $2\pi/3$. The relevant parameters for LH wave accessibility are $(\omega_{pe}/\omega_{ce})^2 = 3.09$ and $n_{\parallel acc} = 3.73$, where $n_{\parallel acc}$ is the minimum value of parallel refractive index required for wave accessibility to the plasma center.

The results of LHRF current profile control at different RF powers are summarized in Table I $(T_{eo} = T_{io} = 1.5 \text{ keV})$ and in Table II $(T_{eo} = T_{io} = 2.0 \text{ keV})$. The current drive figure of merit used in these tables is defined as $\eta_{CD} \equiv \bar{n}_e(10^{20}\text{m}^{-3})I_{RF}(\text{kA}) \times R_o(\text{m})/P_{LH}(\text{kW})$, where $\bar{n}_e$ is the line-averaged density.

|  | Table I |  |  |  | Table II |  |  |
|---|---|---|---|---|---|---|---|
| $P_{LH}(\text{kW})$ | $q_o$ | $I_{RF}(\text{kA})$ | $\eta_{CD}$ | $P_{LH}(\text{kW})$ | $q_o$ | $I_{RF}(\text{kA})$ | $\eta_{CD}$ |
| 0 | 1.00 | 0 | – | 0 | 0.99 | 0 | – |
| 250 | 1.89 | 68 | 0.049 | 200 | 2.18 | 58 | 0.052 |
| 375 | 3.58 | 91 | 0.044 | 250 | 3.26 | 77 | 0.055 |

The results in Tables I and II indicate that $q_o \gtrsim 2$ can be achieved with a lower RF power at the higher electron temperature $(T_{eo} = 2 \text{ keV})$. This is in part because the electron Landau damping is stronger at the higher electron temperature and because the absorption profile is broader and peaked farther from the plasma axis as $T_{eo}$ increases.

Results from the $T_{eo} = 2$ keV case in Table II at $P_{LH} = 200$ kW are shown in Figs. 2(a)-2(d). Figure 2(a) is a plot of $J_{OH}(\psi)$ and $J_{RF}(\psi)$ after the fifth and final iteration between the current drive code and MHD solver. The LHRF current in this case was 58 kA and the ohmic current was 92 kA. The peak in the RF power deposition at $\psi_n \cong 0.35$ is to be contrasted with the peak of $\psi_n = 0.1$ at $T_{eo} = 1.5$ keV and $P_{LH} = 250$ kW.

The resulting $q(\psi)$ profile is shown in Fig. 2(b). The ohmic $q(\psi)$ profile ($P_{LH} = 0$) has also been shown in Fig. 2(b) for comparison (dashed line). The poloidal projection of a single ray trajectory (initial $n_\parallel^o = 4.19$) for this case is shown in Fig. 2(c). Each tick mark along the ray path indicates a 10% decrease in the wave power due to quasilinear electron Landau damping. Finally, the electron distribution function on a flux surface near the maximum in the RF deposition ($\psi_n \cong 0.35$) is shown in Fig. 2(d), plotted as a function of parallel kinetic energy $E = m_e c^2 [n_\parallel/(n_\parallel^2 - 1)^{1/2} - 1]$. It should be pointed out that the slowing down time for the fast electrons in the plateau region of Fig. 2(d) is approximately $\tau_s \gtrsim 3 \times 10^{-3}$ sec (assuming $n_e \simeq 2 \times 10^{19} \text{m}^{-3}$ and $E \simeq 35$ keV at $\psi_n = 0.35$). However, $\tau_s < \tau_L$ so that electrons would be expected to thermalize before spatially diffusing an appreciable distance.

In conclusion, a powerful computational tool has been developed to study lower hybrid current profile control in self-consistent MHD equilibria. Utilizing this model, it has been shown that the $q(\psi)$ profiles [$2 \gtrsim q(\psi) \gtrsim 7$] necessary to access a high beta operating regime in shaped tokamak equilibria can be achieved, via off-axis LHRF current generation with $I_{RF} \gtrsim 0.5 \times I_p$. This model can also be used to study current profile control and bootstrap current generation in reactor configurations where $n_e \gtrsim 1 \times 10^{20} \text{m}^{-3}$ and $T_e, T_i \lesssim 25$ keV.

*Work supported by the U. S. Department of Energy.

**References**

[1] B. Coppi. G. B. Crew, J. J. Ramos, Plasma Phys. Contr. Fusion <u>6</u>, 109 (1981).
[2] M. J. Gerver, J. Kesner, J. J. Ramos, Phys. Fluids <u>31</u>, 2674 (1988).
[3] P. T. Bonoli, R. C. Englade, Phys. Fluids <u>29</u>, 2937 (1986).
[4] M. Azumi, G. Kurita, Proc. 4th Int. Symp. Computing Methods in Appl. Science and Eng. (Paris, 1979) 335.
[5] K. Tani, M. Suzuki, S. Yamamoto, M. Azumi, Report JAERI-M 88-042 (1988).
[6] R. S. Devoto, K. Tani, M. Azumi, 15th Eur. Conf. Contr. Fusion and Plasma Heating, Vol. III (1988) 1055; R. S. Devoto, et al., 16th Eur. Conf. Contr. Fusion and Plasma Physics, Vol. IV (1989) 1295.
[7] M. Porkolab, et al. 12th Int. Conf. on Plasma Phys. and Contr. Fusion Research (Nice, France, 1988) Paper IAEA-CN-50/E-4-8.
[8] J. P. Freidberg, in "Ideal Magnetohydrodynamics" (Plenum, NY, 1987) Ch. 6, p. 111.

**Figure Captions**

Fig. 1. RF power spectra for an eight waveguide, 2.45 GHz, LH grill. Relative waveguide phasings are $\Delta\phi = \pi/3, \pi/2$, and $2\pi/3$. Results were obtained using an LH coupling code [M. Brambilla, Nucl. Fusion <u>16</u>, 47 (1976)].

Fig. 2. Model results for Versator Upgrade parameters ($n_{eo} = 3 \times 10^{19} \text{m}^{-3}$, $T_{eo} = T_{io} = 2$ keV, $B_{to} = 1.0$ T, $I_p = 150$ kA, and $P_{LH} = 200$ kW). (a) Current density vs. $\psi$. (b) $q$ vs. $\psi$ for $P_{LH} = 200$ kW (solid line) and for comparison $P_{LH} = 0$ (dash line). (c) Poloidal projection of ray trajectory ($n_\parallel^o = 4.19$). (d) Electron distribution function versus parallel kinetic energy on a flux surface ($\psi_n \cong 0.35$) near the maximum of the RF deposition profile.

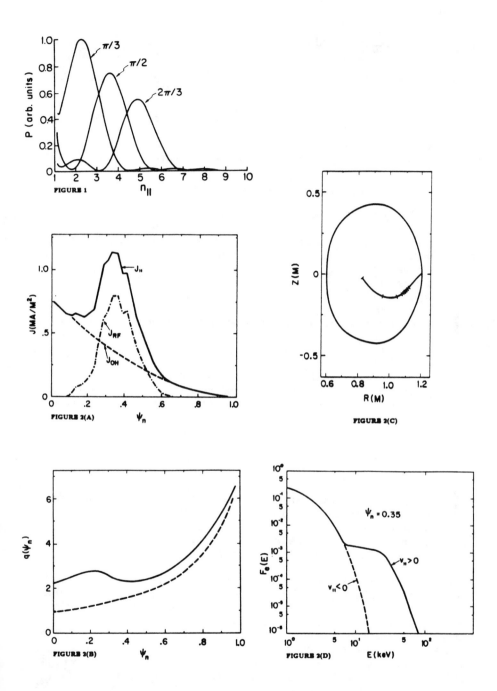

FIGURE 1

FIGURE 2(A)

FIGURE 2(B)

FIGURE 2(C)

FIGURE 2(D)

# Simulation of propagation, absorption and current drive efficiency of Lower Hybrid waves in reactor relevant JET discharges

M Brusati, C Gormezano, S Knowlton(*), M Lorentz-Gottardi(°),
F Rimini

JET Joint Undertaking, CEC, Abingdon, Oxon OX14 3EA

(*) University of Auburn, Alabama
(°) IPP Garching, IPP-EURATOM Association, Garching

## ABSTRACT

A degree of control of the current density profile in JET will be achieved by launching a high directivity narrow spectrum of Lower Hybrid waves at 3.7 GHz. In the preparation of this experiment attention has been focussed on the evaluation of wave absorption and current drive efficiency to be expected in present high performance JET scenarios or reactor relevant scenarios.

The highest slow wave absorption and current drive efficiency are expected in high temperature plasmas with peaked density profiles.

## INTRODUCTION

A Lower Hybrid Current Drive experiment at 3.7 GHz is planned in JET starting late in 1989. The first stage consists of a 16 waveguide array launching a 70% directivity wave pattern with peak $n_\parallel$ ranging from 1.3 to 2.3 into the JET plasma from a main equatorial port, with a power level of 4 MW. A second stage is planned for late 1990 where in excess of 10 MW will be launched from a 32 waveguide array of the multijunction type. A detailed description of the LH plant has been presented recently[1].

When planning the experimental campaign it is important to assess the performance of different JET plasmas as targets, with the aim of maximizing the power absorption in the tail of the electron distribution function and the wave penetration, taking into account the finite poloidal extent of the launcher.

To this purpose the launching of different LH spectra, with power levels up to 10 MW, has been simulated for the different scenarios JET has been operating in, using a flux surface space ray tracing code coupled to a 1D Fokker-Planck code[2]. The plasma magnetic configuration was provided by the JET magnetic data analysis system, and measured temperature and density profiles were taken into account. Simulations have been carried out on a variety of reactor relevant high power ICRH and NBI JET plasmas such as sawtooth stabilized, high current, H modes and pellet fuelled discharges, including wave absorption effects by non thermal ions.

RESULTS

In figure 1 the ray propagation is shown for $n_\parallel$ = 1.8, 2.5 as launched from the outer equatorial plane into a 3 MA, 3.2 T ICRH heated JET plasma during a "monster" sawtooth with medium density ($n_{eo}$ ~ 4.2 x $10^{19}$ $m^{-3}$) and peaked temperature ($T_{eo}$ ~ 10 keV) profiles. Plasmas with peaked profiles provide the best targets for LHCD schemes, as penetration and absorption are enhanced of low $n_\parallel$ waves, thus developing high energy low collisionality electron tails leading to high current drive efficiency.

According to ray tracing calculations, toroidal effects due to the high elongation of JET plasmas induce large deformations in the initial $n_\parallel$ spectrum as it propagates through the plasma and $n_\parallel$-upshift is usually observed after wave reflection at the boundary.

A strong dependence is observed of the computed current drive efficiency $\eta$ on local electron temperature, in agreement with the LHCD theory developed by Fish[3] and with recent results obtained on JT-60[4]. Furthermore the large variation in efficiency observed for different plasmas with comparable average density or temperature values can be ascribed to details in the ray trajectories before absorption takes place and to the spreading of the wave front due to the finite poloidal extent of the launcher.

In figure 2, $\eta$ is shown for different JET plasmas as a function of electron temperature at the peak of absorption and in figure 3 the change in the RF driven current density absorption profile (a) is shown for different values of the electron density (b). Lower Hybrid injection in peaked density pellet fuelled plasmas can increase the operating density range for current drive and lead to off axis RF driven current. In the most favourable cases ("monster" sawtooth plasma with $\langle n_e \rangle$ ~ $3 \cdot 10^{19}$ $m^{-3}$, $I_p$= 3 MA, $B_T$= 3.4 T), it is expected to achieve $\eta$ ~ $1.7 \cdot 10^{19}$ $m^{-2}$ A/W with $n_\parallel$(peak) = 2.3. A similar result is found for $n_\parallel$ = 1.8, where the less effective propagation is balanced by a lower collisionality of the resonating electrons. For 10 MW of launched RF power this leads to a total RF driven current in order of 2 MA. In denser plasmas with peaked density and temperature profiles, such as ICRH heated pellet fuelled discharges ($\langle n_e \rangle$ ~ $4 \cdot 10^{19}$ $m^{-3}$, $n_{eo} \geq 7 \cdot 10^{19}$ $m^{-3}$, $T_{eo}$ ~ 8 keV) reduced wave penetration leads to absorption at lower electron temperature values and lower current drive efficiencies ($\eta$ ~ 1.1 -$1.5 \cdot 10^{19}$ $m^{-2}$ A/W) are predicted.

In the H-mode regime obtained with single null X point configurations, accessibility of the LH wave is reduced due to broader density profiles. Propagation is greatly affected by the up-down asymmetry of the plasma equilibrium and by high $\beta$ effects (fig 4). H modes in the double null configuration present themselves as better high performance target plasmas than the single null ones, mostly due to more peaked density profiles and to up down symmetry.[5]

124

## CONCLUSIONS

Simulations of Lower Hybrid Current Drive scenarios in JET indicate that high efficiencies ($\eta \sim 1.7 \cdot 10^{19}$ m$^{-2}$ A/W) could be achieved in high $B_T$, peaked density, high temperature plasmas, as encountered in sawtooth stabilized and pellet fuelled regimes. Current drive efficiency is affected by changes in density profile and depends strongly on the local value of the electron temperature. In particular the recently observed relation between peaked density profiles and improved confinement regimes can be very beneficial for Lower Hybrid current profile control.

The large variety of reactor relevant scenarios which can be obtained at JET makes the Lower Hybrid Current Drive experiment well suited for ITER related studies, in particular the testing of LHCD schemes in hot plasmas and the damping of LH waves on $\alpha$ particles.

## REFERENCES

1.  S. Knowlton et al, Proc. 14th EPS Conf. on Controlled Fusion and Plasma Physics ('87) Vol. 11D, part III, p. 827.
2.  P. Bonoli, A. Englade, Phys. Fluids, 29, 2937 ('86).
3.  N. Fish, Rev. Mod. Phys., 59 (87) 175.
4.  JT-60 Team, Plasma Phys. and Contr. Fusion ('88) A-I-4.
5.  B. Tubbing et al, Proc. 16th EPS Conf. on Controlled Fusion and Plasma Physics ('89) Vol. 13B, part I, p. 237.

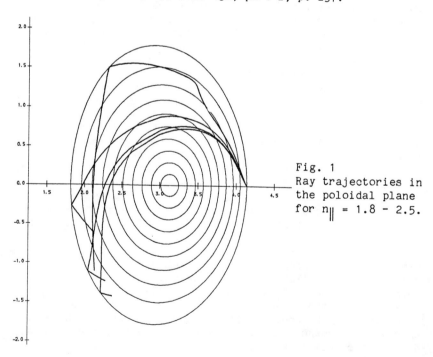

Fig. 1
Ray trajectories in the poloidal plane for $n_{\parallel}$ = 1.8 - 2.5.

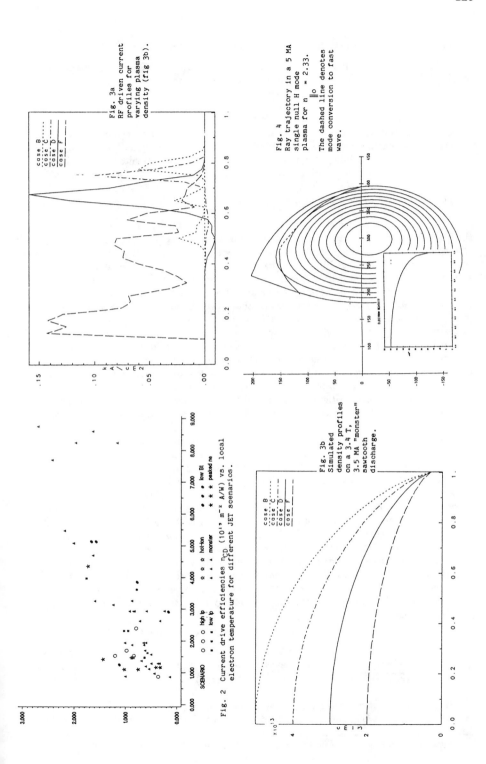

Fig. 3a
RF driven current
profiles for
varying plasma
density (fig 3b).

Fig. 4
Ray trajectory in a 5 MA
single null H mode
plasma for n $\|_0$ = 2.33.

The dashed line denotes
mode conversion to fast
wave.

Fig. 2  Current drive efficiencies $\eta_{CD}$ (10$^{19}$ m$^{-2}$ A/W) vs. local
electron temperature for different JET scenarios.

Fig. 3b
Simulated
density profiles
on a 3.4 T,
3.5 MA "monster"
sawtooth
discharge.

# PULSED LOWER-HYBRID WAVE PENETRATION IN REACTOR PLASMAS

Ronald H. Cohen, Paul T. Bonoli[†], Miklos Porkolab [†] , and Thomas D. Rognlien

Lawrence Livermore National Laboratory

Livermore, CA 94550

## ABSTRACT

Providing lower-hybrid power in short, intense (GW) pulses allows enhanced wave penetration in reactor-grade plasmas. We examine nonlinear absorption, ray propagation, and parametric instability of the intense pulses. We find that simultaneously achieving good penetration while avoiding parametric instabilities is possible, but imposes restrictions on the peak power density, pulse duration, and/or r.f. spot shape. In particular, power launched in narrow strips, elongated along the field direction, is desired.

There is general agreement in the magnetic-fusion community as to the desirability, if not the feasibility, of a steady-state reactor. In a tokamak, achieving this goal requires the sustenance of a non-inductively driven toroidal current. Lower-hybrid (LH) waves are a very appealing choice as a current driver, as LH current drive has an impressive data base, and good theoretical and experimental current-drive efficiency. However, according to conventional wisdom, lower-hybrid current drive doesn't work in the middle of reactor-grade plasmas because accessibility restricts the phase velocity to be low enough that electron Landau damping is severe. But the power expended in Landau damping saturates (due to formation of a quasilinear or nonlinear plateau), allowing burnthrough with enough power. As the absorption formulas below show, the saturated power is in the GW range for a reactor-grade machine such as the International Thermonuclear Experimental Reactor (ITER); such power levels are prohibative if continuous but reasonable as peak levels in a pulsed scheme. Suitable sources of pulsed lower-hybrid power, such as induction-linac-driven relativistic klystrons, are presently under development.

We report here of investigations of several key issues for enhanced LH wave penetration. First is the basic issue of nonlinear absorption of the intense radiation. Because the absorption depends on the width of the parallel-wavenumber ($N_\parallel$) spectrum and the microwave spot shape in the interior of the plasma, ray tracing is important. A ray-tracing code also offers a convenient way of integrating the nonlinear opacity across realistic temperature and density profiles. Finally, because of the wave intensity, it is important to assess the role of parametric instabilities.

We consider first the absorption, which is calculated from the power expended in forming a nonlinear plateau. The result depends on the degree of relaxation of the plateau between successive (as seen by individual electrons) pulses. The limit of most interest (where there is the greatest advantage in pulsing for a fixed average power) is that where the plateau completely relaxes between successive pulses (but the relaxation time is long compared to the pulse duration). In this limit, Maxwellian electrons are incident on the microwave beam; the distribution of emerging electrons is altered only in the band of parallel velocities that can resonantly interact with the waves. Referring to Fig. 1, the power expended by the wave is that utilized in moving the initially Maxwellian electrons from (1) to (2) as they stream through the beam. Hence the rate of decay of the axially averaged (but temporally instantaneous) Poynting Flux $\langle S \rangle$ with path length $s$ is:

$$\langle dS/ds \rangle \cong \left( nmv_t^3/4\pi^{1/2}L_\parallel \right) \exp(-\epsilon_1)H(\Delta\epsilon)$$

where $L_\parallel$ is the r.f. beam scale length along the magnetic field, $\epsilon_1$ is the parallel energy normalized to the electron temperature $T$ at the lower edge of plateau, $\Delta\epsilon$ is the full width of the plateau, $v_t \equiv (2T/m)^{1/2}$, and $H(x) \equiv \left(1 - e^{-x}\right)\left(-1 + x/2\right) + xe^{-x}$; note $H(x) \sim x^3/12$ for small $x$.

If the plateau cannot relax completely between pulses, the peak power required to form the plateau is not as big, since the incident $f$ is partly flattened. The absorption is reduced

---

[†] Permenant address: Massachusetts Institute of Technology, Cambridge MA 02139

approximately by the factor $1 + a\nu_{rep,eff}\tau_{c,eff}$, where $\tau_{c,eff}$ is the time for the plateau to relax, $a \sim 1$, and $\nu_{rep,eff}$ is the average repetition rate experienced by individaul electrons, $\nu_{rep,eff} \equiv \nu_{rep}A_b/A_\psi$ with $\nu_{rep}$ the actual pulse repetition rate, $A_b$ the area spanned by electrons on a flux surface which pass through the r.f. beam during the pulse, and $A_\psi$ the area of the flux surface. Hence, defining a damping decrement $\alpha$ by the relation $\langle dS/ds\rangle = -\alpha\langle S\rangle$, we obtain:

$$\alpha = \frac{nmv_t^3 H(\Delta\epsilon)\exp(-\epsilon_1)}{4\pi^{1/2}L_\parallel\langle S\rangle(1 + a\nu_{rep,eff}\tau_{c,eff})} . \tag{1}$$

In the limit where the plateau relaxes negligibly between pulses, the absorption of *average* power is the same as for continuously applied power; it can be obtained, for example, by calculating the (time-averaged) current and dividing by the current-drive efficiency. That result is reproduced by Eq. (1) in the limit of large $\nu_{rep,eff}\tau_{c,eff}$, provided that one eliminates $S$ in favor of its time-averaged value.

The absorption depends sensitively on the plateau energy $\epsilon_1$ and on the plateau width $\Delta\epsilon$. An upper bound on $\epsilon_1$ is set by the accessibility condition $N_\parallel \gtrsim N_a$, where $N_a = (1 + x^2 - \omega_{pi}^2/\omega^2)^{1/2} + x$ and $x = \omega_{pe}/\omega_{ce}$, the ratio of the electron plasma and cyclotron frequencies. Thus there is strong dependence on $n/B^2$, where $n$ is density and $B$ is the magnetic-field strength. The plateau width, which in the weak-r.f. limit depends only on the spread in $N_\parallel$, now has a lower bound set by single-pass acceleration in the intense r.f. electric field,

$$(\Delta u/\bar{u})_E \cong \left(S_\perp(\text{GW/m}^2)/N_\parallel\right)^{1/4}A^{-1/8}(G/\mathcal{E}_\parallel)^{1/2}(10\,\text{GHz}/f)^{1/2} \tag{2}$$

where $f$ is the wave frequency, $A$ is the ion mass in AMU, and $G \equiv x^{1/2}/(1 - x^{-2})^{1/4}$.

Using (2) in (1), we find the penetration length $L_p \equiv S_\perp/S\alpha$ to vary as $S^{1/4}$ for fixed $u_1$, and to be of the order of a meter for infrequent, 1 GW/m$^2$ pulses with $L_\parallel \sim 5$ m and nominal ITER parameters ($B = 5.3$ T, $n = 0.7 \times 10^{20}$ m$^{-3}$, $T = 25$ keV) (compared to about 10 cm. for linear absorption).

The absorption coefficient (1) accounts only for resonant interaction of electrons with the electric field. Additionally, however, the acceleration of the resonant electrons leads to a charge separation which persists for of order of an electron toroidal transit time, and hence on the pulse timescale, a d.c. electric field which accelerates the electron distribution function in the opposite direction from the resonant acceleration. This effect enhances penetration, but has not yet been adequately quantified.

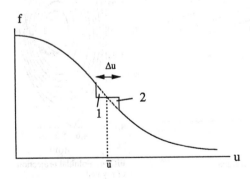

FIG. 1. Distribution function incident on (dashed) and emerging from (solid) r.f. beam vs. parallel velocity

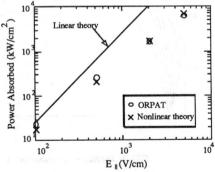

FIG. 2. r.f. power absorpbed for $N_\parallel = 1.7$, $f = 8$ GHz, $T_e = 25$ keV, $L_\parallel = 75$ cm, and $n = 8 \times 10^{19}$ m$^{-3}$

We have compared the absorption predicted by Eq. (1) with results from a multi-particle orbit code ORPAT. Here, Maxwellian electrons are followed for a single pass through a monochromatic, single $N_\parallel$ beam. Fig. 2 illustrates the good agreement with the nonlinear theory, as well as the reduction in absorption compared to the predictions of linear theory (weak r.f. result).

The nonlinear estimates (1) and (2) of the wave absorption and plateau broadening have been incorporated in a toroidal ray tracing code[1] to study the wave penetration problem using realistic plasma profiles in tokamak geometry. The ray tracing calculation[1] has been modified for noncircular equilibria and combined with an MHD equilibrium solver.[2] A rate equation for the wave energy flux is integrated along each ray path and has the form $d\langle S\rangle/dt = -2\gamma\langle S\rangle$, where $2\gamma = \alpha \times$ group velocity.

The density and pressure profiles are taken to be $n_e(\psi) = n_e(0)(1-\psi_n)$ and $p(\psi) = p(0)(1-\psi_n)^2$, where the normalized flux is $\psi_n \equiv (\psi-\psi_0)/(\psi_a-\psi_0)$, $\psi_0 \equiv \psi(0)$, $\psi_a \equiv \psi(a)$, and $T_e(\psi)$, $T_i(\psi)$ are found by taking $p(\psi) = n_e(\psi)[T_e(\psi)+T_i(\psi)]$. The other plasma parameters used were minor radius $a = 1.8$ m, major radius $R_0 = 5.5$ m, elongation $\kappa = 2.22$, triangularity $\delta = 0.61$, $B_0 = 5.3$ T, plasma current $I_p = 18$ MA, effective charge $Z_{eff} = 1.5$, $n_e(0) = 8 \times 10^{19}$ m$^{-3}$, and $T_e(0) = T_i(0) \simeq 33$ keV. The r.f. parameters were $f_0 = 8$ GHz, $N_\parallel^0 = 2.0$, $\Delta N_\parallel/N_\parallel^0 = (\Delta u/u)_0 = 0.05$, $\nu_{rep} = 5$ kHz, power $P_{LH} = (1-50)$ GW, and waveguide-array area $A_B = L_\parallel h = 1$ m$^2$; $h$ is the height normal to field lines. A plot of $N_\parallel$ vs. $\psi$ for five rays launched from a vertical position of $y_0 = +0.5$ m, for $N_\parallel^0 = (1.9, 1.95, 2.0, 2.05, 2.10)$ is shown in Fig. 3. The results indicate that rays which start at the plasma edge ($\psi = 0$) with a spacing in $N_\parallel$ of 0.05 tend to keep this spacing on their first pass into the plasma, but spread substantialy following their closest approach to the magnetic axis. Similar results are found for $N_\parallel(\psi)$ plots made at other vertical positions, $y_0 = (+0.25, 0, -0.25, -0.5)$ m. This property makes it easier to ensure that waves will all penetrate but not overshoot. Note that the nonlinear plateau width (2) is $\simeq 0.1 - 0.13$; there is thus no need to launch a $N_\parallel$ spectrum much narrower. Spatial broadening of the initial waveguide area (r.f. beam) has been found to be minimal. In fact, the r.f. beam area has been found to decrease slightly from 1.0 m$^2$ to 0.8 m$^2$ during the first pass of the r.f. beam into the plasma. The projection of the r.f. beam on each flux surface was found by tracing up to 100 "marker" rays emanating from the waveguide at the plasma surface.

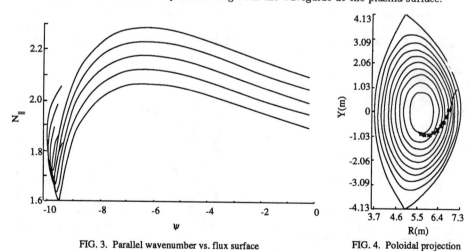

FIG. 3. Parallel wavenumber vs. flux surface

FIG. 4. Poloidal projection of ray path

The best penetration is achieved using long, narrow waveguides. An example of this is shown in Fig. 4 for the parameters given above with $P_{LH} = 1$ GW, $h = 0.1$ m, $L_\parallel = 10$ m and vertical launch point $y_0 = 1.0$ m. Each tick mark along the ray path represents a 10%

decrease in the wave power due to nonlinear damping. The maximum penetration of the ray was to $\psi(x_p) \simeq -10.0$, where $n_e(x_p) \simeq 7.1 \times 10^{19}$ m$^{-3}$ and $T_e(x_p) \simeq 31$ keV (here the magnetic axis corresponds to $\psi \simeq -11.0$). Rays launched from the midplane or below did not get as close to the axis. While a wide launcher results in superior wave penetration, it may be difficult to install in a tokamak. Therefore, we considered a more conventional r.f. beam with $L_\parallel = 1$ m, $h = 1$ m and $P_{LH} = 1$ GW (simple port size in ITER). Then, the maximum penetration distance of the beam was only $\psi(x_p) \simeq -5.0$, $n_e(x_p) \simeq 3.5 \times 10^{19}$ m$^{-3}$, and $T_e(x_p) \simeq 20$ keV. Finally, we note that as $P_{LH}$ is increased from 1 GW to 50 GW with fixed waveguide dimensions of $L_\parallel = 1$ m and $h = 1$ m, the maximum penetration distance increased from $\psi(x_p) \simeq -5.0$ to $\psi(x_p) \simeq -7.0[n_e(x_p) = (3.5 \to 5.0) \times 10^{19}$ m$^-$3 and $T_e(x_p) = (20 - 24.5)$ keV]. The saturation in $x_p$ with increasing r.f. power is mostly due to an increase in the nonlinear broadening of the plateau width $(\Delta u)$.

Parametric instabilities may occur during high power LH pulse injection, especially in the plasma edge region. This may, in turn, restrict the allowed maximum peak power density, and/or maximum pulse length. A detailed theoretical and numerical study of parametric instabilities has been carried out using the previously developed theory and code.[3,4] The results may be summarized as follows. The dominant instability in the edge region $(n_e(a) \gtrsim 3 \times 10^{19}$ m$^{-3}$, $T_e \gtrsim 1$ keV) is a dissipative ion cyclotron quasi-mode, coupled with a lower-hybrid side-band, for incident power densities $S_\perp = P_{LH}/A < 10$ GW/m$^2$. The instability is driven by both $\mathbf{E}_\perp \times \mathbf{B}_0$ drift and parallel $(\mathbf{E}_\parallel)$ drift of electrons. Typical growth rates are $\gamma/\omega_0 \gtrsim 0.02$ for power densities $S_\perp \gtrsim 10$ GW/m$^2$, depending on edge parameters. The lower density and temperature regions $(n_e \sim 10^{19}$ m$^{-3}$, $T_i \sim T_e \sim 100$ eV) are more susceptible to instability, and growth rates of the order of $\gamma/\omega_0 \sim 0.01$ may be achieved with power densities as low as $S_\perp \sim 1$ GW/m$^2$. However, the growth rates decrease rapidly as the temperature and density rise. The growth may be limited by convective losses which limit spatial growth and/or limited pulse length. To limit growth to acceptable values, it may be necessary to cluster waveguide arrays to axial widths of $\Delta z \gtrsim 30$ cm, or heights normal to field lines of $\Delta y \gtrsim 10$ cm, or pulse lengths $\Delta t \gtrsim 10$ ns. Limiting the height is the best choice for good penetration. The waveguide clusters may have to be stacked after suitable spatial separation depending upon pulse length and rep-rate. High rep-rate (50 kHz) may be required to achieve the necessary average power for reactor applications.

In conclusion, we find that the use of pulsed power significantly enhances lower-hybrid wave penetration, and is consistent with avoidance of severe parametric instabilities for suitable choices of launcher shape (narrow strips, elongated along field lines, are best), pulse duration and size. Unresolved issues include nonlinear filamentation, quantification of the charge-separation effect, and the feasibility of long, skinny launching structures.

This work was performed under the auspices of the U.S. Department of Energy by the Lawrence Livermore National Laboratory under contract number W-7405-ENG-48.

## REFERENCES

1. P. T. Bonoli and E. Ott, *Phys. Fluids* **25** 359 (1982).
2. M. Azumi and G. Kurita, Proc. 4th Int. Symp. Computing Methods in Appl. Science and Eng. (Paris, 1979) 335.
3. M. Porkolab, *Phys. Fluids* **20**, 2058 (1977).
4. Y. Takase and M. Porkolab, *Phys. Fluids* **26**, 2992 (1983).

# THE LOWER-HYBRID CURRENT DRIVE AND HEATING SYSTEM FOR THE TOKAMAK DE VARENNES

R. Decoste[a], Y. Demers[b], V. Glaude[c], A. Hubbard[a]
J. Bagdoo[b], G.-A. Chaudron[b], R. Annamraju[d]

Centre Canadien de Fusion Magnétique
Varennes, Québec, Canada, JOL 2PO

## ABSTRACT

A lower hybrid system is under development for current drive and heating on the Tokamak de Varennes. Near steady-state RF current drive and/or transformer recharge should allow long pulse operation of a diverted plasma for up to 30 sec at nearly full machine parameters ($I_p \approx$ 250 kA, $\bar{n}_e \approx 3\times10^{19}$ m$^{-3}$). The system design is for 1 MW of RF power at a relatively high $N_\shortparallel$ (variable around 3.1) in a narrow spectrum (0.4 FWHM) at 3.7 GHz. Design considerations and technical solutions are presented.

## INTRODUCTION

The Tokamak de Varennes (T de V) is a medium size machine presently operated in a Ohmic regime with typical parameters R=0.86 m, a=0.24 m, $B_T$=1.5 T, $I_p$=250 kA, $\bar{n}_e$=3x10$^{19}$ m$^{-3}$ and $T_\infty$=0.8 keV. The planned lower hybrid current drive (LHCD) system should allow long pulse operation of the diverted plasma at nearly full machine parameters for about 30 s. Main areas of research will be LHCD coupling and efficiency, plasma-wall interaction and impurity control with a divertor for long pulses, plasma startup and transformer recharge with the divertor, and auxiliary heating and profile control at high LH wave densities (~1 MW/m$^3$). This paper describes basic considerations and features of the LH system design.

## DESIGN CONSIDERATIONS

The LH system of the T de V will use 2 klystrons for a total power of about 1 MW for 30 s or 1.4 MW for 5 s. The 3.7 GHz frequency should be high enough to avoid density limitations[1] for most cases on the T de V. The antenna, consisting of 2 rows of 8 four-way multijunctions, is designed to launch a narrow spectrum ($\Delta N_\shortparallel$=.4) with an $N_\shortparallel$ adjustable between 2.3 and 3.7 in about 200 ms. Typical spectra, obtained from simulations using a multijunction coupling code[2] are shown in Fig. 1.

Relatively high $N_\shortparallel$ are required because of poor wave accessibily towards the plasma center at low toroidal fields. For example, with $n_{\infty}$=3x10$^{19}$ m$^{-3}$ and

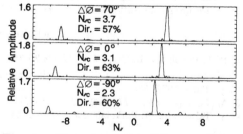

Fig. 1. $N_\shortparallel$ spectra for different phases $\Delta\phi$ between the multijunctions. The directivity represents the relative energy content of the main peak.

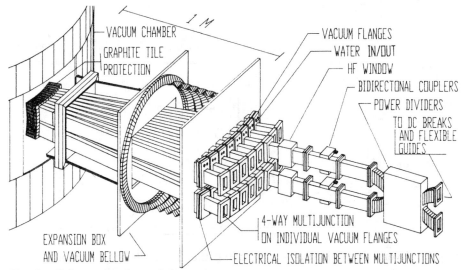

Fig. 2. Schematic view of the antenna.

$B_T$=1.5 T, a $N_{//}$ greater than 3.0 is required to reach the plasma center, proportionally reducing the current drive efficiency[3] ($\eta_{eff}=\bar{n}_e R_{maj} I_p/P_{RF} \propto 1/N_{//}^2$). With $\bar{n}_e$=2x10$^{19}$ m$^{-3}$ and $R_{maj}$=0.86 m, this implies that about 1 MW is required to maintain the plasma current at 200 kA if the LHCD efficiency is 0.35 (10$^{19}$ m$^{-2}$A/W). Such an efficiency is consistent with typical experimental results[4] ($\eta_{eff}$~1 with $N_{//}$~2) and simulations performed with a modified version of the Bonoli-Englade code[5].

Several current drive scenarios are considered for the 30-s pulse duration. In addition to the long pulse condition entirely with RF current drive, an alternating cycle of current drive-enhanced ohmic operation ($N_{//}$~3.5, ~5 s) with low density transformer recharge ($N_{//}$~2, ~5 s) will be examined experimentally. The RF-ohmic part of the cycle would permit higher density operation than pure RF current drive, although at a still higher $N_{//}$. Operation in an H-mode regime, although not yet demonstrated with LH alone, is also a possibility that could require rapid changes of the $N_{//}$ during an experiment.

## DESIGN OVERVIEW

A schematic view of the antenna is shown in Fig. 2 and an overview of the complete RF system is presented in Fig. 3. One klystron can drive 2 rows of 4 multijunctions each, with no phase difference between the top and bottom row. Each multijunction is separated into 4 guides with the length of the dividers (E-plane) adjusted to minimize reflections and increase the directivity at the optimum coupling plasma density. An $N_{//}$ centered at 3.1 with a directivity of 63% is obtained with the built-in 90° phase difference and a 6.5 mm distance between waveguides (including a 2 mm septum). The spectral width $\Delta N_{//}$=.4 (FWHM) is determined by the total grill width of 234 mm (72 mm high), including 2 dummy guides at each extremity. Six high power motor-driven phase shifters can be used to vary the relative phase between adjacent multijunctions from -90° to 70°

132

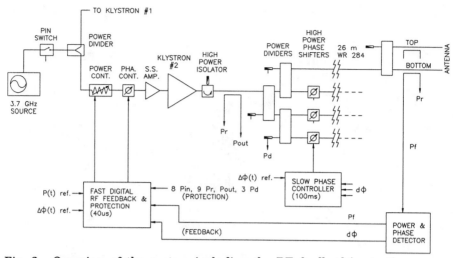

Fig. 3. Overview of the system including the RF feedback/protection.

($2.3 \leq N_{\shortparallel} \leq 3.7$) in less than 200 ms if required. The antenna can be moved ±70 mm with respect to its normal position 20 mm away from the separatrix (R=0.86 m, a=0.26 m).

The multijunctions will be maintained at 200°C with water cooling/ heating to minimize outgassing. With a 1 MW power input, the power density at the grill mouth is about 4.5 kW/cm². Total power losses from the klystrons to the plasma are estimated to be about 15%, including a few % within the antenna (stainless-steel plated or copper). High power isolators should permit operation of the system with up to 10% of the power reflected back towards a klystron. The transmission line uses standard WR-284 waveguide, pressurized with $SF_6$, between the klystron room and the antenna.

The fast digital feedback and protection for the RF system is described in Figs. 3 and 4. The power and phase control of a klystron is based on continuous measurements performed at various points: incident and reflected power on each multijunction, output and reflected power of the klystron, reflected power on the matching loads of the 3 high power dividers next to the klystron and the relative phase of the first multijunction. These measurements are processed by a fast controller during a ~40 µs cycle to determine the power and phase feedback to apply at the input of the klystron, the power balance on the transmission line and the reflection coefficient at each multijunction for protection purposes. Reference waveforms used for the feedback loops can be preprogrammed and/or calculated in real time from plasma parameters ($I_p$, $n_e$, $V_L$, etc.).

A general overview of the control and feedback system is shown in Fig. 4. Both local and remote consoles are available for the user interface and access to the database used to initialize the whole system. Programmable local controllers (PLC) are used for local control, event sequencing, fault detection recording, slow monitoring and phase control of the mechanical high power shifters.

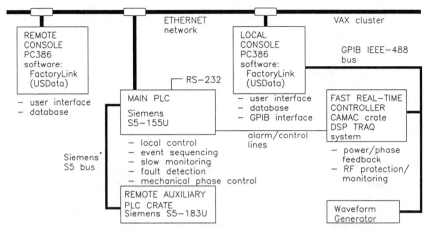

**Fig. 4**. Schematic of the control and feedback system components.

## CONCLUSIONS

The LH system for the Tokamak de Varennes has been designed for reliable and flexible operation in a quasi-continuous regime and should permit significant experiments on long-pulse diverted plasmas. Main features include a narrow spectrum at high $N_z$, that can be changed in real time, and a fast digital control system. The expected commissioning date is early 1992.

## ACKNOWLEDGEMENTS

The authors gratefully acknowledge valuable contributions from J.M. Guay, A. Dubé and L. Vachon for useful insight and technical support, V. Fuchs and I.P. Shkarofsky for the modified LH code results, D. Moreau from the CEA (DRFC-Cadarache, France) for the multijunction coupling code, M.M. Shoucri for accessibility calculations and H.D. Pacher and G.W. Pacher for stimulating discussions and support. The Centre canadien de fusion magnétique (CCFM) is managed and jointly funded by Hydro-Québec, Atomic Energy of Canada Limited (AECL), and the Institut de recherche scientifique (INRS).

## REFERENCES

[a]Hydro-Québec, Varennes, [b]MPB Technologies, Dorval, [c]U. de Montréal, [d]Canatom, Montréal.

1. F. Parlange et.al., Plasma Phys. and Contr. Nucl. Fusion Res. 1986, IAEA, Kyoto, Vol.1, p.563 (1987).
2. D. Moreau et.al., Appl. of RF Power to Plasmas, AIP Conf. Proc. **159**, 135 (1987).
3. N.J. Fisch, Rev. Mod. Phys. **59**, 175 (1987).
4. J.E. Stevens et.al., Nuclear Fusion **28**, 217 (1988).
5. V. Fuchs, I.P. Shkarofsky, R.A. Cairns and P.Bonoli, this conference.

# Simulations of Lower Hybrid Current Drive and Ohmic Transformer Recharge

V. Fuchs, I.P. Shkarofsky, R.A. Cairns[*] and P. Bonoli[†]

Centre canadien de fusion magnétique, Varennes, Québec, Canada J0L 2P0

### Abstract

The Bonoli-Englade code has been modified: a) to allow launching and evolution of narrow Brambilla spectra, b) PID feedback control of the plasma current is added, and c) a new $rf$ current diffusion model is presented. This is applied to the Tokamak de Varennes, in which it is planned to use lower-hybrid current drive to extend operation at full current for the entire duration (30 s) of the toroidal field. The modified code shows that 200 $kA$ of current could be maintained in a steady state with about 1 $MW$ of LH power at 3.7 $GHz$, and $n_o = 4$ x $10^{19}$ $m^{-3}$. With the same power, in an alternating ohmic-$rf$ scenario, the ohmic transformer could be recharged by 2 $Vs$ in about 4 $s$ at a slightly reduced density.

## 1   Introduction

Typical Tokamak de Varennes (T de V) parameters are $R_o = 0.86m$, $a = 0.24m$, and $B_o = 1.5T$, $I_o = 200kA$, $n_o = 4$ x $10^{19}m^{-3}$ and $T_{eo} = 0.8keV$. The planned $rf$ system supplies 1 $MW$ at 3.7 $GHz$ into a 32 waveguide grill producing a Brambilla spectrum, centered at $n_{\parallel} = 3.3$ (determined by the slow wave accessibility) with a narrow width of $\Delta n_{\parallel} = 0.3$[1].

To simulate lower hybrid current drive (LHCD) and transformer recharge with this spectrum, the original Bonoli-Englade[2,3] code was modified to allow an arbitrary $n_{\parallel}$ grill spectrum, a new $rf$ current diffusion equation, feedback stabilization of the plasma current through voltage or power control, and transformer recharging. For an ohmic plasma, control is through the induced voltage, for pure $rf$ current drive, control is through the $rf$ power source, and for $rf$-assisted transformer recharge, it can be either.

We side-launch typically 50 rays from within the Brambilla spectrum. With feedback control at 200 $kA$ and $n_o = 4$ x $10^{19}m^{-3}$, the ohmic equilibrium electron and ion temperatures are 0.8 and 0.5 $keV$, whereas with LHCD, we require 850 $kW$ resulting in 1.6 and 0.8 $keV$, respectively. Recharging the ohmic transformer by 2 $Vs$ in 4 $s$ requires about 1.3 $MW$.

In the next section, we present the new $rf$ current diffusion model equation. In the last section, we provide results on operation scenarios and feedback control.

---

[*]Permanent address: Department of Mathematical Sciences, University of St.Andrews, St. Andrews, Scotland KY16SS9

[†]Plasma Fusion Center, Massachusetts Institute of Technology, Cambridge, MA 02139, U.S.A.

# 2 Diffusion of RF Current

Along with Bonoli *et al.*[2,3] we consider the Fokker-Planck equation

$$0 = \frac{d}{dp_{\parallel}} D_{rf}(r, p_{\parallel}) \frac{dF_e}{dp_{\parallel}} + C(F_e) - \epsilon E_{\parallel} \frac{dF_e}{dp_{\parallel}} - \frac{F_e}{\tau_L} + \Gamma_s \delta(p_{\parallel}). \tag{1}$$

describing the parallel distribution function $F_e(p_{\parallel}) = 2\pi \int_0^{\infty} dp_{\perp} p_{\perp} f(p_{\parallel}, p_{\perp})$. A self consistent equilibrium between $F_e$ and the quasilinear diffusion coefficient $D_{rf}$ is established in an iteration process between solutions of Eq. (1) and power deposition calculations of a toroidally traced slow wave spectrum. The phenomenological loss term $F_e/\tau_L$ represents global loss from a particular flux surface due to radial diffusion. The loss time is given in Ref. 4 and is of the order of the bulk electron confinement time. This loss term is used in place of a spatial diffusion term to make the equation ordinary rather than partial, and thus keep the simulations code within reasonable CPU time limits.

In order to recover the radial diffusion of suprathermals, which affects the $rf$ current density profile, we derive a separate ordinary differential equation in space $r$, by taking the $p_{\parallel}$ - moment of Eq. (1), on the quasilinear plateau. This gives

$$\frac{dJ_{rf}}{dt} = S_{rf} + \frac{1}{r} \frac{d}{dr} \left( r D_J \frac{dJ_{rf}}{dr} \right) - \nu J_{rf}, \tag{2}$$

where $S_{rf}$ is the source term for the $rf$ current density $J_{rf}$, $D_J$ is the current diffusion coefficient, and $\nu$ is a loss rate related to collisional momentum transfer from the tail to the bulk of the distribution function.

The procedure we adopt here is not to calculate the transport coefficients $D_J$ and $\nu$ locally, as in Bonoli's and Englade's original treatment[1], but rather to choose them so that the global conservation predicted by the Fokker-Planck equation is preserved by the current diffusion equation. Our approach is to choose $\nu$ and $D_J$ constant over the cross section for simplicity in such a way that the integrated momentum destruction rate and the diffusion loss in (2) are equal to the corresponding integrated quantities from Eq. (1).

Hence if we integrate Eq. (2) over the cross-section we get

$$\Gamma_{rf} - \nu I_{rf} + 2\pi a D_J \frac{dJ_{rf}}{dr} \Big|_a = 0, \tag{3}$$

where $I_{rf}$ is the total $rf$ current, and $\Gamma_{rf}$ the integrated $rf$ source term, both obtained from Eq. (1). We therefore have to solve Eq. (2) under the boundary conditions ($J' \equiv dJ/dr$).

$$J'_{rf}(0) = 0, \quad J_{rf}(a) = 0, \quad J'_{rf}(a) = \frac{\nu I_{rf} - \Gamma_{rf}}{2\pi a D_J} \tag{4}$$

Since we have three conditions on a second order equation, we have an eigenvalue problem which determines the value of $D_J$ necessary for a self-consistent solution. To find the eigenvalue, we use a bisection method. The procedure is completely automated in the code.

The difference that the new diffusion model makes is illustrated in Fig. 1. These results were obtained before making changes in the code to allow more flexible $n_{\parallel}$-grids for the ray tracing/power-deposition/Fokker-Planck calculations. Better $n_\parallel$-resolution results in a correspondingly larger spread of $J_{rf}$ in configuration space.

# 3 Tokamak Operation Scenarios - Feedback Control

In order to maintain a constant plasma current $I = I_{ref}$, the error (the control variable) $dI = I_{ref} - I$, is fed back to the controlling variable, which is either the induced voltage $V_{OH}$, or the incident $rf$ power, $P_{rf}$, by PID (proportional-integral derivative) control. This is done in the tokamak circuit equation

$$\frac{d}{dt}(LI) + R(I - I_{rf}) = -\frac{d}{dt}(MI_{OH}) \equiv V_{OH}, \qquad (5)$$

where L is the plasma self inductance and M is the mutual inductance between the plasma torus and the ohmic circuit. The PID control is implemented by writing $V_{OH}$ or $P_{rf}$ in the form

$$C_p dI + C_i \int_0^t dI \; dt' + C_d d(dI)/dt \qquad (6)$$

For $I_{rf}$ we assume $I_{rf} \sim P_{rf}$, so that $I_{rf}$ in Eq. (5) can directly be written in the form of Eq. (6).

For an ohmic plasma ($I_{rf} = 0$), I is maintained by $V_{OH} > 0$. For an $rf$ plasma, with no recharge ($V_{OH} = 0$), I is controlled by $P_{rf}$ which provides the necessary $I_{rf}$. With additional transformer recharge ($V_{OH} < 0$), $I_{rf}$ has to be larger, and control is either by $V_{OH}$ or $P_{rf}$, with the other one fixed at a desired value.

As a first example, in Figs. 2a,b we show a controlled I by means of $V_{OH}$ tending to about $-0.2\ V$, at fixed $P_{rf} = 1MW$. Under these conditions, recharge to 2 $Vs$ will therefore take about 10 $s$.

Next we look at two cases, in Figs. 3a,b,c with I controlled through $P_{rf}$ with $V_{OH}$ held constant. The first case with $V_{OH} = 0$ is pure LHCD. The second, with $V_{OH} = -0.5$, shows that about $P_{rf} = 1.4MW$ is necessary for recharge by 2 $Vs$ in 4 $s$.

# 4 Acknowledgments

It is a pleasure to thank members of the CCFM $rf$ group for useful and stimulating discussions in the course of this work. We are deeply indebted to Dr. M. Gavrilovic for informative discussions on the basics of feedback control. This work was supported in part by the CCFM and in part by DOE Contract No. De-AC02-78ET51013. The CCFM is jointly managed and funded by Hydro-Québec, Atomic Energy of Canada Limited, and the institut national de la recherche scientifique.

# References

1  R. Decoste, Y. Demers, V. Glaude, and A. Hubbard, J. Bugdao, G.-A Chaudron, R. Annamraju, this conference.

2  P.T. Bonoli and R.C. Englade, Phys. Fluids 29 (1986) 2937.

3  P.T. Bonoli, M. Porkolab, Y. Takase, and S.F. Knowlton, Nuclear Fusion 28, (1988) 991.

4  H.E. Mynick and J.D. Strachan, Phys. Fluids 24 (1981) 695.

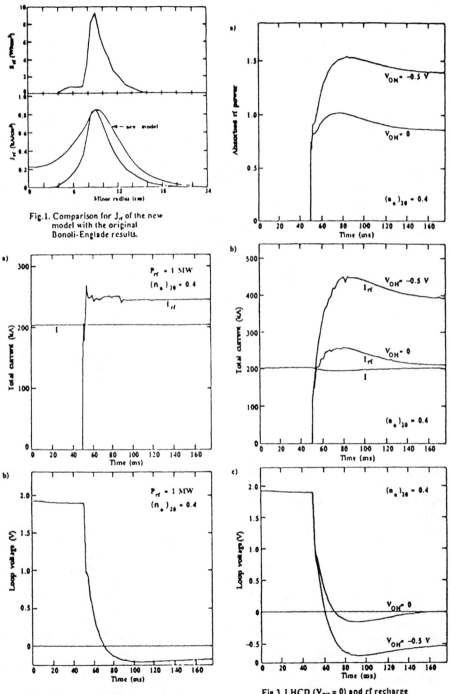

Fig.1. Comparison for $J_{rf}$ of the new
model with the original
Bonoli-Englade results.

Fig.2. Rf recharge controlled by the
induced voltage source.

Fig.3. LHCD ($V_{OH} = 0$) and rf recharge
($V_{OH} = -0.5$) controlled by the
rf power $P_{rf}$.

## FIRST OPERATION OF MULTIJUNCTION LAUNCHER ON JT-60

Y.IKEDA, T.IMAI, K.USHIGUSA, M.SEKI, K.KONISHI, O.NAITO, M.HONDA, K.KIYONO, S.MAEBARA, T.NAGASHIMA, M.SAWAHATA, K.SUGANUMA, K.UEHARA, K.YOKOKURA and JT-60 team
Japan Atomic Energy Research Institute,

### ABSTRACT
The multijunction launcher was designed to improve the directivity of the launched waves with a narrow wavenumber spectrum by dividing the main waveguide into three of secondary waveguides. The coupling agrees fairly well with the theory and the power of more than 2.75 MW has been coupled to a plasma in a few week conditioning. The dependenc of current drive efficiency on the wavespectrum from the multijunction launcher agrees well qualitatively with the Fisch's theory. The current drive efficiency of the multijunction launcher was improved by 40 % compared with the previous conventional launcher on JT-60.

### 1.INTRODUCTION
The development of steady-state current drive would considerably enhance the prospects for a tokamak fusion reactor. Lower hybrid current drive is one of the most promising method to achieve a steady-state tokamak. Another application of lower hybrid current drive is the control of the plasma current profile to improve MHD stability and confinement in tokamaks. So the main objective of LHRF experiment is to optimize the wave spectrum for each item. Recent experimental results have shown that the current drive efficiency was improved by sharpening wave spectrum on PLT[1] and that a good qualitative agreement exists between the Fisch's theory[2] and the experimental data. The theoretical current drive efficiency is given by $\eta_{CD}{}^{th} = K \cdot G$, where $K = (42.45 + 1.954 \cdot Z_{eff})$ and G is defined as follow[3],

$$G = \frac{1}{(N_{\parallel}\min)^2} \cdot \frac{1-(N_{\parallel}\min/N_{\parallel}\max)^2}{\ln(N_{\parallel}\max/N_{\parallel}\min)} \cdot \frac{P_{acc}}{P_{LH}}$$

$N_{\parallel}\min$, $N_{\parallel}\max$ are the minimum and maximum of the effective part of the parallel refractive index spectrum, respectively, and $P_{acc}/P_{LH}$ is the directivity and the accessible power ratio of the launched waves. So a large number of waveguides in toroidal direction are required to launch a narrow wave spectrum $((N_{\parallel}\max/N_{\parallel}\min) \rightarrow 1)$.

A multijunction launcher was proposed in order to increase the number of the waveguide with simplified modification of the launching system[4]. The JT-60 has three LHRF launching systems and has achieved a current drive efficiency of $2.65 \times 10^{19} m^{-2} A/W$ and $2.9 \times 10^{19} m^{-2} A/W$ without and with NBI at $N_{\parallel}$peak of 1.0 - 1.7[5]. And improvement of confinement has been observed at broad plasma current profile by launching the wave spectrum of 1.7 through the conventional launchers, which are composed of a phased waveguides array of 8 columns in the toroidal direction by 4 rows in poloidal direction[6]. We aimed at improving the current drive efficiency and confinement by launching a narrow and highly unidirectional spectrum through a new multijunction type launcher.

We report the overview of this multijunction launcher and the first results of experiments on the JT-60 using this launcher.

## 2.OVERVIEW OF LAUNCHER

From experimental view points, the launching system needs a high controllability of $N_{\parallel}$ to optimize the spectrum. So a large number of waveguides in toroidal direction are required for a launching system. The launching system of the JT-60 features to be separable at 3 m far from the launcher mouth to make it easy to modify the grill part. We modified the grill part from the conventional grill to the multijunction one. The divided number of three in the transmitted waveguide was adopted to sharpen the wave spectrum, and the pitch of the waveguide was constrained by the requirement to connect the conventional launching element. The $N_{\parallel}$peak is given by $30/(f \cdot \delta) \cdot \theta /2\pi$, where f is the frequency(GHz), $\delta$ pitch (cm) and $\theta$ the phase angle. Since We have already had a conventional launcher of $N_{\parallel}$peak of 1.7 for current drive experiment, we selected $N_{\parallel}$peak of 2.2 to obtain more flatten plasma current profile by LHCD. A geometric phase angle $\theta$ of 70° was adopted to launch the $N_{\parallel}$peak = 2.24, when the phase difference between adjacent module was 70°. Since the launching system has 8 modules in the toroidal direction, the wave spectrum can be controllable to select the best value for current drive experiment from 1 to 3.5 with a high directivity as shown in Fig. 1. The multijunction module was manufactured by welding 3 secondary waveguides with geometric phase shifter. A taper waveguide was adopted to match at the geometric phase shifter in each secondary waveguide. The length of the module and the width of the waveguide were limited by the technical requirement of copper plating and welding. The resulting module is shown in Fig. 2. Manufacturing tolerances were ±0.2mm and ±0.2mm on height and width of the waveguide at grill mouth, respectively. The phase shift and power balance were measured on every module and were about 70±10° and 1/3± 1/15, respectively. These discrepancies may be due to the deformation at the welding point in the waveguide. Since the the number of waveguide of 24 in the toroidal direction is much enough to average these discrepancies, the Fourier analysis of the manufactued launcher's spectrum shows that the deformation of the spectrum is little compared with designed spectrum.

## 3. EXPERIMENT

Before the experiment, the launcher was conditioned by only high power RF injection into the vacuum vessel of the JT-60 without plasma, and with Tayor Discharge Cleaning (TDC) plasmas for two days. In the case of without adequate density plasma ( vacuum or TDC plasma ), the temperature rise of about 4.5°C/MW·s at the geometric phase shifter was observed due to a high Voltage Standing Wave Ratio (VSWR) at the secondary waveguides without plasma loading. So the conditioning in this case was performed under high RF field at the secondary waveguides. Following these procedures, a current drive experiment up to 2 MW was carried out after only 30 plasma shots. This conditioning time up to 2 MW was about one-fifth shorter than that of conventional current drive one. We suppose the RF conditioning without adequate plasma is much effective on the multijunction launcher due to high VSWR at secondary waveguides. At present, the maximum injected power reached at 2.75MW.

The measured coupling agreed fairly well with the theoretical one as shown in Fig.3(a). The theory curve assumes an edge density of $5 \times 10^{16}$ $m^{-3}$, a density gradient of $5 \times 10^{19}$ $m^{-4}$ and a vacuum gap of $2 \times 10^{-3}$ m. The temperature rise at the geometric phase shifter shows that a high VSWR stands at high reflection condition in Fig.3(b). This data indicates that a good coupling should be necessary from the view point of not only injection of a high power and designed spectrum into plasmas but also decreasing the rf loss on the grill part in the multijunction launcher.

The current drive efficiency with the multijunction launcher was studied by means of $N_{/\!/}$peak scans from 1 to 2.5. As the plasma current is usually maintained by the feedback system in the JT-60, we determine the current drive efficiency from the data of one-turn loop voltage of approximately zero. Figure 4 shows the current drive efficiency dependent on the $N_{/\!/}$peak at $I$ p = 1 MA. The closed and open data show the experimental $\eta_{CD}$ for the multijunction and the conventional launchers in the same target plasma, respectively. A maximum $\eta_{CD}$ of 2.8 was obtained for the multijunction launcher with the spectrum of $N_{/\!/}$peak = 1.3 at n e = $0.8 \times 10^{19}$ $m^{-3}$ and the $\eta_{CD}$ for the multijunction launcher is about 1.4 time larger than that of the conventional one at $N_{/\!/}$peak = 1.7. This data shows the LH wave with narrow and high directivity spectrum gives a improvement in the current drive efficiency. The reason of the drop in $\eta_{CD}$ at $N_{/\!/}$peak = 1.1 is that wave accessibility limits the minimum spectrum. The solid line in this figure shows the theoretical current drive efficiency of the multijunction launcher calculated from Fisch's theory on the assumption of $Z_{eff}$ = 2 and $N_{/\!/}$acc = 1.3. We also obtained a higher efficiency of 3.4 at a higher plasma current of 1.74 MA at $N_{/\!/}$peak = 1.3. The experimental data qualitatively agreed with the theoretical one. These results clearly show that the fastest wave spectrum with a high directivity, which is accessible into the plasma, is optimum for current drive and also that the launched waves from the multijunction launcher have the calculated spectra as shown in Fig. 1.

## 4. CONCLUSION

A high efficiency current drive of $\eta_{CD}$ = 3.4 have been obtained by using a multijunction launcher. A good agreement between the Fisch's theory and the data from the grill phasing shows that a narrow wavenumber spectrum at the lowest possible $N_{/\!/}$ which can access to the plasma core is the best value to drive a plasma current efficiently. Moreover the multijunction launcher leads to the improved grill conditioning. High availability of the multijunction launcher for current drive have been validated on the JT-60, and therefore such a launcher may be considered for use in the steady-state tokamaks.

## ACKNOWLEDGEMENT

We thank Drs. M.Tanaka, T.Iijima, S.Tamura, Y.Tanaka and M.Ohta for their continued support. We also thank Drs. M. Brambilla and F.X.Söldner for their contribution to the modification of the coupling code and thank Drs. C.Gormezano, A.S.Kaye and G.Tonon for their useful discussion on the multijunction launcher.

REFERENCES
[1] STEVEN,J.,BELL,R.,BERNABEI, S.,et al., Nucl.Fusion 28(1988)217.
[2] FISCH,N.,Phys.Rev.Lett.41(1978)873.

[3]TONON,G.,Plasma Phys. Controll. Fusion 26(1984)145.
[4]NGUYEN,T.K.,MOREAU,D.,Fusion Tech.,2(1982)1381.
[5]USHIGUSA.K.,IMAI,T., et al.in Plasma Physics and Contr. Nuclear Fusion
    Research 1988(Proc. 12th Int. Conf. Nice,1988) Paper IAEA-CN-50/ E-3-1
[6]IMAI,T.,USHIGUSA.K., et al., Nucl.Fusion 28(1988)1341.

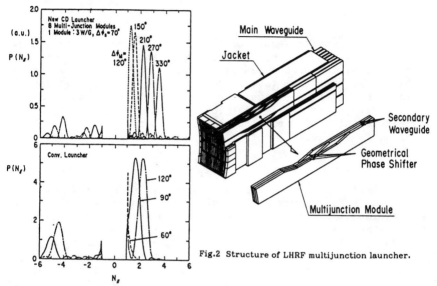

Fig.2 Structure of LHRF multijunction launcher.

Fig.1 $N_{\parallel}$ spectra of a multijunction launcher and
of a conventional current drive launcher.

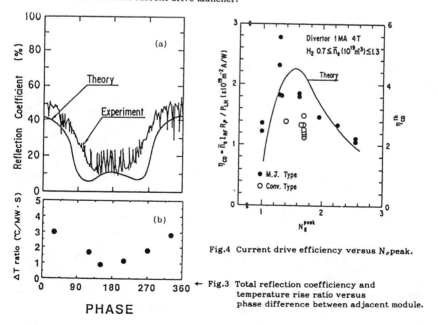

Fig.4 Current drive efficiency versus $N_{\parallel}$peak.

← Fig.3 Total reflection coefficiency and
temperature rise ratio versus
phase difference between adjacent module.

# ELECTRON BEHAVIOR DURING FWCD EXPERIMENT ON JFT-2M

H. Kawashima and JFT-2M group
Japan Atomic Energy Research Institute,
Naka-Machi, Naka-Gun, Ibaraki, 311-02, Japan

## ABSTRACT

Current drive experiments using 200 MHz fast waves have been carried out on the JFT-2M tokamak. Strong electron heating at the plasma center by high-Nz ($7\sim8$) fast waves has been observed. The fast wave heating efficiency is improved when the plasma electrons are preheated by ECH. Indications of fast wave current drive for low values of Nz ($4\sim5$) come from measurements of anisotropic x-ray emission, for which the presence of energetic electrons produced by ECH is essential.

## INTRODUCTION

Recently, fast wave current drive (FWCD) in hot, dense plasmas has been studied both theoretically[1,2] and experimentally[3,4] Fast waves can penetrate into the plasma core and are absorbed mainly by electrons through Landau damping. FWCD experiments at 200 MHz have been carried out in the JFT-2M tokamak. The refractive index (Nz) was chosen to lie in the range $3\sim8$ to permit strong absorption. Measurements of the anisotropy of forward, backward, and perpendicular x-ray emission [5,6] were made in order to check for the presence of fast wave-induced current.

## EXPERIMENTAL SETUP

The target plasma was circular or D-shaped, as defined by inner or outer graphite limiters located in the torus mid-plane. Parameters of the fast wave system, which comprises a 4-loop phased antenna array, are: f=200MHz, Nz=3(8) and $V_{ph}/V_{te}$=30(3) for $\triangle \phi = \pi/2$ ($\pi$), and $P_{max}$=800 kW. X-ray anisotropy measurements were performed using pulse height analyzers. By changing the direction of the plasma current, forward and backward tangential emissions were obtained using one single-channel HPGe system. Perpendicular emissions were measured along a vertical plasma chord with a 5-channel HPGe system[7]

## ELECTRON HEATING BY FAST WAVES

Strong electron heating at the plasma center was observed when fast waves were incident on OH+ECH-preheated plasmas. The temporal evolution of loop voltage (Vl) and electron temperature (Te(r/a=0, +0.4and -0.7)) is shown in Fig. 1. These measurements were obtained with phase difference $\triangle \phi = \pi$ and resultant parallel refractive index Nz=8. Electron temperatures were obtained from soft x-ray pulse height

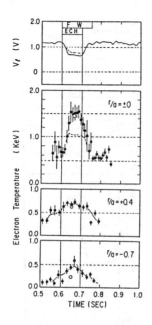

Fig. 1 Time evolution of electron temperature at r/a=0, +4 and -0.7, and of loop voltage during fast wave heating with high Nz. The solid and broken lines correspond respectively to OH+ECH+FW and OH+ECH. Electron temperature is measured using soft x-ray pulse height analysis ( ● ) and Thomson scattering ( ○ ). Plasma current, averaged electron density, and toroidal magnetic field were constant ( Ip=170kA, $\bar{n}_e$ =1.5X10 $^{19}$ m $^{-3}$, and $B_{to}$=1.07T ). Injected power, parallel refractive index and phase difference of the fast waves are : P $_{FW}\sim$400kW, Nz=8, and $\triangle \phi = \pi$, respectively.

analysis (closed circles) and from Thomson scattering measurements (open circles), Te(r/a=0) rises from 1 keV (ECH-heated plasma) to 1.5 keV during the fast wave pulse. Te(r/a=+0.4 and -0.7) also increases but the relative increase is much smaller than that for Te(r/a=0). The loop voltage drops from 0.75 V to 0.65 V as a result of fast wave electron heating. The heating efficiency is $\sim$1.7 X10 $^{19}$m $^{-3}$eV/kW. The absorption efficiency of 20 % is also estimated. These results indicate that fast waves can penetrate deeply into the plasma and can be strongly absorbed at the plasma center, where Te and ne are large. The heating efficiency also depends on Nz and on the electron temperature of the target plasma. These results agree with theoretical predictions.

## ESTIMATION OF FWCD FROM MEASUREMENT OF X-RAY EMISSION ANISOTROPY

Efficient current drive is expected from fast waves with low Nz. The absorbed rf power decreases exponentially with decreasing Nz. We investigated fast wave current drive with $\triangle \phi$ =3/4 $\pi$ and Nz=5, using an OH+ECH preheated, low density target plasma in presence of energetic electrons due to ECH to enhance the interaction between fast waves and electrons. Fig. 2 shows the typical time evolution of plasma parameters. The solid line corresponds to counter-clockwise plasma current, Ip, i.e. the tangential x-ray detector registers forward emission from the plasma electrons. Backward emission is indicated by the broken line. The macroscopic parameters (Vl, $W_{DIA}$ , $\bar{n}e$, $T_{r-ECE}$ ) are obtained without consideration of the Ip-direction. The parallel photon emission in the 8 ~30 keV energy range, presented in the bottom figure, shows a dependence on Ip-direction. Forward emission (closed circles) is about twice as large as backward emission ( open circles ). On the other hand, the perpendicular photon emission is unchanged.

Fig. 3 shows the x-ray energy spectra obtained for each of the 3 viewing directions under the conditions of Fig. 2. The amplifier gain for the tangential detector is reduced and the lower-energy x-ray region (up to 7 keV) is cut off using a thin Al foil in order to bring out the tail component (Fig. 3-(a),(b)). The perpendicular energy measurement corresponds approximately to the thermal region (Fig. 3-(c)). The solid line corresponds to the OH+ECH+FW case, and the broken line to OH+ECH. The broken line indicates ECH production of an energetic electron tail. The tail temperature in broken line shows isotropic heating by ECH since $T_{//F}^* \sim T_{//B}^*$. When fast waves are incident on the plasma, the forward-direction photon counts in the 10 ~30 keV energy range (corresponding to the

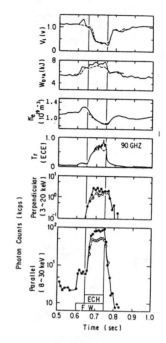

Fig. 2 Time evolution of plasma parameters during measurement of anisotropic x-ray emission. The solid line shows forward emission. Other parameters are : $P_{FW} \sim 400kW$, Nz=8, $\triangle \phi = 3/4 \pi$, Ip=100kA, and $B_{to}=1.07T$.

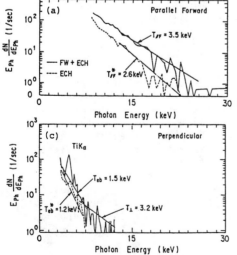

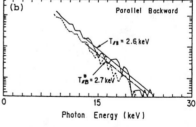

Fig. 3 X-ray energy spectra for forward(a), and backward (b), and perpendicular(c) directions during OH+ECH+FW ( solid line) and during OH+ ECH (broken line).

resonance energy region for the fast waves) increase and $T_{//F}$ increases to 3.5 keV from 2.6 keV. Backward-direction photon counting rates increase slightly and $T_{//B}$ remains unchanged. For the perpendicular measurements, both thermal and tail temperatures ($T_{eb}$, $T_\perp$) rise during the fast wave pulse.

As shown in Figs. 2 and 3, ECH can produce energetic electrons under low-current and low-density conditions ($Ip=100$ kA, $\bar{n}e=1\times10^{19}m^{-3}$). When fast waves are incident on such a plasma, anisotropic x-ray radiation, strongly enhanced in the forward direction, is observed. This observation indicates that the fast waves produce a uni-directional acceleration of resonant electrons, i.e. an indication of the possibility of fast wave-driven current. However appreciable drop of loop voltage due to the fast wave current drive is not observed. It may be the skin time effect. X-ray emission anisotropy is not observed in target plasmas having no energetic electrons.

In the future the electron distribution functions used for comparing the theoretical spectra with the experimental data will be calculated using a bounce-averaged Fokker-Planck equation with a quasilinear diffusion term.

## ACKNOWLEDGMENT

The authors would like to thank members of the JFT-2M operation and heating groups for their technical assistance. Continual encouragement from Drs. S. Simamoto, M. Tanaka and M. Yoshikawa is also gratefully acknowledged.

## REFERENCES

1. K. Theilhaber and A. Bers, Nucl. Fusion, 20, 547(1980).
2. F. Karney and J. Fish, Phys. Fluid, 28, 116(1985).
3. T. Yamamoto et al., in Proc. 12th Int. Conf. on Plasma Physics Controlled     Nucl. Fusion, Nice(1988), IAEA-CN-50/E-4-6.
4. Y. Uesugi et al., in proceeding of16th Europian Conference, 13B-IV, 1259(1989).
5. S. Von Goeler et al., Nucl. Fusion, 25, 1515(1985).
6. H. Kawashima et al., Japan Atomic Energy Research Institute Report, 87-157(1987).
7. H. Kawashima et al., Rev. Sci. Instrum., 59, 1816(1988).

# MEASUREMENT OF SUPRATHERMAL CURRENT CARRIERS AND MOMENTUM RELAXATION BY ELECTRON CYCLOTRON TRANSMISSION

R. K. Kirkwood, I. H. Hutchinson, S. C. Luckhardt, and J.P. Squire
Plasma Fusion Center, MIT, Cambridge, MA 02139

## ABSTRACT

A new diagnostic of suprathermal electron distributions in tokamaks has been developed at Versator II. It is found that the attenuation of obliquely propagating X–mode radiation with $\omega \lesssim \omega_{ce}$ provides a direct measure of the parallel distribution function ($F(p_{\parallel})$). The effects of refraction are eliminated by making a differential attenuation measurement for two beams with equal and opposite launch angle. The asymmetric (current carrying) part of the distribution is then determined. Results show that lower hybrid current drive (LHCD) increases both the number of particles in the 4 to 40 keV range as well as their parallel temperature. The LHCD is then shut off and the population is seen to decay in 1.2 ms ($< \tau$ collisional).

## INTRODUCTION

The production of a population of suprathermal current carriers is essential for the efficient maintenance of toroidal currents by lower hybrid waves [1]. The ultimate efficiency of current drive is determined by the details of the momentum distribution and loss processes of these particles. Calculation of the equilibrium efficiency and momentum distribution relies on knowledge of both the internal wave spectrum and momentum relaxation processes [2]. Several efforts have been made to determine the distribution of suprathermals during LHCD by measurement of the emitted x–ray [3,4] and cyclotron radiation [5,6] spectra. The determination of the distribution function from such measurements requires the assumption of a parameterized model which is fit to the measured spectra. In this paper a technique to directly determine of the parallel distribution of current carriers from electron cyclotron transmission measurements [7] is discussed and results from the current carrying distributions produced by LHCD in the Versator II tokamak are presented.

## CYCLOTRON ABSORPTION THEORY

In this section it is shown that measurements of cyclotron absorption in the vicinity of $\omega = \omega_{ce}$ are directly related to the perpendicular moment of the distribution function of the current carriers ($F(p_{\parallel}) \equiv \int_0^{\infty} p_{\perp} f(p_{\perp} p_{\parallel}) \, dp_{\perp}$), for parameters relevant to LHCD. To understand the wave absorption mechanism consider the condition for wave particle interaction near the cyclotron frequency.

$$\gamma - \omega_{ce}/\omega - N_{\parallel}p_{\parallel} = 0 \qquad (1)$$

Here $\gamma$ is the relativistic factor, $\omega_{ce}$ is the electron cyclotron frequency, $N\|$ is the component of the wave index of refraction parallel to the magnetic field, and $p\|$ is the parallel component of the particle momentum normalized to mc. This condition defines an elliptical contour in momentum space, shown in figure 1, along which the particles must lie in order to interact with the wave. For the present case the particles exist primarily in the region of $|p\|| < .5$ and $|p_\perp| < .25$. In order to interact with these particles the parallel index of refraction is chosen to be 0.4 to and $\omega_{ce}/\omega$ is chosen to be 1.05 to 1.3. The contours for this case are shown as solid lines in figure 1, while dotted lines show the more common case of perpendicular propagation [5,6]. The advantage of the finite $N\|$ contours is that the low $p\|$ region can be accessed more easily. These contours intersect the region of interest nearly perpendicular to the $p\|$ axis which allows a nearly one to one correspondence between $\omega_{ce}/\omega$ and $p\|$. This can be understood from Eq. 1 by recognizing that the doppler term is dominant ($N\|p\| > \gamma-1$) so that the resonance condition is not very sensitive to $p_\perp$.

The absorption rate for waves is determined by the anti–Hermitian parts of the plasma dielectric tensor which are expressed in terms of integrals over these contours [8]. For distributions with $|p\|| \lesssim 1$ the integrands are localized around the low $p\|$ point of intersection of the contour with the $p\|$ axis ($p\|_0$). It has been shown [7] that under conditions reasonable for the distributions in Versator ($p\| >> v_{the}/c$, $k_\perp\rho_e << 1$, $|d(\ln(F(p\|)))/dp\|| < 1/p\|$, and $p_\perp << \sqrt{p\|}$) that the integrals can be expressed in terms of cold plasma factors and the perpendicular moment of the suprathermal distribution function ($F(p\|)$) evaluated at $p\|_0$. An analytic relation between the wave dampening rate and $F(p\|_0)$ results and can be used to diagnose the distribution function. Since the measured parameter is the attenuation of waves traversing the plasma, a ray path of constant $\omega_{ce}$ and $N\|$ is necessary to allow resolution in $p\|$ by way of the resonance condition, Eq. 1. The path of constant major radius, and hence constant B and $k\|$, is provided by a vertical launch in the tokamak as shown in figure 2. Waves are launched along these paths and their attenuation is measured relative to the vacuum case.

In addition to absorptive effects, wave attenuation is produced by the divergence of wave energy induced by refraction. To isolate the absorptive effects attenuation is measured along two ray paths with equal and opposite launch angle and hence $N\|$. The refractive effects arise from the magnitude of $N_\perp$ and $N\|$ which is the same for both rays. The wave particle interaction for the oppositely directed wave takes place with oppositely directed particles. Thus the ratio of the attenuation of the two beams is directly related to the asymmetric part of the distribution $F_a(p\|_0) = 1/2 \, (F(+p\|_0) - F(-p\|_0))$. This is the quantity of interest in current drive experiments since it is the distribution of current carriers.

## RESULTS FROM LHCD PLASMAS

This diagnostic has been used to study LHCD plasmas in the Versator II tokamak. The current drive system is a four wave guide array launching 22 kW of r.f. power at 2.45 GHz. The wave E field is aligned to the toroidal B field to excite slow waves preferentially. The guides are separated in phase by $90^0$ so that the $N\|$ spectrum peaks at 2 and falls to 10% of the peak power at 4.2. The discharge was initiated with inductive drive after which the transformer is open circuited and the plasma current decays with a steady one turn loop voltage of .55 V. The r.f. power is then initiated and the current decay is maintained at 17 kA for the remainder of the r.f. pulse, while the loop voltage falls to $\leq$ .1 V in $\simeq$ 2 ms and remains low. The r.f. power is maintained for 9 ms after which time the current decays and the loop voltage reaches a steady value of .65 V in $\simeq$ 1.2 ms. The cyclotron transmission rates are measured at 5 frequencies between $\omega_{ce}/\omega = 1.05$ and 1.3 once each millisecond from 3 ms before the initiation of the r.f. to 2 ms after the termination. During this time the plasma line averaged density and magnetic field are nearly constant at 3.2 e12 /cm$^3$ and 1.25 T respectively. The time evolution of $F_a(p\|)$ is shown in figure 3. These measurements are averaged over six identical discharges and two time periods. The error bars represent the standard deviation of the six discharges. Increases in the number of particles in response to the r.f. are seen at each energy with the largest proportional increase at the highest energies indicating an increase in parallel temperature. The velocity moment of these distributions provides a measure of the current carried by the particles in this energy range as shown in figure 4. The current carried by the suprathermals rises from an initial value of 7 kA to a peak of 14 kA during the r.f. pulse. Termination of the r.f. causes the current to relax to $\leq$ 3 kA in 1 ms. The time dependence of the distribution in response to the r.f. termination is obtained with greater time resolution by essentially varying the frequency between discharges. The result is shown in figure 5 where each point is an average over 16 independent measurements made in a series of 24 identical discharges. The velocity moment is found to relax with a characteristic time of 1.2 ms. Calculations of the momentum relaxation time are currently being carried out including the effects of the relaxation induced electric field. Preliminary indications are that a finite particle lifetime is necessary to explain the relaxation data.

## CONCLUSIONS

Electron cyclotron transmission measurements in Versator have provided the first direct measure of the energy distribution function of the current carriers during LHCD. Most (62% to 82%) of the plasma current is found to be carried by particles in the energy range 4 to 50 keV though the energy limit due to wave accessibility is much higher (125 keV). Measurements have also determined the time response of the distribution on both the formation and relaxation time scale. The relaxation data is the first measure of the momentum confinement time of the suprathermal current carriers produced during current drive.

## REFERENCES

1. N. Fisch, Rev. Mod. Phys. Vol 59, 175 (1987).
2. P. T. Bonoli, Proc. 7th Top. Conf. on R. F. Power in Plasmas Kissimmee Fl. 85 (1987).
3. S. von Goeler et. al., Nucl. Fus. 25, 1515 (1985).
4. S. Texter et. al., Nucl. Fus. 26, 1279 (1986).
5. K. Kato, and I. H. Hutchinson, Phys. Rev. Lett. 56, 340 (1986).
6. S. K. Guharay, and D. A. Boyd, Nucl. Fus. 27, 2031 (1987).
7. R. K. Kirkwood, et. al. Submitted to Nucl. Fus.
8. I. Fidone et. al. Phys. Fluids 25, 2249 (1982).

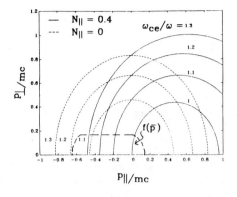

Fig. 1    Resonance contours in $\bar{p}$ space for $N_\parallel = 0$ and 0.4 (oblique launch).

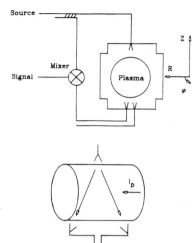

Fig. 2    Experimental apparatus for two beam measurement.

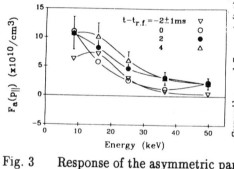

Fig. 3    Response of the asymmetric part of $F(p_\parallel)$ to LHCD.

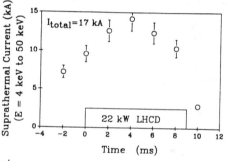

Fig. 4    Suprathermal current measured by cyclotron transmission.

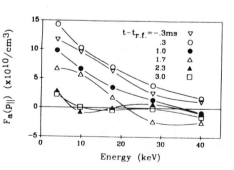

Fig. 5

Relaxation of LHCD induced $F_a(p_\parallel)$.

# SCATTERED DAMPING OF LH FAST WAVES BY DENSITY FLUCTUATIONS IN CCT[*]

K. F. Lai, B. D. Fried and R. J. Taylor
University of California, Los Angeles, CA 90024

## ABSTRACT

Anomalous damping of the LH fast waves has been observed in the UCLA CCT tokamak over a wide range of frequencies and magnetic fields during high power fast wave current drive experiments. In order to identify the wave damping mechanisms, insertible magnetic dipole antennas have been used to launch test waves locally. The damping coefficients of these waves have been determined using detailed poloidal field maps at various distances from the antennas. Lorentzian shaped sidebands have been excited strongly near the fundamental and harmonics of the driver frequencies and correlated very well with the local density fluctuation spectra. The anomalous damping of the fast waves can be attributed to the wave scattering by density fluctuations. The experimental results show good agreement with an approximate wave scattering theory developed by S.N. Antani and D.J. Kaup. Generalization of the scattering theory to a full-wave calculation will be presented.

## INTRODUCTION

Anomalous damping of the lower hybrid fast waves (LHFW) has been observed in the UCLA CCT tokamak during both the high power current drive and small signal test wave experiments[1]. The fast wave damping is 10-100 times more than expected from either Landau or collisional damping. Ray tracing studies[2] have further confirmed the necessity to exist other damping mechanisms to account for the rapid energy loss of the fast waves, especially near the edge of the plasma.

A 3D magnetic probe is used as the major diagnostic of our studies of the waves. Strong sideband generation near the fundamental and harmonics of the RF driver frequencies has been observed. The spectra of the sidebands (Fig.1a) are Lorentzian and correlate very well with the local density fluctuations (Fig.1b) measured by Langmuir probe. At low power levels, the sideband spectra are independent of power and frequencies which suggest scattering rather than parametric processes are responsible for its generation.

Insertible magnetic dipole antennas have been used to launch the fast waves from the center of the tokamak. Because of the finite group angle ($<20°$) of the LHFW, the energy of the waves are essentially guided along the magnetic field lines. The wave damping coefficient can then be determined by detailed poloidal field maps at various distances (Fig.2,3) from the antenna to eliminate geometric distortions.

---

[*]Supported by USDOE Contract DE-FG03-86ER53225.

## SCATTERING THEORIES

There exist at least several different scattering theories[3,4] which may explain the damping of the fast waves. We have chosen the theory developed by S.N. Antani and D.J. Kaup[5] to compare with our experiments mainly because of its simplicity. There are, however, several problems in the original paper that need to be addressed.

An incorrect approximation of $Im(n_\perp) \approx [Im(n_\perp)]^{\frac{1}{2}}$ has resulted in the overestimation of the wave damping by a factor of $10^4$-$10^5$ in the final solution (eqn. 43). Upon correction, we are able to get good agreement between the theory (Fig.4) and the measured damping length (~30cm). Because of the long parallel wavelength of the LHFW, the uncertainty in our measured $n_\parallel$ are quite large and limits the accuracy of our comparison.

At small $n_\parallel$, the fast wave approximation used by the original paper is no longer valid. We have examined the theory in details and found no intrinsic problem to generalize the theory to full-wave calculations. Because of the complexity of the matrix elements, numerical techniques are likely required to solve the full-wave dispersion relation.

In Antani's theory, the wave energy was considered lost once it becomes incoherent due to scattering. This mechanism is similar to the phase mixing process in Landau damping. However the "mystery" is where the energy eventually will end up. A possible answer is the increased density fluctuations of the plasma. This conjecture is consistent with the previous experimental observations by K.L. Wong and P.M. Bellan[6] for the lower hybrid waves.

## CONCLUSIONS

Scattered damping by density fluctuations is a viable explanation for the observed anomalous damping of the LHFW in the current drive experiments. This damping mechanism can prevent the low $n_\parallel$ waves to penetrate to the center of the plasma and be responsible for the plasma filtering of the antenna spectrum. This can mean trouble for the LHFW current drive experiments, if the problem is not addressed properly in the antenna design. There is still many unanswered questions which required better scattering theories and experimental measurements to be resolved in the future.

## REFERENCES

1. K.F. Lai, T.K. Mau, B.D. Fried, R.J. Taylor, Applic. of RF Power to Plasmas (7th Topical Conf., Kissimmee, FL 1987), AIP, New York, 111 (1987).
2. T.K. Mau, K.F. Lai, R.J. Taylor, Applic. of RF Power to Plasmas (7th Topical Conf., Kissimmee, FL 1987), AIP, New York, 207 (1987).
3. P.L. Andrews, Phys. Rev. Let. 54, 2022 (1985).
4. D.R. Thayer and A.N. Kaufman, Bull. Am. Phys. Soc. 33, 2015 (1988).

152

5.  S.N. Antani and D.J. Kaup, Phys. Fluids <u>27</u>, 1169 (1984).
6.  K.L. Wong and P.M. Bellan, Phys. Fluids <u>21</u>, 841 (1978).

FIGURE CAPTIONS

1.  Comparison between RF spectra and low frequency spectra:
    a.  RF sideband spectrum measured by magnetic probe at 21MHz.
    b.  Low frequency density fluctuation spectrum measured by Langmuir probe at 0MHz.
2.  Toroidal field map of fast wave Mod(B) amplitude. Dipole antenna is 45° away toroidally.
3.  Radial wave amplitude at 3 different distances from the dipole antenna.
4.  Comparison between experimental and theoretical attenuation length vs $n_\parallel$ for CCT parameters (f=21MHz, B=230 Gauss, $n=10^{12}cm^{-3}$).

F = 21 MHz **Fig. 1a**          F = 0 MHz          **Fig. 1b**

MOD(B) Profile
Dipole Antenna
$\phi$ = 45°
R = 153.5cm
$\alpha$ = 0°
He
F = 21MHz
B = 230 Gauss

$\beta = -5°$
$\beta = 0°$     **Fig. 2**
$\beta = 5°$

RADIAL POSITION (CM)

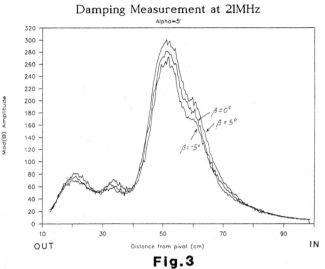

Damping Measurement at 21MHz

Fig.3

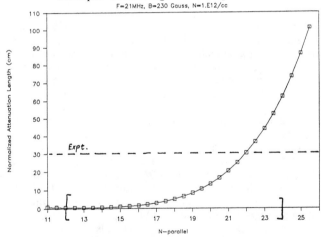

Multiple-Scattering for CCT parameters

Fig.4

# HIGH-POWER TEST OF LOWER HYBRID VACUUM WINDOW

## S. MAEBARA, Y. IKEDA, K. SAKAMOTO, T. NAGASHIMA
Naka Fusion Research Establishment,
Japan Atomic Energy Research Institute,
801-1, Naka, Ibaraki-ken, 311-01, Japan

## H. ARAI, N. GOTO
Tokyo Institue of Technology, Tokyo, 152 Japan

## ABSTRACT

1MW class of RF vacuum window at a frequency of 2GHz is under development for a new LHRF heating system in JT-60 upgrade program. The power capability of a conventional pillbox window, which is used in the JT-60 LHRF heating system, is restricted by the multipactoring discharge at the power level of more than 500kW. In the high power experiment on test stand, it was found that the multipactering discharge is caused by the perpendicular electric field $E_z$ to the ceramic. To suppress the multipactoring discharge at higher power level, a new pillbox window for reducing the perpendicular electric field $E_z$ was designed from a numerical analysis.

## 1. INTRODUCTION

Lower hybrid current drive is one of the most promising method to achieve a steady state operation of tokamaks. A phased array launcher is recognized to be the most effective antenna for launching lower hybrid waves into a tokamak plasmas. Several types of breakdown, however, limit the injection power from the launcher into a plasma. The breakdown at the window is one of the most serious problem from the viewpoint of not only launching high power but also keeping vacuum sealing.

The JT-60 tokamak has three units of LHRF heating systems and the launcher is composed of 32 vacuum waveguides. The same number of pillbox vacuum window are adopted to seal a vacuum at each waveguide, where the power injected through the window is 250 kW for 10 s pulse duration. This vacuum window is the same type of $Al_2O_3$ ceramic as that used in a 1MW klystron which is the RF power source of the JT-60 LHRF heating system. Since the 1MW klystron has 2 output windows, the availability of the launcher's window is twice as high as that of the klystron's one. The serious troubles have not occurred at the vacuum window up to now.

On the contrary, the new LHRF heating system is planned to study current drive with $D_2$ plasmas on the JT-60 upgrade. The number of the waveguide in the launching system should be increased to improve wave spectra by means of using a new multijunction launcher in this new system, and the number of the vacuum window shoud be reduced from 32 to 8 by the economical requirement to simplify the transmission system from the view point of radioactive sealing in $D_2$ operation.

We have been developing the vacuum window to transmit a high power of more than 1 MW at the frequency of 2 GHz band. This paper reports the experimental results from the high power transmission test of the conventional pillbox window of ( Fig. 1 ), which clarifies what is the cause of the power limit of window and presents the design concept of a new type of pillbox window of 1 MW capability .

## 2. HIGH POWER TEST OF CONVENTIONAL VACUUM WINDOW

We performed the high power transmission test of the conventional pillbox window, whose axial length is 56 mm, at the power level of 900 kW, at 2 GHz by using the 1 MW klystron. The schematic of the experimental set up is illustrated in Fig. 2. The tested windows were connected to E-bend rectangular waveguide, which was evacuated up to about $10^{-5}$ Pa. Temperature profile and lightning of the window were monitored using an infrared camera and a photomultiplier, respectively . First, we checked the temperature rise $\Delta T$ of the ceramic due to dielectric loss in the condition of filling $SF_6$ gas at 1 atm in an evacuated part to avoid the breakdown , and then we evacuated the window to perform the transmission test in vacuum. Figure 3 shows the temperature rise of the ceramic in a vacuum condition at an early RF conditioning stage, where f=2 GHz, pulse duration of 1 s with the duty cycle of one tenth. This result was compared with the temperature rise due to dielectric loss, and it was shown a discharge occured on the ceramic at power level of more than 100kW. Here, the temperature increase by the discharge corresponds to ~80°C/s at the power level of 900 kW. Such a sudden increase in the surface temperature will cause the ceramic broken. The discharge was suppressed by the sufficient RF conditioning at the power level of ~500 kW. The infrared camera measurements indicated that the spatial distribution of the temperature has two peak points on the ceramic during discharge as shown in Fig. 4. We measured the electric field pattern in the pillbox window by means of heat sensitive papers, which coincides with the theoretical results[1]. As the results, it was confirmed that the maximum point of the temperature increase of the ceramic corresponds to the maximum point of $E_z$. The discharge is clearly explained by the theory of single surface multipactoring discharge induced by an electric field perpendicular to the ceramic and the space charge[2].

## 3. IMPROVED PILLBOX WINDOW

One method to avoid the RF discharge in the pill box window is to reduce $E_z$. We theoretically found that the edge effect at the junction between the rectangular and the circular waveguide enhanced the perpendicular electric field $E_z$ to the ceramic.[1] Therefore, an oversized pillbox window whose axial length was longer than the conventional one, was designed to reduce the electric field Ez. The axial length of the newly designed pillbox window is determined from the numerical analysis .

Figure 5 shows the numerical results of the reflection coefficienct as a function of the axial length L in the case of 2 GHz. It appears that VSWR of 1.01 in case of L=226 mm is acceptable for the window at the frequency of 2GHz.

The electric field distributions at ceramic for two cases was evaluated by the numerical analysis. Figure 6 shows the numerical results of a perpendicular electric field $E_z$ on the ceramic, where $E_z$ is nomalized that of a $TE_{10}$ mode in the input rectangular waveguide, and Y is the nomalized radius ( 72.5 mm ) of the ceramic. The perpendicular electric field $E_z$ of a new pillbox window is 15 % of conventional one. The numerical analysis also shows that total electric field /E/ on the ceramic of the new pillbox window is decreased by 21.2 % compared with that of the conventional .

Next step, we will carry out a high-power transmission test in vacuum to examinate the availability of the new pillbox window on the resonant ring test-stand, where the power test up to 8 MW can be examined.

## III.SUMMARY

On the high power test, we found that the multipactoring discharge induced by an $E_z$ component is a main cause of RF break-down in the conventional pillbox window. To suppress the multipactoring discharge, we designed a new pillbox window in which the $E_z$ is reduced significantly compared with the conventional one. This will have a capability of power transmission of more than 1 MW at 2 GHz.

## ACKOWLEGEMENTS

We would like to thank to Drs. M.YOSHIKAWA , M.OHTA and M.TANAKA for their continuous encouragements.

## REFERENCES

1. H.ARAI , N.GOTO, et al ., IEEE Trans . on Plasma Science,  PS-14, NO.6  947 (1986).
2. Y.IKEDA et al., to be published IEEE Trans. on Plasma Science.

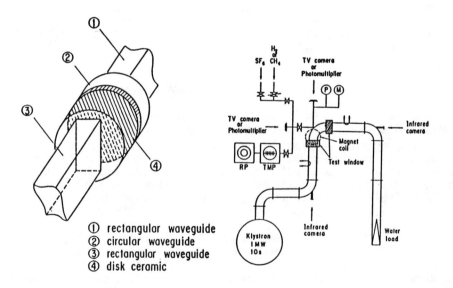

① rectangular waveguide
② circular waveguide
③ rectangular  waveguide
④ disk ceramic

Fig.1 The  structuer of a pillbox window. Fig.2 A shchematic of the experimental setup

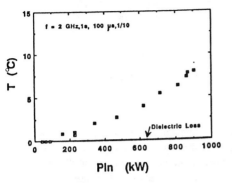

Fig.3 The power dependence of
the temperature rise.

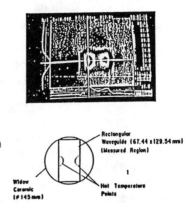

Fig.4 The temperatuer distribution
measured by the infrared camera.

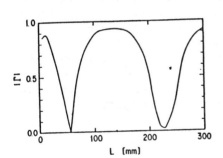

Fig.5 Reflection coefficient as a
function of the L.

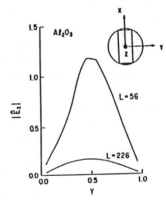

Fig.6 The perpendicular electric feild
distribution $E_z$

# Role of Alpha Particle Damping in Fast Wave Current Drive and Heating

T.K. Mau

Institute of Plasma and Fusion Research
University of California, Los Angeles, Ca. 90024-1600

## ABSTRACT

The impact of energetic alpha particle wave absorption on the range of frequencies for efficient fast wave current drive and heating in a fusion reactor is investigated. The energetic alpha damping decrement is calculated, using a slowing down distribution function, and compared to electron and fuel ion damping over a wide range of frequencies. A combination of strong alpha damping and edge electron absorption in the higher ion harmonic regime limits core fast wave current drive to the lower harmonics (l=2,3).

## INTRODUCTION

Fast wave current drive has been proposed as one of the techniques to achieve steady-state operation in the ITER tokamak reactor.[1] This scheme relies on the wave energy being absorbed by the electrons via the Landau and TTMP processes. Thus, to maximize its efficiency, other competing wave damping mechanisms, such as alpha particle and fuel ion absorption, have to be avoided at all cost. On the other hand, for supplementary heating to ignition, alpha partcles can compete with deuterons for the wave energy because their cyclotron resonance surfaces coincide with each other. In this paper, a physics model is developed to calculate the power damping decrement of the fast wave due to the energetic alpha particles, which is then compared to the electron and fuel ion damping decrements over a wide range of wave and plasma parameters. It is found that the presence of the alpha particles severely limit the range of frequencies in which efficient core current drive can be achieved.

## DAMPING BY ENERGETIC ALPHAS

For simplicity, a 1-D, radial slab plasma model is used to simulate propagation of the fast wave along the equatorial plane in tokamak geometry. The alpha damping decrement $\gamma_\alpha$ along the radial direction is given by:

$$\gamma_\alpha = \frac{P_\alpha}{|V_{g\perp}|U_{emp}} \tag{1}$$

where $P_\alpha$ is the absorbed power density, $V_{g\perp}$ is the group velocity and $U_{emp}$ is the wave energy density. To evaluate $V_{g\perp}$ and $U_{emp}$, the cold plasma fast wave dispersion relation is solved. $P_\alpha$ can be calculated by:

$$P_\alpha = m_\alpha \int d\vec{v} \vec{S} \cdot \vec{v} \tag{2}$$

where $\vec{S} = -\bar{D}_{rf} \cdot \partial F_\alpha / \partial \vec{v}$ is the RF-induced flux in velocity space, $\bar{D}_{rf}$ is the Kennel and Engelmann diffusion tensor[2] due to the Doppler-shifted resonance of the wave with the alphas, and $F_\alpha(\vec{v})$ is the alpha velocity distribution function. In Eq. (2), the integral is over a region of velocity space in which the harmonic alpha resonance condition is satisfied. Noting that $F_\alpha(\vec{v})$ is essentially isotropic provided the RF-induced diffusion is weak compared to collisional diffusion and integrating over $v_\parallel$ in Eq. (2), one obtains

$$P_\alpha = -\frac{2\pi^2 Z_\alpha^2 e^2}{m_\alpha |k_\parallel|} \sum_l \int_0^{v_l} dv_\perp |\vartheta_l|^2 \frac{v_\perp^3}{v} \frac{\partial F_\alpha}{\partial v}\bigg|_{v_\parallel = v_{res,l}} \tag{3}$$

where $|\vartheta_l|$ is given in [2], $v_{res,l} = (\omega - l\omega_{c\alpha})/k_\parallel$ and $v_l = (v_{\alpha 0}^2 - v_{res,l}^2)^{1/2}$, with $v_{\alpha 0}$ being the alpha velocity at birth with an energy of 3.5 MeV. In Eq. (3) the summation over $l$ is for $|1 - l\omega_{c\alpha}/\omega| \le |N_\parallel| v_{\alpha 0}/c$; thus the number of resonances is given by: $\Delta l \sim 2|N_\parallel|(v_{\alpha 0}/c)(\omega/\omega_{c\alpha})$.

Scrutiny of Eq. (3) reveals that the main contribution of absorption comes from the region with $v \sim v_{\alpha 0}$. Furthermore, the damping is dependent on the gradient of $F_\alpha$ with respect to $v$ in the same region. Thus a relatively accurate form of $F_\alpha$ is crucial for the calculation of $P_\alpha$. As such, the model of Kolesnichenko[3] is employed for the present calculations, being given by:

$$F_\alpha(v) = \frac{n_{\alpha 0}}{(\pi^{1/2} u_\alpha)^3} \left\{ \frac{\tau_{p\alpha}}{\tau_{\alpha 0}} f_{\alpha M}(v) + \frac{1}{3} \left( \frac{m_e}{m_\alpha} \right)^{1/2} [1 - f_{\alpha M}(v)] G(v) \right\} \tag{4}$$

where $u_\alpha = (2T_i/m_\alpha)^{1/2}$, $n_{\alpha 0} = \dot{n}_\alpha \tau_{\alpha 0}$, $\dot{n}_\alpha = n_D n_T \langle \sigma v \rangle_{DT}$ is the alpha particle source rate, $\tau_{\alpha 0} (\sim T_e^{3/2}/n_e)$ is the alpha slowing down time, $\tau_{p\alpha}$ is the alpha particle confinement time, $f_{\alpha M}(v)$ is the normalized thermal alpha distribution function and $G(v)$ is given in [3]. In writing Eq. (4), it is assumed that $\tau_{\alpha 0} \ll \tau_{p\alpha}$. This implies that only the thermal alphas are allowed to diffuse, a good approximation under typical reactor conditions. Wong and Ono[4] calculated $\gamma_\alpha$ by the unmagnetized orbit method and their results differ significantly from those reported here, for $\omega_{ci} < \omega \ll \omega_{LH}$.

## RESULTS

A code is written to compare $\gamma_\alpha$ with the electron and fuel ion damping decrements, $\gamma_e$ and $\gamma_{D,T}$ respectively, over a wide range of parameters. Both $\gamma_e$ and $\gamma_{D,T}$ are calculated in a manner analogous to Eq. (1). Electron Landau and TTMP processes are included while mode conversion at the ion harmonics are neglected. In Fig. 1, the damping decrements are shown as a function of D cyclotron harmonics, with $n_e = 0.7 \times 10^{-14} cm^{-3}$, $T_e = 18 keV$, $N_\parallel = 3$ and $\omega_{cD} = 42 MHz$. For this set of parameters, $n_{\alpha t}/n_e = 0.017$ and $n_{\alpha f}/n_e = 0.005$, where $n_{\alpha t}$ and $n_{\alpha f}$ are the total and fast alpha densities respectively. The spike on the left corresponds to the $2\omega_{cD}$ resonance while $\gamma_e$ increases monotonically with frequency at $c/(N_\parallel v_e) = 1.25$. $\gamma_\alpha$ exhibits discrete structure at low frequency, and harmonic overlapping starts to occur above $\omega/\omega_{cD} = 11.6/N_\parallel$ where it decreases with frequency until it is exceeded by $\gamma_e$ at $\omega/\omega_{cD} \simeq 9$. Note that dominant electron damping occurs at $\omega/\omega_{cD} \simeq 2 - 3$ and $15 - 20$, where efficient current drive is possible. In Fig. 2, the damping decrements are compared over a range of $N_\parallel$ from 1.5 to 4.0 at a frequency of 750MHz, all other parameters the same as in Fig. 1. In this case, $\gamma_\alpha$ decreases sharply with $N_\parallel$ while $\gamma_e$ peaks gently at $N_\parallel = 2.5$, probably due to dispersive effects. This result suggests that at the higher harmonics, alpha damping can be made weak compared to electron absorption by increasing $k_\parallel$.

A series of 1-D slab calculations are carried out to simulate launching from the low field side along the equatorial plane. Toroidal effects are taken into account by setting $BR$ and $N_\parallel R$ constant, where B is the toroidal field and R is the major radius coordinate. A set of ITER-like parameters are used: $R_o = 550cm$, $a = 180cm$, $B_o = 5.5T$, and parabolic profiles with $n_{e0} = 10^{14}cm^{-3}$, $\alpha_N = 0.4$, $T_{e0} = 32.4keV$ and $\alpha_T = 0.8$, with $\tau_{p\alpha} = 2sec$. It is found that for all cases at the higher harmonics, either alpha particle absorption dominates or there is little wave penetration to the core when damping by electrons and alphas is strong at the edge. As an example, shown in Fig. 3 is the wave deposition profile for the case with $\omega = 750MHz$ and an incident $N_\parallel$ of 2.5, where the deepest penetration is at $r/a \sim 0.65$. On the other hand, at the low frequency end, core penetration of the wave with dominant electron absorption can be readily obtained. An example is given in Fig. 4 where $\omega = 60MHz$ and $N_\parallel = 2.5$, which results in 80% of electron absorption while 17% and 3% go to the fuel ions and alphas respectively.

For direct fuel ion absorption during the heating phase, low harmonic alpha resonances inside the plasma should be avoided. In the case of $2\omega_{cD}$ heating with $\omega = 84MHz$ and $N_\parallel = 2.5$, alphas absorb 14-48% of the power for $T_{e0}$ of 10-25keV. In conjunction with the 60MHz current drive scenario in Fig. 4, a $2\omega_{cT}$ supplementary heating scheme in the presence of a $^3He$ minority seems appropriate.

## CONCLUSIONS

The role of alpha damping in determining the parameter regime of efficient fast wave current drive and heating has been examined. Strong alpha damping limits the possibility of efficient current drive to the low (l=2,3) and very high harmonic frequency regimes, depending in the latter case on the value of $k_\parallel$. However, for current drive in the core plasma, the low harmonic regime is preferred.

This work is supported by DOE Grant No. DE-FG03-88ER52151.

## REFERENCES

[1] T.K. Mau, D.A. Ehst, R.H. Cohen, Bull. Am. Phys. Soc. 33(1988)1900.

[2] C.F.F. Kennel, F. Engelmann, Phys. Fluids 9 (1966)2377.

[3] Ya.I. Kolesnichenko, Nucl. Fusion 15(1975)35.

[4] K.L. Wong, M. Ono, Nucl. Fusion 24 (1984)615.

## FIGURE CAPTIONS

1. Alpha, electron and fuel ion damping decrements as a function of $\omega/\omega_{cD}$ for $N_\parallel = 3$.
2. Alpha, electron and fuel ion damping decrements as a function of $N_\parallel$ for $\omega/\omega_{cD} = 18$.
3. Wave power deposition profiles for electrons, alphas and fuel ions in ITER plasma with $\omega = 750MHz$ and $N_\parallel = 2.5$.
4. Wave power deposition profiles for electrons, alphas and fuel ions in ITER plasma with $\omega = 60MHz$ and $N_\parallel = 2.5$.

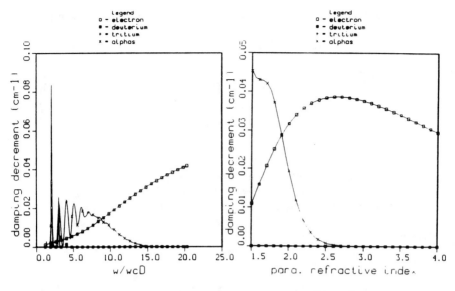

Fig. 1                    Fig. 2

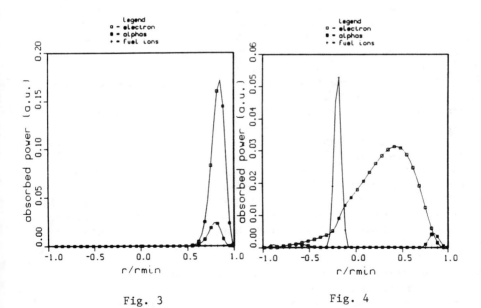

Fig. 3                    Fig. 4

# Fast Wave Current Drive Above the Slow Wave Density Limit

R. McWilliams, D.P. Sheehan*, N.S. Wolf**, and D. Edrich

Department of Physics, University of California, Irvine, CA 92717

## Abstract

Fast wave and slow wave current drive near the mean gyrofrequency were compared in the Irvine Torus using distinct phased array antennae of similar principal wavelengths, frequencies, and input powers. The slow wave current drive density limit was measured for $50\omega_{ci} \leq \omega \leq 500\omega_{ci}$ and found to agree with trends in tokamaks. Fast wave current drive was observed at densities up to the operating limit of the torus, demonstrably above the slow wave density limit.

Following Fisch's[1] prediction, in April 1978 McWilliams experimentally discovered slow wave lower hybrid current drive (LHCD)[2]. Despite the success of slow wave LHCD, there remains a problem with the current drive 'density limit,' a density threshold above which current drive efficiency falls dramatically. Fast waves, on the other hand, might propagate propagate to higher densities and temperatures than the slow wave[3]. Experiments by McWilliams and Platt in the Irvine Torus[4] discovered that a phased array fast wave antenna could drive currents. Recent fast wave current drive results on JIPP T-IIU [5] exhibit many characteristics of slow wave current drive including the density limit. Experimentalists on JIPP T-IIU speculate that their antenna actually coupled to the slow wave, rather than to the fast wave, through parasitic excitation. Goree and Ono have demonstrated parasitic excitation of lower-hybrid slow waves by a fast wave current drive antenna in ACT-I[6].

The experiments were performed on the Irvine Torus, a toroidal device with major radius of 55.6 cm and minor radius up to 10 cm. Graphite limiters reduced the plasma radius to 3.5 cm. For these experiments, $B = 1$ kG . There is no ohmic pulse and, hence, no dc toroidal electric field or poloidal magnetic field from ohmic current. The plasma is produced by electron impact ionization of background gas. The present experiments were performed in helium with plasma densities $n_{eo} \leq 2 \times 10^{12}$ cm$^{-3}$ and bulk electron temperatures $T_{eo} \leq 10$ eV, using up to 3.2 A of ionizing electric current. The toroidally symmetric ionizing electron beams provide a tail population for resonant wave damping and current drive. Beam electron density to bulk electron density was typically $n_{eb}/n_{eo} \approx 10^{-2}$.

Plasma density and bulk electron temperatures were inferred from radially movable Langmuir probes and the propagation angle of lower hybrid waves. Wave-driven currents were measured by means of a small area, single turn, dual loop probe positioned just outside the plasma and oriented to detect $\frac{\delta B_\theta}{\delta t}$, the time rate of change of

* Physics and Eng. Dept., USD, San Diego, CA

** Physics Dept., Dickinson College, Carlisle, PA

the induced poloidal magnetic field. The loops were arranged to discriminate against electrostatic pickup.

The fast wave antenna was excited by application of a 20-200 MHz signal to a sixteen element phased array antenna, as described in reference 4. The slow wave antenna consists of eight coaxial 0.6 cm wide tantalum rings situated equidistantly toroidally outside a thin quartz tube. Similar toroidal vacuum power spectra can be launched by both fast and slow wave antennae. Fast wave and slow wave current drive were compared under similar plasma and wave launching conditions.

Previous tokamak experiments (summarized in references 7 and 8) have studied the slow wave LHCD density limit over one decade in launched wave frequency. The Irvine Torus results extend these data an added order of magnitude by studying $20MHz \leq f \leq 200MHz$. As shown in Figure 1, the Irvine LHCD density limits follow the trend of previous tokamak results. For the present data, the slow wave LHCD density limit was taken to be the density at which current drive efficiency decreased 20% from its maximum value.

Next, waves were launched from the fast wave antenna. The fast wave propagation cutoff density is given approximately by

$$\omega_{pe}^2(cutoff, FW) \approx \omega\omega_{ce}(n_\parallel^2 + n_\theta^2 - 1)^{1/2}(n_\parallel^2 - 1)^{1/2} \qquad (1)$$

We did not expect to observe fast waves for densities below the high $10^{11} cm^{-3}$ range. In fact, fast wave efffects were not observed for $n_e \leq 9 \times 10^{11} cm^{-3}$. However, for densities below $n_e \approx 5 \times 10^{11} cm^{-3}$ the fast wave antenna was observed to launch small amplitude slow waves, which have a dramatically lower cutoff density, given approximately by $w_{pe}^2(cutoff, SW) \approx \omega^2$.

For low densities, the fast wave antenna couples to slow waves (identified by probe measurements) with about 10-30% of the efficiency of the slow wave antenna. At these low densities, the slow waves launched by the fast wave antenna drive electron currents and are subject to the slow wave LHCD density limit just as slow waves launched from the slow wave antenna. Hence, fast wave antennae can be subject to the slow wave LHCD density limit when the antennae are operated in low density plasmas.

When the plasma density is raised sufficiently above the slow wave LHCD density limit, a new regime of current drive appears. For densities high enough that fast wave effects are expected (i.e. densities above wave propagation cutoff), the fast wave antenna drives currents similar to those discovered by McWilliams and Platt[4]. This changing of current drive regimes is shown in Figure 2, which plots driven current versus plasma density for both the slow and fast wave antennae operated at 100 MHz. The slow wave data points (connected by dashed line) show a rapid drop in current at about $2.8 \times 10^{11} cm^{-3}$, followed by current decreasing monotonically with increased plasma density until it falls below detection circuit noise levels.

For densities below about $5 \times 10^{11} cm^{-3}$ the fast wave antenna drives currents with similar density dependency as the slow wave antenna. Above $n_i \approx 9 \times 10^{11} cm^{-3}$, however, the FW antenna driven current displays a new regime of current drive (data connected by solid line). Above this density, in Figure 2, fast wave current drive (FWCD) is observed up to the maximum experimental operating densities. In Figure 2, then, fast wave driven currents are not observed until densities of a factor of 3 greater than the slow wave LHCD density limit are achieved. Additionally, FWCD is seen for densities up to a factor of 6.5 above the slow wave LHCD density limit, this factor being limited by the maximum density achievable in this experiment.

The current drive density limit for the fast wave antenna was compared with the slow wave antenna for $20MHz \leq f \leq 200MHz$, the results shown in Figure 3. The slow wave antenna density limits are given by solid dots. The slow wave components launched by the fast wave antenna and observed only for $n_e \leq 5 \times 10^{11} cm^{-3}$ are shown by triangles and agree well with the slow wave antenna data. Fast wave current drive is indicated by open dots. Below about 80 MHz, the fast wave should be cutoff from propagation and, in fact, below 80 MHz, FWCD is not observed. For 80 MHz and above, however, fast wave current drive is observed up to the maximum operating density for this experiment. When a fast wave current drive density limit was not found, this is indicated by vertical arrows. Experimentally, the magnitude of fast wave driven current was larger at 100 MHz than at 200 MHz, which might be due to wave accessibility considerations for this experiment.

In summary, fast and slow wave lower hybrid current drive (LHCD) was observed in the Irvine Torus. The slow wave LHCD density limit has been observed for $20MHz \leq f \leq 200MHz$, extending the previous tokamak data base by an order of magnitude in wave frequency. At densities below the slow wave LHCD density limit, a fast wave antenna was observed to launch a fraction of power in slow waves and the attendant slow wave LHCD density limit was observed. At densities where fast waves could propagate (which easily could require densities a factor of 3 above the observed slow wave LHCD density limit), fast wave driven current was observed up to densities a factor of 6.5 above the observed slow wave LHCD density limit. Fast wave current drive did not encounter a density limit in these experiments. These results suggest that future tokamak fast wave experiments perhaps should ignore local decreases in driven current, if indicative of SWCD, and press to higher densities where FWCD may flourish.

It is a pleasure to acknowledge Mr. Stacy Roe and Mr. David Parsons for valuable technical assistance. These experiments were carried out under DOE grant DE-FG03-086ER5321 and NSF grant PHY 8606081.

Fig. 1. Slow wave density limit versus frequency for Irvine Torus and various tokamaks (from Ref. 42).

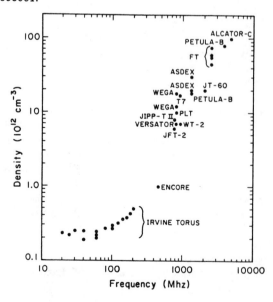

Fig. 2. Wave-driven currents by fast wave antenna (open dots) and slow wave antenna (solid dots) versus plasma density at 100MHz in He plasma. Fast wave antenna drives currents demonstrably above slow wave density limit.

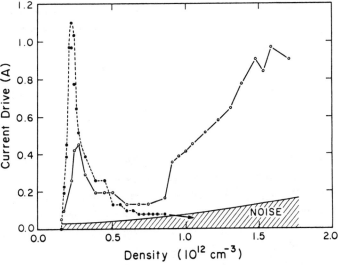

Fig. 3. Density limits versus frequency for fast and slow wave antennae. Solid dots: slow wave antenna; Triangles: presumed due to slow wave component launched by fast wave antenna; Open dots: fast wave antenna. Verticle arrows indicate no density limit was observed.

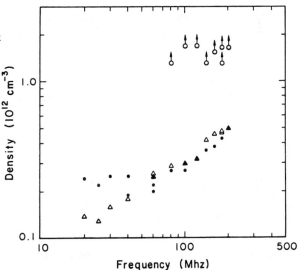

References

1. N.J. Fisch and A. Bers, in Proceedings of the Third Topical Conference on RF Plasma Heating (Caltech, 1978), paper E6.
2. R. McWilliams, L. Olson, R.W. Motley, and W.M. Hooke, Bull. Amer. Phys. Soc. 23, 765 (1978).
3. V.E. Golant, Sov. Phys. 16, 1980 (1972).
4. R. McWilliams and R.C. Platt, Phys. Rev. Lett. 56, 835 (1986).
5. R. Ando, E. Kako, Y. Ogawa, and T. Watari, Nucl. Fusion 26, 1619 (1986).
6. J. Goree and M. Ono, Nucl. Fusion 28, 1105 (1988).
7. J.G. Wegrowe and F. Engelmann, Comm. Plasma Phys. and Cont. Fusion 8, 211 (1984.
8. L.H. Sverdrup and P. Bellan, Phys. Rev. Lett. 59, 1197 (19C7).

# STUDY OF THE PARAMETRIC DECAY INSTABILITIES IN THE 2.45 GHz EXPERIMENT ON ASDEX

V. Pericoli Ridolfini *, K. McCormick, A. Carlson, M. Bessenrodt-Weberpals, ASDEX-team, LH-team

Max-Planck-Institut für Plasmaphysik, EURATOM Association, Garching bei München, Fed. Rep. of Germany

In this paper we present the first results of the study of the parametric decay instabilities (PDI) during lower hybrid (LH) experiment in the ASDEX tokamak with the new 2.45 GHz system [1]. The flexibility of the new grill launcher permits a more strict control of the wave characteristics and therefore a closer comparison with theoretical predictions. In particular we refer to the possibility of carefully shaping the $N_\parallel$ (parallel index of refraction) spectrum of the waves over a large interval, 1.2 - 4.4 with $\Delta N_\parallel \approx 0.7$, and of varying the spatial extension of the LH resonance cones, by means of a partial excitation of the launching grill. Moreover the high frequency, 2.45 GHz, together with the fact that $B_T \leq 2.8T$, rules out any direct RF-ions interaction [2] and allows a clearer study of the effect of these instabilities on current drive (CD) and heating efficiency than before.

## Experimental results

Radiofrequency waves were collected by a small tip antenna inside the vacuum vessel, about 180° toroidally away from the RF grill, and then spectrally analyzed.

The relative position of the collecting antenna and the RF grill should not affect appreciably the shape of the spectra, as observed in a previous work [3].

The plasma parameter space covered in the present study is approximately given by: line averaged plasma density $1.3 \cdot 10^{13} \leq \bar{n}_e \leq 4.7 \cdot 10^{13}$ cm-3, plasma current $300 \leq I_p \leq 420$ kA, toroidal magnetic field $2.2 \leq B_T \leq 2.8$ T, while the LH wave parameters vary in the range: $1.5 \leq N_\parallel \leq 4.4$, RF power $0.3 \leq P_{RF} \leq 1.6$ M.W, grill area $z \cdot y = 34.6 \cdot 22.6$ cm2, (configuration 1,) both upper and lower grill operating, and $36.6 \cdot 11.3$ cm2, (configuration 2,) only one grill operating.

A typical spectrum with well developed instabilities is shown in fig. 1. Its main characteristics are the substantial spectral broadening of the pump, i.e. the injected wave, and the appearence of several satellite lines, or sidebands, shifted from the pump by multiples of the ion cyclotron frequency, $f_{ci}$, evaluated at the torus outer edge. The study is here limited to the pump spectral width and to the first satellite intensity.

## 1) Effect of varying the grill area

Fig. 2 shows two spectra measured during a single shot when the grill configuration was changed

Associazione EURATOM - ENEA sulla Fusione. C.R.E. Frascati  -  Frascati, Roma, ITALY from 1 to 2. In the full area configuration the first satellite is much more developed. Since the power per $cm^2$, i.e. the electrical field, is the same for both cases, this is experimental evidence that the spatial extent of the pump strongly affects the growth of the sideband. The larger is the region where the two waves overlap, the higher is the energy amount that can be coupled to the sideband, since longer is the time of their interaction.

2) Effect of varying the the pump power

Fig. 3 shows for a typical ASDEX shot the spectral broadening of the pump $\Delta f$ and the intensity of the first satellite $I_{PDI}$, as a function of the coupled net RF power $P_{RF}$. The pump width increases approximatively linearly with $P_{RF}$ while the first satellite intensity starts to grow exponentially after a certain threshold and then tends to saturate for higher power.

3) Effect of varying the $N_{\parallel}$ spectrum

Although the influence of the $N_{\parallel}$ value of the waves on PDI has been studied over a rather limited data set, the expected decrease of the threshold power with $N_{\parallel}$ is found.

4) Effect of varying the plasma density

As expected, when the plasma density is increased the pump spectral broadening increases also, while the threshold power, $P_{thr}$, for the onset of the first satellite decreases. That is shown in fig. 4 where the experimentally determined $P_{thr}$ values are compared with the predictions of the code (see below).

5) Effect of varying plasma current, $I_p$, and magnetic field $B_T$

Both pump spectral broadening and first satellite intensity are reduced when either $I_p$ or $B_T$ or both are raised. That agrees with previous observations on FT [3]. Fig. 5 illustrates this behaviour.

Discussion

The interpretation of the experimental results has been performed on the basis of a code already developed for FT LH experiment [3]. The power thresholds are calculated for the decay channel of a LH wave into another LH wave plus an ion cyclotron quasi mode. They are determined by the balance between the instability growth rate and the energy loss due to the convection of the decay products out of the pump resonance cones. All the different possible sidebands are considered, and the minimum $P_{thr}$ value is then chosen. The code has been now extended to take into account all the possible angles of the sideband trajectory in the poloidal plane, previously fixed, and the finite extension of the pump in the poloidal direction, previously assumed infinite. This extension explains the point 1) of the experimental results, as illustrated in fig. 6. Here the threshold power for both grill configurations is plotted versus plasma density. The switching off of one grill corresponds in the figure to a jump from point A, where $P_{RF} > P_{thr,2\ grills}$ (full line), to point B where $P_{RF} > P_{thr,\ 1\ grill}$ (dotted line). For point A, well developed instabilities are expected, and are in fact observed (fig. 2a) while for point B their growth should be prevented (fig. 2b).

The two calculated threshold powers differ less than a factor 2, as it would be expected instead, if they were determined by the electric field magnitude only. This is due to the different trajectories of the LH sideband in the two cases, and consequently to the different times spent within the pump resonance cones. So the primary importance of the convective losses in developing PDI has been directly demonstrated. The quantitative results of the code depend to a great deal on the detailed profile of the edge and scrape off layer density and temperature [3], especially if a shoulder in the density profile can develop just outside the separatrix [4]. So, being at present lacking a full characterization of the edge during the RF phase the best values for the code appeared to be the average ones at the separatrix radius, given in [4]. This can cause some discrepancy with experiment. For example, the calculated PDI growth rate as a function of $P_{RF}$ agrees with data on fig. 3 only within a factor three. However, its scaling with $\bar{n}_e$ is very well reproduced and so are the scalings of $P_{thr}$ with $N_{\parallel}$ and $\bar{n}_e$. The comparison between theory and experiment is shown in fig. 4, as an example for this last scaling . Finally it must be noted that none of these instabilities either prevent or drastically reduce CD or heating of the plasma. However, it is at present under investigation whether they could be responsible for some unexpected variations of the CD efficiency with $\bar{n}_e$ and $I_p$ on ASDEX [5], through modification induced in the pump $N_{\parallel}$ spectrum.

## Figure captions

1) Typical parametric decay spectrum for well developed instabilities.
2) Spectra obtained for the grill configuration 1 a), and 2 b), (see text), $P_{RF} \approx 620 /310$ kW, $N_{\parallel} = 1.8$    $\bar{n}_e = 2.7 \cdot 10^{13}$ cm$^{-3}$    $I_p = 320$ kA    $B_T = 2.17$T.
3) Plot of the pump spectral broadending (full width at -10 db) ($\Delta$), and of the first satellite intensity (o), as a function of the coupled RF power. Same plasma parameter as fig. 2.    $N_{\parallel} = 1.8$.
4) Pump spectral width for $P_{RF} \approx 1$ MW ($\Delta$). Experimental (o) and calculated ($\bullet$) power thresholds versus line averaged density.    $I_p = 420$ kA    $B_T = 2.8$T,    $N_{\parallel} = 1.8$.
5) Pump spectral broadening ($\Delta$) and first satellite intensity (o) versus plasma current. $\bar{n}_e = 2.07 \cdot 10^{13}$cm$^{-3}$  $P_{RF} \approx 950$ kW, $N_{\parallel} = 1.82$. Open symbols $B_T = 2.17$T, full symbols $B_T = 2.8$.
6) Threshold powers, for PDI instabilities (1$^o$ satellite) versus plasma density and $T_e = 80$ eV, for both grills excited, configuration 1, full line, and for only one grill excited, configuration 2, dotted line. Points A and B correspond respectively to the cases in fig. 2 a and 2 b.

[1]  F. Leuterer et al., Proc. 16th Europ. Conf. on Controlled Fusion and Plasma Physics Venice (1989) p. 1287
[2]  J. B. Wegrowe, F. Engelmann, Comments Plasma Physics + Contr. Fus. 8, 211 (1984)
[3]  R. Cesario, V. Pericoli Ridolfini, Nucl. Fus. 27, (1987), 435
[4]  K. McCormick et al., ref. [1] p. 895
[5]  F. Leuterer et al., This Conference

Acknowledgement:
The construction of the 2.45 GHz LH system for ASDEX has been done in cooperation between IPP Garching, ENEA-Frascati and PPPL-Princeton.

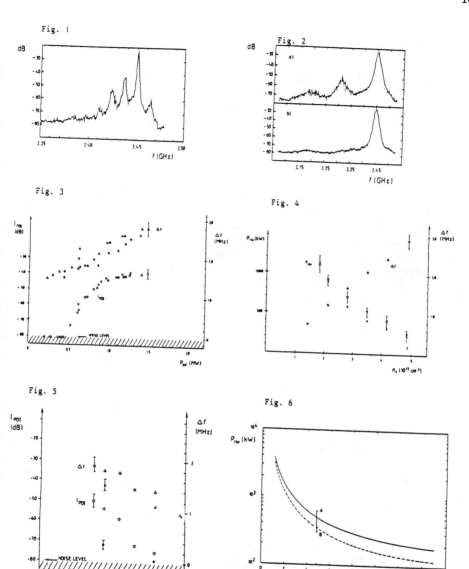

Fig. 1

Fig. 2

Fig. 3

Fig. 4

Fig. 5

Fig. 6

# DESIGN OF NEW LHRF LAUNCHER ON JT-60 UPGRADE

M. SEKI, Y. IKEDA, T. IMAI, K. KONISHI, T. NAGASHIMA and K. UEHARA

Naka Fusion Research Establishment, Japan Atomic Energy Research Institute ( JAERI )
Naka-machi, Naka-gun, Ibaraki-ken 311-01, JAPAN

## ABSTRACT

The JT-60 upgrade is planned to enhance plasma performance of 6 MA divertor $D_2$ plasma. The lower hybrid heating system will also be modified with the launcher to drive the plasma current and to control the plasma current profile efficiently. The new launcher composed of 16 multi-junction modules will be employed at a large horizontal port (0.9 x 0.79 m) to inject around 10 MW of RF power. Each module has a 12- or 18- waveguide divider in the toroidal direction. This new launcher features the controllability of $N_{//}$ spectrum by changing the frequency over the range of 1.74 GHz to 2.23 GHz. The experiment with the new launcher will start in early 1991.

## 1. INTRODUCTION

The plasma current is necessary to maintain the configuration of tokamak plasma, so it is very important that the plasma current is driven by a non-inductive method. Another key point relevant to the reactor grade plasma is to attain a high confinement mode. Therefore we should study the current drive and the active control of plasma parameters. The JT-60 has focussed on these studies, which have been successfully done by using the lower hybrid range of frequencies ( LHRF ).

Moreover the JT-60 upgrade is planned in JAERI, and LHRF will also be modified for the purpose of optimizing the wave spectrum to enhance current drive efficiency. The main objectives of LHRF in the JT-60 upgrade are to attain high current drive efficiency at high temperature $D_2$ plasma and to perform current drive at high density.

The JT-60 upgrade is a poloidal divertor tokamak with a major radius of Rp = 3.4 m , the maximum toroidal field Bt = 4.2 T and a volume of plasma Vp ~ 100 $m^3$. The objectives of JT-60 upgrade are to enhance plasma parameters ( plasma current= 6 MA, temperature =10 ~ 15 keV, density = 7 ~ 10 x $10^{19}$ $m^{-3}$, $D_2$ plasma ) and to study steady state operation.

The experiment of current drive using the multi-junction[1] on the JT-60 led the successful results of the improvement of current drive efficiency by a factor of 1.4 compared with the previous results obtained by using the conventional launcher [2].

We decided that the multi-junction launcher for the JT-60 upgrade should be adopted in order to drive plasma current effectively. The feature of this new LHRF launcher is mainly the controllability of spectrum launched from the grill by changing the frequency.

## 2. OBJECTIVES

The main objective is to study the physical and the technical aspects relevant to the fusion reactor development, so we should demonstrate reactor grade current drive with higher efficiency. Our targets are as follows, high current drive efficiency $\eta_{CD} \sim 4\text{-}5$ ( rf driven current Irf $\sim 4$ MA, density ne $\geqq 1\times10^{19}$ m$^{-3}$) with high temperature($\langle T_e\rangle \geqq$ 3keV ) D$_2$ plasma of the JT-60 upgrade, and verification of current drive at high density with D$_2$ plasmas ( ne $\sim 7 \times 10^{19}$ m$^{-3}$ ).

The target of LHRF for the JT-60 upgrade is shown in Fig. 1 and Table 1 . The coming results on the JT-60 upgrade will be the necessary data base for the design of ITER and/or FER.

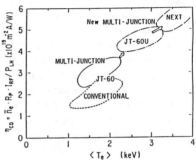

Fig. 1. $< \eta_{CD}$ vs temperature $>$

|  | ACHIEVED | | TARGET |
|---|---|---|---|
|  | OTHERS | JT-60 | for JT-60U |
| RF CURRENTS (MA) | < 0.7 | 2.0 | 4~6 |
| $n_e R I_{RF}$ (MA 10$^{19}$ m$^2$) | < 1.0 | ~8 | ~30 |
| $\eta_{CD}$ (10$^{19}$ m$^3$ A/W) | < 1.5 | 3.4 | ~5 |

Table 1. $\langle$ Target for JT-60U $\rangle$

The current drive efficiency $\eta_{CD}$ has been discussed theoretically[3] and experimentaly[2,4] in order to optimize the refractive index parallel to magnetic field N$_{//}$ .

The theoretical predictions of $\eta_{CD}$ agreed with the experimental results. It is found that the $\eta_{CD}$ increases with decreasing of the peak of N$_{//}$ as long as the accessibility condition N$_{//}{}^{acc}$ is satisfied, and that the narrow N$_{//}$ specturum and the good directivity make the $\eta_{CD}$ higher.

The theoretical $\eta_{CD}{}^{th}$ in the case of $\Delta N_{//}{\to}0$ and the dependence of N$_{//}{}^{acc}$ on the density are shown in Fig. 2. This figure shows that the optimum ranges for low density and high density current drive are 1.3~1.6 and 1.5~2.0, respectively.

The higher N$_{//}$ up to $\sim 4$ may be required for current profile control.

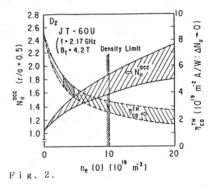

Fig. 2.

$\left\langle \begin{array}{l}\text{Density dependence of Accessibility Condition}\\ \text{and Theoretical Current Drive Efficiency}\end{array}\right\rangle$

## 3. OVERVIEW OF NEW LHRF LAUNCHER

To obtain the objectives of the JT-60 upgrade, we designed the luncher as follows.

The new LHRF launcher will be located on a large horizontal RF port of the JT-60 upgrade vacuum vessel, dimension of which is about 0.9 m (toroidal) x 0.79 m (poloidal) . We will adopt the new multi-junction launcher to get high directivity and sharp spectrum of $N_{//}$ owing to many waveguides arrayed in the toroidal direction. The structure of this multi-junction launcher is shown in Fig. 3. This launcher consists of 4 (toroidal) x 4 (poloidal) = 16 multi-junction modules. A module is drived by a 1 MW klystron. Each module is composed of 12 and/or 18 secondary waveguides in only toroidal direction to simplify the launching structue. The adjacent phase shift in each module is set 60 deg at 2.17 GHz, and the height and the width of the waveguide are 129.54 mm and 10 mm, respectively. Since the width of split wall is 1.5 mm (pitch = 11.5 mm), the peak of $N_{//}$ is about 2. The design effort is made that there are present as many waveguides in the RF port as possible. The geometrical phase shifter with low VSWR over the frequency range of 1.74 to 2.23 GHz is designed in order to control phase shift by changing frequency. The final design of phase setting will be determined after the current drive experiment in JT-60.

Typical $N_{//}$ spectra of the present design are shown in Fig. 4. The spectrum in JT-60 upgrade has more narrow and directive spectrum compared with the previous ones. Its width and directivity are about 0.3-0.4 and 0.95, respectively, and higher current drive efficiency will be expected. The $N_{//}$ spectra in three cases of changing frequency from 1.74 GHz to 2.23 GHz, which is calculated from the multi-junction coupling code, is shown in Fig. 5. We can control $N_{//}$ over the range of 1.3 to 4 by changing frequency and phase difference between adjacent modules corresponding to plasma parameters and experimental objectives.

The power density injected from the launcher will become smaller ( 2.7~4.1 kW ) due to the large RF port. The transmission line connected to the main waveguide of the module has double ceramic windows, and the manifold chamber should be pumped to be kept high vacuum for the purpose of suppressing breakdown and this launcher is baked at around 250°C by circulating high temperature $N_2$ gas. The launcher position can be adjustable shot by shot in order to obtain good coupling property.

We are developing the prototype modules composed of 18 secondary waveguides, which are manufactured by Brazing method and Hot Isostatic Pressing method. A high power test will be carried out on these two method.

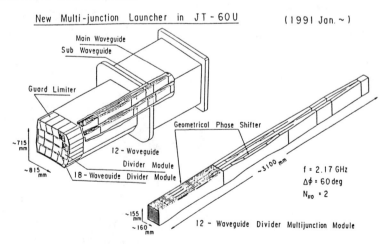

Fig. 3. 〈 4 (toroidal) x 4 (poloidal) x 12 or 18 Waveguide Divider Module 〉

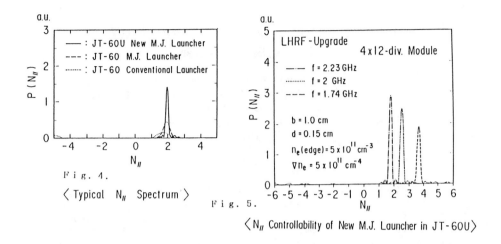

Fig. 4.

⟨ Typical $N_{//}$ Spectrum ⟩

Fig. 5.

⟨ $N_{//}$ Controllability of New M.J. Launcher in JT-60U ⟩

## 4. SUMMARY

This paper described the design of the new LHRF launcher on JT-60 upgrade.    The JT-60 upgrade is aiming at performing the reactor grade current drive by using the new LHRF launcher with high temperature $D_2$ plasma.    This launcher located on the large horizontal port contains 16 multi-junction modules. Each module is composed of 12- or 18- waveguide divider. This new multi-junction launcher is optimized to drive plasma current efficiently. The feature of this launcher is the controllability of $N_{//}$ spectrum by changing the frequency with high directive and sharp spectrum to attain the favorable current drive efficiency.    R & D of manufacturing technique of the module is being carried out.

The experiments with this new multi-junction launcher in the JT-60 upgrade will start in early 1991.

## ACKNOWLEDGEMENT

The authors wish to thank SHI and Toshiba corps., who have contributed to test module. We also wish to express our gratitude to Dr. M. Yoshikawa for his continued encouragement and support.

## REFERENCES

1. T. K. Nguyen, D. Moreau, Fusion tech. 2(1982) 1381.
2. M. Nagami, et al, 16th Europian Conference on Controlled Fusion and Plasma Physics, 1989.
3. T. Imai, et al, Nucl. Fusion 28(1989) 1341.
4. Y. Ikeda, et, al, in this conference

# PENETRABILITY OF LOWER HYBRID WAVES IN A FUSION PLASMA

M. Spada[a,*] and  M. Bornatici[a,b]

a) Physics Department "A. Volta", University of Pavia, 27100 Pavia, Italy;

b) Physics Department, University of Ferrara, 44100 Ferrara, Italy;

*) Present address: The NET Team, c/o Max-Planck-Institut für Plasmaphysik, 8046 Garching bei München, Federal Republic of Germany.

## ABSTRACT

The penetration of lower hybrid waves in a NET-like plasma is investigated by accounting for both (self-consistent) quasi-linear electron Landau damping and (linear) absorption by fusion alpha particles. The waves are treated as magnetized electrostatic electron waves. Numerical results for the profile of both the driven current density and the radial power deposition to both electrons and alphas are presented and discussed. In particular, the fraction of the power absorbed by the alpha particles is evaluated.

## INTRODUCTION

For steady state current drive in reactor-grade plasmas, a possible scenario includes current driven by lower hybrid (LH) waves,[1] for which it is important to ascertain that a significant absorption by alpha particles does not occur.[2] In work reported previously,[3,4] we evaluated the power deposition to the alphas by taking the parallel electron Landau damping either linear[3] or quasi-linear.[4] In the latter case, however, the calculation was not self-consistent, i.e., in evaluating the quasi-linear diffusion absorption of LH waves was disregarded. In this paper we present and discuss numerical results for the absorption profile, as well as for the profile of the driven current density, by accounting for quasi-linear absorption by electrons in a self-consistent way. The approximation made for the wave dynamics, namely to consider the limit of magnetized electrostatic electron waves, is retained.

## OUTLINE OF THE MODEL

The power density dissipated per individual Fourier component in $N_{\parallel}$ is just the spatial rate of change of the (spectral) energy density flux $S_0(N_{\parallel}) \times \exp[-2 \int dx \; (k_e'' + k_\alpha'')(x,N_{\parallel})/\cos\theta(x)]$ so that for a spectrum of $N_{\parallel}$ the power density deposited into either electrons (e) or alphas ($\alpha$) is

$$P_{e,\alpha}(r) = 2 \int dN_{\parallel} \frac{k_{e,\alpha}''(r,N_{\parallel})}{\cos\theta(r)} S_0(N_{\parallel}) \exp[-2 \int_r^a dx \; \frac{(k_e''+k_\alpha'')(x,N_{\parallel})}{\cos\theta(x)}] \quad (1)$$

with $S_0(N_{\parallel})$ the (spectral) energy density flux at $r = a$. Considering a rectangular-like spectrum, one has $S_0(N_{\parallel}) = \mathcal{P}/\Sigma \; \Delta N_{\parallel}$ for $N_{\parallel}^{(min)} < N_{\parallel} < N_{\parallel}^{(min)} + \Delta N_{\parallel}$ and 0 otherwise, with $\Sigma$ the area of the surface relevant to the deposition of the power $\mathcal{P}$ ($\Sigma = 4\pi^2 R_0 r$ for a toroidal surface of major and (effective) minor radius $R_0$ and $r$ ($> 0.4$ a), respectively, and $\Sigma = 0.4 \times 4\pi^2 R_0 a$ if $r < 0.4$ a, a being the tokamak minor radius); $\Delta N_\parallel$ is the width of the LH wave spectrum. The effective wave spatial damping is given by $k''/\cos\theta$, where $k'' \equiv \mathrm{Im}\ \underline{k}\cdot\hat{\underline{v}}_g$ is the imaginary part of $\underline{k}$ along the group velocity $\underline{v}_g$, and $\cos\theta \equiv v_{g\perp}/v_g$ ($= (1 + \omega_{pe}^2/\omega_{ce}^2)^{1/2}\ \omega/\omega_{pe}$, with $v_{g\perp}$ the perpendicular (to the magnetic field) component of the group velocity) accounts for the fact that what is relevant is the wave absorption in the radial (perpendicular) direction (rather than along the group velocity), upon adopting the electrostatic dispersion relation $\omega = \omega_{pe}$ ($1 + \omega_{pe}^2/\omega_{ce}^2)^{-1/2}$ ($k_\parallel/k$).

The absorption of LH waves by the alphas, due to perpendicular unmagnetized Landau damping, for steady-state conditions has been evaluated and discussed in detail elsewhere.[3]

As for the absorption by the electrons, the relevant mechanism is parallel electron Landau damping (ELD); it is evaluated on the basis of the distribution function $F(v_\parallel)$ that is the solution of the steady-state Fokker-Planck (FP) equation describing the balance between the collisional diffusion as well as drag and quasi-linear wave diffusion in (one-dimensional, $v_\parallel$) velocity space (the distribution in $v_\perp$ is assumed to be Maxwellian). Thus, the (spatial) quasi-linear ELD is given by

$$\frac{k''_e}{k''_{e,lin}} \cong \frac{N_\parallel v_t \Gamma/c}{1+N_\parallel v_t \Gamma/c}\ (\frac{c/N_\parallel v_t + \Gamma}{v_\parallel^{(min)}/v_t + \Gamma})\ r^2$$

$$\times \exp\{-[\frac{1}{2v_t^2}\ ((v_\parallel^{(min)})^2 - \frac{c^2}{N_\parallel^2}) + \frac{c/N_\parallel - v_\parallel^{(min)}}{v_t}\ \Gamma]\} \qquad (2)$$

with $k''_{e,lin}$ the linear parallel ELD.[3] In (2), $\Gamma \equiv (v_\parallel/v_t)\ D_c(v_\parallel)/D(v_\parallel)$, where $D_c \equiv (2 + Z_i)\ v_t^5 v_0/v_\parallel^3$ describes the diffusion connected with collisional scattering ($v_t^2 = T/m$; $v_0 = \omega_{pe}^4\ \lambda_{ei}/4\pi n_e v_t^3$, $\lambda_{ei}$ being the Coulomb logarithm). The coefficient $D(v_\parallel)$, describing the diffusion due to the LH waves, is evaluated under the assumption that the shape of the wave spectrum is not significantly affected by the absorption itself. For the rectangular spectrum considered, it is

$$D(v_\parallel) = \frac{A}{v_\parallel^2}\ \frac{\mathcal{P}}{\Sigma\ \Delta N_\parallel}\ \exp[-2\int_r^a dx\ \frac{(k''_e+k''_\alpha)(x,N_\parallel)}{\cos\theta(x)}] \qquad (3)$$

for $c/(N_\parallel^{(min)} + \Delta N_\parallel)$ ($\equiv v_\parallel^{(min)}) \leq v_\parallel \leq c/N_\parallel^{(min)}$, and 0 otherwise, with $A \equiv 4\pi^2(e/m)^2(\omega/\omega_{pe})(c/\omega)(1+\omega_{pe}^2/\omega_{ce}^2)^{-1/2}(1-\omega^2/\omega_{pe}^2-\omega^2/\omega_{ce}^2)^{-1/2}$ Note that the r.h.s. of (2) tends to 1 in the linear regime ($\Gamma \to \infty$), whereas it tends to 0 in the plateau regime ($\Gamma \to 0$).

## NUMERICAL ANALYSIS AND DISCUSSION

The radial profile of the absorption is now evaluated numerically for a profile of the electron density and temperature given, respectively, by $n_e(r) = (n_{e0} - n_{ea}) [1 - (r/a)^2]^{\gamma_n} + n_{ea}$, and $T_e(r) = (T_{e0} - T_{ea}) [1 - (r/a)^2]^{\gamma_t} + T_{ea}$. For the numerical results that follow we take the profile indices $\gamma_n = 0.5$ and $\gamma_t = 1$; the central and edge-density $n_{e0} = 10^{14}$ cm$^{-3}$ and $n_{ea} = n_{e0}/10$, respectively; the edge-temperature $T_{ea} = 0.5$ keV, whereas the central temperature $T_{e0}$ is either 30 keV or 15 keV; $Z_i = Z_{eff} = 1.3$. Furthermore, the major and minor radius are $R_0 = 630$ cm and a = 205 cm, respectively, and the magnetic field is 60 kG. As for the LH parameters, we consider the frequencies f(GHz) = 4.6, 5, 5.5 and a rectangular spectrum of width $\Delta N_{||} = 0.1$ and $N_{||}(min) = 1.75$ (> $N_{||}(acc)(r/a = 0)$). The results from one set of calculations are shown in Fig.s 1 for $T_{e0} = 30$ keV and in Fig.s 2 for $T_{e0} = 15$ keV, for 20, 60 and 100 MW of LH power. More specifically, the radial profile of the absorbed power density, given by (1), is shown in Fig.s 1a and 2a, and the profile of the corresponding driven parallel current density is shown in Fig.s 1b and 2b. The notable feature is that the power deposition and driven current are restricted to the outer half (in r) of the plasma, the LH penetration improving somewhat as the incident LH power increases and/or the temperature decreases, in either case the ELD tending to be weaker. The figure of merit $\tilde{\eta}$ (= $n_e^* I R_0/\mathcal{P}$, where $n_e^*$ is the density in the point in which the deposition of the power is maximum, and I is the driven current; note that the incident power is used in the figure of merit) for $T_{e0} = 30$ keV (15 keV) and f = 4.6 GHz is 0.13 (0.09), 0.11 (0.07) and 0.10 (0.06) $\times 10^{20}$ A/W·m$^2$ for $\mathcal{P}$ = 20, 60 and 100 MW, respectively, with a maximum driven current of 2.19 MA for $T_{e0} = 30$ keV and 100 MW of LH power. An enhancement factor $\approx 2$ for $\tilde{\eta}$ should be considered since this model is one-dimensional.[5]

The fraction of the injected LH power absorbed by the $\alpha$-particles is given in the following table for $T_{e0} = 30$ keV and, within brackets, for 15 keV, for three different values of both the LH frequency and power:

| | $\mathcal{P}$(MW) = 20 | 60 | 100 |
|---|---|---|---|
| f(GHz) = 4.6 | 9.3 % (12.5 %) | 17.8 % (22.1 %) | 23.7 % (28.4 %) |
| 5 | 6.2 % (10.8 %) | 14.0 % (19.5 %) | 19.6 % (25.4 %) |
| 5.5 | 0.0 % (7.1 %) | 0.6 % (15.0 %) | 2.5 % (20.5 %) |

Note that the power deposition to the alphas is greater for $T_{e0} = 15$ keV than for $T_{e0} = 30$ keV, despite the fact that $n_\alpha(T_{e0} = 15$ keV) < $n_\alpha(T_{e0} = 30$ keV). This results from the improved penetration of the LH waves at lower temperatures.

In conclusion, with the model assumptions made and for the limited parameter range considered, the LH wave absorption by the alphas is found not to be large, in part as a consequence of the limited penetration of the LH waves, due to the strong peripheral ELD.

## ACKNOWLEDGEMENTS

Useful discussions with F. Engelmann and A. Nocentini are gratefully acknowledged. This work was supported by EURATOM and the Ministero della Pubblica Istruzione of Italy.

## REFERENCES


1. R. S. Devoto *et al.*, Proc. of the 16th Europ. Conf. on Controlled Fusion and Plasma Physics, Venice (1989), Vol. 13B Part IV, p. 1295.
2. P. T. Bonoli and M. Porkolab, Nucl. Fusion **27**, 1341 (1987).
3. M. Spada and M. Bornatici, in *Theory of Fusion Plasmas*, Proc. of the Joint Varenna-Lausanne Int. Workshop, Chexbres, Switzerland, 1988 (Ed. Compositori, Bologna), p. 673.
4. M. Spada and M. Bornatici, Proc. of the 16th Europ. Conf. on Controlled Fusion and Plasma Physics, Venice (1989), Vol. 13B Part III, p. 1181.
5. C. F. F. Karney and N. J. Fisch, Phys. Fluids **22**, 1817 (1979).


## FIGURE CAPTIONS

Fig. 1a Profile of the power density (1) deposited in both electrons (full line) and α-particles (dashed line) for $T_{e0}$ = 30 keV, f = 4.6 GHz and various values of the incident LH power.

Fig. 1b Profile of the driven current density for the electron absorption shown in Fig. 1a.

Fig. 2a The same as Fig. 1a for $T_{e0}$ = 15 keV.

Fig. 2b The same as Fig. 1b for $T_{e0}$ = 15 keV.

*Fig. 1a* r/a

*Fig. 1b* r/a

*Fig. 2a* r/a

*Fig. 2b* r/a

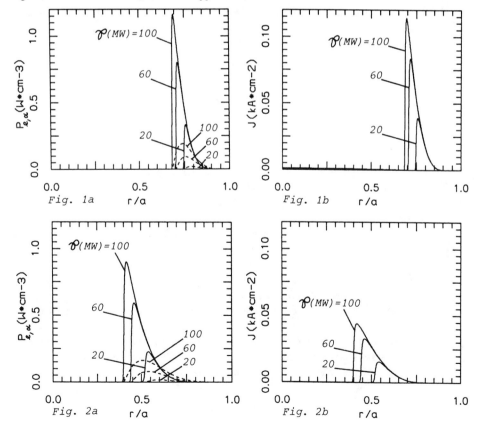

# WAVE PROPAGATION IN A TOKAMAK AT LOWER HYBRID FREQUENCIES[1]

Donald C. Stevens and Harold Weitzner

Courant Institute of Mathematical Sciences

New York University, New York, New York 10012

## ABSTRACT

Maxwell's equations with the cold plasma dielectric tensor are solved to study wave propagation and absorption in a tokamak at lower hybrid frequencies. A resonance other than conventional lower hybrid resonance is exhibited and is shown to be extremely sharp. Numerical limitations of this technique are discussed.

## INTRODUCTION

In an earlier paper[1], we described a code developed to solve Maxwell's equations with the cold plasma dielectric function in axisymmetric geometry. Here, we explore some properties of numerical solutions of the system for a modest sized tokamak such as Petula-B for incident waves at lower hybrid frequencies. We exhibit a strong, sharp resonance different from the usual lower hybrid resonance and we show the sensitivity of this resonance to parameters and antenna position. We discuss numerical limitations on the computability of these solutions for finer numerical grids, larger tokamaks, and higher frequencies.

## THE MODEL

We take $\Lambda = c/\omega$ and solve Maxwell's equations in the form

$$\Lambda \nabla \times \mathbf{E} = -i\mathbf{B} ,$$

$$\Lambda \nabla \times \mathbf{B} = i(\sigma/(i\omega) + \mathbf{I})\mathbf{E} = i\kappa \cdot \mathbf{E} ,$$

where $\kappa$ is the cold plasma dielectric tensor to which we have added a small imaginary multiple of the identity matrix, $i\epsilon\mathbf{I}$, to resolve resonances. In cylindrical coordinates $(r, \theta, z)$, we assume $\mathbf{E} = \mathbf{E}(r, z)e^{iM\theta}$ and $\mathbf{B} = \mathbf{B}(r, z)e^{iM\theta}$. With these coordinates the system reduces to two second order partial differential equations for $rE_\theta$ and $rB_\theta$. To determine $\kappa$ we taken an exact solution of the Grad-Shafranov tokamak equilibrium equations, a Sol'viev equilibrium, so that all toroidal effects are included.

In this geometry[2] the only resonances occur at particular points on the surfaces $\kappa_\perp(\kappa_\perp B_T^2 + \kappa_\parallel B_p^2) = 0$, where $\kappa_\perp = 0$ is the usual lower hybrid resonance, and in a non-periodic geometry the other resonance would occur at $\kappa_\perp + \kappa_\parallel = 0$. $B_T$ and $B_p$ are the toroidal and poloidal components of this equilibrium magnetic field. The reduction from the six Maxwell equations to two second order equations involves algebraic elimination of four unknowns, and this elimination becomes singular on particular curves, referred to in Ref. 1 as $\Delta(r, z) = 0$. We consider that any energy density peaks of the solution on these curves are spurious and we discuss this point below.

---

[1]This work was supported by the U.S. Department of Energy, Grant No. DE-FG02-86ER53223.

We select parameters roughly comparable with Petula-B and we select $r_o = .72M$, $a = .17M$, $B_o = 2.8T$, $\bar{n}_e = .6 \times 10^{19}M^{-3}$, $I_p = 150kA$, and $\omega = 1.3GHz$. We scan in the parameter $M$ which we allow to take non-integral values. We could easily have scanned on other, more physically reasonable parameters, but our scan readily shows the phenomena we seek. We take a particularly simple antenna consisting of one point on the boundary on which we give a unit value to $B_\theta$, elsewhere on the boundary $B_\theta = 0$. We set $E_\theta = 0$ on the boundary. We calculate $E_\theta$ and $B_\theta$ wave fields in the interior of the domain. We evaluate the energy absorption rate in the plasma and the complex Poynting vector on the boundary near the antenna; at other boundary points the Poynting vector vanishes. Energy conservation holds within a few percent. A measure of the strength of the resonance is the ratio of the real part of the Poynting vector to its absolute value. Off resonance the ratio is almost zero. In plots of this resonance parameter as a function of position, we parametrize the outside and inside of the tokamak by $0 \leq p \leq 1$ and $2 \leq p \leq 3$, respectively. The top and bottom of the tokamak are $1 \leq p \leq 2$ and $3 \leq p \leq 4$, respectively. In the plots of wave fields and energy absorption contours, the outside and inside of the tokamak are at the top and bottom of each frame, respectively. In the field and energy absorption contours two ovals dotted appear. The outer dotted oval corresponds to $\Delta = 0$, the inner dotted oval to the resonance $\kappa_\perp B_T^2 + \kappa_\parallel B_p^2 = 0$. No lower hybrid resonance appears in the system.

## RESULTS

Figures 1 and 2 have contours of $E_\theta$ and $B_\theta$ fields and Fig. 3 has contours of energy density for $M = 45.742$ which is near a resonance. Figures 4-6 have these contours for $M = 45.74587$, a resonance value. The energy absorption at resonance is fairly clearly focussed on a few particular points on the resonant surface, but not on the full surface. Figures 7 and 8 have energy density contours at other resonances. All these solutions are for $\epsilon = 10^{-6}$. Figure 9 shows a scan of the resonance parameter as a function of $M$ for $\epsilon = 10^{-4}$. The peaks of energy absorption broaden and their heights decrease toward 0 as $\epsilon$ increases, and the peaks sharpen, their heights approach 1, and the mode pattern does not change as $\epsilon$ decreases. Figure 10 plots the resonance parameter as a function of antenna position with $M = 42.0411632$ and $\epsilon = 10^{-6}$.

We have not developed satisfactory means to deal with the numerical singularities of $\Delta = 0$. Several of Figures 1-8 show spurious peaks at points near the surface $\Delta = 0$. The calculations for these figures were done with grids of $41 \times 41$. At still finer grids this problem gets worse. These difficulties, together with the practical limitations of working with extremely large matrices, limit the ability of calculations such as these to represent larger tokamaks at higher frequencies which automatically require much finer grids. Nonetheless these computations clearly show the novel resonance in tokamaks with its extreme sharpness and sensitivity to antenna position.

## REFERENCES

1. Lower Hybrid Wave Propagation in a Tokamak, D.C. Stevens and H. Weitzner, in Applications of Radio-Frequency Power to Plasmas, Seventh Topical Conference, Kissimmee, FL (1987), AIP Conference Proceedings 159, eds. S. Bernabei and R.W. Motley, p. 203.

2. Wave Propagation Based on the Cold Plasma Model, H. Weitzner, CIMS-NYU Report MF-103.

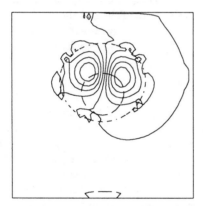

Fig. 1  $E_\theta$  M=45.742

Fig. 2  $B_\theta$  M=45.742

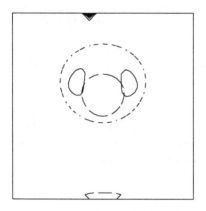

Fig. 3  Energy Density  M=45.742

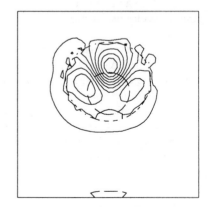

Fig. 4  $E_\theta$  M=45.74587

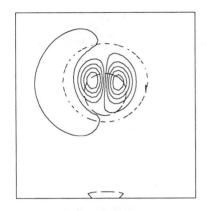

Fig. 5  B$_\theta$  M=45.74587

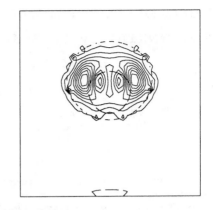

Fig. 6  Energy Density  M=45.74587

Fig. 7  Energy Density  M=45.89329129

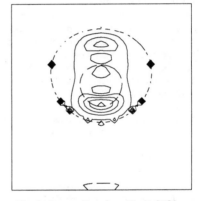

Fig. 8  Energy Density  M=41.0782

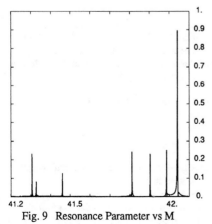

Fig. 9  Resonance Parameter vs M

Fig.10  Resonance Parameter vs Antenna Position

# BROADENING OF LOWER HYBRID WAVE SPECTRA BY MAGNETIC RIPPLE; SIMULATION OF EXPERIMENTAL RESULTS AND APPLICATION TO NET.

*J.-G. Wégrowe*

*The NET Team, c/o Max-Planck-Institut für Plasmaphysik,
D-8046 Garching bei München, Federal Republic of Germany*

## ABSTRACT

Various mechanisms have been proposed to explain the commonly observed '$N_{//}$-upshift' of lower-hybrid waves (LHW). This term was coined to describe the fact that, to account for the magnitude of the absorbed power and of the driven current, in many experiments one has to assume that the spectral power density spectrum extends to much lower phase velocities (larger $N_{//}$) than expected from the launched spectrum.

As proposed recently[1], the magnetic ripple (and also magnetic islands) acting like a diffraction grating upon the lower-hybrid waves, could be a quite efficient mechanism for this effect; it may also explain the lack of penetration of LHW sometimes observed in the ion regime since the broadening is found to increase with $(k_\perp/k_{//})$, and thus with density.

In this paper, we describe the physical mechanism involved and present numerical simulations of ASDEX lower-hybrid current drive experiments performed by means of a 1D Fokker-Planck code. In this experiment, a strong broadening of the launched wave spectrum is needed to explain the results[2]. Ripple-broadening permits a satisfactory simulation of ASDEX shots in various conditions.

This mechanism of spectral broadening, unlike others (e.g. edge bouncing) is active even when total absorption is expected during the first pass through the plasma, which is the case in thermonuclear conditions. Simulation using NET or ITER burn phase parameters show that, if this effect is actually operative, a strong degradation of the wave penetration results.

## I - INTRODUCTION.

A number of possible explanations for the occurrence of the upshift has been proposed: edge bouncing of the the rays[3], strong density fluctuations[4], non linear effects in caustics[5], effect of Parail-Pogutse instability[6], ponderomotive effects[7], diffraction effects[8], ray stochasticity[9] and wave scattering on toroidal inhomogeneities like the magnetic ripple[1], to be described below. However, to our knowledge, no experimental evidence exists as yet for supporting one or the other specific mechanism.

It is however mandatory to understand the reasons for the spectral broadening when considering the use of LHW to a Tokamak reactor since maintaining a narrow wave spectrum is essential for a good penetration.

## II - THE 'RIPPLE-BROADENING' MECHANISM.

The basic mechanism invoked is the diffraction of the LHW on the regular ondulations of the magnetic field which act as a grating.

Although the spread of the sidebands due to this diffraction is very small (ripple wavelength assumed to be large as compared to the parallel wavelength, i.e $K << k_z$), the modulation depth and thus the distorsion of the wavefront can be quite large due to the large ratio of $k_\perp$ to $k_{//}$. This effect has been computed in

/1/; we shall give below a less rigourous derivation of its magnitude to illustrate the process.

We assume in the following a slab geometry (Fig. 1). The total unperturbed magnetic field is in the direction z (we neglect the rotational transform), the radial direction is r, and y represents the poloidal direction.

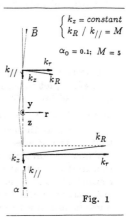

Fig. 1

The magnetic field with ripple is characterized by the angle $\alpha$ of the total field with Oz ($\alpha_0$ is half the peak-to-peak ripple): and by the wavelength of the perturbation: $2\pi R/m$, large as compared to the parallel wavelength of the wave.

$$\alpha(z) = \alpha_0 \cos(mz/R) \qquad (1)$$

We consider one of the wavelets of the impinging wave spectrum: $\vec{E}(N_{//}) \simeq \hat{k}\, e^{i(k_z z + k_y y + k_r r)}$ , whereby $\vec{k}_y + \vec{k}_r = \vec{k}_\perp$ and the quasi electrostatic approximation is made: $\vec{E}_{k_z} \propto \hat{k} E^0_{k_z}$.

The dispersion relation applicable locally in the frame attached to the magnetic field yields $k_\perp = M(k_{//}$ , plasma parameters and frequency$)$ namely,

$$M = \frac{\omega_{pe}}{\omega}\frac{1}{\sqrt{1 + \omega^2_{pe}/\omega^2_{ce} - \sum_i(\omega^2_{pi}/\omega^2)}} \gg 1 \qquad (2)$$

M , i.e. the plasma parameters, are taken to be constant along z. Now to compute the local values of $k_{//}(z)$ and $k_\perp(z)$ corresponding to the initial wavelet, we write that $k_z$ and $k_y$ are constant along the rippled field:

$$k_z(z) = k_{//}\, cos\alpha - k_R\, sin\alpha = k_{//}\, cos\alpha - \sqrt{k^2_\perp - k^2_y}\, sin\alpha \qquad (3)$$

i.e., by using $k_\perp = Mk_{//}$:

$$k_{//}(z) = \frac{k_z cos\alpha + sin\alpha\sqrt{M^2 k^2_z - (cos^2\alpha - M^2 sin^2\alpha)k^2_y}}{cos^2\alpha - M^2 sin^2\alpha} \qquad (4)$$

For simplicity, consider the case $k_y = 0$:

$$k_{//}(z) = \frac{k_z}{cos\alpha - M sin\alpha}; \quad k_r(z) = k_z\frac{M cos\alpha + sin\alpha}{cos\alpha - M sin\alpha} \qquad (5)$$

These formulae show the strong modulation of the wave numbers due to the anisotropy (M$\gg$ 1) (as illustrated above).

Assuming furthermore $\alpha \ll 1$ and $M\alpha \ll 1$, we get:

$$k_r(z) \simeq k_z M(1 + M\alpha) = k_{r0} + M^2\alpha k_z \qquad (6)$$

(where $k_{r0} = Mk_z$ is the unperturbed radial wave number). Computing now the wave spectrum $\vec{F}_{(r_0 + dr_0)}(k'_z)$ from the perturbed phase distribution after a radial propagation $dr_0$, we obtain,

$$\vec{F}_{(r_0+dr_0)}(k'_z) \simeq \int_{-\infty}^{+\infty} \hat{k} e^{-i(k'_z - k_z)z + ik_{r0}\, dr_0}\, e^{ik_z dr_0 M^2\alpha} dz \qquad (7)$$

replacing $\alpha$ by its expression as a function of z, and using a well known Bessel identity, one has:

$$\vec{F}_{(r_0+dr_0)}(k'_z) \simeq e^{ik_{r0}\, dr_0}\sum_\infty^\infty \hat{k}_n J_n(q)\delta\left(k'_z - k_z - n\frac{m}{R}\right) \qquad (8)$$

with: $q = M^2 k_z m\alpha_0\, dr_0/R$. For $dr_0 = 0$, one recovers the original wavelet (delta function), but already after a short radial propagation($q \simeq$ n, i.e: $dr_0 = nR/M^2 mk_z\alpha_0$), the power has been transferred on the side-bands around $k_z \pm nK$. In other words, the original power spectrum is gradually shifted up- and downwards at an approximate rate:

$$\frac{\Delta k_z}{k_z} = \frac{mM^2\alpha_0}{R}dr_0$$ (9)

(as obtained in /1/ by more sophisticated means). To estimate the magnitude of this effect, let us consider the following set of parameters (ASDEX): $f = 1.3$ GHz; $n_e = 2\ 10^{19}m^{-3}$; $B = 2.5$ T; gas: D; $R = 1.65$m; $\alpha_0 = 5.10^{-3}$; $m = 16$ (coils). This gives: $M = 30$ and $\Delta N_{//}/N_{//} = 0.44 dr_0^{[cm]}$.

Note that the relative broadening is proportional to the density. Equ.(4) shows that a strong effect is also to be expected when $k_y \simeq k_\perp$ (neighbourhood of the whispering gallery: although there the ripple is much less, this effect may contribute to penetration beyond the predictions of ray-tracing). Finally it must be remarked that besides the ripple, also other kinds of regular perturbations can give rise to a spectral broadening. (magnetic perturbation in islands[10,1], density modulation due to islands, etc...).

## III - SIMULATION OF ASDEX RESULTS

ASDEX LHW current-drive experiments have been simulated by means of the 1-D Fokker-Planck code LOCH[11]. The code uses a cylindrical geometry, ray-tracing was not implemented (note that the required spectral broadening effects largely exceed the spectral variations due to classsical toroidal propagation).

The launched power spectrum is assumed to have the shape:

$P(N_{//}) = \left\{\frac{sin\left[\pi(N_{//}-<N_{//}>/\Delta N_{//})\right]}{\pi(N_{//}-<N_{//}>/\Delta N_{//})}\right\}^2 \frac{1}{(N_{//}^2-1)^{2/3}}$, where $<N_{//}>$ and $\Delta N_{//}$ are given by the grill geometry and phasing. Spectral broadening by magnetic ripple is implemented as follows: at each radial step, before performing the FP calculation, the power $P_r(N_{//})$ of each spectral components is distributed equally in two sidelines respectively shifted by $\Delta N_{//} = \pm N_{//}\ \frac{m\delta(r)dr}{R}\left[\frac{N_\perp}{N_{//}}\right]^2 cos\Phi$ (where $m$ is the number of toroidal field coils, R and a the major and minor radii, $\delta$ the magnetic ripple amplitude at radius r for which we take $\delta(r) \simeq \delta(a)\left[\frac{R+r}{R+a}\right]^{m-1}$, and $\Phi$ the angle between radial and poloidal components of the wave vector (here assumed to be given by $\Phi = \pi|a - r|/4a$). $N_\perp$ is computed by the warm dispersion relation.

Analytical fits of the measured density and temperature profiles are used in the simulations.

The experiments have been made at various LHW input power, maintaining constant the plasma current [12]. The comparison between code and experiment[11] for 3 shots with null, co- and counter-acting electric field are summarized in the following table:

**TABLE I**

| shot | experiment ($I_p = 300$ kA) | | | | | code | | | |
|------|------|------|------|------|------|------|------|------|------|
| | $n_e(0)$ $[10^{19}m^{-3}]$ | $T_e(0)$ $[keV]$ | $P_{inj}$ $[kW]$ | $V_{loop}^\Omega$ $[V]$ | $V_{loop}^{LH}$ $[V]$ | $\delta(a)$ $[\%]$ | $V_{loop}^{LH}$ $[V]$ | $I_{driven}$ $[kA]$ | $I_{tot}$ $[kA]$ |
| 18468 | 0.80 | 2.81 | 375 | 0.86±0.2 | 0.0±0.2 | 0 | 0.09 | 3 | 37 |
| 18468 | 0.80 | 2.81 | 375 | 0.86±0.2 | 0.0±0.2 | 2 | 0.09 | 260 | 297 |
| 18470 | 0.96 | 2.19 | 90 | 1.09 ±0.2 | 0.41 ±0.22 | 2.2 | 0.70 | 162 | 305 |
| 18466 | 0.72 | 1.96 | 920 | 1±0.2 | -0.25 ±0.2 | 1.8 | -0.22 | 391 | 303 |
| B=2.2T; $Z_{eff}$=3; $<N_{//}>$=1.90; $\Delta N_{//}$=0.955; f=1.3 GHz | | | | | | | | | |

Note that without any broadening of the launched spectrum, the driven current fails by 2 orders of magnitudes to explain the experimental level. On the contrary, a satisfactory modelling of the experimental data can be achieved with this model, using a ripple value of the order of 2%. It must be noted however that this value is about twice the value of ASDEX. Considering the rough approximations used to compute the broadening, this gross agreement encourages to investigate further the possible responsibility of the proposed mechanism in the spectral broadening effect. Fig. 2 represents the wave power spectra during inwards propagation for the same shot with and without spectral broadening.

## IV - SIMULATION FOR NET PARAMETERS.

LHW are considered in the Next-Step devices for current ramp-up, and, in combination with other methods, for plasma and current initiation and for steady-state current drive during burn. In the latter case, the LHW does not generally penetrate beyond mid-radius. One condition to achieve an acceptable penetration is to keep the wave spectrum as narrow as possible (the quasi-linear absorption coefficient being roughly proportional to $\Delta N_{//} < N_{//} >$).

Fig. 3 shows the dramatic effect in NET II[13] on the wave penetration of the ripple broadening modelled as above, for a realistic value of the edge ripple. The parameters used in this run are the following:
$R = 6.3$m, a $= 2.05$m, $\kappa = 2.2$, $B_0 = 6$T, $Z_{eff}=1.5$, I=15MA, $n_e(0)= 1.05 10^{20} m^{-3}$ (profile: square root of parabola), $T_e(0)= 30$keV (profile: parabola).
f= 5.6GHz. We have chosen a very narrow initial wave spectrum, close to the accessibility limit: $< N_{//} > = 1.62$, $\Delta N_{//} = 0.1$ (full width of 1st. lobe).

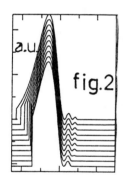

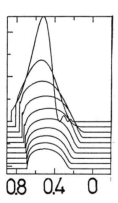

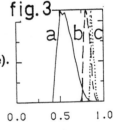

Figure Caption:
Fig. 1: Geometry.
Fig. 2: Spectral power density as a function of $v_{//}/c = 1/N_{//}$ for equidistant radii (top=plasma edge, bottom=centre). Top figure without spectral broadening; bottom figure $\delta = 2\%$.(ASDEX)
Fig3: Radial driven current density profiles in NET II as functions of r/a.
a) $\delta = 0$, b) $\delta = 0.5\%$, c) $\delta = 1\%$.

/1/ J.-G. Wégrowe, Proc. 14th Eur. CCPP, Madrid 1987 III, 911
/2/ IAEA INTOR-Related Specialists Meeting on Non-Inductive Current-Drive, Garching, FRG, Sept. 1986, NET-PM-86-003 and IAEA-TECDOC-441, Vienna, 1987.
/3/ P. Bonoli and R.C. Englade Phys. Fluids, 29, 2937 (1986)
/4/ P. Bonoli, E. Ott, Phys. Fluids, 25(2), 359, Feb 1982.
/5/ E. Barbato, A. Cardinali, F. Santini, Proc. of the 4th. Int. Symp. on Heating in toroidal Plasmas, Roma, It, (1984)
/6/ see e.g. J. Vacklavik et al. Plasma Physics 25, 1283, (1983)
/7/ R. Croci and E. Canobbio; Z. für Naturf. A-42,10, (1987), 1067.
/8/ G.V. Pereverzev, JETP Lett.,44,9,(1986), 549
/9/ D. Moreau et al., 15th. Eur. Conf. on CFPP, Dubrovnik (1988),III-995 and , 16th. Eur. Conf. on CFPP, Venezia, (1989),III-1169
/10/ F. Romanelli et al., Proc. 14th Eur. CCPP, Madrid 1987 III, 903
/11/ J.-G. Wégrowe and G. Zambotti, 16th. Eur Conf. on CFPP, Venezia, (1989),IV-1247
/12/ F. Leuterer et al., 12th. Eur Conf. CFPP, Budapest (1985),II-240
/13/ The NET Team, R. Toschi, IAEA-CN-50/G-1-1, 12th. Int. Conf. PPCF, Nice, F, Oct. 1988.

CHAPTER 3

ION CYCLOTRON RANGE OF FREQUENCIES

# ICRF and ICRF plus Neutral Beam Heating Experiments on TFTR

P. Colestock, A. Cavallo, W. Dorland, J. Hosea, G. Greene, G. Hammett, H. W. Hendel, B. Howell, K. Jaehnig, R. Kaita, S. S. Medley, C. K. Phillips, A. L. Roquemore, G. Schilling, J. Stevens, B. Stratton, D. Smithe, A. Ramsey, G. Taylor, J. R. Wilson, S. J. Zweben and the TFTR Group, Princeton University, W. Gardner and D. Hoffman, ORNL

## I. Introduction

An ICRF heating program is currently underway on TFTR to explore the potential for improving the confinement and enhancing the fusion gain of large-scale, hot tokamak plasmas. It is expected that the ability of ICRF waves to penetrate to the core of large, dense plasmas and deposit energy locally near the cyclotron resonance layer will lead to optimal heating rates and improved fusion gains, especially at densities where present neutral injection energies are insufficient for adequate penetration. One of the goals of these experiments, hence, is to explore the characteristics of the wave heating at comparatively high densities. A further enhancement of the global confinement in the so-called supershot regime may also be expected, since in this case the central electrons are only weakly heated by the injected ions, while the centrally-generated resonant ions typically can be made sufficiently energetic to couple virtually all of their energy to the electrons via collisions.

Another important goal of this work is the study of the physical effects in the near field of the antennas and the determination, as well as control, of the processes limiting the antenna power density. This effort is directed toward the improvement of the antenna operating characteristics for application to the CIT antenna design.

In these experiments, up to 2.8 MW at 47 MHz has been applied with an end-fed antenna, consisting of a pair of adjacent striplines. Plasma heating experiments have been carried out in both the minority hydrogen and minority $He^3$ regimes. In the former case, the addition of deuterium neutral beam injection permits the possibility of second harmonic resonant beam ions. The study thus far has been focussed on the heating of low density and comparatively high temperature target plasmas, which, in the presence of sufficient auxiliary power give rise to the strongly-peaked profiles and enhanced energy confinement characteristic of supershots. The effects of antenna phasing have also been studied in these experiments leading to the observation of significant visible emission and impurity production under certain pathological conditions. In this paper we review the experiments in both the rf only and rf plus neutral beam injection heated plasmas. We present an anlysis of the transport physics, including a model for the wave deposition, resonant ion slowing down and 1-D transport, both with and without an injected fast ion species. In addition, we report detailed observations of the antenna-edge plasma interaction and the effects of antenna phasing. Finally, we discuss plans and prospects for future ICRF experiments on TFTR.

## II. Ohmic Target Plasma ICRF Heating Results

The heating regime studied most extensively in this work has been minority H or $He^3$ in a $He^4$ plasma. Power levels up to 2.8 MW have been applied to these discharges with a pair of current straps driven out-of-phase and up to 1.8 MW when driven in-phase. The dependence of the operating characteristics on phasing is significant and will be discussed more below. Details of the antenna design are described elsewhere[1]. The effects of the minority H heating on the bulk discharge parameters are shown in Fig. 1. An increase in the total plasma stored energy, determined magnetically, is accompanied by a pronounced increase in the central electron temperature at the rate of ~ 2 eV/kW/$10^{19}$ m$^{-3}$. In these cases, it is significant that the radiation from high Z ions during the rf pulse is negligible, while the low Z radiation rate actually falls, relative to the ohmic plasma. Associated with these effects is a slight increase in density, as shown. The results for He$^3$ minority heating were qualitatively similar. However the electron heating rate and the global stored energy increase was about 30% lower than that of the H minority case. The incremental stored energy as a function of the applied power is shown in Fig. 2. Analysis of these experiments has been carried out using a 1-D time independent transport code including a full-wave model for the wave deposition and an isotropic slowing down model for the rf-induced fast ions. Details of this work are presented elsewhere.[2] As expected, strong fast ion tails are generated which, in the case of the H minority, are estimated to give up 90% of their power collisionally to the electrons.

As a consequence of the strong central electron heating, it is possible to directly estimate the central rf power density from the slope of the sawtooth oscillations of the electron temperature. A comparison of the electron heating rate with the predictions of the full wave deposition analysis indicates that the measured central power densities fall 3 - 4 times below the theoretical values. Similar deposition profile broadening has been observed in previous experiments[3,4]. A feasible explanation of these observations is the possibility of rapid mixing of the fast ions within the q=1 surface during the sawtooth

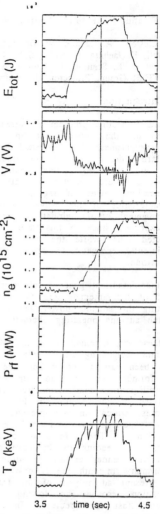

Fig. 1 Time evolution of H minority heating in a He$^4$ plasma

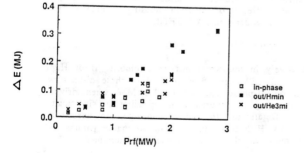

Fig. 2 Incremental stored energy for H minority in- and out-of-phase and He$^3$ minority out-of-phase.

crash.[5-7]    Profile measurements indicate a flattening of the central electron temperature and a slight decrease in the central density.    According to the model[5], the sawtooth crash reconnects the magnetic field lines out to approximately    $\sqrt{2}$ times the q=1 radius.    If the fast ions are then instantaneously redistributed within this volume, the power available for central electron heating is decreased to a level consistent with the sawtooth slope, as shown in Fig. 3(a).    Such a physical picture is further supported by the off-axis heating case shown in Fig. 3(b), where no central tail is expected

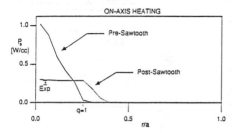
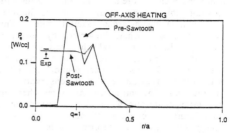

Fig. 3(a)    Theoretical electron heating power density profiles for minority H heating with and without sawtooth mixing model.    Experimental value on-axis is shown.

Fig. 3(b)    Theoretical electron heating power density profiles as in 3(a) for off-axis case.

without a mixing effect.    Further evidence for the sawtooth mixing is inferred from off-axis charge exchange measurements that show high energy fluxes inverted with the central sawteeth, as shown in Fig. 4.    Other mechanisms for fast ion transport within the q=1 surface have been investigated, but have been found to be too small to explain these observations.[8]

Fig. 4 Deuterium charge exchange flux at 136 keV for an H minority case with 110 keV neutral beam injection.    The detector sightline is 30 cm off-axis (q=1 at r=15 cm).    An increase in the flux occurs at the sawtooth crash.    Reference levels are specified by the diagnostic neutral beam as shown.

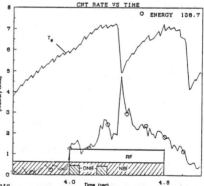

## III.    RF plus Neutral Beam Heating Experiments

A series of experiments was also carried out with rf power added to low density discharges, dominated by neutral injection heating, in an effort to determine the viability of improving the so-called "supershot" performance with ICRF.    In these experiments, up to 7 MW of CO or COUNTER injected deuterons at a primary energy of 110 keV was injected into the discharge prior to the rf pulse, as shown in Fig. 5.    The heating effects were found to be roughly additive, though the largest stored energy increase was observed with fully CO injection. The greatest effect of the additional rf power was to produce a central electron temperature increase, consistent with the slowing down of the resonant ions. The higher electron temperature, in turn, produced an increase in the beam ion slowing down time resulting in an increase of the D-D beam-target reaction rate. This was offset somewhat, however, by the 10% density increase, which reduced the beam-beam reaction rate. Good agreement with the 1-D model was found, consistent with these neutron emission observations. In the case of minority hydrogen heating, the beam deuterons were also observed to be significantly accelerated above the beam energy by the rf, as described below.

Modelling of these experiments was carried out with the same model used above in the rf only case, but with the additional feature of beam ions that may be resonant with the rf waves. The simulation of the tails of both minority hydrogen and beam deuterons was done with a bounce-averaged Fokker-Planck code, including the spatial dependence of the rf power deposition. The comparison of the measured beam ion distribution with this theory is shown in Fig. 6. The fit to the deuteron distribution

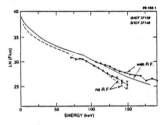

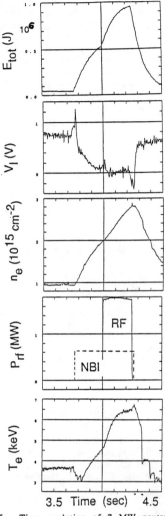

Fig. 6 Deuteron charge exchange measurements of beam slowing down spectrum with and without ICRF. Solid curve is theoretical fit to second harmonic-heated CO -injected beam ions above the beam energy (110 keV).

Fig 5. Time evolution of 7 MW neutral beam injection plus 2 MW ICRF in the H minority regime into a $He^4$ target plasma.

above the beam energy is found by adjustment of the hydrogen concentration, which controls the power partitioning between hydrogen and deuterium. The power absorbed by the deuterons can be directly determined by an integration of these spectra, yielding approximately 20% of the total rf power in the deuterons. This value is in good agreement with the theoretical prediction from the full wave damping model and a 3% minority ion concentration, provided the elevated beam energies are included in the damping calculation. We note that both ion species are measureably anisotropic, and contain a sizable fraction of trapped particles whose turning points lie in the resonance layer. Such a picture is consistent with the measurements from the vertically-oriented charge exchange analyzer array that indicate a pronounced increase in flux as the resonance layer approaches the viewing line, as shown in Fig. 7. Similar results are obtained by edge particle detectors that respond to high energy

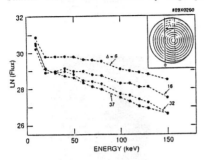

Fig. 7    Charge exchange hydrogen distributions during a scan of the resonance layer location relative to the detector sightline for an H minority heated case. The spatial dependence is consistent with an accumulation of trapped particle with turning points in the resonance layer.

protons, as shown in Fig. 8. In these results, the escaping proton flux is observed to be highly sensitive to the applied rf power as well as to the location of the resonance layer, relative to the detector. An investigation of this flux using single particle orbit analysis reveals that the high energy protons must be receiving a large increment in perpendicular energy (~ 100 keV) in a single transit of the resonance layer in order to be able to intercept the detector. Although the total flux of such particles is small, their behavior violates the quasi-linear assumption commonly applied to rf-induced velocity diffusion, and warrants further study.

## IV.   Antenna Effects

As indicated above, significant differences in heating efficiency were noted, depending on whether the antennas were driven in- or out-of-phase. Associated with in-phase drive, bright visible emission was observed near the ends of the antennas, as shown in Fig. 9, at an applied rf power of 1.1 MW. The intensity of this emission was greatest at a toroidal field of 3.4 T (corresponding to 2.5T at the antenna) and was associated with a large increase in the titanium content of the plasma. The only source of titanium in these experiments is believed to be the TiC coated Faraday shields. It is noted that the out-of-phase drive showed neither a significant impurity increase nor any visible emissivity over the full range of applied powers (up to 2.8 MW).

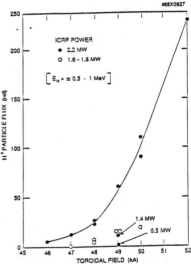

Fig. 8    Escaping proton flux during the toroidal field scan of Fig. 7. Measurements indicate a strongly power dependent lost particle flux at high energy.

194

Detailed measurements of the loading have also been carried out and compared to a one dimensional coupling model that includes the possibility of standing waves both in the radial and toroidal directions.[9-11] Using measured density and temperature profiles and adjusting the minority concentrations accordingly, a good fit to the dependence of the loading on density can be obtained for both antenna phasings, as shown in Fig. 10 In the in-phase case, radial standing waves form between the mode conversion layer and the antenna, resulting in the widely-spaced peaks shown. A second group of standing waves formed from the rf power that has tunnelled through the mode conversion zone and reflected from the inside vessel wall is believed to be responsible for the split peaks observed. In order to produce this feature theoretically, a hydrogen concentration of <1% was required, and this value is roughly consistent with the deposition model fit described above. This value, however, is considerably lower than the edge hydrogen concentration, measured spectroscopically. Moreover, significant high energy tail components were required to match the peak-to-valley ratio observed. It is interesting to note that while the slope of the observed background loading increase with density is well-described by the theory, the best fit requires the introduction of an offset value added to the loading resistance, as indicated in Fig. 10(a). This value is approximately 30% of the average loading. In the out-of-phase case, the loading increases similarly with density, however, no such offset is required to fit the observed data. In this case, the observed cutoff density (where the loading curve intercepts the $n_e$ axis) suggests a dominant $k_{||}$ component of 7 m$^{-1}$, in rough agreement with expectations. In addition, the out-of-phase case exhibits a series of small loading peaks as a function of density, as indicated in Fig. 10(b). Both the spacing and peak-to-valley ratio of these modes is reproduced by the theoretical model as described above, with no other changes in assumptions or profiles, provided the toroidal connection length is included in the model. Thus, in this case, the modes appear to be weak toroidal standing waves rather

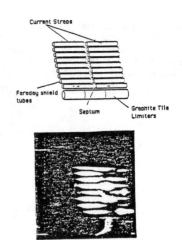

Fig. 9 Visible light emission and corresponding schematic view of portion of the antenna during in-phase operation, 1.3MW.

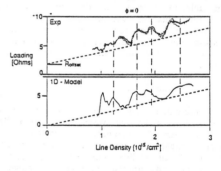

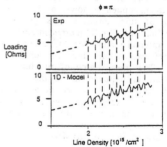

Fig. 10 Comparison of antenna loading measurements and 1-D coupling model for the H minority case (a) in-phase (b) out-of-phase.

than local radial standing waves, as occurs in the in-phase case. It is also noted that the spacing between potentially excited poloidal modes does not correspond to any of the spacings observed here. Thus we conclude that the antenna has a narrow poloidal mode spectrum, in agreement with the predictions of 2-D coupling models.[1,2]

## V.   Conclusions

In ICRF heating experiments at up to 2.8 MW on TFTR, we have successfully heated the plasma core and the results can be described well by the theoretical model.   In particular, the minority heating mode can be used to heat electrons on axis via fast ion slowing down, resulting in a significant increase in the central electron temperature. It is expected that such a situation can improve the neutral beam heating case, especially at high densities, by virtue of the localized central deposition.   Good agreement beween theory and experiment regarding details of the ion distributions has also been obtained, including the power partitioning between deuterons and protons,   the energy dependence, and the effects of sawteeth on the fast ion profiles. Observations of the antennas during operation indicate a clear advantage in out-of-phase operation and suggest that local   edge phenomena may be responsible for the lower observed central heating efficiency in the in-phase case.   A detailed comparison of the antenna loading measurements with a 1-D theory reproduces much of the observed behavior, suggesting that both radial and toroidal standing waves occur, in reasonable agreement with the mode conversion theory and suggestive of a well-collimated antenna spectrum.

Future plans for the TFTR experiments include the installation of an additional 4-5 MW in the 40-80 MHz range.   Emphasis will be placed on enhancing the performance of neutral beam heated   plasmas, as well as the exploration of heating at higher densities, i. e. in conjunction with pellet injection experiments.   In addition, the potential for possible stabilization of MHD modes by the generation of a large fast ion population will be investigated.   Of particular importance in these experiments will be the monitoring of effects observed on the antennas and the clarification of the dependence of the coupling and heating results on the antenna phasing.

Work supported under USDOE Contract # DE-AC02-76-CHO3073

References:

1.   J. C. Hosea, et. al., This conference

2.   C. K. Phillips, G. W. Hammett, et. al., This conference

3.   E. Mazzucato, R. Bell, et. al., Proc. 10th International Conference on Plasma Physics and Controlled Nuclear Fusion Research,( London), IAEA, Vienna, Austria, (1985)   Vol. 1, p. 433

4.   V. P. Bhatnagar, J. G. Cordey, et. al., Joint European Torus, JET-p(88)51, Sept. 1988

5.   B. B. Kadomtsev, Sov. J. Plasma Physics, 1 (1975), p.389

6.   S. Itoh, K. Itoh, et. al., Proc. 14th European Conf. on Controlled Fusion and Plasma Physics, Madrid, (1987), 11D, p. 1204

7.  L. -G. Eriksson and T. Hellsten, Proc. 16th European Physical Society, Venice, (1989), p. 1077

8.  L. Chen V. Vaclavik and G. W. Hammett, Nuclear Fusion, **28**, (1988), p. 389

9.  Y. Lapierre, Chauffage Cyclotronique Ionique dans le Tokamak de Fontenay aux Roses, Ph. D. Thesis, L'Universite Paris-Sud, (1982), University Microfilms, Ann Arbor

10.  G. J. Greene, P. L. Colestock, et. al., This conference

11.  A. L. McCarthy, V. P. Bhatnagar, et. al., Proc. 15th European Conference on Controlled Fusion and Plasma Heating, Dubrovnik, (1988), Vol. II, p 717

12.  D. N. Smithe, P. L. Colestock, R. J. Kashuba and T. Kammash, Nuclear Fusion, **27**, 8 (1987) p. 1319

# HAMILTONIAN FORMULATION OF THE ICRF HEATING DYNAMICS IN TOROIDAL MAGNETIC FIELD CONFIGURATION.

A. Becoulet, D.J. Gambier, J-M Rax, A. Samain

Association Euratom CEA sur la Fusion, CEN Cadarache, France 13108.

## ABSTRACT

For the ICRF wave particle coupling, it is only when the Random Phase Approximation (RPA) is valid that a diffusion process takes place, from which particle heating results. In the prospect of coming studies of suprathermal population dynamics, i.e. low collisionnality regime, only intrinsic stochasticity above its threshold may legitimate the RPA. It is then essential that ICRF stochastic heating of passing and trapped particles be reviewed. Finally, two suprathermal ICRF schemes are investigated, in which the criterion and efficiency of enhanced fusion rate and fast wave current generation are given.

## ICRF HEATING ISSUE AND METHOD OF ANALYSIS

In essence, ICRF is a particular case of wave particle resonant coupling. It occurs, along the guiding centre trajectory, when the particle gyrophase matches the wave phase: $\omega = -\omega_{ci}(r_G(t))$. From a ballistic point of view, the effect of the particle wave interaction leads to an integration of the phase $\phi(t)$ along the trajectory, i.e. $\int \exp\{i\phi(t)\}\, dt$ ; $\phi(t) = \omega t + \phi_{ci}(t)$. This integral to be non trivial requires that the phase is stationnary around some resonance points. For ions trajectories in the Tokamak magnetic configuration, resonances occur at times $\{t^*\}$ for which the phase may be developped as : $\phi(t) = \phi(t^*) + v_{//}\nabla_{//}(\omega_{ci}(t^*))\,(t-t^*)^2/2$. However heating, to take place, implies that in between two successive interactions, a decorrelation mechanism be at work so that each resonant interaction leads to an independent result. This is realised when the phase has diffused by at least 1 rd in the corresponding time interval, that is : the successive values taken by $\phi(t^*)$ are random. The net result of each interaction is simply : $\int \exp\{i(v_{//}\nabla_{//}\omega_c)\tau^2/2\}\,d\tau = \sqrt{\dfrac{2\pi}{iv_{//}\nabla_{//}\omega_c}}$. Therefore ICRF heating results from both a resonant interaction and a diffusion process. In this case the Quasi Linear theory is applicable and the evolution of the distribution function, under the effect of the RF perturbation, is governed by a Fokker-Planck equation.

In present Tokamaks, the decorrelation mechanism, insuring the RF diffusion in the velocity space, is produced by Coulombian collisions. However, diffusion is rapidly stopped due to

198

the $E^{-3/2}$ collisionality scaling, and other decorrelation mechanisms must be sought. The analysis is conducted within the Hamiltonian formalism where the action-angle dynamics is formally simple. In this frame, resonance overlapping may occur producing the necessary randomisation of the phase and leading to diffusion in the phase space.

Precisely, in an axisymmetric unperturbed tokamak magnetic configuration, the particle trajectory is integrable and described by a set of three action-angle variables $\{J_k, \Phi_k\}$. Its hamiltonian $H_0(J_k)$ depends only on the action variables and the motion is quasi-periodic : the actions remain constant and the angles rotate linearly ($\Phi_k = \omega_k(J_k)t + \Phi_{0k}$). Practically, the effect of a low frequency ICRF perturbation is to disrupt this adiabatic behaviour in regions of the phase space where an efficient coupling (i.e. resonances) between fields and particle occurs. In order to identify those regions, the wave hamiltonian perturbation is to be expanded as a Fourier series :

$$\delta H = \sum_N h_N \exp(in_k\Phi_k + i\omega t) + \text{complex conjugate} ; N = \{n_k\}$$

Resonant interaction takes place when, for a given $N$, $h_N \neq 0$ and the phase of the perturbation is stationnary along the unperturbed trajectory, i.e. $\Omega = n_k\omega_k(J_k) + \omega = 0$. The latter defines a resonance surface in the phase space $J = \{J_k\}$; $J_R$. Let assume $h_N$ hermitian and almost constant along $N$, then $\delta H = s(J) \cos(n_k\Phi_k+\omega t)$; $s(J)=h_N+h_N^*$, and the time evolution of $J$ is given by the Hamilton equation : $dJ/dt = N s(J) \sin(n_k\Phi_k+\omega t)$. The motion occurs along the vector $N$ and may be represented by the set : $J = J_R + N\lambda$; $\varphi = n_k\Phi_k+\omega t$ with $d\lambda/dt = s(J) \sin(\varphi)$.

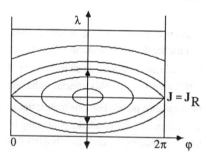

Fig.1 : Island in the $(\lambda, \varphi)$ phase space

Introducing the curvature of the unperturbed Hamiltonian : $\alpha = n_k n_1 \dfrac{\partial^2 H_0(J_R)}{\partial J_k \partial J_1}$ , taken on the resonance surface $J_R$ , it is possible to construct an invariant $W = \frac{1}{2}\alpha\lambda^2 + s(J_R) \cos\varphi$. The motion, in the $(\lambda, \varphi)$ space, is typical of an island like structure (Fig.1), defining a trapped (resp. circulating) domain for particles verifying $|W| < s(J_R)$ (resp. $|W| > s(J_R)$ ).

Considering now a family of resonance surfaces $\{J_{Ri}\}$, nearly parallel so that $N_i \approx N$, the distance between two nearby surfaces is $\Delta\Omega_R = (\Delta n_k)\omega_k$. Correspondingly, the width of the island in $\Omega$ is of order $\Delta\Omega_I = 4\sqrt{s\alpha}$ . If $\Delta\Omega_R < \Delta\Omega_I$, then overlapping of island starts appearing and the regular behaviour of perturbed trajectories is destroyed, creating stochastic domains in the neighbour of resonance surfaces[1]. Stochasticity is insured if $S = \Delta\Omega_I/\Delta\Omega_R$ is much larger than 1.

The overall effect of stochasticity is that : (i) the perturbed trajectory is no longer confined around the resonance surface and may explore a large domain of the phase space, (ii) the asssociated phase is chaotic. Therefore in a phase space domain where resonance with the wave is effective ($h_N \neq 0$) and the density of resonance surfaces is large, the process associated to the particle wave interaction is diffusive, thus validating the Quasi Linear Theory (QLT).

The method of analysis amounts to select the set of action-angle variables relevant to the physical situation, calculate the respective positions of resonance surfaces and the associated $h_N$'s, derive the curvature of the unperturbed Hamiltonian, evaluate the stochasticity parameter and finally construct a Poincaré mapping to check the onset of stochasticity.

## ACTION-ANGLE VARIABLES FOR PASSING AND TRAPPED PARTICLES

The ion trajectory is assumed to be described within the frame of the adiabatic theory, with the guiding centre $x_G$ labelled by $\{R,r,\phi,\theta\}$ and the gyromotion by $\{\rho_c,\phi_c\}$ : $R,r,\rho_c$ are respectively the major, minor and gyro radii; $\phi,\theta$ are angles around the major axis and the magnetic axis and $\phi_c$ is the gyro angle.

The constant set $\{J_k\}$ specifies the unperturbed particle trajectory (the $J_k$'s analytic expressions being written in Table I). The set $\{J_k\}$ is in correspondance with the 3 adiabatic invariants $\{\mu,H,\mathcal{M}\}$ : $\mu = \dfrac{mv_\perp^2}{2B}$ ; $H = \frac{1}{2} mv_{//}^2 + \mu B$; $\mathcal{M} = e\Psi_P + mv_\varphi R$, with indices $\perp$ and $//$ standing for perpendicular and parallel directions to the total magnetic field, $\Psi_P$ the poloidal magnetic flux and $v_\varphi$ the velocity in the toroidal direction. On the other hand, the set $\{\Phi_k\}$ resulting from the integration of the unperturbed trajectory for a given set $\{J_k\}$, specifies : (i) the gyrophase $\phi_c$ through $\phi_1 = \phi_c - \Theta(\mu,H,\mathcal{M})$, (ii) the drift surface in the poloidal and toroidal directions through $\phi_2$ and $\phi_3$ respectively (Table I).

Table I : action angle variables for deeply passing and trapped particles ( index 0 for value on the magnetic axis, $\Phi_T$ toroidal magnetic flux function, $\bar{\theta}$ maximum amplitude for trapped particle ($\theta=\bar{\theta} \sin(\phi_2)$), $\mathcal{J}_0$ zeroth order Bessel function).

| Passing particle | Trapped particle |
|---|---|
| $J_1 = -\dfrac{m}{e}\mu$ ; $\omega_1 = -\dfrac{eB_0}{m}$ ; $\phi_1 = \phi_c + \dfrac{r\omega_1}{R_0\omega_2}\sin(\theta)$ | $J_1 = -\dfrac{m}{e}\mu$ ; $\omega_1 = -\dfrac{eB_0}{m}[1 - \dfrac{r}{R_0}\mathcal{J}_0(\bar{\theta})]$ |
| $J_2 = e\Phi_T(\dfrac{J_3}{e}) + qmR_0v_{//}$ ; $\omega_2 = \dfrac{\omega_3}{q}$ ; $\phi_2 = \theta$ | $J_2 = \frac{1}{2} mR_0^2q^2\bar{\theta}^2\omega_2$ ; $\omega_2 = \omega_b = \dfrac{v_\perp}{qR_0}\sqrt{\dfrac{r}{2R_0}}$ |
| $J_3 = e\Psi + mR_0v_{//}$ ; $\omega_3 = \dfrac{v_{//}}{R_0}$ ; $\phi_3 = \phi$ | $J_3 = e\Psi$ ; $\omega_3 = \omega_d = \dfrac{qmv_\perp^2}{2erR_0B_0}$ ; $\phi_3 = \phi - q\theta$ |

The value $\omega_1(J) = d\phi_1/dt$ is the time average value along the trajectory of $\omega_c(x_G)$ and the difference $\omega_c - \omega_1$, of order $\omega_c r/R$, exhibits a large variation rate. Because of the axisymmetry, the relation of $\theta$ and $\phi$ in terms of $\phi_2$ and $\phi_3$ is written : $\theta = \varepsilon\phi_2 + \hat{\theta}(\phi_2)$; $\phi = \phi_3 + q(\bar{\Psi})\hat{\theta}(\phi_2)+ \hat{\phi}(\phi_2)$; where $\hat{\theta}(\phi_2)$ and $\hat{\phi}(\phi_2)$ are $2\pi$-periodic functions of $\phi_2$ ($\varepsilon = 1$ for passing particles, resp. 0 for trapped particles and $q(\bar{\Psi})$ is the safety factor). Therefore, the trajectory is completed within the time duration $2\pi/\omega_2 : \omega_2(J) = d\phi_2/dt$, allowing to restrict the study over this time period.

## HEATING OF PASSING PARTICLES

In the ICRF scheme the electromagnetic power is carried by a fast wave polarised in the extraordinary mode. In the plasma core, the wave is a compressionnal Alfven wave with an electric field and a wave vector predominantly perpendicular to the magnetic field : $k_\perp \approx \omega/c_A$. The pulsation $\omega$ is imposed by the RF generator and the coupling launcher allows a discret toroidal mode spectrum in $k_\phi : k_\phi = N/R$.

A prerequisite for an effective heating[2] requires the components $h_N$, associated to the resonance $\Omega = 0$, to be different of zero. The compressional wave can be described with the gauge $[A_\perp, \psi=0]$ so that the perturbed Hamiltonian is of the form : $\delta H = - e \, v_\perp A_\perp(r+r_c)$. This last expression, when Taylor expanded around $r$, takes the form of a Fourier series in $\exp\{ip\phi_c\}$ :

$$\delta H = \sum_p h_p(\mu, H, r) \exp(ip\phi_c + i\omega t) + \text{c.c.}$$

If we now compare to the action angle expression of $\delta H$ and consider one toroidal mode N, it is clear that for $n_3 \neq N : h_N = 0$. Furthermore as for a given set $\{J_k\}$ the trajectory is completed when $\Phi_2$ is varied within $2\pi$, it is sufficient to Fourier transform $\delta H$ integrating over $\Phi_2$ to derive $h_N$, or equivalently to perform an integration over time over a period $2\pi/\omega_2$ :

$$h_{n_1=p,n_2,N} = \frac{1}{2\pi} \int_0^{2\pi} \delta H \exp(-in_k\phi_k - i\omega t) \, d\phi_2 = \frac{\omega_2}{2\pi} \int_0^{2\pi/\omega_2} h_p \exp(- ip\Theta - i(n_2\omega_2 + N\omega_3)t) \, dt$$

As mentioned, $\Theta$ has a variation rate ( $d\Theta/dt = \omega_c - \omega_1 \approx \omega_c r/R$ ) typically much larger than the one of $h_p$. Therefore the integral will be non vanishing if there exists along the trajectory points where the phase is stationary, namely when :

$$p\omega_c(r^*) = p\omega_1 + n_2\omega_2 + N\omega_3$$

The number of resonances involved is then defined by the number of $n_2$ verifying the previous relation; over the passing trajectory $\omega_c(r^*)$ varies by $2\omega_1 r/R$ so $\Delta n_2 = 2p \, \omega_1/\omega_2 \, r/R$ and $n_2 \approx pr\omega_c/R\omega_2$ (Table II). The integration is then performed as usual : the phase, developped around $x_G^* = x_G(t^*)$, is assumed to have diffused in between two successive resonances, so that each resonance ($\theta^* = \{\theta_1^*, \theta_2^*\}$) participates independently to the final result[2] :

$$\left| h_{p,n_2,N} \right|^2 = \sum_{\theta^*} \left| \frac{\omega_2}{2\pi} \right|^2 \left| \frac{2\pi}{pv_{//}|\nabla_{//}\omega_c(r^*)|} \right| \left| h_p(\mu, H, x_G^*) \right|^2$$

when the $h_p$'s variations along the path of integration are larger than the characteristic scale of the saddle point; $\tau_c = 1/\sqrt{pv_{//}|\nabla_{//}|\omega_c}$. Applied to the fundamental resonance ($p=1$), it yields :

$$\left|h_{1,n_2,N}\right| = \frac{ev_\perp}{\omega\sqrt{\pi}\sin(\theta^*)\,\Delta n_2}\left|E\right| \; ; \text{ with } \left|h_1(\mu,H,r)\right| = \frac{|e|\,v_\perp}{2\omega}\left|E\right|$$

Table II : characteristic frequencies $v = \omega/2\pi$ as functions of temperature for a helium-4 plasma with $B_0 = 4$ T, $q = 2$, $r = 0.5$ m, $R_0 = 2.5$ m, $\bar{\theta} = 1$ rd ( $n_e = 10^{20} m^{-3}$, E = 100 V/cm ).

| | | passing particles $[v_{/\!/}(\theta=0) = v_\perp(\theta=0)]$ | | | trapped particles | | |
|---|---|---|---|---|---|---|---|
| $v_{/\!/}(\theta=0)$ (keV) | $n_2$ | $v_1$ (MHz) | $v_2$ (MHz) | $v_3$ (MHz) | $v_1$ (MHz) | $v_2$ (MHz) | $v_3$ (MHz) |
| 10 | -276 | - 30 | 0.02 | 0.04 | - 25 | 0.02 | 0.003 |
| 800 | -30 | - 30 | 0.20 | 0.40 | - 25 | 0.20 | 0.24 |
| 3500 | -14 | - 30 | 0.40 | 0.80 | - 25 | 0.43 | 1.20 |

A decorrelation process is efficient when it destroys the regular motion displayed by the island like structure. Collisions essentially induce a diffusion of $J_2$ : $\delta J_2 \approx J_2 \sqrt{t/\tau_{coll}}$, and diffusion will results if, after a duration $\Delta t$, and before a revolution is completed; $\Delta t\, d\varphi/dt = 1$, $\Delta t < 2\pi/\omega_2$. Now $d\varphi/dt \approx n_2\partial\omega_2/\partial J_2\,\delta J_2$ and then $\Delta t \approx [\sqrt{\tau_{coll}}\,R/r\omega]^{2/3}$, which amounts to the condition : $(qr/\rho_c)^2 qR/\lambda_{coll} = S_{coll} \gg 1$, verified for T « 20 keV for the Table II parameters with $T_e = T_i$.

In the case of diffusion induced by intrinsic stochasticity, S must exceed 1. To derive it, the curvature of the unperturbed Hamiltonian must be explicited, i.e $\alpha = n_k n_l\,\partial\omega_k/\partial J_1$ which here takes the form $\alpha \approx (N+n_2/q)^2/mR_0^2$. As in this case $\Delta\Omega_R = \omega_2$, S becomes :

$$S = \frac{4q}{v_{/\!/}}\left(N + \frac{n_2}{q}\right)\sqrt{\frac{2ev_\perp|E|}{m\omega\sqrt{\pi}\sin\theta^*\,\Delta n_2}}$$

For T < 10 keV (Table II), $S_{coll}/S>1$ and collisions are preponderant, then for $T<T_S = 800$ keV; S>1 stochasticity dominates and for $T>T_S$ diffusion ceases to be effective. However, due to the N antenna spectrum, the stochasticity parameter may be largely increased. For non integer q value, the resonant surfaces for families $\{1,n_2,N\}$ and $\{1,n_2,N'\}$ are interposed and the distance in between two successive surfaces is effectively reduced. As for example $N'=N+1$, the distance may be made less than $\omega_2$ when $|\Delta n_2+q|<1$. The onset of the stochastic regime is confirmed performing a Poincaré mapping, here plotting the successive $\lambda$ values versus $\phi_2$ (Fig.2). Then RPA is valid over a large temperature domain, provided the N antenna spectrum is sufficiently rich.

## CYCLOTRON RESONANCE EFFECT ON TRAPPED PARTICLES

Along a trapped particle trajectory, the interaction with the ICRF field may occur at most four times. As long as the banana tip $(\theta=\bar{\theta})$ is far away from the interaction zone $(\theta^*)$, each saddle

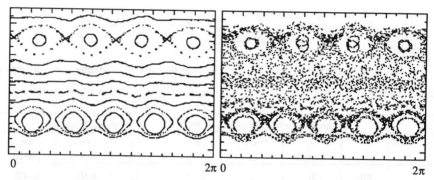

0                                2π 0                                2π

Figure 2 : Poincaré mapping of $\lambda$ versus $\Phi_2$ ( the combination $\phi_1 + N\phi_3 + \omega t$ fixed ) : (l) one resonant N=20 ($n_2$=-4,-5),500 keV $\alpha$-particles and 25V/cm; (r) resonant families N=20 ($n_2$=-4,-5) and 21 ($n_2$=-6) q=1.5.

point contributes independently to the final $h_N$ result and its derivation is identical to the previous case. This is true when the distance to the resonant zone is greater than the width of the resonance itself, i.e. when $|\bar{t} - t^*| \gg \tau_c \approx \sqrt{\dfrac{2\pi}{pv_{//}\nabla_{//}\omega_c(t^*)}}$. However when the banana tip becomes close to the resonance $v_{//}\to 0$, the two consecutive saddle points are linked together. To overcome this difficulty, the phase must now be developped around $\bar{\theta}$ instead of $\theta^*$. This leads to the $h_N$ expression :

$$|h_{1,n_2,N}|^2 = \sum_{\theta} |\omega_2|^2 \left(\frac{2}{\ddot{\omega}_c(\bar{\theta})}\right)^{2/3} |Ai(z)|^2 |h_1(\mu,H,\bar{x}_G)|^2$$

where $z = \dfrac{|\omega_c(\bar{\theta}) + \omega|}{(\ddot{\omega}_c(\bar{\theta})/2)^{1/3}}$ is the Airy function argument and $\ddot{\omega}_c(\bar{\theta}) = -\dfrac{v_{//}(\theta=0)\omega_2 r}{qR_0^2}\omega_c(\bar{\theta})\sin(\bar{\theta}) > 0.$

The S value is of the same functional form as before. In term of diffusion the two situations may be compared for the same $v_{//}(\theta=\theta^*)$, then the stochastic parameters S are almost equal but they correspond to different total particle energies $E_T \approx E_P(1+R/r)/2$. Therefore the energy range for the RPA validity is appreciably extended in the case of fast trapped particles.

## SUPRATHERMAL ICRF FUSION RATE ENHANCEMENT SCHEME

The quasi linear absorption of ICRF power is at the centre of experimental studies of the suprathermal population dynamics. As it produces a fast ion tail, a multiplication of the fusion power may result, steming from the wave induced enhancement of the fusion rate.

Above the stochasticity threshold, the wave particle interaction can be viewed as an incoherent sum of quantum kicks. In this approach, the QLT is equivalent to the classical limit (h=0) of the quantum population balance. Bearing this in mind, if at a time t=t' one ICRF photon of energy $hv_c$ is absorbed by a resonant ion of velocity $v_r$ and mass $m_r$, the ion receives a kick in velocity $\Delta v_r = hv_c/m_r v_r$ and a ion hole pair is created in the distribution function. The collisional relaxation of this ion hole excitation is governed by the distribution function f solution of :

$$\frac{\partial f(v,v_r,t)}{\partial t} - C(v).f(v,v_r,t) = \frac{[\delta(v-v_r+\Delta v_r) - \delta(v-v_r)]}{4\pi v^2}\delta(t-t')$$

where C is the linearized fast ion collision operator. The number of ion-hole excitations per unit time is given by $W_a/m_r v_r \Delta v_r$ where $W_a(v_r)$ is the absorbed ICRF power. Taking then the classical limit of h=0 corresponding to the quasi linear regime, the steady state modification of the ion distribution may be expressed in terms of the absorbed power and the inverse of the collision operator (i.e. the Green function):

$$f(v,v_r) = -\frac{W_a(v_r)}{m_r v_r}\frac{\partial}{\partial v_r}\left(C^{-1}(v)\frac{\delta(v-v_r)}{4\pi v^2}\right)$$

This allows to derive an expression of the wave induced fusion power $W_f$ steming from the reaction between resonant ions and the cold plasma target of density $n_t$. The efficiency of this process is defined as the ratio of the fusion to the RF absorbed power and takes the form :

$$\eta(v_r) = \frac{W_f}{W_a} = \frac{\tau_s\sigma_{rt}(v_r)v_r^2 Q_{rt}n_t}{m_r(v_r^3 + v_c^3)}$$

with $\tau_s$ the slowing down time and $v_c$ the critical velocity[3], $\sigma_{rt}$ and $Q_{rt}$ the standard cross section and thermonuclear energy. For the $D(t,n)^4He$ reaction, the scaling of this quasi-linear suprathermal power multiplication efficiency indicates that a value of $\eta$ above unity is well within the experimental reach. However even in the QL regime, this scenario is limited because it involves only a small population.

In conclusions when conditions for significant suprathermal power coupling (S>1) are fulfilled, power amplification takes place in a reacting plasma. The enhancement effect is limited and is only of interest for present fusion physics study.

## ELEMENTS OF FAST WAVE CURRENT DRIVE BY TTMP

In contrast to Low Hybrid waves, the fast wave, in a situation of no ion cyclotron resonance, does not suffers from an accessibility problem which is advantageous in a current drive scenario. The FWCD scheme aims at sustaining current by both an unbalanced N antenna spectrum and a direct absorption of the ICRF power on electrons, through parallel Landau damping or TTMP[4]. We focus on TTMP with the perturbed Hamiltonian $\mu\delta B$ ($n_1=0$) and its effect on passing electrons. For sake of simplicity the derivation is conducted for non relativistic electrons as the relativistic extension gives the same stochasticity parameter.

In this resonant scheme, the QL transfer of parallel momentum is insured by an efficient overlapping of the toroidal N modes. Therefore, the main resonance corresponds to $n_1 = n_2 = 0$, and $\Omega = 0 = \omega + N\omega_3$. The distance between two resonant surfaces $\Delta\Omega_R$ is $\omega_3 = v_{//}/R$ and resonant electrons have their parallel velocity verifying $v_{//,res} = -\omega R/N$. Accordingly (Table I), the associated $h_N$ is $\mu b_N$ when $\delta B$ is Fourier analysed : $\delta B = \sum_N b_N e^{i(\omega t + N\phi)}$, while the Hamiltonian curvature is $\alpha = N^2/mR_0^2$. Then the stochasticity parameter becomes, with $\mu$ the pitch angle :

$$S = \frac{4\omega R}{v_{//}} \frac{v_\perp}{v_{//}} \sqrt{\frac{b_N}{B_0}} = \frac{4\omega R}{v} \frac{\sqrt{1-\mu^2}}{\mu^2} \sqrt{\frac{b_N}{B_0}}$$

For resonant electrons at a given N, S varies linearly with $v_\perp$ and it is only when $v_\perp$ is large that $S>1$ (typically in a D-plasma with Table II parameters, $v = 19$MHz and $N = 5$; $T_\perp > 37$keV). This limits the quasi linear domain in pitch angle ($\mu < 0.5$ for the above parameters). The poloidal $n_2$ antenna spectrum is then essential to produce FWCD as for non integer q value, the distance between adjacent resonant surfaces is seriously decreased. This allows for a wider quasi linear domain accessible to FWCD and correspondly larger pitch angles.

The final question is that of the current drive efficiency. For $S>1$, the quasi-linear theory and its quantum kick picture are valid and the excitation-relaxation reasoning leads to the Fisch efficiency result for total velocity v :

$$\eta(v,\mu) = \frac{v^2}{(Z+5)} \frac{3\mu^2+1}{\mu}$$

For the FWCD, due to the boarder $S=1$, the QL coupling is limited to a portion of the $(v_\perp, v_{//})$ space where v is large and $\mu$ small. On the opposite, for the LHCD case, the stochastic zone is simple and $\mu = 1$ is usually taken. In practice the two schemes suffer from coupling limitations, however of different nature. Nevertheless, for a given phase velocity v, the efficiency ratio $[\eta_{FW}/\eta_{LH} = (3\mu^2+1)/4\mu^3]$ appears to favor the FWCD scheme.

## CONCLUSIONS

For ICRF coupling in the low collisionality regime, the QL diffusion is induced by intrinsic stochasticity, provided the N toroidal antenna spectrum is rich. ICRH then results and produces fast ion tails, which in turn may enhanced the fusion reaction rate. The Hamiltonian formalism allows to address the question of FWCD : the TTMP current drive is made possible by the poloidal antenna spectrum. It is limited within a domain a small pitch angles, but presents no other accessibility problems, which is favorable to the scheme.

## REFERENCES

1.      B.V. Chirikov, Physics Report 52, 265 (1979).
2.      D.J. Gambier, A. Samain, Nucl. Fusion 25 (3), 283 (1985).
3.      J.G. Cordey, M.J. Houghton, Nucl. Fusion 13, 215 (1973).
4.      D. Moreau, J. Jacquinot, P.P. Lallia, JET Rep. JET-P(86)15, 65 (1986).

# PHYSICS OF HIGH POWER ICRH ON JET

D.F.H. Start, V.P. Bhatnagar, G. Bosia, D.A. Boyd[1], M. Bures,
D.J. Campbell, J.P. Christiansen, J.G. Cordey, G.A. Cottrell,
G. Devillers, L.G. Eriksson[2], M.P. Evrard[3], J.A. Heikkinen,
T. Hellsten, J. Jacquinot, O.N. Jarvis, S. Knowlton, P. Kupschus,
H. Lean[4], P.J. Lomas, C. Lowry, P. Nielsen, J. O'Rourke,
G. Sadler, G.L. Schmidt[5], A. Tanga, A. Taroni, P.R. Thomas,
K. Thomsen, B. Tubbing, M. von Hellermann, U. Willén[2]

JET Joint Undertaking, Abingdon, Oxon, OX14 3EA, UK
[1] University of Maryland, College Park, Maryland USA
[2] Chalmers University, Gothenburg, Sweden
[3] LPP-ERM/KMS, EUR-EB Association 1040 Brussells Belgium
[4] Culham Laboratory, Abingdon, Oxon, OX14 3DB, UK
[5] Princeton Plasma Physics Laboratory, New Jersey USA

## ABSTRACT

Ion Cyclotron Resonance Heating (ICRH) experiments have been
carried out in a wide range of JET plasmas. A high confinement
region in the plasma core has been discovered during on-axis heating
of pellet fuelled discharges. Sawtooth-free periods and improved
heating efficiency have been achieved in high current plasmas by
heating during the current rise. Combined ICRH and neutral beam
injection (NBI) in double null X-point plasmas has yielded D-D
fusion rates up to $2 \times 10^{16} s^{-1}$ and electron and ion temperatures of
11keV and 17keV respectively. Non-thermal fusion reaction rates
from $(He^3)D$ minority ICRH have been reproduced by a model which
predicts $Q \sim 0.7$ for $(D)T$ minority ICRH in JET. RF modulation
studies have yielded power deposition profiles and heat transport
coefficients inside the q=1 surface. Alignment of the antenna
screen to the magnetic field minimises nickel impurity release.

## INTRODUCTION

ICRH on JET is provided by 8 antennae situated on the median
plane on the low field side of the machine. Each antenna consists
of two vertical strip lines with centres separated by 0.31m toroidally and carrying currents with either zero (monopole) or $\pi$ (dipole)
phase difference. Also, the phases between antennae can be imposed
to provide a phased array for future fast wave current drive
experiments. This array has already proved useful as a 'super
dipole' with zero phasing between strip lines in each antennae but $\pi$
phasing between adjacent antennae. The resulting $k_{||}$ spectrum
produces a coupling resistance and nickel impurity release intermediate between those with the normal dipole and monopole phasings.
The system operates in the 23MHz to 57MHz frequency range with
a generator power capability of 30MW. So far a maximum of 18MW has
been coupled to the plasma, the principle limitation being breakdown
in the vacuum lines which have been redesigned for the next operat-

ing period. An automatic matching system which controls the frequency can track rapid changes in antenna/plasma coupling due to density variations or radial eigenmode excitation. This fast system is being augmented by a slow system which controls the tuning stubs and which will almost completely automate the matching process. The Faraday screens of the antennae are water cooled which has allowed long pulse operation to be demonstrated with 6MW of power being coupled for 20 sec with no deleterious effects. However, eventually four antenna have developed leaks as a result of mechanical failure of supporting blocks due to disruptions. These blocks are now strengthened. New beryllium screens (September 1989) will allow long pulse operation to resume.

This paper is a summary of the principal results obtained with ICRF during 1988. In the next section we discuss edge effects, specifically nickel impurity release and parametric decay of the fast wave. In section 3 we present measurements of the power deposition profile, the direct electron heating fraction and the electron thermal diffusivity inside the $q = 1$ surface from RF modulation studies during Monster sawteeth. In section 4 we report on the achievements of on-axis ICRH in high performance scenarios, involving pellet fuelling, RF heating in the current rise and in double null X-point discharges. Finally in section 5 we use a model, validated by (He$^3$,D) fusion yield experiments, to predict non-thermal (D,T) Q values for JET with (D)T ICRH minority heating.

## EDGE EFFECTS: NICKEL IMPURITIES AND PARAMETRIC DECAY WAVES

Previous experiments have shown that nickel is released only from the screens of energised antennae suggesting that local RF fields play a role[1]. However the perpendicular electric field of the fast wave is probably not responsible since the influx does not correlate with eigenmode excitation[2]. On the other hand, non-alignment of the screen elements with the magnetic field B allows a parallel component, $E_{\parallel}$, to penetrate into the edge plasma and possibly accelerate ions into the screen to release nickel by sputtering. The screen elements on JET are aligned to within $\sim 5°$ under normal operating conditions. On-axis (H)D heating experiments at $I_p = 2MA$, $B_T = 2.1T$ used both normal and reversed toroidal field directions which gave screen angles of 5° and 25° to B. The intensity of the NiXXVI line was three-fold enhanced by this increase in angle. Moreover, the coupling resistance, electron temperature and incremental confinement time were all reduced by 30% in the reversed field case. Such reduced performance implies a smaller power fraction coupled to the fast wave with perhaps increased excitation of the slow wave as the screening of $E_{\parallel}$ becomes less effective. Either this or near field effects (sheath rectification) could cause nickel influx.

Probe measurements in the plasma edge have detected parametric decay of the fast wave into two slow waves or into a Bernstein wave and a quasi-mode. An example is shown in Fig. 1. Spectra were recorded at several RF power levels in a 3MA, 3.15T discharge containing a mixed D, He³ plasma with H-minority ions. The RF frequency was 48MHz to give on-axis heating, the phasing was monopole and the toroidal field direction was normal. As the RF power is increased, peaks emerge which can be identified as quasi-modes for deuterium and He³ with their corresponding Ion Bernstein Waves (IBW). In this example any decay to the slow waves at 24MHz is obscured but has been seen when the IBW branch is forbidden.

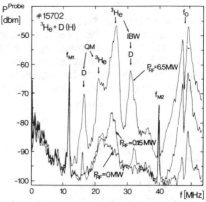

Fig. 1. Parametric Decay Waves in JET Edge Plasma

## RF MODULATION DURING MONSTER SAWTEETH

The response of the central electron temperature, $T_e(0)$ to 4Hz square wave modulation of central He³ minority ICRH between 6MW and 8.5MW is shown in Fig. 2. The discharge parameters were $I_p$ = 2MA, $B_T$ = 3.4T, $n_e(0)$ = 4.2x10¹⁹m⁻³ and the He³ concentration was 5% of the electron density. The sharp change in the derivative of $T_e(0)$ at the RF switch up (switch down) is due to the modulated direct electron heating. Direct heating occurs through mode conversion, electron Landau damping or TTMP; indirect electron heating arises from friction with the minority ions. The magnitude of the discontinuity in $\partial T_e(0)/\partial t$ gives a modulated direct heating power density of 50kW/m³ in the plasma centre. The amplitude profile of the 4Hz component of $T_e$ is almost gaussian with a peak value of 200eV on- axis and with a width of 0.3m. The phase delay between the $T_e$ oscillation and the power modulation is 82 ± 5° in the centre and increases monotonically towards the limiter. The phase and amplitude profiles were analysed using a simple electron heat diffusion model to obtain a thermal diffusivity 1.5 < $\chi_e$ (m²s⁻¹) < 3.0, a gaussian power deposition width of 0.18 ± 0.04m and a direct heating fraction of 10 ± 3%. This localised heating agrees well with self consistent full wave/Fokker-Planck calculations.

With lower He³ concentration (2%) and higher average power, (10MW compared with 7.2MW), the $T_e$ response is quite different to that above. The amplitude profile is hollow at 4 Hz modulation but peaked at 16Hz (Fig. 3). A hollow profile suggests off-axis heating (although the minority resonance layer was central) except that the phase delay is least on-axis. These data can be reproduced, however, if there is a depletion of the minority heating on-axis as the power is increased, such as could occur if the width of the minority heating profile were modulated. The required fractional change in the profile width is about 4% which is similar to the calculated oscillating minority energy fraction and could perhaps arise from a modulation of the banana orbits. This interpretation is clearly not unique and another possibility has appeared recently from studies of the minority Fokker Planck equation with a modulated power source. The phase of the power transferred to the electrons is found to be delayed by $\sim \pi/2$ at low and high energies but is advanced by $\sim \pi/2$ at energies around the tail temperature. The data could then be readily explained if this latter component dominates the heating on-axis due to orbit effects.

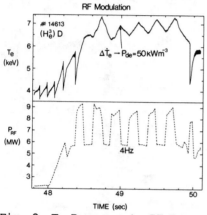

Fig. 2. $T_e$ Response to RF Power Modulation

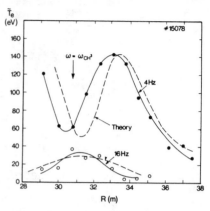

Fig. 3. Radial Profile of Modulated $T_e$

## HIGH PERFORMANCE SCENARIOS

PELLET FUELLED DISCHARGES  Plasmas with peaked density profiles produced by pellets have been successfully heated with H-minority ICRF without the density pump-out observed in previous experiments. The time evolution of plasma parameters in a 3MA, 3.3T discharge is shown in Fig. 4.  Pellets of 2.7mm diameter were injected at 41.5s and 42s followed by a 4mm pellet at 43s which produced a central density, $n_e(0) = 9 \times 10^{19} m^{-3}$. The 4mm pellet was closely followed by 5MW of neutral beam power and $\sim$ 13MW of ICRF hydrogen minority heating, which raised $T_e(0)$ to 12keV at 44.25s when a sudden collapse

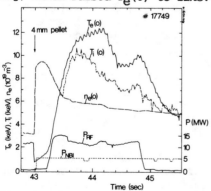

Fig. 4. Time Evolution of Pellet Fuelled Plasma

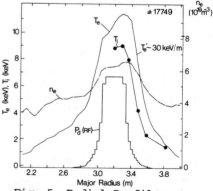

Fig. 5. Radial Profiles at t = 44.2 sec

occurred. The central ion temperature, $T_i(0)$, from doppler broadening of NiXXVII radiation, initially increased to 10keV at 43.8s and slowly decayed thereafter. The NBI allowed $T_i(r)$ profiles to be measured by charge exchange spectroscopy[3] as shown in Fig. 5 for t = 44.2 sec. Also shown are LIDAR profiles of $T_e$ and $n_e$, the latter remaining peaked although the central density is much less than that just after pellet injection. The $T_e(r)$ and $T_i(r)$ profiles are extremely peaked even by comparison with those during monster sawteeth and the strong gradients imply enhanced confinement. For diffusive heat loss the electron power balance at minor radius a in cylindrical geometry can be written

$$P_e = 4\pi^2 Ran_e(a)\chi_e(a) \left.\frac{\partial T_e}{\partial r}\right|_a$$

where $P_e$ is the power input to electrons within the radius a, R is the major radius and $\chi_e$ the thermal diffusivity. According to ray tracing, most of the power is deposited inside a = 0.35m (see Fig. 5). From the $T_e$ gradient and the calculated ICRF power flow to the electrons (69%) we estimate $\chi_e \sim 0.8m^2/s$. This value is a factor 2-3 less than those obtained for monster sawteeth. This estimate is confirmed by detailed transport calculations[4] which find that the improvement extends over the inner half radius where $\chi_e \sim 0.8m^2/s$ and $\chi_i \sim 0.6m^2/s$. Outside this region the values of $\chi_e$ and $\chi_i$ are higher by factors of 2 and 4 respectively.

Statistical comparisons of this improved confinement regime with non-peaked density profile discharges (including plasmas with monster sawteeth) show that $T_i(0)$ and $T_e(0)$ are factors of 2 and 1.4 higher, respectively, for the same value of $P_{total}/n_e(0)$. Also the D-D fusion reaction rate is a factor of 4 larger, but the global energy confinement is only increased by 20%. The fusion parameter $n_D(0)T_i(0)\tau_E$ reaches $2\times10^{20}keVm^{-3}s$ which is comparable with values in good H-mode plasmas. The strong pressure gradients produced bootstrap currents of the order of 0.8 MA[5] which, together with the initial cooling by the pellet, tend to broaden the current profile: equilibrium analysis gives a value of q(0) close to 1.5 prior to the crash. After the crash there is a strong increase in n = 2 MHD activity with an odd poloidal mode number, m ≥ 3. Theoretical studies suggest that ballooning modes[6] or modes with intermediate n values[7] ('infernal' modes) could cause the collapse.

CURRENT RISE HEATING  ICRH hydrogen minority heating of deuterium plasmas has been applied on-axis during the current rise of 5 MA and 6 MA discharges in order to heat the plasma before or shortly after the onset of sawteeth. For $I_p$ = 5MA the retarded current penetration delayed the first sawtooth collapse by up to 1.2 s. This collapse occurred for q(0) ∿ 1 according to polarimetry data. With 11 MW of RF power the central electron temperature reached 10.5 keV for $n_e(0)$ = $6\times10^{19}m^{-3}$ just prior to the collapse. The temperature profile was highly peaked, $T_e(0)/\langle T_e\rangle \sim$ 3-4 and the discharges were particularly clean with $Z_{eff} \sim$ 2. The 6 MA discharges exhibited small sawteeth before the heating pulse which, for $P_{RF}$ > 6 MW, produced a monster sawtooth as the current reached 5 MA. Current rise heating is the only way monster sawteeth have been produced in such high current discharges. Fig. 6 shows the larger values of $n_e(0) \cdot T_e(0)$ during the current rise compared with those obtained

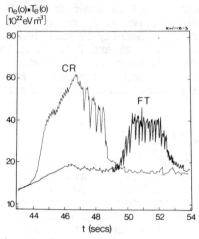

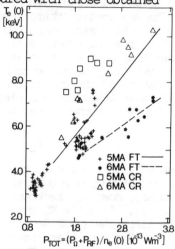

Fig. 6. Current Rise and Flat Top Heating for $I_p$ = 5MA

Fig. 7. $T_e(0)$ During Current Rise and Flat Top Heating

during the flat top. This improvement stems from the enhanced $T_e(0)$. A statistical comparison of current rise and flat top values of $T_e(0)$ plotted against $P_{total}/n_e(0)$ is shown in Fig. 7 for both $I_p$ = 5 MA and $I_p$ = 6 MA. Similar comparisons for other parameters lead to the following summary of the enhancements obtained with current rise heating.

1) A 50% increase in $T_e(0)$ for equal values of $P_{total}/n_e(0)$.
2) 10-20% improvement in $T_i(0)$ in both 5 MA and 6 MA plasmas.
3) An increase in the global energy confinement by typically 15% due to the suppression of the sawteeth.
4) Twofold enhancement of the D-D fusion rate in 5 MA plasmas partly due to second harmonic heating of deuterium.

The best values of the fusion parameter achieved so far is $n_D(0)T_i(0)\tau_E = 1.65 \times 10^{20}$ $(m^{-3}keVsec)$. This scenario appears to be the most suitable for ICRH (D)T operation as discussed in section 5.

DOUBLE NULL X-POINT DISCHARGES Combined NBI ($\leq$ 21MW) and ICRH ($\leq$ 11MW) experiments have obtained D-D reaction rates up to $2 \times 10^{16} s^{-1}$ in 3MA, 3.2T double null X-point deuterium plasmas. This configuration was chosen for good density control through the pumping action of the X-point carbon tiles and for good matching of the plasma boundary to the RF antenna curvature[8]. The deuterium beam energy was $\sim$ 80keV and the 48MHz, on-axis ICRH used hydrogen minority ions. The pre-heating target density was $\sim$ 1.5x10[19]$m^{-3}$ but increased during the heating pulse, thereby contributing to a non-steady neutron production, principally from beam-plasma inter actions, which reached a maximum for $n_e(0) \sim 3 \times 10^{19} m^{-3}$. Values of $T_e(0)$ and $T_i(0)$ up to 11keV and 17keV respectively were achieved. A plot of reaction rate versus NBI power, in Fig. 8, shows the enhancement with combined heating compared with beam-only cases. TRANSP code simulations show that this enhancement is not solely due to the improved plasma parameters, notably $T_e$, when RF is applied. This conclusion implies an acceleration of the beam ions by second harmonic heating although neutron spectroscopy reveals the presence of only a weak tail above 80keV.

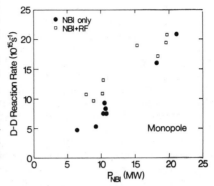

Fig. 8. D-D Fusion Rate for NBI + RF and NBI Only

During these experiments, several H-modes were formed and in one case (# 18773) the limiter clearance was only 1cm compared with 3cm normally required for H-modes. This discharge was fuelled with three 2.7mm pellets and then heated with 10MW of ICRF and 15MW of NBI during which a 3MA to 3.5MA current ramp was applied. The confinement time was ∿ 0.6s (twice the Goldston L-mode value), the energy content was 11MJ, $T_e(0)$ and $T_i(0)$ were close to 10keV and the D-D reaction rate reached $1.9 \times 10^{16} s^{-1}$. The proximity of the plasma gave good coupling ($R_c$ ∿ 6Ω) and the radiated power rise rate was low (∿ 5MWs$^{-1}$) compared with previous NBI + RF heated H-modes (∿ 10MWs$^{-1}$) indicating a low influx of nickel. In Fig. 9 the energy content, W, is compared with values for NBI alone and NBI + RF in 3MA H-modes. The upper and lower

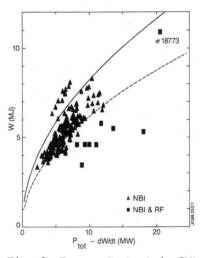

Fig. 9. Energy Content in 3MA H-modes

curves are 2 x Goldston scaling for $I_p$ = 3.5MA and 3MA, respectively. This type of discharge will be investigated further during the next experimental campaign.

## NON THERMAL FUSION FROM (D)T MINORITY ICRH

Experiments using He$^3$ minority ICRH in deuterium plasmas have yielded up to 60kW of fusion power from He$^3$-D fusion reactions[8]. The results, particularly the scaling with RF power, have been simulated using both a Stix model[9] and a self consistent full wave/-Fokker Planck treatment[10]. In this section we describe predictions of the Stix model for the D-T fusion yield from fundamental ICRF heating of deuterium in tritium JET plasmas. The plasma parameters are based on those attained in high performance scenarios with;

a) pellet injection at 3MA, b) current rise heating at 5MA and c) monster sawteeth at 3MA. In the model, the profiles of $n_e(r)$ and $T_e(r)$ were parabolic to a power $\gamma$ which was determined by fitting experimental data. Central values, $n_e(0)$ and $T_e(0)$, were extrapolated to high power levels ($\sim$ 20MW) using the measured offset linear scaling laws of the type $T_e(0) = \alpha + \beta P_{tot}/n_e(0)$ as shown in Fig. 7. We also assume $T_i = T_e$ which is not fully justified experimentally. However, $T_i$ approaches $T_e$ in the pellet fuelled discharges which have central densities close to that for which the (D)T scheme is optimum. The RF power absorption was investigated by ray tracing for f = 25MHz, $B_T$ = 3.55T and dipole antennae phasing. For example, with 30% deuterium minority, the pellet fuelled discharge parameters gave 80% single pass absorption on deuterium, 17% direct electron heating (TTMP + ELD) and 3% was unabsorbed. In the Stix model the power deposition was taken as gaussian with a width of 0.2m. The predicted fusion yields are plotted against $n_e(0)$ in Fig. 10 for 20MW of power coupled to the minority ($P_{RF} \sim$ 24MW), $n_D/n_e$ = 15% and 30%, $Z_{eff}$ = 2, and off-axis heating ($r \sim$ 0.3m). The fusion yield peaks for densities close to $1\times10^{20}$m$^{-3}$ when the minority tail temperature is optimum ($T_{tail} \sim$ 140keV). For this case of fixed $Z_{eff}$ the pellet fuelled discharges and the current rise heating give similar values of $Q_{RF}(= P_{Fusion}/P_{RF}) \approx$ 70%. However,

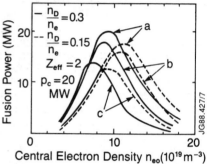

Fig. 10. Predicted Fusion Power for (D)T Minority ICRH

only the current rise heating achieved $Z_{eff}$ = 2. With pellets $Z_{eff}$ was about 3 for maximum $T_e(0)$ and this further dilution reduces $Q_{RF}$ to 50%.

## SUMMARY

ICRH has made a major contribution to high performance JET plasmas. Its strongly localised nature, as verified by modulation studies, allowed heating well within the central enhanced confinement region in peaked denisty plasmas and has generated sawtooth-free periods during the current rise phase of 5MA and 6MA discharges. Previously encountered problems of ICRF heating of H-modes (low coupling, impurity influxes) appear to be absent in a new scenario allowing 1cm plasma/limiter separation in double-null X-point discharges. Further investigation of this scenario, together with a) (H)D and (He³)D experimental simulations of the (D)T high Q scheme, b) fast wave and minority ion current drive studies and c) investigation of synergism with combined RF and NBI, will all be essential elements in the near term development of ICRH on JET.

## ACKNOWLEDGEMENTS

It is a pleasure to acknowledge the assistance of all our colleagues in the JET team. Particular thanks go to the tokamak operating team, the RF and NBI operating teams and to the members of the diagnostic groups contributing to these measurements.

## REFERENCES

1. M. Bures et al, Plasma Physics and Cont. Fusion 30(1988) 149.
2. J. Jacquinot et al, Plasma Physics and Cont. Fusion 30(1988)1467.
3. M. von. Hellermann et al, 16th Eur. Conf. on Contr. Fusion and Plasma Physics, Venice 1989.
4. V.P. Bhatnagar et al, ibid.
5. P. Stubberfield et al, ibid.
6. R.M.O. Galvão et al, ibid.
7. L.A. Charlton et al, Proc. of 1989 Int. Sherwood Theory Conf., Texas 1989, paper 2C27.
8. B. Tubbing et al, 16th Eur. Conf. on Fusion and Plasma Physics, Venice 1989.
9. D.A. Boyd et al, to be published in Nuclear Fusion.
10. L.G. Eriksson et al, Nuclear Fusion 29(1989)87.

# SPECTRAL SHAPING AND PHASE CONTROL
# OF A FAST-WAVE CURRENT DRIVE ANTENNA ARRAY*

F. W. Baity,  W. L. Gardner,  R. H. Goulding,
D. J. Hoffman, and P. M. Ryan
Oak Ridge National Laboratory, Oak Ridge, TN  37831-8071

## ABSTRACT

The requirements for antenna design and phase control circuitry for a fast-wave current drive (FWCD) array operating in the ion cyclotron range of frequencies are considered.  The design of a phase control system that can operate at arbitrary phasing over a wide range of plasma-loading and strap-coupling values is presented for a four-loop antenna array, prototypical of an array planned for the DIII-D tokamak (General Atomics, San Diego, California).

The goal is to maximize the power launched with the proper polarization for current drive while maintaining external control of phase.  Since it is desirable to demonstrate the feasibility of FWCD prior to ITER, a four-strap array has been designed for DIII-D to operate with the existing 2-MW transmitter at 60 MHz.

## SPECTRAL SHAPING

The spectrum launched by an array of side-by-side loop antennas  depends on the dimensions, number, and spacing of the loops; the mutual coupling between loops; the relative phasing between loops; and the ratios of currents in each loop.  The geometry of the loops is driven by considerations of maximizing the magnetic flux linked to the plasma, independent of any concern for driving plasma current.  The number of loops will be controlled by the available space inside the tokamak.  The mutual coupling between loops can be varied by the use of slotted septa between loops.

The spectra produced both with and without solid septa between loops are compared in Fig. 1 for the geometry of the DIII-D four-loop array.  The calculation was based on equal currents in all loops and $90°$ phasing between loops.  With no septa between loops, approximately 81% of the power is launched with $k_{\parallel}$ in the current drive peak; with full septa, only about 48% of the power is launched in the current drive peak.  The effect of the plasma  absorption on the spectrum was not taken into account.  This must be done for calculations of the current drive efficiency.[1]

*Research sponsored by the Office of Fusion Energy, U.S. Department of Energy, under contract DE-AC05-84OR21400 with Martin Marietta Energy Systems, Inc.

## IMPEDANCE MATCHING CIRCUIT

For the DIII-D demonstration all four loops are driven from a single transmitter, but the circuit described here applies to any array with a pair of loops fed from the same rf source. For optimum spectral control it is necessary to have equal current magnitudes on each loop. Likewise, arbitrary phasing between loops is required for optimal $k_{\parallel}$ shaping. A practical system will also be stable over a range of plasma-loading resistance with high coupling between loops ($KQ \geq 1$, where $K$ is the coupling factor between loops).

A schematic drawing of the circuit is shown in Fig. 2. The loops forming the array are fed in pairs by unmatched tees. Each loop is prematched by a single stub tuner to obtain the value of the input admittances at the tee that, for a given value of plasma loading, results in equal currents in each array element and produces an impedance match at the inputs to the tee. The phase between loops of each pair is controlled by an additional phase shifter in one leg of the tee.

With a coupled, lossy transmission line model, it can be shown[2] that, for the four-element array with currents that are equal in amplitude but not in phase, to a good approximation the real parts of the current strap input impedances scale as follows:

$\mathrm{Re}\,(Z_{in1}) \propto R - K_1 \omega L \sin \theta$

$\mathrm{Re}\,(Z_{in2}) \propto R + (K_1 - K_2)\,\omega L \sin \theta$

$\mathrm{Re}\,(Z_{in3}) \propto R - (K_1 - K_2)\,\omega L \sin \theta$

$\mathrm{Re}\,(Z_{in4}) \propto R + K_1 \omega L \sin \theta$

where $R$ is the resistance per unit length; $L$ is the inductance per unit length, $K_1$ is the coupling constant (ratio of $M/L$, where $M$ is the mutual inductance per unit length) between loops 1 and 2 and between loops 3 and 4; $K_2$ is the coupling constant between loops 2 and 3; and $\theta$ is the (equal) phasing between neighboring loops. The small coupling between nonadjacent loops was ignored. If the currents into both elements in a pair are balanced, equal powers into the two tees will produce equal currents in all loops. In addition, this relation will hold independently of the relative phase and of the values of $K_1$ and $K_2$. Thus, pairing elements 1 and 4 together and elements 2 and 3 together appears to be the most favorable of the possible choices for a parallel feed arrangement.

The transmission line parameters for the DIII-D current drive antenna design were calculated by using a 2-D magnetostatic code[3], and the mutual inductances were measured on a small antenna array having a geometry similar to the DIII-D design. The strap length in the calculations was taken to be 0.4 m. Figure 3 shows the results of a calculation in which the resistance per unit length of the strap is varied, simulating changes in resistive plasma loading. The calculations were done for a frequency of 60 MHz, and the tuning element lengths were chosen to obtain equal currents and 90° phase difference between straps for 6-$\Omega$/m resistive loading. The current amplitude ratios remain constant for a wide range of loading values, although the phasing of the currents changes. Changing the length of the lines in legs 1 and 3 between the tees and the tuning stubs can alter the behavior of the circuit so that the phasing is stable except for very low loading values to changes in resistive loading, whereas the amplitudes are not stable (see Fig. 4).

In comparison, Fig. 5 shows the situation in which the input impedance of each array element is matched individually to 50 $\Omega$ (for 6-$\Omega$/m loading) at the generator side of the single stub tuner. The forward power in all four legs is chosen to be the same for this case, as it would be if hybrid splitters were used to divide the power between legs. Note that the current ratios are far from unity, producing a significant degradation in the directionality of the $k_\parallel$ spectrum produced. Of equal importance is the fact that, with the same amount of power applied, peak voltages are significantly higher for this case than for the equal-current case. This is due to unequal voltages on the straps, which correspond to the unequal currents, with some lines having higher voltages and some lower voltages than is the case with equal currents. The result is shown in Fig. 6, which shows the maximum voltage encountered in the four lines between the stub tuners and the array elements as the load varies for the cases of equal forward power and equal currents. For 2-MW input power and 4-$\Omega$/m resistive loading, the peak voltage for the case of equal forward power is 10 kV higher than seen in the equal-current case. This difference could be critical if it were necessary to operate with only light loading.

## CONCLUSIONS

In general, spectral shaping is improved by increasing the coupling between antennas, but phase control becomes more difficult. With the phase control system previously described, it is possible to operate at arbitrary phasing between antennas at coupling values considerably above critical coupling. These features allow both the power-handling capability of the antenna array and the $k_\parallel$ spectrum generated to be optimized to the fullest extent possible.

## REFERENCES

1. M. J. Mayberry et al., "60 MHz Fast-Wave Current Drive Experiment for DIII-D," paper presented at this conference.
2. P. U. Lamalle, A. M. Messiaen, and P. E. Vandenplas, "ICRH Antenna Array Analysis and Resulting Problem of Generator Matching," *IEEE Trans. Plasma Sci.*, **PS-15** (1), 60–69 (1987).
3. G. L. Chen et al., "Resonant Loop Antenna Design with a 2-D Steady-State Analysis," AIP Conference Proceedings, 159, 382-385 (1987).

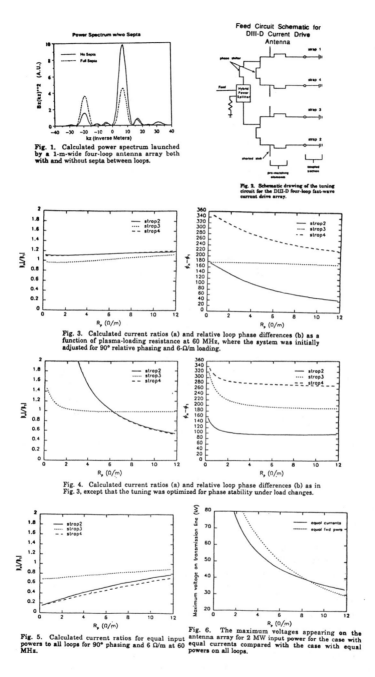

**Fig. 1.** Calculated power spectrum launched by a 1-m-wide four-loop antenna array both with and without septa between loops.

**Fig. 2.** Schematic drawing of the tuning circuit for the DIII-D four-loop fast-wave current drive array.

**Fig. 3.** Calculated current ratios (a) and relative loop phase differences (b) as a function of plasma-loading resistance at 60 MHz, where the system was initially adjusted for 90° relative phasing and 6-Ω/m loading.

**Fig. 4.** Calculated current ratios (a) and relative loop phase differences (b) as in Fig. 3, except that the tuning was optimized for phase stability under load changes.

**Fig. 5.** Calculated current ratios for equal input powers to all loops for 90° phasing and 6 Ω/m at 60 MHz.

**Fig. 6.** The maximum voltages appearing on the antenna array for 2 MW input power for the case with equal currents compared with the case with equal powers on all loops.

# FAST WAVE CURRENT DRIVE MODELING FOR ITER AND PROSPECTS FOR A NEAR-TERM PROOF OF PRINCIPLE EXPERIMENT*

D. B. Batchelor, E. F. Jaeger, M. D. Carter, and D. W. Swain
Oak Ridge National Laboratory
Oak Ridge, Tennessee 37831-8071

## Introduction

It is widely recognized that a key element in the development of an attractive tokamak reactor, and in the successful achievement of the mission of ITER is the development of an efficient steady state current drive technique. An important figure of merit in this context is the quantity

$$\gamma = \frac{n_e}{10^{20}m^{-3}} I_{RF} \frac{R}{P_{RF}} = 1.6 \times 10^{-3} T_e \ \tilde{j}/\tilde{p} \tag{1}$$

Ehst[1] has provided a simple algebraic expression for $\tilde{j}/\tilde{p}$ obtained as a fit to Fokker-Planck results of Karney and Fisch[2] employing a momentum conserving collision operator but not including trapped particle effects

$$\tilde{j}/\tilde{p} = \frac{8u^2}{5 + Z_{eff}} + 2 + \frac{12(6 + Z_{eff})}{(5 + Z_{eff})(3 + Z_{eff})} + \frac{2D}{Z_{eff}u} \tag{2}$$

where the factor $D = 8.09$ if the dominant damping mechanism is electron TTMP. For parameters projected for the steady state technology phase of ITER values of $\gamma$ greater than 0.5 are easily achieved for current driven near the magnetic axis. With $Z_{eff} = 2.2$, $\tilde{j}/\tilde{p}$ given by Eq (2) has a minimum at $u = 1.5$ which at a central temperature of $T_e(0) = 33$kev gives $\gamma = 0.35$ from Eq (1). If such values are actually achievable the fast wave current drive would appear to be a very attractive scheme for ITER. The two primary issues to be resolved are:

1. Can an acceptable antenna be designed to produce a $k_\parallel$ spectrum with most power in the phase velocity range $v_e < v_{phase} < c$ ?

2. Will the RF power be efficiently absorbed by electrons in the desired velocity range without unacceptable parasitic damping by fuel ions and $\alpha$ particles?

To address these issues we have employed the ORION code which gives a 3D full-wave solution of the fast wave equations in tokamak geometry along with power absorption and current drive profiles. This 3D code realistically models the launched antenna spectrum with radial focusing, includes toroidal eigenmode effects (multiple pass absorption) if present, and includes the effects of ion cyclotron harmonic and $\alpha$ particle damping.

## ITER FWCD Antenna Array

The results presented here are based on a preliminary antenna array design being developed by J. J. Yugo of the Fusion Engineering Design Center and by members of the ORNL Fusion Energy Division Plasma Technology Section. The array consists of 12 individually phased antennas with centers separated

* Research sponsored by the Office of Fusion Energy, under Contract No. DE-AC05-84OR21400 with Martin Marietta Energy Systems, Inc.

toroidally by 45 cm. Each antenna consists of 5 poloidally stacked straps to a total poloidal extent of 300 cm. The straps are 22 cm wide, 2 cm thick and are recessed from the first wall by about 8 cm with an additional 10 cm separation between the back of the current strap and the conducting backplane of the antenna enclosure. The antennas are separated by a slotted isolating septum of 2 cm thickness. The model presented here does not include the image currents due to the septa. The model is however being extended to include this effect. Engineering details of the antenna design and its integration into the ITER device will be presented separately by J. J. Yugo.

## Description of the 2D Full Wave Model

The 2D full wave code ORION[3,4,5] solves the vector wave equation for the fast wave fields

$$\nabla \times \nabla \times \mathbf{E} - \frac{\omega^2}{c^2} K \cdot \mathbf{E} = i\omega\mu_0 \mathbf{J}_{ant} \qquad (3)$$

where $K$ is the dielectric tensor. We use a reduced warm plasma description for the plasma response in which $K$ is the warm plasma dielectric tensor expanded to 2nd order in Larmor radius. Maxwellian electrons, deuterium and tritium components as well as slowing down alpha particles are included. For the values of $k_\parallel$ and $k_\perp$ required we approximate $k_\parallel$ by $k_{toroidal}$ and obtain $k_\perp$ from the fast wave root of the local warm plasma dispersion relation. In this way warm plasma damping effects (cyclotron damping, Landau damping and TTMP) are included and the effect of the presence of the Bernstein wave on the fast wave is approximated, while maintaining a 2nd order differential system. The comparison of this reduced kinetic description with the full 6th order equations (including Bernstein waves) is discussed in ref. 6. From Eq (6) two components of $\mathbf{E}$ perpendicular to $\mathbf{B}$ are solved while $\mathbf{E}_\parallel$ is obtained from the parallel equation as a perturbation. From $\mathbf{E}$ the local rate of energy deposition is obtained as $\dot{W} = \text{Re}\{\mathbf{E}^* \cdot \mathbf{J}_p\}/2$. The wave parallel phase velocity is $v_{phase} = \omega/k_\parallel$ where we approximate $k_\parallel = N_T R_{maj}$. Then the current driven by the power in a given toroidal mode is

$$J(N_T) = \text{sgn}(N_T) \frac{19.19 \times 10^{18} T_e(0)}{2\pi R_{maj} n_e(0) \ln\Lambda} \quad \tilde{j}/\tilde{p} \quad \dot{W}(N_T) \qquad (4)$$

where $\tilde{j}/\tilde{p}$ is given by Eq(1) with $u = \omega/k_\parallel v_{th}$. However for $v_{phase} > c$, $J(N_T)$ is taken as zero and to account in a coarse way for trapped particle effects, $J(N_T)$ is also taken as zero for $v_{phase} < v_e$. The total driven current is then the sum over $N_T$.

## ITER Parameters

For purposes of these calculations we have taken ITER parameters as given for the steady state technology phase described in ref. 7. Specifically $R_{maj} = 5.5$m, $R_{min} = 1.8$m, $\mathbf{B}_0 = 5.3$T, $I_p = 18$MA, $\kappa = 2.0$, $\delta = 0.4$. There is an additional 10 cm scrape off region between the plasma and the antenna, and since our models do not allow recessed antennas at this time, we take the wall radius to be at the backplane location, 10 cm beyond the antenna. We have assumed a peak temperature of 33 kev, $\langle T_e \rangle \sim 18$kev, and peak density $n_e(0) \sim 0.7 \times 10^{20}$, $\langle n_e \rangle = 0.47 \times 10^{20}$ where $\alpha_T = 1.0$, $\alpha_n = 0.5$. Assuming a Troyon g factor of 2.9, with these parameters the $\beta$ limit is given by (see ref. 8.) $\beta_{max} = gI_p/R_{min}B_0 = 5.5\%$. The thermal component is (using the notation of ref. 8) $\beta_{th} = 0.4n_{20}T_{10}(1+n_i/n_e)/B_0^2 \simeq 2.5\%$, while the fast alpha component is $\beta_{f\alpha} = 0.29(n_{DT}/n_e)^2(T_{10} - 0.37) \simeq 0.6\%$. Thus $\beta_{total} \simeq 3.1\% < \beta_{max}$. Also in accordance with the guidelines of ref. 8, we have taken $Z_{eff} = 2.26$ appropriate to a 10% thermal $\alpha$ fraction.

## ITER Results

Fig (1a) shows the spectrum of absorbed power (solid curve, the dashed curve is the absorption spectrum unweighted by the antenna spectrum) expressed in terms of $k_\parallel$ on the magnetic axis, for a relative phasing between elements of $-0.45\pi$. This phasing was found to give maximum efficiency for this configuration. The spectrum is seen to be well peaked about $n_\parallel \simeq -2.45$ with very little power propagating in the opposite direction. The net efficiency is $I/P_{Tot} = 0.131$ A/W ($\gamma = 0.33$) where $P_{Tot}$ is the total power launched into the plasma [i.e. the integral of Fig(1a)]. Fig (1b) shows contours of power deposition in the midplane of the antenna obtained from the ORION code. The circular region around the axis is TTMP/ELD on the electrons whereas the vertical strip to the left is ion absorption at the tritium 2nd harmonic. In this case 89% is absorbed by the electrons and 11% is absorbed by tritium. Essentially none is absorbed by $\alpha$ particles. Fig(1c) shows the flux surface average absorbed power profile. The driven current profile follows the electron absorption profile closely.

## Possibilities for a FWCD Proof-of-Principle Experiment

A critical need in this field is for a clean proof-of-principle experiment on a large hot tokamak. A very attractive candidate is D-III-D where a proposal has been made and a collaborative effort has begun between GA and ORNL to design an appropriate antenna array and perform such experiments. With 2MW of ECH power, it is anticipated that the high central electron temperatures needed for good electron absorption and high current drive efficiency can be obtained on D-III-D. For these calculations we have assumed a peak temperature of 7 kev and peak density $n_e(0) \sim 0.2 \times 10^{20}$, $R_{maj} = 1.67$m, $R_{min} = 0.63$m, $\mathbf{B}_0 = 1.0$T. The antenna consists of 4 current straps of width 11 cm and center-to-center separation of 23 cm. With ORION toroidal eigenmodes are seen in $k_\parallel$ which make the current drive efficiency somewhat sensitive to the array phasing. In this case optimum efficiency was obtained with $\Delta\phi = \pi/2$. The net current drive efficiency was $I/P_{Tot} = 0.42$A/W which gives $\gamma = .093$.

## Conclusions

These calculations suggests that it is indeed possible to obtain the needed spectral control to efficiently drive current in an ITER scale device. The requirements on central temperature for high values of $\gamma$, while stringent, are common to all current drive strategies. Such high electron temperatures assure absorption of the fast waves by electrons. Clearly to proceed with confidence to an ITER scale current drive system based on fast waves, a convincing experiment is needed on a major existing tokamak. Our calculations indicate that such a proof-of-principle experiment is feasible on D-III-D.

## References

1. D. A. Ehst and K. Evans, Nuclear Fusion 27,1267 (1987).

2. C. F. F. Karney and N. J. Fisch, Phys. Fluids 28, 116 (1985).

3. E. F. Jaeger, D. B. Batchelor, H. Weitzner, and J.H. Whealton Computer Phys. Communications 40,33 (1986).

4. E.F. Jaeger, D.B. Batchelor, and H. Weitzner, "Global ICRF Wave Propagation in Edge Plasma and Antenna Regions with Finite $\mathbf{E}_{\parallel}$" submitted to Nucl. Fusion.

5. D. B. Batchelor, E. F. Jaeger and P. L. Colestock, "Ion Cyclotron Emission form Energetic Fusion Products in Tokamak Plasmas- a Full-Wave Calculation" accepted Phys. Fluids.

6. E. F. Jaeger, D. B. Batchelor and H. Weitzner, Nucl Fusion 28, 53 (1988).

7. ITER Concept Definition, Report ITER-1, Oct. 1988.

8. N. A. Uckan and ITER Physics Group, "ITER Physics Design Guidelines," ITER-TN-PH-8-7, Jan. 1989.

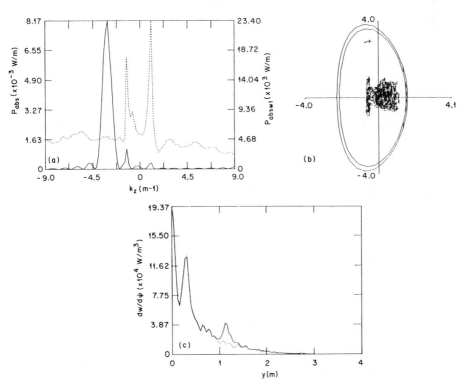

Fig. 1.(a) Absorbed power weighted by antenna spectrum (solid) and unweighted (b) Contours of power deposition (c) Flux surface average of power absorbed, total (solid), electrons (dashed).

# ICRF ANTENNA RESISTANCE AND THE H-MODE :
# EXPERIMENTAL AND THEORETICAL PARAMETER DEPENDENCE
# IN ASDEX AND OPTIMIZATION FOR ASDEX UPGRADE

M.Brambilla, J.-M. Noterdaeme, H.-P. Burkhart[1], F. Hofmeister
Max-Planck-Institut für Plasmaphysik, Euratom Association
D-8046 Garching bei München, Fed.Rep. Germany

## ABSTRACT

The antenna resistance in ASDEX was calculated using the FELICE 3-dim antenna code with the measured temperature and density profiles and the plasma position. The change of antenna resistance at the H-mode transition was modelled by varying the density gradient at the boundary. The theoretical estimates are in reasonable agreement with the measured values. During the H-phase, the excitation of eigenmodes complicates the picture.

The magnitude of the sudden decrease of the antenna resistance at the transition is experimentally observed to be a weak function of plasma position and density, in agreement with theoretical estimates.

The geometrical dimensions of the antenna and the phasing of its conductors also influence the magnitude of this decrease at the transition. The code is used to analyse the sensitivity of the antenna resistance of ASDEX Upgrade in this respect.

## INTRODUCTION

It has been shown on several machines (ASDEX, JFT-2M, JIPPT-IIU, JET) that ICRF can be used to obtain the H-mode. However, the change of antenna resistance at the H-transition, and the resulting technical difficulties, has caused some concern about the compatibility of H-mode and ICRF heating. In this paper we investigate how we can minimize this reduction of antenna resistance at the transition.

Firstly we show, by detailed comparison between the antenna resistance, calculated with the FELICE[1] code, and the experimentally measured values, for a particular shot on ASDEX, that the calculated resistance reasonably reflects reality.

Then we analyse experimentally the influence of the plasma parameters on the magnitude of decrease of the antenna resistance at the transition. The general trends are checked against the theoretical expectations and found to be in good agreement. This gives us further confidence in the theoretical model. It also indicates how (by proper choice of the plasma parameters) one can minimize, for a given antenna, the reduction in antenna resistance at the transition.

Finally, we rely on the calculated values to optimize the geometrical parameters of the ASDEX Upgrade antenna, to minimize this reduction in antenna resistance at the H- mode transition.

## EVOLUTION OF THE ANTENNA RESISTANCE

For a particular discharge in ASDEX, the antenna resistance[2] was calculated, using the electron temperature profiles, and the density profiles provided every 17 ms by the Thomson scattering system. The ion temperature was assumed to be equal to the electron temperature. The plasma position, measured by magnetic loops was taken into account since it plays an important role in the coupling through the antenna-plasma distance. The plasma position (d) is measured with respect to R =1.65 m, with

---

1 Ludwig-Maximilians-Universität, München.

a = 0.4 m. Antenna protection limiters are located at R = 2.12 m, the central conductor at R = 2.14 m and the return conductor at R = 2.215 m. Fig. 1 shows the experimental traces for the discharge for which the resistance was modelled. The resistance was calculated for different e-folding lengths of the boundary density (L=5, 2.5, 1.25, 0 cm). The L-mode values are well reproduced with 5 cm > L > 2.5 cm and taking $n_\varphi$ >10 (the contribution of low $n_\varphi$ is largely overestimated by the slab model when eigenmodes are present). The change in resistance at the transition to the H-mode is modelled at $t_1$ (last L-mode profile), by changing the boundary density gradient to L=1.25 and 0 cm (Fig. 2). During the H-mode narrow peaks are observed in the antenna resistance. They correspond to eigenmodes[3]. Since the experimental antenna resistance is sampled every 4 ms, and the profiles, on which the calculated resistance is based, only every 17 ms, it is not clear whether the discrepancy observed between measured and calculated resistance during the H-mode is due to this difference in sampling times, or to other reasons ( overestimation by the model of the amplitude of the eigenmodes, larger sensitivity to the correct position)

Fig. 2. Comparison of the calculated and experimental antenna resistance for the discharge shown in Fig. 1.

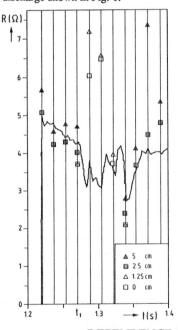

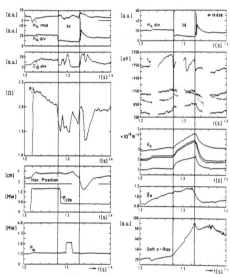

Fig. 1. Experimental traces for a particular shot with an H-transition. Note that the transition occurs when the electron temperature at the edge rises due to the sawtooth. The sharp decrease of the antenna resistance at the transition leads to an increased voltage in the transmission lines, and shut down of one generator (because of arcing). The antenna resistance is given for one half-antenna.

## DEPENDENCE ON PLASMA PARAMETERS

The decrease of the antenna resistance at the H-mode transition shows experimentally a dependence on a number of parameters (similar observations were made, using low power coupling tests, on D-III-D[4]). It decreases with decreasing antenna-plasma distance (Fig. 3) and increases with increasing line-averaged density (Fig 4). Both trends are in agreement with theoretical estimates (the large scatter of the experimental points in the figures is because the data are not from single parameter scans). The relative decrease of the antenna resistance at the transition also tends to

224

increase with increasing neutral injection power. Thus, in order to reduce the jump at the H-mode transition, one should work at small antenna-plasma distance, low density, and low NI power. The agreement of the calculated trends with the experimentally observed ones further increases confidence in the model.

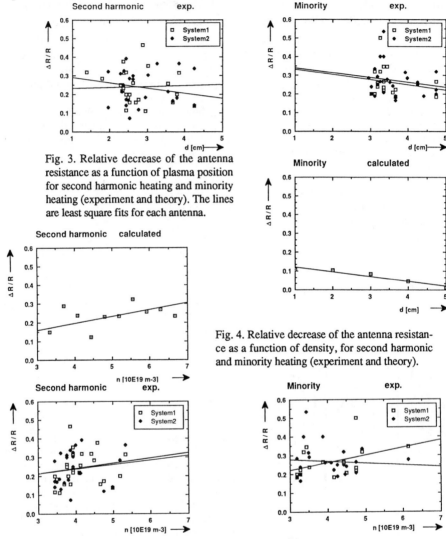

Fig. 3. Relative decrease of the antenna resistance as a function of plasma position for second harmonic heating and minority heating (experiment and theory). The lines are least square fits for each antenna.

Fig. 4. Relative decrease of the antenna resistance as a function of density, for second harmonic and minority heating (experiment and theory).

OPTIMIZATION OF THE ANTENNA GEOMETRY

The code was finally used to optimize the antenna geometry for ASDEX Upgrade. The antenna has two toroidally spaced conductors, which can be operated in-phase or out-of-phase. A model density profile was taken with the central density $n_0 = 6 \times 10^{19}$ m$^{-3}$ and separatrix density $n_s = 2 \times 10^{19}$ m$^{-3}$. The separatrix is located at R = 2.13 m. The protection limiters are located at R = 2.18 m, the antenna is modelled with the central conductors at 2.225 m, the return conductor at 2.29 m. The sensitivity of the antenna resistance to the H-mode transition was analysed with respect to the width of the conductors (w), the distance between them (gap), and the phasing. The edge gradient was varied between 4 cm and 0 cm to model the H-transition. With both conductors in phase ($\varphi = 0^0$) and a gap > 15 cm, the L-mode antenna resistance increases, the H-mode resistance decreases with increasing gap width and it is possible to obtain an antenna resistance that is not sensitive to the H-mode transition (Fig. 5). For the preferred $\varphi = 180^0$ configuration the antenna resistance in the L-mode is much larger than in the $\varphi = 0^0$ case and decreases with increasing gap (Fig. 6). The H-mode resistance is only slightly different from the $\varphi = 0^0$ case but increases with increasing gap. The difference between L and H could be minimized if a very large gap (> 30 cm) were possible, and more easily so for wide conductors.

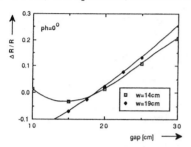
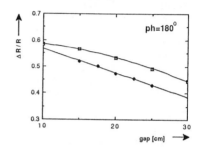

Fig. 5. Relative change of the antenna resistance.

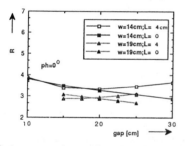
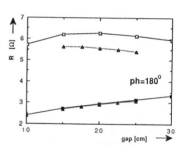

Fig. 6. Antenna resistance for ASDEX Upgrade, as a function of distance between the two antenna conductors (conductor widths 14 cm and 19 cm) for different values of the density gradient at the boundary and different phasings. The value for one conductor is double the value in the figure.

## CONCLUSION

The FELICE 3-dim code reproduces reasonably well the experimental values of the antenna resistance. Experimentally and theoretically, plasma parameters can be found which minimize the decrease of the antenna resistance at the H-mode transition. The ASDEX Upgrade antenna is, for in phase operation, insensitive to the transition.

## REFERENCES

1. M. Brambilla, accepted for publication in Plasma Physics and Contr. Fusion.
2. Y. Ogawa et al., Europhys. Conf. Abstr. (Proc. 16 th. Eur. Conf. on Controlled Fusion and Plasma Physics, Venice, 1989), Eds. S.Segre, H.Knoepfel, E.Sindoni, Vol 13B, Part III, 1085.
3. F. Hofmeister et al., Fusion Technology (Proc. 14th. Symp., Avignon, 1986), CEC, Vol.1, 801.
4. M.J. Mayberry et al., AIP Conf. Proc. (Proc. 7th Conf. on Appl. of RF Power to Plasmas, Kissimee, 1987), Eds. S. Bernabei, R.W. Motley, Vol. 159, 278.

# ION ENERGY AND PLASMA MEASUREMENTS
# IN THE NEAR FIELD OF AN ICRF ANTENNA*

J. B. O. Caughman, II, and D. N. Ruzic
University of Illinois, Urbana, IL 61801

D. J. Hoffman
Oak Ridge National Laboratory, Oak Ridge, TN 37831

## ABSTRACT

Plasma properties and ion energies have been measured in the near field of an ICRF antenna to determine the effects of rf fields in a magnetized plasma sheath on the energy of ions incident on the surface of the Faraday shield. A resonant loop antenna with a two-tier Faraday shield was used on the RF Test Facility at Oak Ridge National Laboratory (ORNL). The magnetic field near the antenna is ~2 kG, and the plasma density is ~$10^{11}$ cm$^{-3}$ with an electron temperature of 6-10 eV. The time-varying floating potential was measured with a capacitively coupled probe, and the time-averaged electron temperature, electron density, and floating potential were measured with a Langmuir probe. Both probes were scanned poloidally in front of the antenna, parallel to the current strap. Diagnostics for measuring ion energies included a gridded energy analyzer located directly below the antenna. Measured ion energies are compared with predictions from a computational model for determining the energy and angular distribution of ions incident on a surface in a magnetized plasma sheath with a time-varying plasma potential.

## INTRODUCTION

The interaction of the rf fields near an ICRF antenna with the surrounding plasma is important in understanding the generation of impurities from the antenna. Changes in the plasma parameters that result from the application of power from the antenna must be measured to help explain this interaction. Several of these plasma parameters have been measured within 1 cm of the surface of the Faraday shield tubes of a single-strap ICRF antenna in the RF Test Facility at ORNL. The antenna used in the experiment (shown in Fig. 1) was a resonant loop antenna operated at 42 MHz. The rf power was varied up to 60 kW, and the target plasma was generated by ~16 kW of 10.6-GHz ECH with a background hydrogen gas pressure of (1-3) × 10$^{-4}$ Torr.

## EXPERIMENT

The time-varying floating potential was measured with a capacitively coupled probe that was scanned in front of the antenna. The probe, described in Ref. 1, was calibrated at the rf frequency (42 MHz). A Langmuir probe scanning in the same area as the capacitive probe was used measuring the time-averaged electron temperature, electron density, and floating potential. The Langmuir probe was terminated on a small dc and rf load and thus measured the time-averaged current as

---

*This research was supported by the Magnetic Fusion Energy Technology Fellowship Program administered by Oak Ridge Associated Universities for the U.S. Department of Energy and by U.S. DOE Contract No. DE-AC05-84OR21400 and Subcontract No. 19X-SB359V with Martin Marietta Energy Systems, Inc.

a function of applied probe bias voltage. The ion saturation region of the measured probe characteristic appears to follow a $(V_{bias})^{1/2}$ dependence. With this dependence assumed, the time-averaged ion saturation current was calculated and subtracted from the measured current to give the time-averaged electron current. The electron temperature was then calculated from the lower portion of the $Ie$-$V$ curve to avoid problems associated with response of a Langmuir probe in an rf plasma.[2] The electron density was calculated by measuring the ion current well into the saturation region and then corrected by using the LaFramboise method.[3] The time-averaged floating potential was taken from the time-averaged current measurement and then corrected for self-bias due to rf.[1,2]

A gridded energy analyzer located ~7 cm below the antenna measured the distribution of ion energies incident on a grounded surface in a magnetized rf plasma. Since the magnetic field was parallel to the surface and to the biasing grids of the analyzer, the analyzer was thin ($\leq 1$ mm thick) so that the ion energies perpendicular to the magnetic field could be measured.

## RESULTS

The capacitive probe results indicate that the floating potential oscillates at the rf frequency and can reach values of up to 300 V p-p for an antenna current of ~400 A. The rf floating potential, normalized by the antenna current, is shown in Fig. 2 for various rf powers and gas pressures. The plasma loading and the plasma density were lower for the 0.1-m Torr case. While the precise scaling with the rf electric field has not yet been quantified, the potential generally increases with increased antenna current and plasma loading. The value of the potential is fairly constant in the poloidal direction, parallel to the current strap, and generally follows the magnetic field pattern of the antenna instead of the voltage distribution on the current strap. This result indicates that the potential formation is caused mainly by the electromagnetic fields and not by the electrostatic fields.

The electron temperature in front of the antenna increases with increased rf power. Without rf, the electron temperature is 6-10 eV. With rf, the electron temperature $T_e$ increases to values above of 60 eV for an rf power of ~25 kW. It appears to be higher closer to the antenna surface. The electron density ~1 cm in front of the antenna is $(3-6) \times 10^{10}$ cm$^{-3}$ and generally decreases when rf is applied. The density decreases closer to the antenna surface. The time-averaged floating potential at ~1 cm in front of the antenna increases from ~5 V without rf to over 70 V with ~25 kW of rf power.

The ion energy distribution measured with the energy analyzer shows an increase in the ion energies hitting a grounded surface during rf. Figure 3 shows the measured perpendicular ion energy distribution for rf powers of 0-25 kW. The energy distribution is peaked at 5-15 eV without rf and broadens to higher energies with increased rf power. Ion energies above of 300 eV have been measured with ~25 kW of rf power. This increase in ion energies will lead to increased erosion of the antenna surfaces. The net result of these measurements is that the electron temperature, plasma potential, and ion impact energies generally increase with rf power.

## DISUSSION

This experiment was designed to test rf-plasma interactions near the antenna with rf fields and antenna conditions similar to those found in high-power rf experiments on confinement devices. The antenna voltages and currents in these experiments were 50-100% of those that would be expected in a tokamak. For example, at ~25 kW, the peak antenna voltage was ~20 kV and the antenna current was ~500 A. While most of the rf power in a tokamak will be absorbed in the resonance zone, the power and fields must pass through the low-density near-field area of the antenna. The amount of power deposited in the near field is not known exactly, but the power levels absorbed in our experiments are reasonably close to those expected. Some of this power appears to be coupling to the electrons and increasing their energy. This is consistent with theoretical predictions of electron heating in a rf sheath at the Faraday shield.[4,5] The electron temperature clearly increases with the rf electric field and rf power. Since the antenna used in this experiment has only one current strap, the effects of phasing between adjacent straps on the electric field structure, the plasma density, and the electron temperature near the antenna were not studied.

The increase in the ion energies measured with the energy analyzer is consistent with an increased sheath potential due to an increase in the electron temperature and in the fluctuating plasma potential. A computational model of a magnetized rf sheath currently in development,[6] shows that the energy of ions incident on a grounded surface will increase with increased electron temperature to values consistent with those measured with the energy analyzer. The model shows that the distribution is peaked near the time-averaged plasma potential. The time-averaged plasma potential is an input to the model and is taken to be the time--averaged floating potential plus $2.5T_e$ (Ref. 7).

## CONCLUSIONS

Experiments have shown that large rf plasma potentials exist in front of the antenna. The potentials are caused by the rf power from the antenna and seem to follow the magnetic field pattern of the antenna. These large potentials cause an increase in the ion energies near the antenna surface and could increase the amount of erosion and impurity generation from the antenna. Although other mechanisms might exist for the observed increase in ion energies, this increase appears to be at least partially due to the increase in the sheath potential caused by the increase in the electron temperature.

## REFERENCES

1. J. B. O. Caughman, II, D. N. Ruzic, and D. J. Hoffman, J. Vac. Sci. Technol. A, (accepted for publication May/June 1989).
2. N. Hershkowitz, M. H. Cho, C. H. Nam, and T. Intrator, Plasma Chem. Plasma Process. 8 (1), 35 (1988).
3. P. M. Chung, L. Talbot, and K. J. Touryan, Electric Probes in Stationary and Flowing Plasmas: Theory and Application (Springer--Verlag, New York, 1975).
4. G. A. Emmert, Bull. Am. Phys. Soc. 33 (9), 1875 (1988).
5. F. W. Perkins, Princeton Plasma Physics Laboratory Report PPPL-2571.
6. J. B. O. Caughman, II, D. N. Ruzic, and D. J. Hoffman, Bull. Am. Phys. Soc., 33 (9), (1988).
7. R. Chodura, Phys. Fluids 25, 1628 (1982).

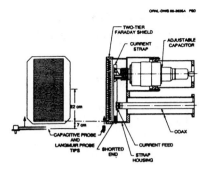

Fig.1 Front view and sectional view of the resonant loop antenna.

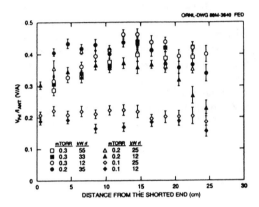

Fig.2 RF floating potential, normalized by the antenna current.

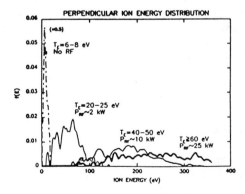

Fig.3 Measured perpendicular ion energy distribution for rf powers of 0–25 kW.

# MODE CONVERSION AT
# THE HIGHER ION CYCLOTRON HARMONICS

S.C. Chiu, V.S. Chan, and R.W. Harvey
General Atomics, San Diego, CA 92138-5608

M. Porkolab
Massachusetts Institute of Technology, Cambridge, MA 02139

## ABSTRACT

It has been demonstrated that mode conversion of fast waves to ion Bernstein waves can be calculated from a reduced second order differential equation for the wave fields rather than the 4th order equations used in earlier studies near the ion-ion hybrid resonance and the second harmonic resonance. Here the underlying justification of the method is discussed. It is shown that the method works for high harmonic resonances and an analytical formula for the tunneling coefficient is derived. The result is a generalization of a previous result obtained by Ngan and Swanson and is applicable when $k_\perp \rho_i$ is large. Recently, there is interest in using fast waves for current drive at high ion cyclotron harmonics frequencies in tokamaks. Generally, the fast wave will encounter ion cyclotron harmonics within the plasma cross-section. For efficient current drive, the minimization of the mode conversion processes sets restrictions to the choice of frequencies and magnetic fields. This is discussed using the derived formula.

## I. INTRODUCTION

The use of the second order Budden equations to study the mode conversion problem near the cold ion-ion hybrid resonance of a multi-ion species tokamak plasma was well known for a long time.[1-3] Mode conversion in tokamak plasmas at the second and higher ion cyclotron harmonics was first studied using higher order wave equations.[4] In the last few years, there was renewed interest in using the Budden equation to study mode conversion at the second harmonics.[5-9] The motivation for reducing the order of the wave equation is that the second order equation is considerably simpler analytically and numerically. In this paper, we extend this method to the case of higher ion cyclotron harmonics. In Section II, the justification of the method is clarified and a Budden equation is derived. An expression for the tunneling coefficient which generalizes a previous calculating using higher order equations[10] is obtained. We then apply the expression to investigate the constraint set by mode conversion for fast wave current drive at high harmonics.

## II. BUDDEN TUNNELING AT HIGHER ION CYCLOTRON HARMONICS

In Ref. 5, it was shown that in the case of pure second harmonics of fast waves (FW), the correct Budden equation is obtained by substituting the cold plasma value, $k_{\perp F}$, of the perpendicular wave number of the fast wave into the warm plasma terms. Two points were clarified. First, the neglect of the odd derivative terms can be justified by requiring $(n_z^2 + c^2/2V_A^2) \gg (\beta_i \zeta_0/4)(\zeta_0^2/\mathcal{R}_0)$, where $n_z$ is the parallel refractive index, $V_A$ is Alfvén speed, $\beta_i$ is ion beta value, $\zeta_0 = \omega/k_\parallel V_{Ti}$, $\mathcal{R}_0 = \omega R_0/c$, $R_0$ being major radius of the tokamak. Second, replacing the $k_\perp^2$ in warm terms by the cold value is heuristically appealing in the asymptotic sense when the warm plasma effect is viewed as a source term for the FW branch. It was not clear what the small parameter is. On closer examination of the procedure, it becomes apparent that in using the cold FW value, $k_{\perp F}$, there is an underlying assumption that neither the FW electric field nor its first derivative may vary considerably on passing through the singular layer. This is equivalent to assuming that the tunneling parameter, $\eta$, must be small, where $\eta$ is related to the transmission coefficient $T$ by $|T|^2 = \exp(-\eta)$; that is, there is an assumption that the tunneling layer is "thin." At the second harmonic it may happen that $\eta$ is not small so that this procedure may not be very accurate in predicting $\eta$ for large major radius, $R$, and high beta machines, although this inaccuracy is not very important because at those parameters, the transmission will be rather small anyway. However, it is generally true that $\eta$ will decrease as the harmonic number becomes higher, so that we can expect that this procedure will be rather accurate for higher harmonic mode conversion processes.

We assume that the plasma and wave parameters are such that they are sufficiently far from the FW–SW (slow wave) mode conversion region. Then the FW is described by the wave equations

$$
\begin{pmatrix} \epsilon_{11} - n_\parallel^2 & \epsilon_{12} \\ -\epsilon_{12} & \epsilon_{22} - n_\parallel^2 + \frac{\partial^2}{\partial x^2} \end{pmatrix} \begin{pmatrix} E_x \\ E_y \end{pmatrix} = 0 \quad , \tag{1}
$$

where $x$ is the "radial" coordinate, $\epsilon_{ij}$ are the dielectric tensor elements including warm plasma effects, and $x$ is normalized in $k_0 = \omega/c$. The perpendicular wave number of the cold FW is given by

$$
n_{\perp F}^2 \equiv c^2 k_{\perp F}^2 \big/ \omega^2 = \left[ \frac{(S - n_\parallel^2)^2 - D^2}{S - n_\parallel^2} \right] \quad , \tag{2}
$$

where $S$ and $D$ are the cold plasma dielectric functions defined in Ref. ll. Using the procedure just described, and with the notations

$$
K_{11}^{(H)} = \sum_\alpha \frac{\omega_{p\alpha}^2 \xi_0}{\omega^2 \mu_\alpha} e^{-\mu_\alpha} \ell_\alpha^2 I_{\ell\alpha}(\mu_\alpha) Z(\xi_{\ell\alpha}) \quad , \tag{3}
$$

$$
K_{22}^{(H)} = K_{11}^{(H)} + \sum_\alpha \frac{2\omega_{p\alpha}^2 \xi_0}{\omega^2} e^{-\mu_\alpha} \mu_\alpha \left( I_{\ell\alpha} - I_{\ell\alpha}' \right) Z(\xi_{\ell\alpha}) \quad , \tag{4}
$$

$$
K_{12}^{(H)} = \sum_\alpha \frac{\omega_{p\alpha}^2 \xi_0}{\omega^2} e^{-\mu_\alpha} \ell_\alpha \left( I_{\ell\alpha} - I_{\ell\alpha}' \right) Z(\xi_{\ell\alpha}) \quad , \tag{5}
$$

where $\omega_{p\alpha}$ is ion-plasma frequency for the $\alpha$-th specie, $\mu_\alpha = \frac{1}{2} k_{\perp F}^2 \rho_\alpha^2$, $I_{\ell\alpha}$ are Bessel functions, $Z$ is plasma z-function, and

$$
\xi_{\ell\alpha} = \frac{\omega - \ell_\alpha \Omega_\alpha}{k_\perp V_{T\alpha}} \quad , \tag{6}
$$

is the resonance factor at the $\ell_\alpha$-th resonance, one arrives at a Budden-like equation:

$$
\frac{\partial^2 E_y}{\partial x^2} + \left[ n_{\perp F}^2 + \frac{(S - n_z^2)^2 K_{22}^{(H)} - 2D(S - n_z^2) K_{12}^{(H)} + D^2 K_{11}^{(H)}}{(S - n_z^2)^2} \right] E_y = 0 \quad . \tag{7}
$$

Here, terms of higher order in $K_{ij}^{(H)}$ are neglected because they are small when the inequality mentioned at the beginning of the section is satisfied. Only resonant terms are included in $K_{ij}^{(H)}$. When the dissipation is neglected and the large argument expansion is used for the $z$-functions, Eq. (7) becomes the Budden equation. Thus, using the $1/R$ dependence of $B$, we have $\xi_{0\alpha} Z(\xi_{l\alpha}) = - R_0/x$, where $R_0$ is the major radius, and

$$K_{ij}^{(H)} = -\chi_{ij}^{(H)} \frac{R_0}{x} \quad , \tag{8}$$

where

$$\chi_{11}^{(H)} = \sum_\alpha \frac{\omega_{p\alpha}}{\omega^2} \frac{e^{-\mu-\alpha}}{\mu_\alpha} \ell_\alpha^2 I_{l\alpha}^2(\mu_\alpha) \quad , \tag{9}$$

$$\chi_{12}^{(H)} = \sum_\alpha \frac{\omega_{p\alpha}^2}{\omega^2} e^{-\mu_\alpha} \ell_\alpha \left(I_{l\alpha} - I_{l\alpha}'\right) = \sum_\alpha \chi_{12\alpha}^{(H)} \quad , \tag{10}$$

and

$$\chi_{22}^{(H)} = \chi_{11}^{(H)} + \sum_\alpha \frac{2\mu_\alpha}{\ell_\alpha} \chi_{12\alpha}^{(H)} \quad . \tag{11}$$

The tunneling parameter is then

$$\eta = \pi \mathcal{K} / n_{\perp F} \quad , \tag{12}$$

where

$$\mathcal{K} = \frac{\omega}{c} R_0 \left\{ \chi_{22}^{(H)} - \frac{2D}{(S - n_z^2)} \chi_{12}^{(H)} + \frac{D^2}{(S - n_z^2)^2} \chi_{11}^{(H)} \right\} \quad , \tag{13}$$

Equation (13) is a generalization of the tunneling parameter derived in Ref. 10. For a single ion-specie plasma, at $\omega = \ell\Omega_i$, we may approximate $k_{\perp F} \simeq \ell\omega_{pi}/c$. Assuming $\frac{1}{2} k_{\perp F}^2 \rho_i^2 < \ell$, using series expansion for $I_{l\alpha}$, etc., we obtain

$$\eta_\ell \simeq \pi \frac{\omega_{pi}}{c} R_0 \frac{(\ell - 1)}{2(\ell - 2)!} \left(\frac{\ell^2 \beta_i}{4}\right)^{\ell-1} e^{-\ell^2 \beta_i/2} \quad . \tag{14}$$

When $e^{-\ell^2\beta_i/2} \to 1$, Eq. (14) reduces to the expression in Ref. 10. For tokamak parameters, Eqs. (12) and (13) or Eq. (14) are more appropriate because $\ell^2\beta_i/2$ can exceed unity.

There has been considerable interest in driving steady-state current by FW at high harmonics.[12-14] The FW, on penetrating to the plasma core, will encounter high harmonic resonances. The current drive efficiency will be adversely affected if the transmission $|T|^2$ is significantly less than unity. For a given density, $R_0$ and $\beta_i$, the requirement that $\eta_\ell$ be small poses a lower limit on the frequency. In Fig. 1, we plot $\eta_\ell$ as a function of $\ell = \omega/\Omega_i$ as given by Eq. (14) for a hydrogen plasma of density $n = 1.25 \times 10^{14}$ cm$^{-3}$ and temperature 15 keV, with $\beta_i$ as a parameter. For $\beta_i = 3\%$, $\ell > 6$ is appropriate, while for $\beta_i = 5\%$, $\ell > 10$ is needed.

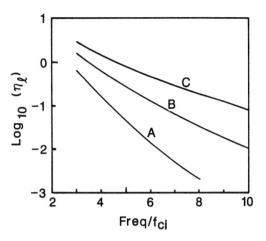

FIG. 1. Tunneling parameter, $\eta_\ell$, for ion Bernstein wave mode conversion as a function of frequency for $\beta_i = 3\%$ (A), $\beta_i = 5\%$ (B), and $\beta_i = 7\%$ (C). Plasma density $= 1.5 \times 10^{14}$ cm$^{-3}$, plasma temperature $= 15$ keV.

## III. SUMMARY

We have obtained a general formula, Eqs. (12) and (14), for the tunneling parameter of FW at arbitrary ion harmonics. This formula is a generalization of a formula previously obtained in Ref. 10 by Ngan and Swanson. Application of this formula gives the lower limit of the frequency of fast waves needed to drive current in tokamaks at high harmonics.

## IV. ACKNOWLEDGMENT

This is a report of work sponsored by the U.S. Department of Energy under Contract No. DE-AC03-89ER53277.

## V. REFERENCES

1. D.G. Swanson, Phys. Rev. Lett. **36**, 316 (1976).
2. F.W. Perkins, Nucl. Fusion **17**, 1197 (1977).
3. J. Jacquinot, B.D. McVey, J.E. Scharer, Phys. Rev. Lett. **39**, 88 (1977).
4. D.G. Swanson, Phys. Rev. **28**, 2645 (1985).
5. S.C. Chiu, Phys. Fluids **28**, 2808 (1985).
6. A. Kay, R.A. Carns, C.N. Lashmore-Davies, in Proc. 13th Euro. Conf. on Contr. Fusion and Plasma Heating (European Physical Society, 1986) p. 93.
7. C.N. Lashmore-Davies et al., Phys. Fluids **31**, 1614 (1988).
8. D.N. Smithe et al. Nucl. Fusion **27**, 1319 (1987).
9. E.F. Jaeger, D.B. Batchelor, H. Weitzner, Nucl. Fusion **28**, 53 (1988).
10. Y.C. Ngan, D.B. Swanson, Phys. Fluids **20**, 1920 (1977).
11. T.H. Stix, The Theory of Plasma Waves, McGraw-Hill, New York (1962).
12. K.L. Wong, M. Ono, Nucl. Fusion **24**, 615 (1984).
13. P.L. Andrews, D.K. Bhadra, Nucl. Fusion **26**, 897 (1986).
14. S.C. Chiu, V.S. Chan, R.W. Harvey, M. Porkolab, General Atomics Report GA-A19534 (1989), submitted to Nucl. Fusion.

# ANALYTIC STUDIES OF ICRF HEATING

## C. Chow and A. Bers

*Plasma Fusion Center, Massachusetts Institute of Technology, Cambridge, Massachusetts 02139*

## V. Fuchs

*Centre Canadien de Fusion Magnétique, Varennes, Québec J0l 2P0, Canada*

## ABSTRACT

We obtain the scattering coefficients for minority and second harmonic ICH D-H and D-T-$^3$He scenerios. A straight forward $k_\perp \rho_i$ expansion is not sufficient to reproduce the D($^3$He) dispersion relation. We obtain properly corrected dielectric tensor elements and give results for various cases.

## I. INTRODUCTION

In ICRF minority and second harmonic heating power is injected on a fast Alfvén wave (FAW) which propagates to a resonance layer in the center of the plasma where it couples to an ion-Bernstein wave (IBW). Power is transmitted, mode converted and reflected for low field incidence.[1,2] Damping is present on both the FAW and the IBW. This coupling-damping problem has been solved with various numerical codes.[3-6] Here we generalize a method to obtain the scattering coefficients in closed form.[7] We apply this method to D-H and D-T-$^3$He heating in arbitrary species concentration ratios.

## II. THE REDUCED SYSTEM OF EQUATIONS

We start from the standard zero-electron-mass, Vlasov-Maxwell local dispersion relation in ICRF where the dielectric tensor elements have been expanded to first order in $(k_\perp \rho_i)^2$. In the coupling region we assume slab geometry in which the tokamak toroidal magnetic field is directed along the $z$ coordinate, and its gradient is along $x$. The gradient scale length is $R_0$, the tokamak major radius. The dispersion relation normalized to the Alfvén velocity $c_A$ is then

$$N_\perp^4 K_1 + N_\perp^2 (K_0 - 2\lambda_N K_1) - 2\lambda_N K_0 + \lambda_N^2 = 0 \tag{1}$$

where $c_A^2 = B_0^2/\mu_0 n_1 m_1$, $N_\perp = c_A k_\perp/\omega$, $K_0$ and $K_1$ are functions of $x$ derived from the expanded dielectric elements and $\lambda_N$ is a parameter depending on $N_\parallel$.[7] The IBW dispersion relation is given by $K_0 + K_1 N_\perp^2$. Using this we write (1) in the coupled form

$$(2\lambda_N - N_\perp^2)(K_0 + K_1 N_\perp^2) = \lambda_N^2. \tag{2}$$

Coupled differential equations immediately follow

$$F'' + 2\lambda_N F = \lambda_N \Phi, \tag{3a}$$

$$(K_1 \Phi')' - K_0 \Phi = -\lambda_N F, \tag{3b}$$

where $F$ is the FAW amplitude and $\Phi$ is the IBW amplitude. These equations can be solved analytically to obtain the transmission coefficient $T$ and the high field mode conversion coefficient $C_H$. $T$ has also been solved analytically independently.[9]

Letting $(K_1\Phi')' \simeq -K_1 N_c^2\Phi$ which amounts to treating the IBW as a driven wave at $N_\perp = N_c$, the FAW $N_\perp$ at the coupling point, gives[8]

$$F'' + [2\lambda_N - \lambda_N^2/(K_0 + N_c^2 K_1)]F = 0, \tag{4}$$

from which $T$ and $R$ are obtained. Given $R$ the low field incidence conversion coefficient $C_L$ is found. Currently (4) is solved numerically for $R$.

## III. D-H HEATING

The dispersion elements for this case given in ref. 7 were limited to very low concentrations of H. The following elements are correct for arbitrary concentrations of H:

$$K_0 = -\frac{1}{3} - \frac{\eta}{8} + \frac{\eta}{4\alpha_2}Z\left(\frac{x}{\alpha_2 R_0}\right), \quad K_1 = \frac{\beta_D}{4\alpha_1}Z\left(\frac{x}{\alpha_1 R_0}\right),$$

$$\lambda_N = \frac{1}{3} + \frac{\eta}{4} - N_\parallel^2,$$

where $\alpha_i = N_\parallel v_{ti}/c_A$, $v_{ti} = 2T_i/m_i$. We label the ions D and H with 1 and 2 respectively. From these elements we obtain the scattering coefficients. For instance

$$T = \exp[-\pi\sigma N_c^3/(1 + \eta/2 + N_\parallel^2)^2]$$

where

$$N_c^2 = \lambda_N \frac{1 + \eta/2 + N_\parallel^2}{1/3 + \eta/8 + N_\parallel^2}, \quad \sigma = (\eta + \beta_1 N_c^2)R_0\omega/4c_A,$$

and $\sqrt{\beta_i} = v_{ti}/c_A$.

## IV. D-T-$^3$HE HEATING

For the case when all 3 species are present the tensor elements are

$$K_0 = -\left(\frac{9}{7} + \frac{\theta}{2} + \frac{3\eta}{8} + N_\parallel^2\right) + \frac{3\eta}{4\alpha_3}Z\left(\frac{x}{\alpha_3 R_0}\right),$$

$$K_1 = \frac{3}{8}\frac{\theta\beta_2}{\alpha_2}Z\left(\frac{x}{\alpha_2 R_0}\right) + \frac{54}{35}\beta_1 - \frac{3}{8}\frac{\eta\beta_3}{\alpha_3}Z\left(\frac{x}{\alpha_3 R_0}\right),$$

$$\lambda_N = \frac{3}{7} + \frac{3\eta}{4} + \frac{\theta}{2} - N_\parallel^2.$$

where $\eta = n_3/n_1$ and $\theta = n_2/n_1$. The labels 1,2 and 3 refer to the species D,T and $^3$He respectively. We are able to consider arbitrary concetrations of each species. For simplicity, however, we express the above elements in a form which requires a non-zero concentration of D. This yields

$$T = \exp(-\pi\sigma N_c^3/\nu^2)$$

where

$$\sigma = \left(\frac{3}{4}\eta + \frac{1}{4}\theta\beta_1 N_c^2 + \frac{13}{20}\eta\beta_1 N_c^2\right)R_0\omega/c_A,$$

$$N_c^2 = \lambda_N \nu / \left( \frac{9}{7} + \frac{\theta}{2} + \frac{3\eta}{8} + N_\parallel^2 \right),$$

$$\nu = 3 + \frac{3}{2}\eta + \frac{\theta}{2} - N_\parallel^2.$$

For $\theta = 0$, $K_1$ above will go to zero between the cutoff and the harmonic. The dispersion relation (1) will then break down. We correct for this by using

$$K_1 = \frac{13}{20} \frac{\eta \beta_1}{\alpha_3} Z \left( \frac{x}{\alpha_3 R_0} \right).$$

This was obtained by expanding $K_1$ in $x$ around the coupling point.[10] The new element does not affect $T$ but is important for the other coefficients. Results for D-T($^3$He) and D($^3$He) are shown in figures 1 and 2.

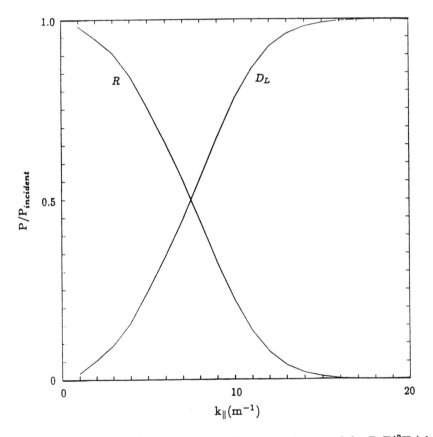

FIG. 1. Power transfer coefficients and the powers dissipated for D-T($^3$He) heating with CIT type parameters. Coefficients other that $R$ and $D_L$ (L refers to low-field side) are too small to appear on the graph. $R_0 = 2.1$m, $f = 95$MHz, $B_0 = 10$T, $n_e = 6 \times 10^{20}$m$^{-3}$, $T_0 = 20$Kev, $\eta_D = \eta_T = .45, \eta_{^3He} = 0.05$. All concentrations are with respect to electron density.

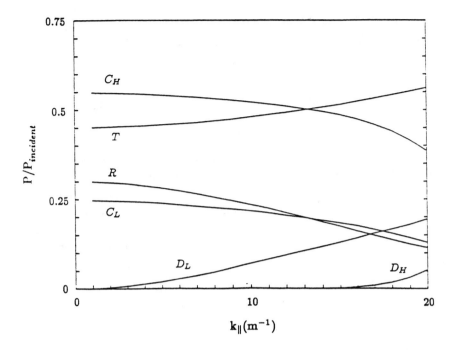

FIG. 2. Power transfer coefficients $T, R, C_H, C_L$, and the powers dissipated, $D_H$ and $D_L$ for D($^3$He) heating with Alcator C mod type parameters. $R_0 = 0.64$m, $f = 80$MHz, $B_0 = 9$T, $n_e = 5 \times 10^{20}$m$^{-3}$, $T_0 = 2$Kev, $\eta_{3He} = 0.05$. (L and H refer to low-field and high-field side respectively)

## REFERENCES

1. T.H. Stix, Nucl. Fusion **15**, 737 (1975)
2. D.G. Swanson, Phys. Rev. Lett. **36**, 316 (1976)
3. D. Smithe, P. Colestock, T. Kammash, and R. Kashuba, Phys. Rev. Lett. **60**, 801(1988).
4. E.F. Jaeger, D.B. Batchelor, and H. Weitzner, Nucl. Fusion **28**, 53 (1988).
5. K. Imre and H. Witzner, submitted to Plasma Phys. Controlled Fusion.
6. H. Romero and J. Scharer, Nucl. Fusion **23**, 363 (1987).
7. V. Fuchs and A. Bers, Phys. Fluids **31**, 3702 (1988).
8. C. Lashmore-Davis, V. Fuchs, G. Francis, A.K. Ram, A. Bers, and L. Gauthier, Phys. Fluids **31**, 1641 (1988).
9. G. Francis et. al. in AIP Conf. Proc. 159,1987,pp 370-373, and Bull. Am. Phys. Soc. **33**, 2115(1988).
10. C. Chow, V. Fuchs, and A. Bers, In preparation.

# SCATTERING OF AN ICRF MAGNETOSONIC WAVE BY PLASMA DENSITY TURBULENCE.*

*Daniel R. Cook and Allan N. Kaufman,*
*Lawrence Berkeley Laboratory and Department of Physics,*
*University of California, Berkeley, California 94720.*

## ABSTRACT

A fast ICRF magnetosonic wave, launched into a tokamak plasma, scatters off turbulent density fluctuations in the plasma edge. We use cold-fluid theory to calculate the angular distribution of the scattered wave and find it to be predominantly perpendicular to the incident wavevector for second harmonic majority heating. We calculate the mean free path and find it to be large compared to the size of tokamak devices. Therefore, scattering of ICRF magnetosonic waves by density turbulence is an utterly negligible effect.

## INTRODUCTION

The incident magnetosonic wavevector has $k_\parallel \ll k_\perp$ and experimental data indicate that the density turbulence has $k_\parallel \approx 0$. We therefore in our model let $k_\parallel = 0$ for all wavevectors, incident, scattered, and turbulent. The scattering occurs in the plane perpendicular to $B_o$. Experimental data indicate that the correlation length of the density turbulence is small compared to the magnetosonic wavelength. The density turbulence fluctuates slowly compared to the frequency of the magnetosonic wave, so the turbulent scatterers look stationary to the incident wave, and the scattering is elastic. In the wave field quantities which follow, an $e^{i\omega t}$ time dependence is to be understood. We found many useful ideas which helped us with this calculation in Ishimaru's fine book, Wave Propagation and Scattering in Random Media[1].

*Work supported by US DOE under contract No. DE-AC03-76SF00098 and by US DOE Magnetic Fusion Science Fellowship, administered through Oak Ridge Associated Universities.*

We begin with a few basic symbols and relations:

$$\psi(x,y) \equiv \tilde{B}(x,y)/B_o, \qquad \mu(x,y) \equiv n_t(x,y)/n_o, \qquad \mathbf{k'} \equiv \mathbf{k}_{scat} - \mathbf{k}_o,$$

$$\psi(\mathbf{x}) = \psi_{inc}(\mathbf{x}) + \psi_{scat}(\mathbf{x}), \qquad n = n' + \tilde{n}, \qquad \mathbf{k}_{scat} \equiv k_o\hat{r},$$

$$\psi_{inc}(\mathbf{x}) = \psi_o e^{i\mathbf{k}_o\cdot\mathbf{x}}, \qquad n' \equiv n_o + n_t, \qquad k_o^2 \equiv \omega^2/C_A^2,$$

$$C_A^2 \equiv B_o^2/4\pi M_i n_o, \qquad \Omega_i \equiv eB_o/M_i c, \qquad \lambda_o \equiv 1/k_o,$$

$$Re\left[e^{-i\omega t}\tilde{B}(x,y)\right], \qquad \text{magnetic field of the wave}$$

$$n_o, \qquad \text{uniform background density}$$

$$\tilde{n}, \qquad \text{density variation due to wave}$$

$$n_t, \qquad \text{turbulent density fluctuation}$$

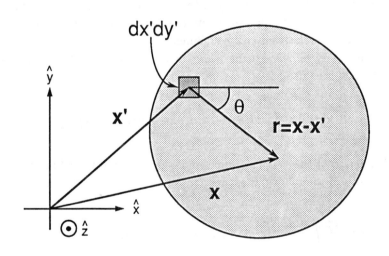

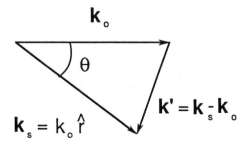

From the equations of two–component cold fluid theory, we obtain the following wave equation:

$$(\nabla^2 + k_o^2)\psi(x,y) = \left\{-k_o^2\mu + \left[\nabla\mu - i\frac{\omega}{\Omega_i}(\hat{z} \times \nabla\mu)\right] \cdot \nabla\right\}\psi(x,y)$$

We make the Rayleigh–Born Approximation on the wavefunction in the above equation by letting $\psi_{scat} \ll \psi_o$. Next we perform a far field approximation on the Green's Function to obtain the following scattered wavefunction:

$$\psi_s(x) = \sqrt{\frac{k_o}{8\pi}}\left(i\hat{k}_o - \frac{\omega}{\Omega_i}\hat{z} \times \hat{k}_o\right) \cdot \nabla_x \int d^2x' \; \mu(x') \; \frac{e^{ik_o[\hat{k}_o \cdot x' + |x - x'|]}}{|x - x'|^{\frac{1}{2}}}$$

The spectral density $S(k'; x')$ is the Wigner Function of the relative density turbulence, $\mu(x')$. It represents the *frequency– integrated wavevector content* of the relative density turbulence at position $x'$. The $\langle \; \rangle$ indicates a time average over a time which is long compared to the timescale of the density turbulence.

$$S(k';x') \equiv \int d^2s \; e^{-ik' \cdot s} \left\langle \mu(x' + \frac{1}{2}s) \; \mu(x' - \frac{1}{2}s)\right\rangle$$

We construct the Wigner Function of the scattered wave function and use it to obtain the following expression for the scattered wave intensity:

$$\langle|\psi_s(x)|^2\rangle = \frac{k_o^3}{8\pi}\int \frac{d^2x'}{r} \; g(\theta) \; S(k';x') \; |\psi_o|^2, \qquad g(\theta) \equiv \cos^2\theta + \left(\frac{\omega}{\Omega_i}\right)^2 \sin^2\theta$$

The strength of the density turbulence, $\langle\mu^2(x')\rangle$ is related to the spectral density and to the wavenumber spread $k_n$ of the density turbulence as follows:

$$\langle\mu^2(x')\rangle = \int \frac{d^2k'}{(2\pi)^2} \; S(k';x') \equiv \frac{k_n^2}{(2\pi)^2} S(k' \to 0; x'),$$

$$k_n^2 \equiv \int d^2k' \; \frac{S(k';x')}{S(k' \to 0; x')}$$

We obtain the scattering cross section density from the scattered wave intensity and use it to construct the mean free path,

$$\frac{1}{l(\mathbf{x}')} \equiv \int d\theta \, \sigma(\mathbf{x}'; \theta),$$

which may be expressed in terms of the wavenumber spread $k_n$ of the density turbulence as follows:

$$\frac{\lambda_o}{l} = \frac{\pi^2}{2} \left[ 1 + (\frac{\omega}{\Omega_i})^2 \right] \langle \mu^2 \rangle \left( \frac{k_o}{k_n} \right)^2$$

Using Ritz et al.[2] data from TEXT we obtain the following:

$k_o \approx 0.1 \ cm^{-1}$,     at the plasma edge

$k_n \approx 6 \ cm^{-1}$,     spread in turbulent wavenumbers

$\mu \approx 0.2$,     turbulence strength at the plasma edge

$\omega/\Omega_i = 2$,     second harmonic heating

$\lambda_o/l \approx 10^{-4}$,     from above equation

$l \approx 1 \ km$,     **mean free path**

## CONCLUSION

Two dimensional cold–fluid theory predicts a mean free path large compared to the size of tokamak devices. Scattering of incident ICRF magnetosonic waves by turbulent density fluctuations is **utterly negligible**.

## REFERENCES

1.    A. Ishimaru, Wave Propagation and Scattering in Random Media (Academic Press, N.Y., 1978), p. 329.
2.    C.P. Ritz et al., Nucl. Fusion 27, 1125 (1987).

# PERPENDICULAR ION ACCELERATION DURING ICRF, LHRF AND ECH EXPERIMENTS IN TOROIDAL PLASMAS IN THE UCLA CCT DEVICE

J.D. Evans, G.J. Morales and R.J. Taylor
University of California, Los Angeles, CA 90034

## ABSTRACT

Measurements of the fast neutral flux in the CCT (Continuous Current Tokamak) device in ICRF-, LHRF-, and ECH-produced plasmas show a common feature: fast ion tails ($E_i/T_i \approx 100$) with large ratios of perpendicular-to-parallel energy ($E_\perp/E_\parallel \approx 100$). Perpendicular fast ion production is particularly strong during fast wave RF operation, regardless of driver frequency, but enhanced ion generation has been observed at $\omega = n\Omega_i$, up to $n = 7$. Superthermal ion tails ($E_i/T_i \approx 20$) are also present in high-harmonic ICRF-heated tokamak discharges, but the ion distribution function relaxes towards isotropy in pitch angle due to improved ion confinement. In all of the above plasmas, the perpendicularity of the fast ion distribution increases with energy.

## INTRODUCTION

A wide variety of RF heating and current drive experiments have been conducted on the UCLA CCT (Continuous Current Tokamak) device in the ion-cyclotron, lower-hybrid and electron-cyclotron range of frequencies, in both tokamak and non-tokamak plasmas. Further description of the experiments can be found in the literature[1,2] and at this conference[3]. With the lone exception of electron-cyclotron-heated tokamak discharges, fast (tail) ion populations have been observed in all of these plasmas using a fast neutral spectrometer of the stripping-cell type. This spectrometer is mounted in such a way that the viewing angle with respect to the toroidal magnetic field **B** can be continuously varied, so that the neutral flux versus spectrometer viewing angle with respect to **B** can be measured. An interesting feature of the data thus obtained is that the fast ion populations ($E_i/T_i \geq 20$) are highly anistropic in pitch angle in all of the above plasmas. This feature is most striking in the non-tokamak cases, where the half-angle of the neutral flux with respect to the perpendicular direction $\approx 3°$ typically, which decreases as the ion energy increases. The auxiliary-heated tokamak fast neutral flux is perpendicular to a lesser degree, probably due to the fact that the ratios of the collisional relaxation rates to the charge-exchange ion loss rate is larger in tokamaks than in non-tokamak discharges (because of the larger degree of fractional ionization of neutrals in tokamaks). These ions are present throughout the ion-cylotron/lower-hybrid range of frequencies, but are most apparent during fast wave excitation at ion cyclotron harmonic frequencies. These results suggest that a physical mechanism present in plasmas is responsible for partial conversion of RF wave energy to perpendicular ion energy, the efficiency of this process increasing as the RF frequency approaches a resonant frequency of the plasma.

# EXPERIMENTAL DESCRIPTION

All of the experiments described here were performed on the UCLA CCT device (R = 1.5m, a = 0.4m). In the RF-heated tokamak cases, the parameters were: B = 2.5kG, $I_p \approx$ 40kA, $n_e \approx 2 \cdot 10^{12} cm^{-3}$, $T_e \approx 300eV \approx 3T_i$, in hydrogen. Presently, there are 4 $E_\perp B$ fast-wave antennas, one $E_\parallel B$ "Bernstein wave" antenna, and 3 microwave-oven-type magnetrons (f = 2.45GHz) coupled to the machine. A complete description of the antenna system is presented at this conference[3]. A 1-30 MHz RF-generator was used to supply RF power to the fast-wave antennas. In both tokamak and non-tokamak cases, $P_{rf} \approx$ 20kW. In the RF-only case, a steady-state, low-density, low-temperature plasma ($n_e \approx 3 \cdot 10^{11} cm^{-3}$, $T_e \approx 15eV \approx 3T_i$) at B = 0.8kG and f = 7MHz (f $\approx 6\Omega_i$) was produced by energizing one antenna. In the fast-wave LHRF-heated tokamak case, an 80 MHz RF-generator (f $\approx 20\Omega_i$) was used. For the ECH-produced plasma, a magnetron (f = 2.45GHz, X-mode polarization, outside launch) was operated at $P_{\mu w} \approx$ 2kW. For this plasma, B = 0.8kG, $n_e \approx 2 \cdot 10^{10} cm^{-3}$, and $T_e \approx 10eV \approx 3T_i$.

A scannable fast neutral spectrometer system (fig. 1) was developed and used for measuring the fast neutral flux versus viewing angle. The spectrometer is coupled to the chamber with a flexible vacuum bellows and mounted on circular tracks so that the spectrometer viewing angle can be continuously varied from -20° to +20°, where 0° is taken to be the direction perpendicular to B for convenience. Since the magnetic field lines are not concentric with the pivot point of the spectrometer (located on the outer wall), the angle of trajectory of the sampled neutrals (i.e. ion pitch angle $\chi$) is different from the viewing angle $\theta$. The relationship between $\theta$ and $\chi$ at a given major radius R (in general not equal to the outer wall major radius $R_w$) is:

$$\sin\chi = (R_w/R)\sin\theta \qquad (1)$$

Since $\chi$ depends on $\theta$ and R in general ($\theta \neq 0$), the spectrometer samples fast neutrals from charge-exchanged ions with a range of ion pitch angles. This difficulty may be removed by obtaining knowledge of the local fast neutral flux as a function of major radius. This was achieved in the non-tokamak cases by mounting a "scoop tube" apparatus on the spectrometer, which consists of a metal tube and a slideable vacuum seal. The spectrometer is then mounted on linear tracks, so that the tube may be scanned in and out of the plasma in order to measure the total line-integrated neutral flux versus major radius. The derivative of this curve with respect to major radius then yields the local neutral flux as a function of radius. In this way, the flux in the ECH-case was found to come from a narrow region about the upper hybrid resonance layer $R_{uh}$. In this case, $\chi$ is then easily found from (1) by substituting R = $R_{uh}$, and the neutral flux vs. ion pitch angle follows from the data. In the non-tokamak RF case, the flux was almost uniformly radially distributed, so that the flux vs. pitch angle must be evaluated point-by-point from the data. This "scoop tube" procedure is impossible in tokamaks, since no material object can probe the plasma core without disruption.

## EXPERIMENTAL RESULTS

Figure 2 illustrates the dependence of the 800 eV fast neutral flux on viewing angle for all of the plasmas. Flux vs. angle data for the RF-produced discharge, the RF-heated tokamak plasma, and the purely ohmic tokamak shots are displayed in figure 2. There is a gradual increase in anisotropy (perpendicularity) as one looks at the ohmic, RF + tokamak, and RF-only discharges. For comparison, the expected dependence of fully isotropic flux on viewing angle is included (solid curve). The fact that this curve fits the ohmic tokamak data so well indicates that the perpendicular and parallel ion temperatures are approximately equal. The anisotropy present in the other cases shows that the perpendicular ion energy is higher in the RF + tokamak case, and that the fast ions are almost completely perpendicular in the RF-produced plasma. Figure 3 shows the fast neutral flux vs. angle in the RF-only case, for 3 different values of neutral energy. Here, there is an increase in perpendicularity as the energy increases. Figure 4 contains the ECH-only data, which is similar to the RF-only data, except that the flux peaks are somewhat narrower in the ECH case.

In order to assess the importance of ion-cyclotron-harmonic effects, the following experiment was performed. A narrow outer edge target plasma ($n_p \approx 5 \cdot 10^{11} cm^{-3}$, $T_e \approx 30eV \approx 3T_i$, width $\approx 10cm$) was produced by energizing the $E_\parallel B$ antenna. One of the $E_\perp B$ fast-wave antennas was then used to couple low power ICRF to the plasma at $P_{rf} \approx 1kW$, f = 3.65 MHz. The toroidal field was pulsed up to B=3kG and allowed to ramp down in such a way that the ion-cyclotron harmonic number varied in time from 1 to 5 in the edge-localized plasma during the course of the discharge. Figure 3 shows the dependence of the 500 eV fast neutral flux at $\theta = 0°$ and at $\theta = 10°$ on field (hence on ion cyclotron harmonic number). Peaks in the neutral flux are observed at $\omega = n\Omega_i$, for n = 1...5, during this discharge. The half-angle for the neutral flux is at $\theta \approx 10°$ in this case, possibly due to the higher ion temperature and fractional degree of ionization than in the previous non-tokamak experiments. Subsequent experiments (not shown here for lack of space) have demonstrated neutral flux peaks for n = 6 and 7 as well.

## CONCLUSIONS

Perpendicular ion acceleration is observed in a wide variety of RF-heated toroidal plasmas. This phenomenon is particularly apparent in non-tokamak plasmas for which $\omega = n\Omega_i$. These results suggest that a physical mechanism inherent in plasmas is responsible for the conversion of RF wave energy to perpendicular ion energy. The exact nature of this mechanism remains to be determined.

## REFERENCES

[1]  R.J. Taylor, J. Nucl. Mat. 145-147, 700-3 (1987).
[2]  K.F. Lai et al., AIP Conf. Proc. 159, 111-4 (1987).
[3]  R.J. Taylor, et al., this conference.

245

## FIGURE CAPTIONS

[1]  CCT scannable fast neutral spectrometer schematic.
[2]  800 eV neutral flux vs spectrometer viewing angle.
[3]  Neutral flux vs viewing angle in RF-plasma.
[4]  Neutral flux vs viewing angle in ECH-plasma.
[5]  500 eV neutral flux vs B (ion cyclotron harmonic number).

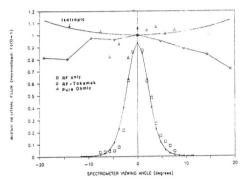

**Fig 2**

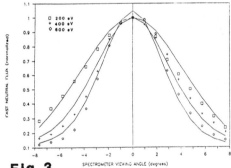

**Fig 3**

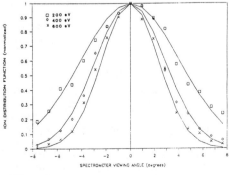

**Fig 4**

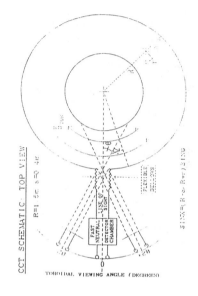

**Fig. 1**

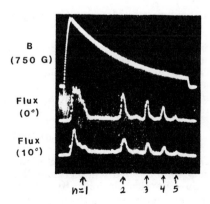

**Fig 5**

# ALCATOR C-MOD ICRF ANTENNA AND MATCHING CIRCUIT*

S. N. Golovato, M. Porkolab, Y. Takase, H. L. Holcomb

M. I. T. Plasma Fusion Center, Cambridge, MA 02139

## Abstract

Alcator C-Mod will be a compact, high field, high density, divertor tokamak. Two FMIT transmitters will supply 4 MW of power in 1 sec pulses at 80 MHz for ICRF heating. Fast wave minority heating experiments are planned in D($^3$He) at 8 T and D(H) at 5.5 T. The first antenna will have a single current strap inside a box structure, which will be movable radially. The antenna will be inserted through a side port, making the rf power density on the antenna surface $\sim$2 kW/cm$^2$ at 2 MW. The antenna will be center-tapped for mechanical strength and have a double layer Faraday screen tilted along the field lines. The antenna geometry was chosen to maximize power coupling assuming voltage-limited operation. A wide antenna with slotted box sides appears the best design, and 10 $\Omega$ of loading is required to couple 2 MW of power at a voltage limit of 40 kV. Matching is achieved by choice of the drive point to a resonant circuit formed by the antenna and a loop of transmission line outside of the vacuum and by tuning elements in the transmission line to the transmitter.

## I. Introduction

Alcator C-Mod is a compact (R=66 cm, a=20 cm), high field (9 T), high density ($\bar{n}$=1-10×10$^{20}$ m$^{-3}$, elongated ($\kappa$=1.8-2), divertor tokamak. Up to 3 MA of toroidal current will provide a hot ($T_e \sim T_i \sim$ several keV) ohmic plasma and good confinement for high energy minority ions. Fueling will be by gas puffing and/or pellets. Fast wave heating experiments are planned in the minority heating regime for $^3$He in a deuterium plasma at 8 T and D(H) at 5.5 T. The ICRF heating system for the Alcator C-Mod tokamak will provide 4 MW of power in 1 sec pulses at 80 MHz from two FMIT transmitters. The first antenna to be installed in Alcator C-Mod will have a single current strap and will be movable radially. The antenna will be inserted fully assembled through a side port. The goals with this antenna will be to study coupling and heating in high density, shaped, diverted discharges. The initial experiments will establish the radial location of the antenna relative to the separatrix, the best plasma shape, and the disruption speed. This knowledge will aid in designing a second, optimized antenna which will likely have two phased current straps and require internal assembly.

## II. Antenna and Matching Circuit Design

The basic features of the design are a center-tapped current strap for mechanical strength, the widest current strap that can be fit through the port opening, slotted sides on the antenna box, and a double layer Faraday screen tilted along the field lines. The antenna is driven from each end 180° out of phase. The antenna is modular in design such that the box sides, current strap, Faraday screen elements, and protection limiters are separately replaceable. The antenna configuration is shown in Fig. 1. The mechanical and thermal analyses of the antenna structure are discussed in a companion paper.[1]

The requirements of the structural support against disruption forces and the narrow port opening restrict the size of the two coaxial feeds to the current strap to 5" OD. The 5" line will have a characteristic impedance of $\sim$ 30$\Omega$. The feeds to the antenna ends are connected together outside of the vacuum region to form a closed loop which is three wavelengths long. By forming this resonant loop, the antenna can be fed at a low impedance point in the loop while maintaining the desired out of phase excitation.[2] Between the resonant loop and the transmitter, a phase shifter and stub tuner are used

* work supported by U. S. DOE Contract DE-AC02-78ET51013

to match the input impedance of the loop to the 50 $\Omega$ transmission line. The resonant loop and matching circuit are shown in Fig. 2. At the tranmitters, three coaxial switches will be used to allow flexibility in operation of the ICRF system. It will be possible to connect the driver ($\sim$130 kW), the final power amplifier (Eimac X2242, $\sim$2 MW), or a low power source or network analyzer to the antenna without any reconnecting of components. Tuning of both the driver and the final amplifier into the dummy load will also be possible.

The choice of a wide antenna relative to the size of the cavity was made by analyses which assumed that the power limiting factor would be the peak voltage. It may be that the limit will be the high power density on the antenna surface, but the constraint that the antenna be inserted through the port opening sets the power density independent of the particular antenna geometry. Similar codes by Mau[3] and Ryan[4] study the effects of cavity geometry both on the coupling to the plasma and on the antenna inductance. The coupled power in voltage-limited operation may be written $P = (1/2\omega^2)(R/L^2)V^2$ where R is the plasma loading on the antenna and L is the antenna inductance. Maximizing $R/L^2$ couples the highest power at the voltage limit. Lower inductance reduces the voltage on the antenna for a given current ($I=V/\omega L$). Typically, geometrical effects that increase R also increase L making the design a trade-off between these two parameters. Results of the Ryan code are shown in Fig. 3 for a variety of geometries. It can be seen that a wide strap and open box sides maximize $R/L^2$, with a deeper box having a weaker effect. In practice, opening the box sides is accomplished by slotting the sides to allow the rf magnetic flux to penetrate. The case indicated by the circled point on the top curve was chosen, with a wide current strap and the sides slotted halfway. Opening the sides completely both weakens the structure and might allow neutral gas to build up in the box.

Another code using transmission line models for the antenna and matching system determines the peak voltage and its position in the system for a given plasma loading. The most difficult part of the antenna to model is the inductance because of the complicated geometry of the current strap and box, including the radial feeds. A mock-up of the antenna geometry was constructed and laboratory measurements of the inductance were used to correct the code calculation. The correction to the code model for a bare antenna (no cavity sides or Faraday screen) were only 15-20%. Using the geometry chosen from Fig. 3 in the transmission line code gives the important prediction that for a voltage limit of 40 kV, 10 $\Omega$ of loading are required to couple 2 MW from the antenna. This loading level is consistent with predictions from the Brambilla code.[5] A voltage limit of 40 kV has been achieved in experiments but the power density at 2 MW ($\sim$ 2 kW/cm$^2$) is still about a factor of two higher than was achieved experimentally in Alcator-C at high densities ($n_e \gtrsim 3 \times 10^{20} cm^{-3}$).[6] However, with divertor operation and pellet fueling in Alcator C-Mod, steeper edge profiles are expected. This may facilitate operation at higher power densities than in Alcator-C.

The transmission line analysis is also used to choose the drive point in the resonant loop. Figure 4 shows the chosen operating point, which reduces the VSWR that must be tuned out by more than a factor of two while still maintaining out-of-phase excitation of the antenna. The sensitivity of the tuning to changes in plasma loading was studied. The VSWR remains below 1.5 for changes in resistive loading of up to 40%. However, only 2% changes in the reactive loading can be tolerated. A double layer Faraday screen was chosen, in part, to reduce the sensitivity to changes in reactive loading, at the expense of some reduction in resistive loading.

A vacuum test stand with a full-scale C-Mod side port will be available for antenna testing prior to installation in C-Mod. Vacuum conditioning techniques for reaching the maximum voltage standoff will be established. The effect of bake out on the antenna structure will be tested as well as the ease with which the antenna can be tuned. The control and data acquisition system that will be used during tokamak operation will be developed and tested while the antenna is on the test stand. The results of the test stand work, along with knowledge gained from initial experiments will feed into the design of the second antenna. An antenna design incorporating two current straps fed through a single port will be pursued.[1]

## III. Control and Diagnostics

The instrumentation and control requirements for the ICRF system fall into two categories. First are the control and monitoring on a shot-to-shot timescale, which will be done by a PLC (programmable logic controller). Second are diagnostics that monitor conditions during a shot through CAMAC. The PLC system will control movement of the antenna and tuning elements, moniter status, and provide safety interlocking. Diagnostics such as thermocouples on the antenna assembly and pressure in the coaxial line will be monitored by the PLC. The user interface and status display for the PLC will be an IBM PC with color monitor using commercial software. Appropriate design of the display will allow the PLC to flag problems or dangerous situations.

Data from diagnostics on a fast time scale during a shot will be acquired by CA-MAC. The most important of these will the antenna loading. The loading will be measured in three different ways in the antenna system. There will be two bi-directional couplers, one between the resonant loop and the tuning elements to measure the mismatch that must be tuned out and one at the transmitters to measure the matching to the 50 $\Omega$ line. By measurement of the forward and reflected voltage and their relative phase, the complex reflection coefficient can be determined. There will also be two sets of voltage probes spaced along a half wavelength to measure the VSWR. Knowing both the VSWR and the position of the voltage maximum allows evaluation of the complex impedance at any point in the line. One probe set will be in the resonant loop to determine the input impedance of the antenna. Another will be between the resonant loop and the tuning elements to determine the input impedance of the resonant loop, supplying a redundant measure of the mismatch to be tuned out. The better this can be characterized, the fewer shots will be required to tune the antenna, a very important consideration on a device with a 20 minute shot cycle. Finally there will be magnetic pickup loops in the antenna box to measure the antenna current directly. The combination of these measurements with an appropriate antenna model will provide a measure of the load resistance.

Emphasis will be placed on diagnostics of the edge plasma during ICRF. There will be at least three sets of RF probes distributed around the torus to look at surface wave excitation. Each set will include 2-3 orthogonal magnetic probes and a high frequency electrostatic probe. Sets of Langmuir probes will measure the RF perturbation of the edge temperature and density profiles. A periscope will allow observation of the front of the antenna during the plasma pulse. Impurity generaton will be studied spectroscopically both at the edge and in the core plasma. There will also be a $CO_2$ laser scattering diagnostic to look at RF effects in the plasma core, such as Bernstein wave excitation near the mode conversion layer.

## References

[1] Y. Takase et al., this conference.

[2] J. R. Wilson et al., in *Proceedings of the 7th Topical Conference on Applications of Radio-Frequency Power to Plasmas*, Kissimmee, FL (AIP, New York, 1987), p. 294.

[3] T. K. Mau et al., *IEEE Trans. Plasma Sci.* **PS-15**, 273 (1987).

[4] P. M. Ryan et al., in *Proceedings of the 15th European Conference on Controlled Fusion and Plasma Heating*, Dubrovnik, 1987, part II, p. 795.

[5] M. Brambilla et al., *Nucl. Fusion* **28** 549 (1988).

[6] T. D. Shepard, Ph.D. Thesis, MIT, 1988.

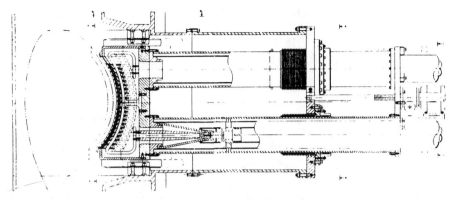

Figure 1. Antenna Configuration

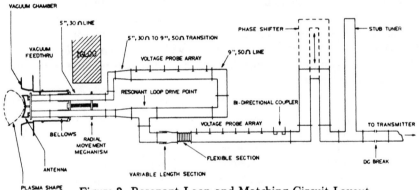

Figure 2. Resonant Loop and Matching Circuit Layout

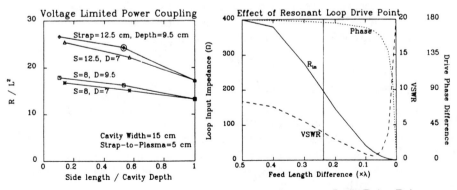

Figure 3. Antenna Geometry Effects          Figure 4. Resonant Loop Drive Point

# ICRF ANTENNA DESIGNS FOR CIT AND ALCATOR C-MOD*

R. H. Goulding, F. W. Baity, P. L. Goranson, D. J. Hoffman, P. M. Ryan,
D. J. Taylor, and J. J. Yugo

Oak Ridge National Laboratory, Oak Ridge, TN 37831-8071

## ABSTRACT

An ion cyclotron range of frequencies (ICRF) launcher for the Compact
Ignition Tokamak (CIT) has been designed. This launcher incorporates four current
straps in a $2 \times 2$ configuration. The current straps consist of end-fed loops that are
grounded in the middle. An antenna similar in geometry, size, and feed configuration
to a single strap of the CIT launcher will be built for use on Alcator C-Mod. The
design must provide maximum power levels of 4 MW/port for CIT and 2 MW/port for
C-Mod, pulse lengths of 5–10 s for CIT and 1 s for C-Mod, and power densities up to
$2 \text{ kW/cm}^2$. The design uses a Faraday shield consisting of Inconel rods with
mechanically attached graphite tiles; the shield and the current strap are cooled by
radiating to a gas-cooled backplane. A feed configuration compatible with the end-fed
antenna design has been developed and features tunability in three bands in the range
65–130 MHz. It uses an external resonant loop with integral tuning elements. It has
been designed to maximize power handling capabilities, minimize space requirements,
and facilitate remote handling.

## INTRODUCTION

The initial ICRF system for CIT will provide 10 MW of power to the plasma,
and consideration is being given to increasing this to 25 MW. The 25 MW system
would utilize a total of 16 radio frequency (rf) power units feeding 32 current straps in
8 ports. The proposed antenna is an end-fed loop, grounded in the middle, with
matching elements in an external resonant loop outside the vacuum boundary. The
antennas are of modular, compact construction for installation and removal through the
midplane port. Remote maintainability plus the reactor-like operating environment
have a major impact on the design of the launcher for this machine. An antenna that is
being designed for Alcator C-Mod will have a single strap and will be similar in size,
geometry, and feed configuration to a single strap of the CIT launcher.

## SYSTEM OVERVIEW

The baseline CIT antenna design consists of four end-fed loop antennas in a
single port (101.6 cm $\times$ 37.5 cm), which form one module as shown in Figs. 1 and 2.
The antenna is tuned and impedance matched with capacitive elements in an external
resonant loop outside the vacuum boundary. RF power is provided by 16 modified
surplus Fusion Materials Interaction Tests (FMIT) rf transmitter units, with Varian
X–2242 tetrodes or an improved equivalent, for an output power of 2 MW per
transmitter at 95 MHz.

*Research sponsored by the Office of Fusion Energy, U.S. Department of Energy,
under contract DE-AC05-840R21400 with Martin Marietta Energy Systems, Inc.

## LAUNCHER MECHANICAL DESIGN

The antenna system interfaces with the feed line at a constant-impedance vacuum feedthrough with a brazed alumina dielectric separating the pressurized 9-in. transmission line from the evacuated 5-in. antenna feed line. Figures 1 and 2 show the details of the side and top views of the baseline antenna configuration.

Each antenna strap has a center ground, which is securely bolted to the cavity back wall. This ground connection also reacts any disruption loads on the current strap into the wall. The antenna position within the port is adjusted by twin threaded shafts. Vacuum bellows, metal seals, and the brazed alumina dielectric in the feedthrough provide the vacuum boundary.

A remotely actuated flange is used to couple the impedance-matching and transmission line components to the antenna. The impedance-matching assembly for all four straps is disconnected as a unit from the antenna by an automated sequence of operations, and the assembly can then be removed by the overhead crane and remote manipulator.

Disruption forces are reacted into the vacuum vessel through keyways at the top and bottom of the antenna module. Acceleration of the antenna module through a 0.32-cm gap between the keyway and the module has been calculated to cause an impact load a factor of nine times larger than the applied load. Such a force magnification is unmanageable and requires eliminating the gap in the keyway and incorporating elastic materials as shock absorbers to decelerate the antenna. The solution of this design problem will await modeling of the disruption loads.

The Faraday shield is constructed of copper-plated Inconel rods (0.5 in. o.d.) in a staggered double row. Graphite tiles are mechanically attached to the plasma side of the tubing. The shield is cooled by radiation to the cavity back wall which is cooled with nitrogen gas at 200–350° C between shots.

A one dimensional model of the Faraday shield has been used to evaluate the temperature evolution of the graphite tiles and Inconel rods. The model includes a time-dependent heat flux and models cooling by radiation and conduction to the surrounding structure. For an rf power density through the Faraday shield of $1.0 \text{ kW/cm}^2$, $50 \text{ W/cm}^2$ is assumed to be deposited in the Faraday shield for 10 s. Plasma radiation and neutrons deposit $30 \text{ W/cm}^2$ during 4 s of the plasma burn and $150 \text{ W/cm}^2$ during 6 s of field ramp-up where the plasma is limited by the outboard wall. For this case the peak graphite temperature is 1220° C, and the peak Inconel temperature is 667° C. The peak temperature in the Inconel is reached approximately 3.2 s after the end of the shot.

## FEED SYSTEM ELECTRICAL DESIGN

The feed system for the CIT antenna has been designed to meet several objectives, including pulse lengths of 10 s, power levels of 1 MW per strap (2 MW for C-Mod), phasing of currents in toroidally adjacent straps at 0° or 180°, and tuning between 65 and 130 MHz. It uses an external resonant loop with integral tuning elements for each current strap, and is a modification of a Princeton design, used for an ICRF antenna now installed on TFTR.

The system, shown schematically in Fig. 3, is band tunable, with three bands approximately 10 MHz wide between 65 and 130 MHz. It can be adjusted to allow for variations in plasma resistive and reactive loading, changes in coupling between straps due to plasma effects, and changes in phasing of toroidally adjacent current straps. The resonant loop is matched to 50 $\Omega$ at the feed point, minimizing current and voltage amplitude on the feed lines.

Each poloidally adjacent strap (Fig. 3) is fed through a hybrid power splitter. The splitters are fed by separate transmitters with equal power to each splitter and with the forward voltage phase difference between splitters fixed at either $0^\circ$ or $180^\circ$.

## CIRCUIT MODELING

The circuit has been modeled using lossy transmission lines with distributed coupling between lines 1 and 3, and between lines 2 and 4, assumed constant along the line; coupling between other line combinations is ignored. For the results shown, the characteristic impedance of the current straps (neglecting coupling) is assumed to be 60 $\Omega$, and the phase velocity to be 0.82 c. These values are typical of similar antennas that have been built previously, such as the ORNL antenna for TFTR.[1] The resonant loop transmission lines are assumed to have a characteristic impedance of 30 $\Omega$ and a phase velocity of 1.0 c.

Figure 4 shows the capacitances required to match to 50 $\Omega$ at the tee as a function of frequency for a $180^\circ$ phasing case. The operating bands are also shown. The center frequencies of these bands can be modified by changing the line lengths between the straps and capacitors. The values of $C_3$ and $C_4$ (not shown) correspond to those of $C_1$ and $C_2$ for $180^\circ$ phasing.

Figure 5 shows the sensitivity of the match at the input to tee 1 as the capacitances are varied over a range of $\pm 10\%$ at the same frequency. Capacitors $C_3$ and $C_4$ have relatively little effect on the match, which suggests that an iterative technique can be used to tune the antennas in practice. Figure 6 shows contours of constant maximum input power per strap as a function of resistive loading and the feed point $\alpha$ (see Fig. 3) for f = 95 Mhz. For 6-$\Omega$/m loading, the maximum power is approximately 1.8 MW. This calculation assumes a maximum voltage of 50 kV on the resonant lines and maximum rms currents of 1200 A on the resonant lines and 1100 A on the capacitors.

## CONCLUSIONS

An ICRF heating system designed for CIT features a Faraday shield with mechanically attached graphite tiles, which is cooled by radiating to a gas-cooled back plane. A tuning and matching system has been developed with a tuning range of 65–130 MHz in three bands that can operate at power levels greater than 1 MW with 6-$\Omega$/m loading.

## REFERENCES

1. D. J. Hoffman et al., "Coupling of ICRF Power with the TFTR ORNL Antenna," Bull. Am. Phys. Soc. **33**, 2094 (1988).

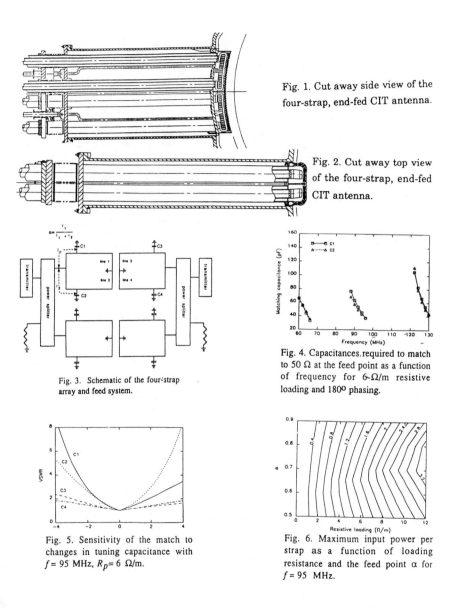

Fig. 1. Cut away side view of the four-strap, end-fed CIT antenna.

Fig. 2. Cut away top view of the four-strap, end-fed CIT antenna.

Fig. 3. Schematic of the four-strap array and feed system.

Fig. 4. Capacitances required to match to 50 $\Omega$ at the feed point as a function of frequency for 6-$\Omega$/m resistive loading and 180$^\circ$ phasing.

Fig. 5. Sensitivity of the match to changes in tuning capacitance with $f = 95$ MHz, $R_p = 6$ $\Omega$/m.

Fig. 6. Maximum input power per strap as a function of loading resistance and the feed point $\alpha$ for $f = 95$ MHz.

# ICRF COUPLING ON TFTR USING THE PPL ANTENNA

G.J. Greene, P.L. Colestock, J.C. Hosea, C.K. Phillips,

D.N. Smithe, J.E. Stevens, J.R. Wilson, W. Gardner,[†] and D. Hoffman[†]
Princeton Plasma Physics Laboratory*
Princeton, New Jersey 08543

## ABSTRACT
Coupling of the PPL ICRF antenna to the TFTR plasma is experimentally measured as $R_c = P/I^2$, where $P$ is the net power dissipated and $I$ is the RF current at a point in the resonant loop. The relation of $R_c$ to the equivalent antenna loading resistance is investigated using a transmission line model that includes the antenna structure and the feedthrus. Coupling has been experimentally characterized for a variety of discharge conditions including $H$ and $^3He$ minority, $D$ and $^4He$ majority plasmas. Effects of antenna phasing, $D$ neutral beam injection, RF power level, plasma density and position are discussed. Distinct and reproducible eigenmodes in the loading are observed in $H$-minority, $D$-majority plasmas during the density rise accompanying neutral beam injection. A 1-D wave propagation model has reproduced the general structure of the modes. For $0-\pi$ toroidal phasing, the modes arise from radial reflections from both the resonance-absorption layer and the inner wall of the tokamak; for $0-0$ phasing, the modes result from toroidal interference.

## I. INTRODUCTION
The loading of the TFTR ICRF antennas due to the tokamak plasma is a quantity of fundamental importance in assessing antenna performance. The magnitude of the loading relative to the vacuum case gives a measure of the efficiency of energy transfer to the plasma. The loading also determines the maximum power that can be applied to the antenna. Characterization of the dependence of the loading on plasma parameters is of importance in designing the RF impedance matching system and can yield information about the wave propagation and absorption processes taking place within the tokamak.

This paper presents studies of antenna coupling performed with the PPL antenna during the 1988 TFTR experimental run. Initial experiments[1] revealed a significant difference (30%) in ICRF heating efficiency depending on the toroidal phasing of the antenna loops, and these observations motivated a careful analysis of the associated antenna-plasma coupling.

## II. EXPERIMENTAL ARRANGEMENT
The PPL antenna consisted of two poloidally-directed current straps, located on the low-field side of the vessel at Bay 'M' and covered with a Faraday shield.[2] Each strap was 10 cm wide and extended poloidally $\pm 21$ degrees about the equatorial plane with respect to the minor axis. The center of each strap was grounded to the backplane, and the straps were driven at each end. The centers of the two current straps were separated toroidally by 33 cm, and a slotted septum between the loops permitted significant mutual coupling. For data presented here, the face of the Faraday shield was fixed at a major radius of 3.62 m, and the generator frequency was 47 MHz.

Transmission line loops outside the vessel connected the upper and lower ends of each current strap, forming an 'inner' loop that was resonant at 47 MHz. The tap points of the two inner loops were joined with transmission line sections to form a third, 'outer' loop, whose length was chosen so that the antenna current straps were toroidally either in phase or out of phase. The RF current was measured with couplers, located near the high-current point in each inner loop, that consisted of shielded magnetic loop probes built into the wall of the transmission line. A dual directional coupler measured the net dissipated power to the system.

## III. TRANSMISSION LINE MODEL
Antenna coupling is typically measured experimentally as an 'equivalent coupling resistance', given by $R_c = P/I^2$, where $P$ is the net power dissipated by an antenna loop and $I$ is the (rms) current flowing at a point of minimum voltage in the transmission line that feeds the antenna loop.[3] Theoretical modeling of the antenna loading generally assumes a current distribution on the antenna elements and calculates the dissipated power, $P_a$. The loading resistance is then obtained from $R_a = P_a/I_a^2$, where $I_a$ is the (rms) current at the antenna loop midplane. Although the coupling resistance and the antenna loading resistance are often compared, they

---

[†] Oak Ridge National Laboratory

* This work supported by U.S. DOE Contract No. DE-AC02-76-CHO-3073.

are not equivalent. It is important to model the antenna and the loading diagnostic as an entire system in order to clarify the relation between $R_c$ and $R_a$.

The model used in this work is shown in Fig. 1. Each of the radiating current strap sections (labeled AUL, AUR, etc.) is modeled as a uniform distributed lossy transmission line with a complex characteristic impedance $(Z_a)$, phase velocity $(V_a)$, distributed series resistance, $R_s$, and inductance, $L_s$. The antenna loading is defined here as $R_a = R_s S_a$, where $S_a$ is the length of an up-down pair of poloidal radiating current straps (79 cm).

The sections of current strap (labeled SW) that connect the radiating elements to the vacuum feedthrus are modeled as uniform strip-lines, and the non-uniform vacuum feedthrus (labeled FT) are modeled as a sequence of five sections with different impedances and phase velocities. The sections labeled T1 - T5 are coaxial transmission lines, and the boxes marked C1 and C2 represent the RF current couplers. Coupling between the antenna loops is represented by mutual inductances between each up-down and left-right pair of elements. The induced voltage sources (VUL, VLL, etc.) represent the contribution from currents in the associated poloidal and toroidal pair elements. The presence of plasma is modeled as a change in $R_s$.

Transmission line theory allows reduction of the model to a collection of linear equations which are solved numerically. The model was used to calculate the coupling resistance $R_c$ that would be experimentally measured for a given antenna loading $R_a$; the ratio is plotted as a function of $R_c$ in Fig. 2. The value of the ratio is slightly larger than 2 and is relatively constant over most of the range characteristic of typical TFTR operation $(R_c \sim 3 - 10\ \Omega)$, but the ratio increases rapidly as $R_c$ decreases (due to the background losses in the transmission line elements). The ratio $R_c/R_a$ does not vary significantly with small changes in the mutual inductances or with the antenna toroidal phasing, but does exhibit some dependence on the antenna distributed inductance (which is thought to decrease slightly when the plasma is present). The corresponding curves associated with a reduction in the distributed inductance of 10% and 20% from the vacuum value are also shown in Fig. 2. For a 10% reduction in $L_s$ and $R_c = 5\ \Omega$, the change in $R_c/R_a$ is about 5%. Thus, within certain limits, the measured coupling resistance can be taken to be linearly proportional to the antenna loading resistance.

The transmission line model can also be used to calculate the efficiency of energy transfer to the plasma. The ratio of the power dissipated by the plasma to the net power supplied at the feed point is plotted in Fig. 2 as a function of $R_c$. It is seen that for $R_c > 3\ \Omega$, more than 90% of the supplied power is absorbed by the plasma.

## IV. RESULTS

The variation of coupling resistance with total applied RF power for a series of shots with $0 - \pi$ toroidal phasing is shown in Fig. 3. The discharge parameters were $I_p = 1.1$ MA, $\overline{N_e \ell} = 4.5 \times 10^{15}$ cm$^{-2}$, $R_{plasma\ edge} = 3.57$ m, majority species = $^4$He, minority species = H. The value of $R_c$ is shown for each of the two antenna loops, calculated at a time near the end of the RF pulse, and the corresponding increment in central chord-averaged density due to the RF is also shown. The change in loading is approximately proportional to the RF power, increasing by some 40% as the power rises to 1.5 MW. The RF-induced density increment also appears to be proportional to the applied power but with a threshold of $\sim 150$ kW.

The dependence of the coupling on the plasma position was investigated by applying RF power to a discharge in which the plasma size was ramped. The corresponding waveforms are shown in Fig. 4 and demonstrate that the loading is proportional to the edge density (measured as a vertical chord average at a major radius of 3.47 m). For this shot, $P_{RF} = 100$ kW, majority $= ^4$He, minority = H, and the antenna toroidal phasing was 0-$\pi$. The central line-average density rose from 3.79 to 3.86 $\times 10^{15}$ cm$^{-2}$ during the RF pulse but was not directly related to the loading as it exhibited no discontinuity at $t = 1.95$ s, the time at which the plasma growth halted.

Antenna loading during deuterium neutral-beam heated discharges was investigated for both 0-0 and 0-$\pi$ toroidal phasing. For the 0-0 phasing case, large fluctuations were seen in the loading during the beam pulse. The variations were reproducible but their shape changed significantly with beam power. Plotting the coupling resistance as a function of the central line-averaged density during the beam pulse (Fig. 5a) demonstrates that the fluctuations are in fact global-density dependent modes. For these data, $P_{RF} = 1.0$ MW, $P_{NB} = 3.1 - 8.1$ MW, majority $=$ D, minority $=$ H. The loading exhibits five distinct peaks as the density increases. The largest peak-to-valley variations are some 45% of the mean value, and the increase of the mean with density can be approximated as an offset-linear function with a slope of $\sim 3.0 \times 10^{-15}\ \Omega$ cm$^2$ and an intercept of $\sim 1.5\ \Omega$ at zero density.

Global density-dependent modes were also observed with $0 - \pi$ toroidal phasing, but the structure was quite different. Figure 6a plots the loading as a function of density for two shots with slightly different beam powers. For these shots, $P_{RF} = 1.6$ MW, $P_{NB} = 6.8, 7.1$ MW, majority = D, minority = H. The mode amplitude is much smaller than for the $0 - 0$ phasing case and decreases as the density rises. The average loading may also be approximated as an offset linear function of density, with a slope of $\sim 3.5 \times 10^{-15}\ \Omega\,cm^2$ and an intercept of $\sim -2.2\ \Omega$. The relatively uniform mode separation ($\Delta \overline{N_e \ell} \sim 1 \times 10^{14}\ cm^{-2}$) is much smaller than the mode spacing for the case of 0-0 toroidal phasing ($\Delta \overline{N_e \ell} \sim 6 \times 10^{14}\ cm^{-2}$).

Although the slope of the curve of mean loading versus density is similar for the 0-0 and 0-$\pi$ antenna phasings, the offset between the two amounts to nearly 4 $\Omega$. At a given density, the loading for the 0-0 phasing is significantly higher than for the 0-$\pi$ case. This result is of particular interest because of the lower heating efficiency observed with 0-0 phasing. The increased loading in that case may represent the excitation of an undesired mode or surface wave.

A 1-D wave model that includes mode conversion[4] has been used to calculate the expected antenna loading for the experimental cases shown in Figures 5a and 6a. The general form of the calculated loading curve shown in Fig. 5b is that of three broad peaks, the first two of which appear to be split. There is good general agreement with the experimental data, which also shows the peak splitting. Examination of the calculated wave fields shows that the three broad peaks result from reflection of the incident wave off the resonance-absorption layer at the tokamak center. The splitting of the peaks arise from waves that have tunneled through the absorption layer and reflect off the inner wall of the vessel. The predictions of the model for 0-$\pi$ antenna phasing are shown in Fig. 6b. Although there is a phase difference between the calculated and measured modes, the peak spacing and the overall rise with density are approximately correct. Analysis of the model has shown that the mode structure in this case arises from toroidal interference rather than from the purely radial resonances respsonsible for the structure in the 0-0 case.

In order to reproduce the peak-to-valley ratio of the modes in the 0-0 phasing case with the 1-D model, however, it was found necessary to postulate that a constant background resistance had been added to the experimental data. This observation supports the idea that the larger loading and lower heating efficiency for the 0-0 phasing results from an additional deposition mechanism unrelated to fast wave generation.

Data from two shots with different minority species (and consequently different toroidal fields) but similar plasma characteristics ($P_{RF} = 1.5$ MW, $P_{NB} = 0$, 0-$\pi$ phasing, $\overline{N_e \ell} = 5 \times 10^{15}\ cm^{-2}$, majority = $^4$He) are compared in Fig. 7. The coupling resistance for the $^3$He-minority case exhibits large-amplitude, narrow, dense modes throughout the RF pulse, while the H-minority case has a relatively smooth characteristic (consistent with the previous data at high density). The mean value of the loading is similar in the two cases. The coupling for three $^3$He-minority discharges with different densities is shown in Fig. 8 as a function of central line-averaged density. The peaks are evidently modes and they occur in regular pairs, the second being narrower than the first. Clarification of the physical nature of these modes is a subject of current investigation.

## REFERENCES

1. J.R. Wilson, et al., Proc. 12th Int. Conf. on Plasma Physics and Controlled Nuclear Fusion Research (IAEA, Nice, 1988), paper IAEA-CN-50/E4-1.
2. J.R. Wilson, et al., Proc. 7th Topical Conf. on Applications of RF Power to Plasmas (AIP, New York, 1987), p. 294.
3. K. Theilhaber, et al., Nucl. Fusion **24**, (1984) 541.
4. A.L. McCarthy, et al., Proc. 15th Eur. Conf. on Controlled Fusion and Plasma Heating (Dubrovnik, 1988), Vol. II, p. 717.

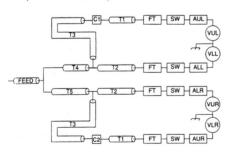

FIG. 1. Schematic of transmission line model of PPL antenna system.

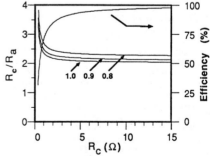

FIG. 2. Ratio of coupling resistance to antenna loading, for three values of $L_s$, and the power transfer efficiency.

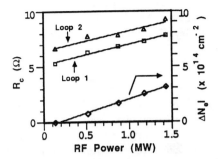

FIG. 3. Variation of coupling resistance of each loop with RF power, for 0-π phasing. Also plotted is the density increment due to the RF power.

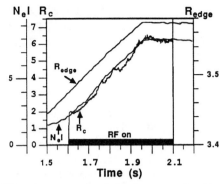

FIG. 4. Antenna coupling during plasma expansion. Also plotted are the edge density and the plasma edge position. Units are: $R_{edge}$ - m, $R_C$ - ohms, $N_e\ell$ - $10^{15}$ cm$^{-2}$.

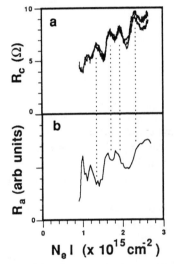

FIG. 5. a) Coupling versus chord averaged density for 0-0 antenna phasing. Four shots with different beam power are overlayed. b) Loading from the 1-D wave model.

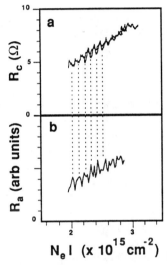

FIG. 6. a) Coupling versus chord averaged density for 0-π antenna phasing. Two shots with similar beam powers are overlayed. b) Loading from the 1-D wave model.

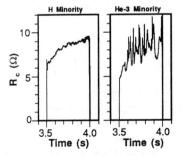

FIG. 7. Comparison of loading for H and He-3 minority species and 0-π antenna toroidal phasing.

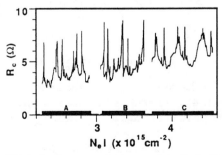

FIG. 8. Loading versus chord averaged density for He-3 minority species and 0-π phasing. Three shots with different density ranges are displayed.

# ANALYSIS OF CHARGE EXCHANGE MEASUREMENTS DURING ICRF AND ICRF+NBI HEATING IN TFTR

G. W. Hammett, W. Dorland, R. Kaita, S. S. Medley, D. N. Smithe, P. L. Colestock
Princeton Plasma Physics Laboratory, P.O. Box 451, Princeton NJ 08543 USA

## ABSTRACT

Charge-exchange measurements during ICRF-only and combined ICRF (2 MW) and NBI (7 MW) heating have been made with vertically-viewing mass-resolving analyzers. We have observed an enhancement of the deuterium spectrum above the beam injection energy, providing evidence of second harmonic acceleration of beam ions by the ICRF waves. The measured spectra are in reasonable agreement with the predictions of a comprehensive ICRF+NBI code, given the measurement error bars and the simulation uncertainties due to the hydrogen concentration, RF power profile, and effects of sawteeth. In the ICRF-only case, we present measurements from a toroidal field scan which show that the flux of 150 keV H-minority ions varies by a factor of 7 and is largest when the resonance layer is over the detector sightline. Future experiments are proposed to extend the results obtained here.

## ICRF MODELLING

The ICRF code we used to model the charge exchange measurements consists of two coupled parts. The first part is a 3-D poloidal mode expansion code[1,2] which solves a contracted second-order equation for the fast wave (including fundamental and second harmonic ion damping, electron damping, and mode conversion). The second part is a bounce-averaged quasilinear Fokker-Planck code[3] (including nonzero $k_\parallel$ and the full Bessel functions for nonzero $k_\perp \rho$) which calculates the minority ion distribution function $f(E, \mu, r, t)$. Beam-beam collisions are modelled with as collisions with an equivalent Maxwellian. The neutral density profile $n_0(r)$ is predicted with the FRANTIC package, and used in the calculation of the line-integrated charge-exchange spectrum. We found it was important to include the Shafranov shift when calculating the spectra, so that the detector sightlines would pass close enough to the hot center. It is important to have enough energy grid points in the Fokker-Planck code to accurately predict the shape of $f$ above the injection energy. Our upwind numerical scheme approximates the change in $f$ between two adjacent grid points by $\exp{-\delta E/T_{tail}} \approx 1/(1 + \delta E/T_{tail})$. We used 250 equally spaced grid points from 0 to 200 keV, so the simulated slope should be within 10% of the true slope for tail temperatures over 4 keV (which is satisfied in the center of our plasmas).

## SECOND HARMONIC HEATING OF THE DEUTERIUM BEAM

The charge-exchange detector used in these experiments (described in more detail in Ref. 4) is located in the basement below TFTR, and views vertically upwards at a major radius of $R_{maj} = 2.97$ m. Fig. 1 shows the measured deuterium charge exchange spectra for the case of NBI-only (2 co-injection and 1 counter-injection beams) and for the case of combined ICRF and NBI. In the beam-only case, there is insufficient signal (compared to the background) above 150 keV, while in the beam+ICRF case the tail is measurable out to 190 keV. Drawing a straight line through the data points at 110 and

150 keV yields a slope equivalent to 11 keV for the beam only case, rising to 18 keV for the beam+ICRF case. For the NBI only case, the central plasma parameters were $T_e$ = 5.1 keV, $T_i$ = 18. keV, $n_e$ = 2.1 × $10^{19}$ /m$^3$, 0.19 m Shafranov shift, $Z_{eff}$ = 3.1, and $P_{inj}$ = 7.2 MW. For the NBI+ICRF case, the central plasma parameters were $T_e$ = 5.2 keV (this was half-way through a large sawtooth which went 15% higher at its peak $T_e$), $T_i$ = 14. keV, $n_e$ = 2.3 × $10^{19}$ /m$^3$, 0.26 m Shafranov shift, $Z_{eff}$ = 3.1, and $P_{inj}$ = 7.1 MW, $P_{rf}$ = 2 MW. Other plasma parameters which remained the same in both cases were: outer flux surface at Rmaj = 2.59 m, a=0.94 m, $B_t$ = 3.279 T, and $I_p$ = 1.2 MA, ICRF frequency $f_{rf}$ = 47.0 MHz, and neutral beam injection energy $E_{inj}$ = 102-104 keV. The ratio of the hydrogen to the *thermal* deuterium was assumed to be 15% (as measured by $H_\alpha/D_\alpha$ spectroscopy near the edge) throughout the plasma, resulting in $n_H/n_e$ = 3.9% near the center of the plasma because of the impurity and beam densities. The simulated spectra are normalized (reflecting the uncertainty in the the neutral density) to match the lower energy data around 100 keV. Other than this normalization there were no free parameters in the simulations. The predicted enhancement of neutrons by the ICRF-driven D tail is small (under 20%). Experiments on JET[5] with higher ICRF power have shown that substantial enhancement of the neutron yield can be achieved, and they found good agreement between measurements and theoretical predictions.

Fig. 1 shows that the data and theory agree fairly well for the beams-only case. The measurement for the beams+RF case shows a statistically significant tail compared with the predictions if the ICRF-beam interaction were ignored. However, the observed tail is less than predicted if the effects of the second harmonic acceleration due to the ICRF are included. Several possible explanations for this are explored in Fig. 2. One is that the power profile (or the Shafranov shift) may be slightly different than calculated. The predicted RF power absorbed by the deuterium is 0.2 W/cm$^3$ at r=13 cm (the minor radius tangency point of the detector) but drops to 0.04 W/cm$^3$ at r=23 cm (the minor radius tangency point of a vertical detector at R=307 cm). Fig. 2 shows that the predicted tail varies quite significantly over this small distance. Another possible explanation is related to the large sawtooth event which occurred 0.2 s before the measurement was taken, which is modelled in Fig.2 simply by running the Fokker-Planck code only for 0.2 s instead of the full 0.45 s needed to reach equilibrium. We found that the predicted spectra was surprisingly not very sensitive to the hydrogen concentration, but in any case one might think that the hydrogen concentration may be lower than the 3.9% assumed here because of the strong central deuterium fueling by the beam.

Fig. 3 shows measurements from a case with counter-NBI only, and counter-NBI+ICRF heating. For the counter-NBI only case, the central plasma parameters were $T_e$ = 3.6 keV, $T_i$ = 5.1 keV, $n_e$ = 2.6 × $10^{19}$ /m$^3$, 0.17 m Shafranov shift, and $Z_{eff}$ = 4.0. For the counter-NBI+ICRF only case, the central plasma parameters were $T_e$ = 4.8 keV, $T_i$ = 7.5 keV, $n_e$ = 3.7 × $10^{19}$ /m$^3$, 0.20 m Shafranov shift, and $Z_{eff}$ = 3.6. Other plasma parameters are the same as before. Counter-injection beams by themselves produce rather poor heating results. Also, the observed tail falls below the predicted tail. The reason for this is not clear. If one arbitrarily ignores beam-beam collisions (which have a large effect on the tail temperature because the background plasma is so cold), then somewhat better agreement is found. Although the fit is not as nice as in

Fig. 1, we again find that to explain the measured tail during counter-beams+ICRF we need to invoke the presence of ICRF beam acceleration.

At large energies, where the charge-exchange cross-section is beginning to drop, the signal drops below the background (as measured by masked detectors). One must be careful in measuring this background and subtracting it off, or one can see what appears to be significant tail even in the absence of ICRF. While our results in Figs. 1-3 are certainly consistent with acceleration of the beam ions by the ICRF, the observed tail is significantly above the no-RF expectation only at the highest energies where the background subtraction is beginning to get important. Our measurements could be improved by lowering the injection voltage, so that we have more room to see a tail, and by higher RF power (which will be available in this summer's experiments) to produce a larger tail.

## RESONANCE LAYER SCAN

RF-only experiments were conducted in a $^4$He plasma to avoid deconditioning the limiter. Fig. 4 shows the results of a resonance layer scan during H minority heating. These spectra have been corrected for the difference in the cross-sections for protons charge-exchanging with helium neutrals instead of deuterium neutrals. Note that the flux of 150 keV hydrogen rises by a factor of 7 as the resonance layer moves toward the detector. There are several ways in which these preliminary results may be improved. We plan to do a more complete resonance layer scan to look on both sides of the detector sightline, to see if the signal falls again as the resonance layer moves beyond the detector position. We plan to do the measurements in a deuterium majority plasma, which should increase our signal by a factor 10. And we are investigating the possibilities of using these techniques to place upper-bounds on the radial diffusion of fast ions.

## ACKNOWLEDGEMENTS

We thank J.C. Hosea, J.R. Wilson, A.T. Ramsey, A.L. Roquemore, and other members of the TFTR ICRF and diagnostics group for making these measurements possible. This work was supported by U.S. DOE contract No. DE-AC02-76CH03073.

## REFERENCES

[1] D.N. Smithe, P.L. Colestock, R.J. Kashuba, T. Kammash, Nucl. Fus. **27** (1987) 1319.

[2] D.N. Smithe, C.K. Phillips, G.W. Hammett, and P.L. Colestock, this conference.

[3] G.W. Hammett, Ph.D. Thesis, Princeton University (1986), available from University Microfilms, Ann Arbor MI 48106 (USA).

[4] A.L. Roquemore et.al., Rev. Sci. Instrum. **56**, 1120 (1985).

[5] G.A. Cottrell, et.al., in "Proceedings of the Seventh Topical Conference on Applications of Radio-Frequency Power to Plasmas" (Kissimee, FL, 1987), (AIP Conference Proceedings **159**, New York, 1987).

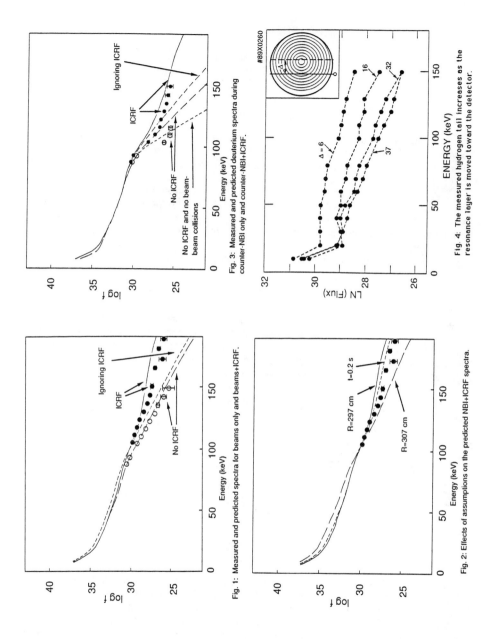

Fig. 1: Measured and predicted spectra for beams only and beams+ICRF.

Fig. 2: Effects of assumptions on the predicted NBI+ICRF spectra.

Fig. 3: Measured and predicted deuterium spectra during counter-NBI only and counter-NBI+ICRF.

Fig. 4: The measured hydrogen tail increases as the resonance layer is moved toward the detector.

# AN ALGORITHM FOR THE CALCULATION OF ICRH IN TOKAMAKS

B. M. Harvey and R. A. Cairns
Department of Mathematical Sciences, University of St Andrews,
St Andrews, Fife, KY16 9SS, U.K.

## ABSTRACT

A computational scheme for the calculation of 3D full-wave solutions to Maxwell's equations in a plasma described by Vlasov's equation and under conditions typical of Ion Cyclotron Resonance Heating experiments, is developed for tokamaks with elliptical cross section.

The method used is based on the work done by Smithe, Colestock, Kashuba and Kammash on a tokamak with circular cross section. The problem is reduced to a series of 2D ones by expressing the perturbing fields as series of toroidal modes. The assumption of toroidal symmetry for the plasma allows each toroidal mode to be solved separately.

For the cross section, elliptical coordinates are used

$$x = a \sinh(u)\sin(v) \qquad y = -a \cosh(u)\cos(v)$$

and the perturbing fields are expressed as Fourier series in v. These 'poloidal' modes are coupled, not only by variations in the magnetic field, plasma density and temperature, but also by the form of the curl operator in elliptic coordinates. However, the coupled differential equations in u can be solved numerically for a finite number of modes with acceptable results.

## INTRODUCTION

The physical model is of an inhomogeneous plasma in a toroidal magnetic field bounded by a perfectly conducting wall. The wall has an elliptical crossection similar to that of J.E.T. although the triangularity of J.E.T. is not included in the present treatment. The antenna is a current sheet of the form

$$\vec{J} = \frac{\delta(u-u_{ant})s(v)p(\phi)}{hh_{\phi}} \vec{e}_v \tag{1}$$

$$h_{\phi} = R_o + a \sinh(u) \sin(v) \qquad h = a(\sinh^2(u) + \sin^2(v))^{1/2}$$

In general, the elements of the exact dielectric tensor would contain the well known homogeneous plasma terms, and in addition, terms due to the spatial dependence of the equilibrium distribution function and magnetic field. Although these parameter gradient terms have been shown, in slab geometry[2,3,4], to have significant effects on the propagation of the Bernstein mode, it has also been shown[2,3,4] that the propagation of the fast mode is almost completely unaffected by these terms.

Restricting attention to the fast mode, for the case of minority heating the dielectric tensor can be modelled by the cold plasma expression, whereas for the case of harmonic heating the tensor can be modelled by the homogeneous hot plasma tensor, with $k_{\perp}^2$ replaced by the differential operators. Since only the fast wave solutions to this system are being considered it is logical to replace these operators with the fast wave value $k_f^2$. This is an approximation that has been used in slab models with results[5,6,7] which compare well with those obtained by the computationally much more expensive approach of retaining the differential operators.

## WAVE EQUATIONS

In the ion cyclotron range of frequencies, the conductivity parallel to $\vec{B}_0$ is much greater than that perpendicular to $\vec{B}_0$ and so $E_\phi$ is suppressed. Considering only $E_v$ and $E_u$ the dielectric tensor (equation 7 of ref 1) is obtained for the nth toroidal mode.

$$K_{1,1} = K_{2,2} = K_\perp \qquad K_{1,2} = -K_{2,1} = iK_x$$

$$K_x = \frac{-\omega_{pe}^2}{\omega\omega_{ce}} + \sum_{ions} \frac{h_\phi\omega_{pi}^2}{2nV_{thi}\omega}(Z(\frac{h_\phi(\omega-\omega_{ci})}{nV_{thi}}) - Z(\frac{h_\phi(\omega+\omega_c)}{nV_{thi}}) + \frac{k_f^2 V_{thi}^2}{2\omega_{ci}^2}(Z(\frac{h_\phi(\omega-2\omega_{ci})}{nV_{thi}}) - Z(\frac{h_\phi(\omega+2\omega_{ci})}{nV_{thi}}))) \quad (2)$$

$$K_\perp = 1 + \sum_{ions} \frac{h_\phi\omega_{pi}^2}{2nV_{thi}\omega}(Z(\frac{h_\phi(\omega-\omega_{ci})}{nV_{thi}}) + Z(\frac{h_\phi(\omega+\omega_c)}{nV_{thi}}) + \frac{k_f^2 V_{thi}^2}{2\omega_{ci}^2}(Z(\frac{h_\phi(\omega-2\omega_{ci})}{nV_{thi}}) + Z(\frac{h_\phi(\omega+2\omega_{ci})}{nV_{thi}}))) \quad (3)$$

Following the method of ref 1 the evolution equations for $E_v$ and $H_\phi$

$$\frac{d}{du}(hE_v) = k_0^2 \frac{h^2}{h_\phi} \frac{i\omega\mu_0}{k_0^2}h_\phi H_\phi + \frac{d}{dv}(\frac{1}{\overline{K}_\perp}\frac{d}{dv}(\frac{i\omega\mu_0}{k_0^2}h_\phi H_\phi) - i\frac{\overline{K}_x}{\overline{K}_\perp}hE_v) \quad (4)$$

$$\frac{d}{du}(\frac{i\omega\mu_0}{k_0^2}h_\phi H_\phi) = i\frac{\overline{K}_x}{\overline{K}_\perp}\frac{d}{dv}(\frac{i\omega\mu_0}{k_0^2}h_\phi H_\phi) - (\overline{K}_\perp - \frac{\overline{K}_x^2}{\overline{K}_\perp})hE_v \quad (5)$$

$$\overline{K}_\perp = h_\phi(K_\perp - (\frac{n}{k_0 h_\phi})^2) \qquad \overline{K}_x = h_\phi K_x \qquad k_0 = \frac{\omega}{c}$$

are expanded in 'poloidal' components and then solved using a modified version of a code developed by Smithe, Colestock, Kashuba and Kammash[1].

$$\frac{dF_m}{du} = \sum_{n=-\infty}^{\infty}((a_{m-n} - md_{m-n}n)H_n + mc_{m-n}F_n) \quad (6)$$

$$\frac{dH_m}{du} = \sum_{n=-\infty}^{\infty}(c_{m-n}nH_n - f_{m-n}F_n) \quad (7)$$

$$hE_v = \sum_{m=-\infty}^{\infty} F_m e^{imv} \qquad \frac{i\omega\mu_0}{k_0^2}h_\phi H_\phi = \sum_{m=-\infty}^{\infty} H_m e^{imv} \qquad \frac{1}{\overline{K}_\perp} = \sum_{m=-\infty}^{\infty} d_m e^{imv} \qquad \frac{\overline{K}_x}{\overline{K}_\perp} = \sum_{m=-\infty}^{\infty} c_m e^{imv}$$

$$\bar{K}_\perp - \frac{\bar{K}_x^2}{\bar{K}_\perp} = \sum_{m=-\infty}^{\infty} f_m e^{imv} \qquad \frac{k_0^2 h^2}{h_\phi} = \sum_{m=-\infty}^{\infty} a_m e^{imv}$$

## THE CODE

Clearly, for scalars and vectors to be single valued functions of position, on the line u=0 any scalar or any $\phi$ component of a vector must be an even function of v while any u or v component of a vector must be an odd function of v. A set of independent initial value problems satisfying this condition are integrated from the origin to the wall without including the antenna jump condition.

$$H_m(u_{ant}^+) = H_m(u_{ant}^-) - s_m P_n \tag{8}$$

$$s_m = \frac{1}{2\pi} \int_0^{2\pi} s(v) \, e^{-imv} \, dv \qquad P_n = \frac{1}{2\pi} \int_0^{2\pi} p(\phi) \, e^{-in\phi} \, d\phi$$

Then a 'driven' solution, with zero field inside the antenna, and satisfying the jump condition at the antenna, is integrated to the wall. Finally, a combination of the undriven solutions is found which, on the wall, cancels the $E_v$ from the driven solution. Adding this combination to the driven solution satisfies the wall condition, $E_v=0$, and so solves the full problem.

Care must be taken in the selection of suitable initial conditions for the 'undriven' solutions. Pure poloidal 'modes' become very rapid functions of position near the foci of the ellipse because the scale factor h tends to zero. In an exact solution, with an infinite number of modes, the resulting spikes would cancel but for realistic computation, combinations of poloidal modes which are better behaved are required. Fortunately such combinations (Mathieu functions) were developed in the late 19th[8] century for modelling oscillations of elliptically bound membranes. More recently efficient numerical methods for the calculation of the required Fourier series have been developed[9].

Using odd and even Mathieu functions as the initial conditions for $E_v$ and $H_\phi$ (figure 1) avoids the spurious focal spikes.

figure 1: Contour plots of real($E_u+iE_v$)

## RESULTS

Since the motivation for these developments was the considerable ellipticity of J.E.T., parameters typical of J.E.T. (semi-major axis 2.1m, semi-minor axis 1.25m, major radius 3.09m, $T_i = T_e$ = 5keV, peak density $n_i = n_e = 2.72 \times 10^{-13} cm^{-3}$) were used in a series of runs, with a toroidal mode number of 15, to examine the behaviour of the plasma loading as a function of frequency (figure 2). Not typical of J.E.T. was the composition of the plasma, a 50/50 mix of deuterium and tritium.

Although the code can handle minority heating, only heating at the deuterium harmonic was used in these runs.

As the changing frequency moves the resonance through the standing wave set up by the antenna, the antenna loading rises and falls significantly, as would be expected. Less easily predicted is the focusing and defocusing of the heating effect. While the x coordinates of the heated regions are determined simply from the applied frequency, the y coordinates are not. In figures 3a-3c the heated regions not only move to lower magnetic fields as the frequency is reduced, they also coalesce to form one central hot spot.

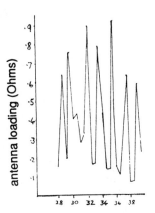

Frequency (MHz)
Figure 2.

Figure 3 contour plots of power deposition

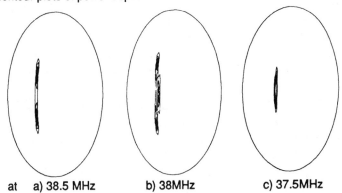

at   a) 38.5 MHz     b) 38MHz     c) 37.5MHz

Acknowledgment-We would like to thank Dr P. L. Colestock for providing a copy of the code of Smithe, Colestock,Kashuba and Kammash[1].

## REFERENCES

(1) D. N. Smithe, P. L. Colestock, R. J. Kashuba and T. Kammash, Nucl. Fusion 27, 1319 (1987)
(2) B. M. Harvey, Ph.D. thesis (1988)
(3) D. G. Swanson, Phys. Fluids 28, 2645 (1985)
(4) H. Romero and J. Scharer, Nucl. Fusion 27, 363 (1987)
(5) C. N. Lashmore-Davies, V. Fuchs, G. Francis, A. K. Ram, A. Bers and L. Gauthier, Phys. Fluids 31, 1614 (1988)
(6) A. Kay, R. A. Cairns and C. N. Lashmore-Davies, Plasma Phys. Controlled Fusion 30, 471 (1988)
(7) P. L. Colestock and R. J. Kashuba, Nucl. Fusion 23, 763 (1983)
(8) E. L. Mathieu, Journal de Math. XIII, 137 (1868)
(9) G. Blanch, J. Math. Phys. 25, 1 (1946)

# HIGH-POWER TESTING OF THE FOLDED WAVEGUIDE*

G. R. Haste and D. J. Hoffman

Oak Ridge National Laboratory, Oak Ridge, TN 37831-8071

## ABSTRACT

The folded waveguide, an alternative launcher for large plasma devices, is under development at Oak Ridge National Laboratory (ORNL) using a full-scale 80-MHz model. In the folded waveguide, rf fields exist over a surface area, which increases the waveguide's multipactoring susceptibility. Edges with small radii of curvature enhance the field strength and breakdown susceptibility.

Previous high-power tests demonstrated steady-state operation at 8 kW, limited by structure heating, and pulsed operation at 15 kW, limited by discharges. Recent modifications have led to operation at 30 kW for 1-s pulses, limited by the transmitter. The maximum electric field of 22 kV/cm is comparable to that expected for a 2-MW launcher with plasma loading.

Measurements of a loop antenna and a folded waveguide show that typical electric fields in the internal structure of the folded waveguide are about half those for the loop. High-power testing will be carried out with a higher-power transmitter and operation in the RF Test Facility (RFTF). The device will be operated in vacuum, with a magnetic field, in the presence of a plasma, and at still higher powers.

A folded waveguide array is being considered for current drive on the International Thermonuclear Engineering Reactor (ITER).

---

* Research sponsored by the Office of Fusion Energy, U.S. Department of Energy, under contract DE-AC05-84OR21400 with Martin Marietta Energy Systems, Inc.

# INTRODUCTION

A full-scale model of a folded waveguide launcher[1,2] has been tested at 80 MHz. This model, shown in Fig. 1, was originally built of aluminum and tested at low power. To prepare for higher-power tests, the

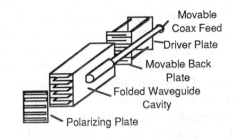

Fig. 1.  Folded waveguide coupler.

aluminum pieces were copper plated. Low-power testing was done at atmospheric pressure; high-power testing required vacuum operation.

## LOW-POWER TESTS

The basic principles of operation can be demonstrated at low power. The resonant frequency for the lowest-order mode is determined by adjusting the movable back plate so the cavity is one-half guide wavelength long. The position of the movable coaxial feed line determines the matching, and this position also depends on the Q. These variations are shown in Figs. 2 and 3.

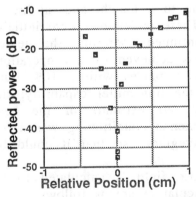

Fig. 2.  Matching as a function of coax contact point.

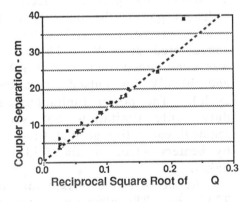

Fig. 3. Variation of optimum coupler position with Q.

Figure 2 shows the matching as a function of the position of the coax line, and Fig. 3 shows the variation of the optimum position as the Q is varied by placing absorbent plastic sheets (Eccosorb) at various locations in front of the folded waveguide. The separation between the coupler and the back plate varies as $Q^{-1/2}$, in agreement with a simple calculation.

The Eccosorb plastic sheets were also used to compare the behavior of the folded waveguide with that of a loop antenna having nearly the same dimensions. Typical electric fields for the loop antenna (e. g. between the strap and the Faraday shield) were roughly twice those for the folded waveguide (between the vanes). If the typical electric field determines the launcher capability, the limiting power for a folded waveguide would be about 4 times that for a loop antenna.

## HIGH-POWER TESTS

One of the principal threats to the folded waveguide concept is multipactor discharges. The electric fields range between zero, at the ends of the waveguide, to their maximum value along the central vane and cover a large surface area. The conditions are therefore favorable for multipactoring over some large areas. High-power testing in vacuum was designed to meet that threat.

Multipactoring did indeed occur with high-power operation, as shown by pressure increases with rf power, increases in reflected power, the inability to tune to a match, and, at higher powers, emission of visible light. These indications were initially encountered at 7 W. Through continued operation, the device was conditioned such that 8 kW was achieved for cw operation, and 30 kW for pulsed operation (1-s pulses with a 10% duty cycle). The power limitation was the transmitter capability; the duty cycle limitation was set by our estimate of the heat removal capability at that time. Antenna conditioning was aided by heaters added to the folded waveguide, so that it could be baked.

At 30 kW the maximum electric field is calculated to be 22 kV/cm. The elements of this calculation are as follows.

The voltage applied to central vane at the feed point is equal to the voltage on the center coax conductor, which is determined by the power and the coax impedance.

The maximum voltage on the central vane is the maximum of a sine curve with the applied voltage at the feed point, and zeros at the ends.

The electric field is the maximum voltage along the central vane, divided by the vane-to-vane spacing, times a field enhancement factor, calculated to be roughly 3, due to the curvature of the vane tips.

A 2-MW launcher, with vane tip radii 3 times larger than the present device and with its Q reduced due to plasma loading, would have electric fields comparable to those in the present, unloaded device at 30 kW.

## FUTURE GOALS

The folded waveguide will soon be mounted on the RFTF for more extensive testing. There it will encounter magnetic fields, higher gas pressure, a plasma and a microwave environment, and a higher-power rf source. The effects, both singly and in concert, will provide a more severe and realistic test of the concept.

Goals to be pursued at low power, with smaller (and therefore higher-frequency) models, include determining the effects of holes in the sides of the folded waveguide for improved vacuum conductance from the cavity interior, using a pair of folded waveguides side by side with arbitrary phasing for spectrum control, and investigation of alternative coupling schemes to introduce power into the folded waveguide.

An array of folded waveguides has been considered for current drive on ITER. If phase control for separate elements of that array is at least as good as for loop antennas, then a comparable spectrum can be launched. The advantages of the folded waveguide array are the increased power capability and, perhaps, additional strength to resist disruption forces.

[1] T. L. Owens, IEEE Trans. Plasma Sci. PS-14 (6), 934 (1986).
[2] T. L. Owens, G. L. Chen, G. R. Haste, and P. M. Ryan, AIP Conf. Proc. 159, 298 (1987).

# MEASUREMENTS OF ICRF FIELDS NEAR A MODEL ANTENNA*

N. Hershkowitz, R. Majeski, T. Tanaka and T. Intrator
University of Wisconsin-Madison, Madison, WI 53706

## ABSTRACT

RF B fields are measured near the Faraday shield of a model loop antenna in a vacuum and in the presence of a plasma. The axisymmetric plasma is produced in the central cell of the Phaedrus-B tandem mirror. Results are found to be in good agreement with the predictions of the code ANTENA.

## INTRODUCTION

This paper presents measurements of the rf B fields near the Faraday shield of a model antenna in the presence and absence of plasma. The current strap is a 60° loop oriented in the poloidal (azimuthal) direction. The purpose of these experiments is to bench mark codes which have been and will be developed for describing ICRF in tokamaks. Although measurements are carried out in the central cell of a tandem mirror plasma, the plasma density and temperature near the antenna are comparable to those present in tokamak edge plasmas. Although $B_o$, the dc magnetic field is much lower, the plasma density and radius of the Phaedrus-B plasma is sufficient to support multiple fast magnetosonic waves.

## EXPERIMENT

Measurements have been made of the rf B fields near the plane of the single layer Faraday shield of a low power model antenna (see Fig. 1). Measurements were carried out in vacuum (in air), with and without a Faraday shield and in the presence of a plasma. Plasma experiments employed the central cell plasma of the Phaedrus-B tandem mirror[1]. Plasma is produced for up to 20 msec by one or more dual half turn antennas(operated at power levels ~ 200 kW). The Phaedrus-B central cell was operated with the axial (toroidal) magnetic field equal to $B_o$ = 900 ± 45 gauss over the central 2 m.

The model antenna was located at + 75 cm with respect to the central cell midplane. The radial positions of the antenna strap, Faraday shield, and central cell limiters were 19 cm, 17 cm, and 17 cm respectively. The radial variation of a representative Phaedrus-B central cell plasma, obtained with

Langmuir probes, is given in
Fig. 2. The density is relatively
flat out to approximately 10 cm and
then falls to the value at the
shield. Peak plasma densities are
$n \sim 3 - 6 \times 10^{12}$ cm$^{-3}$ with $n \geq 5 \times$
$10^{11}$ cm$^{-3}$, 3 cm from the front
plane of the Faraday shield.
Representative electron
temperatures are 20 eV on axis and
40 eV near the plasma edge.

Measurements were made with a
"B dot" magnetic probe (with three
orthogonal coils[2]. The model
antenna (shown in Fig. 1) consists
of a 60° current loop, width = 10
cm, operated at 3.5 MHz (2.7 $\Omega_i$)
with $I_{ant} \sim 100$ A p-p. Magnetic
probes were scanned in the radial
(r) and "toroidal" (z) direction
and the model antenna itself was
scanned in the poloidal (θ)
direction.

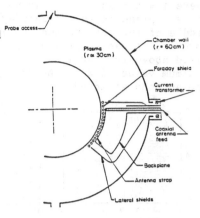

Phaedrus-B test antenna sketch

Fig. 1. Schematic of the model antenna.

Vacuum data were supplemented with data from an identical model
antenna operated in air. In this paper we give vacuum data both
with and without Faraday shields and for several Faraday shield
configurations. Plasma data all have a Faraday shield present.

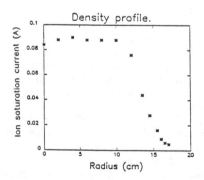

Fig. 2. Radial
density profile
determined with a
Langmuir probe.

EXPERIMENTAL RESULTS

Vacuum Data

Magnetic field data are presented in Fig. (3) in the form of
mod $B_j$ contours in planes parallel to the the Faraday shield,

where j = z,r and θ. It is not surprising that the maximum field
in each plane above the Faraday shield was located above the
middle of the antenna and was in the toroidal direction.
However, it was surprising that the strength of the poloidal
field above the corners of the antenna was comparable to the
maximum value of B . The data given in Fig. 3 show that $B_{\theta max}$ at
the antenna strap edges is ~ 0.4 of the rf $B_{z max}$.

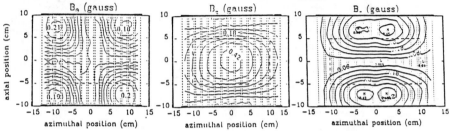

Fig. 3. Mod B contours in a plane 1.5 cm from the antenna
Faraday shield in a vacuum.

Plasma Data
    Mod-B contours in the presence of plasma are quite similar to
those obtained in a vacuum. Details of the the radial variation
of the three components of $B_{rf}$ at the center, corner and edge of
the antenna strap are given in Fig. 4. Effects associated with
wave propagation begin to be apparent at 5 cm from the Faraday
shield.

## DISCUSSION

    Ko et al.[3] have shown that the ARGUS code can be used to
model the vacuum fields of this experiment and that very good
agreement is obtained. The code self-consistently determines the
variation of the current density in the current strap in the
toroidal direction and also accurately describes the currents in
the back plane and Faraday shield.
    We have employed the azimuthally symmetric ANTENA code[4] to
model the near fields in the presence of plasma. Results from
ANTENA are compared to experimental data in Fig. 4.
    Good qualitative agreement was found for the radial
variations of the three components of B for plasma radii greater
than 5 cm (i.e. out to 12 cm from the Faraday shield). Measured
B fields for r < 5 cm tended to be somewhat larger than ANTENA
predictions. It is not yet clear whether this discrepancy is
real or if it is associated with magnetic probe perturbations to
the plasma. Both the code and data show that the plasma has
little effect on the rf B fields within 5 cm of the Faraday
shield.
*Work supported by U.S. DOE Grant DE-FG02-88ER53264.

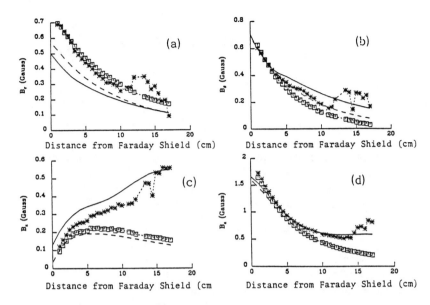

Fig. 4.    Components of B versus r (measured from the
Faraday shield), above the corner (a,b,c) and the
center (d) of the antenna strap determined
experimentally (data) and predicted by ANTENA (lines).
Dashed traces and squares correspond to vacuum fields.

## REFERENCES

1. R. Breun, et al "Radial Transport in the Phaedrus-B Tandem
   Mirror," 12th Int'l Conference on Plasma Physics and
   Controlled Nuclear Fusion Research, Nice, France, to be
   published in IAEA-CN-50/H-1-2, Vienna (1989).
2. T. Intrator, S. Meassick, J. Browning, R. Majeski, J.R.
   Ferron, N. Hershkowitz, to be published in Nuclear Fusion,
   International Atomic Energy Agency (1989).
3. W. Grossmann, K. Ko, A. Drobot, R. Majeski, T. Tanaka, N.
   Hershkowitz, "3D Modeling, Simulation and Evaluation of ICRF
   Antennas," presented at the Sherwood Theory Conference,
   April 3-5, 1989, San Antonio, TX.
4. B. McVey, "ICRF antenna coupling theory for a cylindrically
   stratified plasma," NTIS Document No. DE85004960 (National
   Technical Information Service, Springfield, VA, 1984).

# ICRF HEATING ON TFTR WITH THE ORNL ANTENNA*

D. J. Hoffman, W. L. Gardner, and P. M. Ryan
Oak Ridge National Laboratory, Oak Ridge, Tennessee 37831

G. J. Greene, J. C. Hosea, J. R. Wilson, and J. E. Stevens
Princeton Plasma Physics Laboratory, Princeton, New Jersey 08542

## ABSTRACT

Initial ion cyclotron range of frequencies (ICRF) heating experiments on TFTR began in the summer of 1988. Although we were in the commissioning stage for much of the equipment, some plasma coupling measurements were made in the fall. This paper is focused on the results from the Bay L antenna.

## INTRODUCTION

The Bay L antenna comprises a movable array of two loops (designated straps 3 and 4) installed in a 60- by 90-cm port. It was principally designed to compare, in plasma, a graphite Faraday shield to thin coatings, but it has additional design features[1] that, in combination with those of the Bay M antenna, were intended to help in the future selection of design options for ICRF antennas. These options include wall slotting vs closed walls, active vs inertial shield cooling, internal vs external matching, vacuum vs $SF_6$ as the high-voltage medium, and copper vs silver as the conducting medium. Although most of these options have not yet been fully tested, some, such as slotting, have been seen to make a significant difference in coupling.[2]

The plasma response to the Bay L antenna has shown that each strap can effect central heating. Despite the limited operation (a few days of running) and the use of $SF_6$ instead of vacuum on the capacitors, the antennas were powered to 600 kW on either single strap and 500 kW on the combined pair. The phased launching is tentatively comparable[2] with that obtained on Bay M, but lower power makes it necessary to record incremental plasma performance instead of absolute response.

## ANALYSIS OF COUPLING

The measured power of 600 kW per strap was achieved for an average power density of $\approx 400$ kW/cm$^2$ flowing through the antenna to the $2 \times 10^{13}$ cm$^{-3}$ plasma located 4 cm from the face of the antenna. This was seen to be the limit of the antenna at the time, so the circuit needed to be fully analyzed to ascertain whether low-voltage breakdown in the structure or low coupling was responsible. Previous circuit models[3] using lumped elements implied that the voltage was low ($\approx 30$ kV). During the vacuum opening, a fully distributed, mutually coupled, lossy coax model was devised for the Bay L antenna (Fig. 1). This model has the expected spatial hyperbolic functions, but has two additional terms that are the result of the mutual inductance between the two straps. The mutual coupling between the two straps is

*Research sponsored by the Office of Fusion Energy, U.S. Department of Energy, under contract DE-AC05-84OR21400 with Martin Marietta Energy Systems, Inc.

included only on the poloidal radiating element. The circuit is reduced to finding appropriate coefficients on the coupled legs from the following equations, assuming that the load, capacitance, and inductance per unit length of each strap are given by R', C', and L', respectively:

$$V_3 = a_1\cosh(\Gamma_+x) + a_2\sinh(\Gamma_+x) + a_3\cosh(\Gamma_-x) + a_4\sinh(\Gamma_-x) \tag{1}$$
$$V_4 = a_1\cosh(\Gamma_+x) + a_2\sinh(\Gamma_+x) - a_3\cosh(\Gamma_-x) - a_4\sinh(\Gamma_-x) \tag{2}$$
$$I_3 = (a_1/Z_+)\sinh(\Gamma_+x) + (a_2/Z_+)\cosh(\Gamma_+x) + (a_3/Z_-)\sinh(\Gamma_-x) + (a_4/Z_-)\cosh(\Gamma_-x) \tag{3}$$
$$I_4 = (a_1/Z_+)\sinh(\Gamma_+x) + (a_2/Z_+)\cosh(\Gamma_+x) - (a_3/Z_-)\sinh(\Gamma_-x) - (a_4/Z_-)\cosh(\Gamma_-x) \tag{4}$$

If the mutual coupling is given by k, then

$$\Gamma_\pm = \sqrt{\{[R' + j\omega L' \bullet (1 \pm k)] \bullet [j\omega C']\}}, \quad Z_\pm = \sqrt{\{[R' + j\omega L' \bullet (1 \pm k)]/[j\omega C']\}}.$$

At x = 0 (the input voltage), $V_3 = a_1 + a_3$, and $V_4 = a_1 - a_3$. The other two coefficients can be obtained by relating the transformed capacitances at the other end of the two straps.

Circuit parameters were determined through exhaustive correlation of implicit measurements (capacitance values, resonances, and Q to yield circuit parameters), explicit measurements (time domain reflectometry for these parameters), and two-dimensional magnetostatic calculations (to confirm the measurements). Table I lists values of every element for strap 3 of the Bay L antenna. Figure 2 shows capacitance vs load per unit length for strap 3's parameters with strap 4 detuned.

## IMPLICATIONS OF THE CIRCUIT ANALYSES

These calculations confirmed that the Bay L antenna had been conditioned to well beyond 60 kV with vacuum in the capacitors and was limited to ≈48 kV with 15 psig of $SF_6$ in the capacitors. At 50 kV, the average internal electric field is 43 kV/cm, while that at the ends of the capacitor cans is 1.3 MV/cm. This average field is consistent with the breakdown strength of $SF_6$ at 15 psig.

A more important implication of the distributed model is that the loading is lower than predicted by lumped models. For TFTR discharges of $2 \times 10^{13}$ cm$^{-3}$ plasma located 4 cm away, the model showed 2.1 Ω/m for strap 4 and 1.5 Ω/m for strap 3, representing relative loads of ≈6.1 and ≈4.5, respectively. This imbalance in loading, unknown during phased operations, complicated analyses. Measurements of the port and antenna location, made during the vacuum opening, show that strap 3 was recessed ≈6 mm from strap 4 due to the port's misalignment. Flux linkage models[3] predict that the recess should reduce the load by 14%, so the imbalance is not completely understood. At this position, the respective power limits were 600 kW and 390 kW. The power in each strap was limited by the voltage handling limits of the $SF_6$. When reactive loading is included, both antennas operated at ≈50 kV, the conditioning voltage. Moving the antenna in 8 mm resulted in a significant gain in power handling (550 kW to 590 kW) and a 37% gain in loading (vs 28% predicted by the flux linkage model). Thus, the antenna maintained its voltage capabilities despite the increased particle flow.

The model showed two other trends in loading. First, the loading was linearly proportional to the line-average density for ohmic discharges. Increasing the edge density (e.g., with neutral beam injection into the discharge) raised the loading. However, this has not yet been quantified, since a systematic study was not made. The second trend is that the loading per strap with two straps is greater than single-strap loading. Both in-phase and out-of-phase coupling benefited from this effect. The trend was complicated by the unbalanced loading on the two straps.

## PLASMA RESPONSE

Because the rf power was ≈1/6 of the combined rf and ohmic power in the discharge, the plasma effects could clearly be seen, but some statistical variations could confuse results. Generally, the heating occurred at the resonance zone. Little impurity introduction was observed. No surface interactions at the face of the antenna, as seen on the Bay M antenna in phase, were detected.

One trend observed on the Bay L antenna, as with the Bay M antenna, is that generally the in-phase operation resulted in ≈60% of the beta increase observed with out-of-phase operation (Fig. 3). On Bay L, when individual straps were powered, the beta increases were comparable to the in-phase rate. Each single strap had the expected heating rate. Because both in-phase and single-strap configurations have some spectral power near the wave numbers of zero, and because the single-strap and out-of-phase configurations are peaked at higher wave numbers than the in-phase configuration, it is suspected that the near-zero spectral wave power is not being usefully transmitted to the core.

## FUTURE GOALS AND CONCLUSIONS

The conclusions of the first runs held some encouragement. The antennas could be operated at good power densities (400 kW/cm$^2$) and at high voltages (50 kV). The medium in the capacitor was seen to be the voltage limit. Finally, at the 500-kW level, the best core heating was achieved with out-of-phase operation. Single-strap or in-phase operation resulted in a 40% reduction of efficiency.

Since the detailed analyses led to the conclusion that the loading was low, it is important to try to increase the antenna voltage. Near-term plans include final confirmation of the loading on TFTR. During the vacuum break, the vacuum system for the capacitors has been upgraded so that the capacitors can be pumped by a turbomolecular pump during the magnetic pulse. If the plasma voltages continue to be the same as the conditioning voltages achieved with vacuum in the capacitors, the power should increase by more than 67%. We will also try to facilitate loading by moving the antenna 8 mm closer to the plasma. Finally, higher-density plasma scenarios in TFTR are planned, including pellet-fueled discharges.

## REFERENCES

[1] D. J. Hoffman, et al., in Proc. 15th Eur. Conf. Controlled Fusion and Plasma Heating, Dubrovnik, 1988, Vol. 12B, Part II, pp. 770–3 (1988).

[2] J. Hosea, et al., Bull. Am. Phys. Soc. **33** (9), 2094 (1988).

[3] D. J. Hoffman, et al., Bull. Am. Phys. Soc. **33** (9), 2094 (1988).

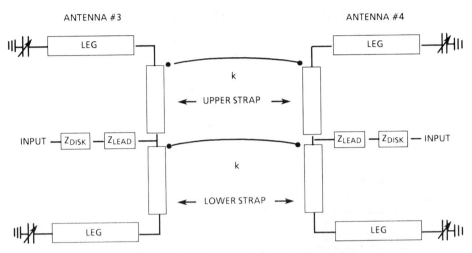

Fig. 1. Circuit diagram of the Bay L antenna.

Table I. Circuit parameters for strap 3
(Tap point = 0.5927, mutual coupling = 0.02005)

|  | Strap | Leg | Disk | Lead |
|---|---|---|---|---|
| Impedance ($\Omega$) | 42.0 | 75.0 | 17.15 | 60.0 |
| Relative phase velocity (%) | 66.2 | 99.9 | 34.3 | 100.0 |
| Vacuum losses ($\Omega$/m) | 0.34128 | 0.010 | 0.010 | 0.010 |
| Length (m) | 0.6775 | 0.4243 | 0.0254 | 0.20 |

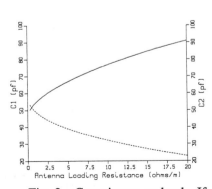

Fig. 2. Capacitance vs load. If the antenna is matched, the capacitance pairs can be used to infer plasma loading.

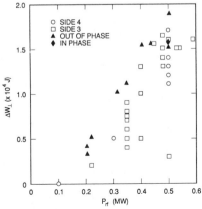

Fig. 3. Change in stored perpendicular energy vs rf power for various phasings.

# ICRF ANTENNA DESIGNS AND RELATIVE PERFORMANCES ON TFTR*

J. Hosea, P. Bonanos, P. Colestock, G. Greene,
S. Medley, C. Phillips, J. Stevens, J. Wilson and TFTR Group
Princeton University, Princeton, NJ 08543

W. Gardner, D. Hoffman, and D. Swain
Oak Ridge National Laboratory, Oak Ridge, TN 37831

## ABSTRACT

Two compact antenna designs, each having two toroidally separated current straps, are being used to perform high power ICRF experiments on TFTR. For the modest density regimes ($\bar{n}_e \leq 3 \times 10^{13}$ cm$^{-3}$) studied in the 1988 TFTR run, the end-fed slotted-wall design delivered up to 2.8 MW (source limited) and the off-center fed, solid-wall design delivered up to 0.6 MW (coupling/configuration limited). Both antennas exhibited similar heating efficiencies which were optimum (~ 30% greater on average) for out-of-phase current strap excitation. In-phase strap operation of the first antenna design at relatively high power resulted in a dramatic localized sputtering of the antenna shield TiC surface, suggesting the interaction of Bernstein waves with local impurity resonances. Out-of-phase operation of this design exhibited no such interaction and gave a high loading value consistent with 6 MW capability with future two-source operation even at $\bar{n}_e$ ~ 2 x $10^{13}$ cm$^{-3}$. Lower than expected antenna loading for the second antenna design has been found to account for the power limit which will be eased during the upcoming run period by placing this antenna closer to the plasma and further enhancing its voltage standoff capability to extend its operation to the multimegawatt power level.

## TFTR ICRF ANTENNAS

Two compact antennas have been developed jointly by ORNL and PPPL to support the ultimate delivery of 10 MW of source power to the TFTR plasma. These antennas are designated by their location on TFTR -- Bay M (6 MW design) and Bay L (4 MW design) indicated in Fig. 1 -- and have the principal characteristics noted in Table I.[1-3] The designs have been chosen to highlight the effects on performance of the launched wave spectrum, shield outgassing and impurity effects, and of variations in matching and conditioning.

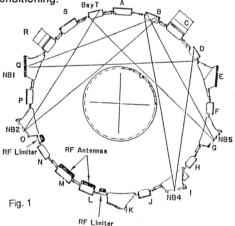

Fig. 1

Table I

|  | Bay L | Bay M |
|---|---|---|
| Design Power (MW) | 4 | 6 |
| Pulse Length (Sec) | 2 | 2 |
| Box Wall/Septum Configuration | Solid | Slotted |
| Design Shield power density (kW/cm$^2$) | 1.2 | 1.3 |
| Shield Material | Graphite | TiC |
| Shield Protective Bumpers | Graphite | Graphite |
| Number of current straps (toroidally) | 2 | 2 |
| Strap Feed Point | Central | End |
| Matching (Capacitors) | Internal | External |

## PLASMA HEATING

Plasma heating experiments have been performed for relatively low density regimes on TFTR. Typical conditions of interest here are $\bar{n}_e \sim 2 \times 10^{19}$ m$^{-3}$, R = 2.62 m, a = 0.96 m, Helium, H-minority, $B_T$ = 3.3 T.

The Bay M and Bay L antennas have been operated at 47 MHz to power levels up to 2.8 MW and 0.5, respectively, for single-source, two-strap excitation. In addition, 0.6 MW has been delivered to one strap of the Bay L antenna during conditioning and power limit tests. The out-of-phase incremental heating efficiency for the Bay M antenna ($\sim$ 2 eV/kW/10$^{19}$ m$^{-3}$) is observed to approach the ohmic level and the in-phase heating is found to be $\sim$ 30% less efficient on average.[4] Although the measurement accuracy is reduced at the lower powers delivered by the Bay L antenna, the same heating trends are observed[5] and the best out-of-phase incremental heating value observed thus far is also $\sim$ 2 eV/kW/10$^{19}$ m$^{-3}$ (Fig. 2a). Although conclusive experiments await higher power operation, the $\Delta T_e$ for the Bay L antenna is typically less than the value for the Bay M antenna (Fig. 2b), suggesting that hydrogen gas may be evolving from the cooled, graphite-coated Faraday shield.

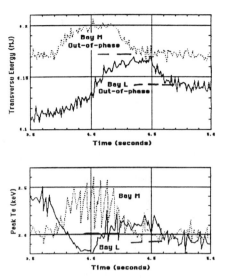

Fig. 2  Bay L and Bay M antenna heating at P$_{RF}$ = 0.5 MW. P$_{RF}$ pulse is 0.5 sec beginning at 4 sec: Bay L, 3.7 sec: Bay M.

The impurity effects during the ICRF experiments have been on the whole benign.[6] The one exception has been for in-phase strap excitation of the Bay M antenna for which Ti, the shield material, is introduced into the discharge in minute quantities (up to P$_{RF}$ = 1.8 MW).

## BAY M ANTENNA SHIELD-PLASMA INTERACTION FOR IN-PHASE EXCITATION

One primary reason for choosing different shield materials for the two antennas was to permit exploration of the shield-plasma interaction characteristics and, in particular, investigation of the shield impurity release mechanism(s) observed on JET[7,8] and JT-60.[9] Accompanying the Ti (shield material) release for in-phase strap excitation of the Bay M antenna, we have observed a dramatic localized shield-plasma sputtering interaction as indicated in Fig. 3. Precise measurements have revealed that the protective graphite bumpers fail to shadow the shield by 1 mm over the entire poloidal extent. As a result, the shield-plasma interaction is expected to be stronger than for well shadowed designs. However, the appearance of sputtering zones at the top and bottom of the shield only (at a major radius R) and only for in-phase operation suggests the possibility that long wavelength Bernstein wave excitation with subsequent damping at the harmonic resonances of partially ionized carbon or He$^+$ is responsible for the interaction.[10,11]

280

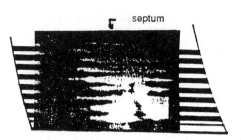

Fig. 3 Close-up of bottom interaction zone for Bay M in-phase excitation.

## ANTENNA LOADING

The Bay M antenna has been designed to optimize loading: minimal shield-strap separation of 8 mm, zero shadowing (- 1mm) by the side protective tiles, and slotted walls. As a result, the loading, even at densities as low as $n_e \sim$ $2 \times 10^{19}$ m$^{-3}$, is ~ 40 times the vacuum value when the straps are jointly fed from one generator as indicated in Fig. 4. Shown there are the loci of loading before and after the two quarter-wave transform sections of the transmission system (prior to matching to the generator). This level of loading is sufficient to support 6 MW operation at the 50 kV standoff level which also has been achieved at lower $n_e$.

However, it should be noted that the loading exhibits a reactive change with increasing load (change in the polar angle is equivalent to a frequency upshift caused by a reduced strap reactance due to image currents in the plasma). Therefore, future full power operation using two sources will require that the reactive loading as well as the resistive loading be tracked to match the plasma conditions.

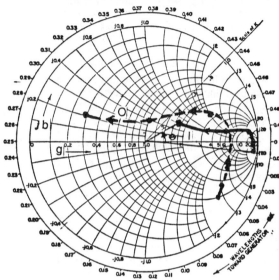

Fig. 4 Bay M loci before λ/4 transformers (I) and before matching (0) for $\bar{n}_e$ increasing to $2 \times 10^{19}$ m$^{-3}$.

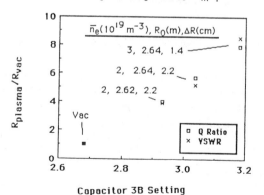

Fig. 5 $R_{pl}/R_{vac}$ load ratio obtained for one strap of Bay L antenna (Bay M side) from VSWR and Q measurements versus bottom capacitor setting. (ΔR is is limiter-antenna separation.)

The Bay L antenna has been designed to accentuate exclusion of gas from the box and for a relatively wide frequency tuning range: the walls are solid, the protective bumper tiles are designed to shadow the shield face by 5 mm, and the shield-strap separation is 15 mm. Consequently, the loading is expected to be reduced by about a factor of 2 from that of the Bay M antenna. In fact, recent analysis[5] and measurements (Fig. 5) support the conclusion that the loading for the initial run conditions at low density on TFTR was about a factor of 5 below the Bay M antenna level, and that this relatively low value was responsible for the observed power limit.

## PLANS

During the coming run period on TFTR, two-source operation will be employed to reach the design power levels or power limits for both antennas over a wide range of plasma conditions including those for the supershot, L-mode, and pellet regimes. The Bay L antenna power capability will be improved by moving it 8 mm closer to the plasma edge, enhancing the capacitor voltage standoff with vacuum above the 45 kV level obtained with $SF_6$ gas and by increasing the plasma edge density. Thus the relative performance of the differing features of the antennas will be quantified at the multimegawatt level to support antenna design choices for future devices, both compact (Alcator C, CIT, Ingitor) and large scale (ITER, NET).

Ultimately, both antennas will be modified to permit full design power operation in the high neutral beam power supershot regime for which the edge power flux is very large and the edge plasma density is relatively low.

*This work supported by US Department of Energy Contract No. DE-ACO2-76-3073.

## REFERENCES

1.  J.R. Wilson et al., Proc. 7th Topical Conf. on Applications of RF Power to Plasmas (AIP (New York, 1987) p. 294.
2.  D.J. Hoffman et al., Ibid., p. 302.
3.  D.J. Hoffman et al., Proc. 15th European Conf. on Controlled Fusion and Plasma Heating (Dubrovnik, 1988) Vol. II, p. 770.
4.  J.R. Wilson et al., Proc. 12th Int. Conf. on Plasma Physics and Controlled Nuclear Fusion Research (IAEA, Nice, 1988) paper IAEA-CN-50/E4-1.
5.  D.J. Hoffman, et al, This conference.
6.  J.E. Stevens et al., This conference.
7.  J. Jacquinot et al., Proc. 11th Int. Conf. on Plasma Physics and Controlled Nuclear Fusion Research (IAEA, Kyoto, 1986) IAEA (Vienna, 1987) p. 449.
8.  M. Bures et al., Plasma Physics and Controlled Fusion Research 30 (1988) 449.
9.  H. Kimura, TFTR Antenna Design Review 1986; H. Kimura et al., Proc. 14th European Conf. on Controlled Fusion abd Plasma Physics (Madrid, 1987) Vol. II, p. 857.
10. J. Hosea, Proc. Course and Workshop on Applications of RF Waves to Tokamak Plasmas (Varenna, 1985) Vol. II, p. 666.
11. S. Puri, Phys. Rev. Lett. 61 (1988) 959.

# ICRH in Large Tokamaks with Poloidal Magnetic Field*

Kaya Imre**, Harold Weitzner and D.C. Stevens
*Courant Institute of Mathematical Sciences, NYU, New York, NY 10012*

D.B. Batchelor
*Oak Ridge National Laboratory, P.O. Box Y, Oak Ridge, TN 37831*

## ABSTRACT

The poloidal component of the equilibrium magnetic field is shown to alter fundamentally the wave propagation characteristics across ion cyclotron resonance layers. The local theory, or perpendicular stratification approximation, which ignores this field component is replaced by a nonlocal theory for which the field equation is an integrodifferential equation. We give a few exact general results, and sample numerical results for a typical CIT application.

## I. INTRODUCTION

The wave propagation within slightly inhomogeneous plasmas in large toroidal devices usually can be studied within the geometrical optics approximation. However, it is well-known that this model fails for waves propagating near resonance surfaces, such as surfaces on which the wave frequency is a non-zero multiple of the ion cyclotron frequency of any of the ion species present. We have applied boundary layer analysis[1] to study the wave propagation across ion cyclotron resonance layers in large toroidal devices. In such devices, typically the equilibrium magnetic field has both toroidal and poloidal components, with the toroidal field much larger than the poloidal field. The perpendicular stratification approximation assumes that the poloidal field is negligible and that the equilibrium magnetic field only varies in a direction perpendicular to itself. In our previous studies we have adopted this assumption, and obtained field equations valid within the resonance region, which are ordinary differential equations. This appoximation can only be justified when the effects of the poloidal field is negligible.[2] The purpose of this paper is to investigate the case when the perpendicular stratification approximation fails due to the poloidal field effects.

## II. BASIC FORMULATION

We study the poloidal field effects on wave propagation across ion cyclotron resonance layers in large toroidal devices by using a boundary layer analysis similar to those used in previous local analyses. We introduce $\varepsilon = n_c v_{th}/c$ as an expansion parameter, where $n_c$ is the cold plasma refractive index at resonance $(\mathbf{r} = 0)$. Within the Nth harmonic resonance layer we introduce the stretched variable $\xi$ by the relation

and we define

$$\xi\, N\delta\omega = \omega - N\frac{ZeB(\mathbf{r})}{mc}, \quad \delta = \frac{c}{\omega n_c}\frac{|\nabla B(0)|}{B(0)},$$

$$a = \mathbf{B}_p\cdot\nabla B(n_c^2\, v_{th}\,\omega)\,/(c\,|\nabla B|)^2.$$

Very approximately, in a near circular tokamak, $a = (\varepsilon/\delta)\,(1/q)\,(h/R)$, where $q$ is the safety factor, $h$ is the distance from the equatorial plane, and $R$ is the major radius. For CIT, $0 \le |a| \le 2$. We solve the Vlasov equation within the boundary layer, and calculate the current $\mathbf{J}$. Only for $|a| \ll 1$ are poloidal field effects negligible, and local perpendicular stratified theory applicable. Otherwise, $\mathbf{J}$ is a nonlocal functional of the induced field, and the associated field equation is an integrodifferential equation, whose solution can be shown to match the geometrical optics modes. Using $(+,-,\|)$ representation, and eliminating $E_-$ within Maxwell equations, we obtain

*Work supported by U.S.D.O.E. Grants No. DE-FG02-86ER53233 and DE-FG02-ER53223.*
**Permanent Address: College of Staten Island, CUNY*

$$(D^2 + 1)E_+ + (D^2 + 1 + \frac{\tilde{\varepsilon}_{--}}{\tilde{\varepsilon}_{++}}) \frac{1}{\tilde{\varepsilon}_{++} + \tilde{\varepsilon}_{--}} \varepsilon^H_{++} E_+ = 0,$$

where

$$D = K \, d/d\xi, \; K = d\xi/d[k_c x](0), \; \tilde{\varepsilon}_{\pm\pm} = \varepsilon^c_{\pm\pm} - n^2_{\parallel},$$

and where $\varepsilon^c$ is the cold plasma dielectric tensor at resonance. The operator $\varepsilon^H_{++}$ is to be obtained from Vlasov equation for each specific problem. We study three distinct problems of practical interest: (i) minority fundamental resonance, (ii) second harmonic resonance, and (iii) ion-ion hybrid resonance. In the first case, we obtain

where

$$\varepsilon^H_{++} E_+ = i \frac{X_F}{a_F \delta} \int_{-\infty}^{+\infty} d\xi' \, F\left(\frac{|\xi - \xi'|(\xi + \xi')}{2a_F}\right) \exp[i\bar{n}_F(\xi - \xi')] \, E_+(\xi'),$$

$$F(\lambda) = \frac{1}{2} \int d^3u \, \frac{G_0(u)}{|u_\parallel|} \exp\left(\frac{i\lambda}{|u_\parallel|}\right), \quad \bar{n}_i = \frac{\varepsilon_i n_\parallel}{\delta a_i n_c}, \quad X_i = \left(\frac{\omega_{pi}}{\omega}\right)^2,$$

and $G_0$ is the normalized equilibrium distribution function. The subscript F denotes the ion species at the fundamental resonance.

For the second problem of second harmonic resonance we obtain a fourth order integrodifferential equation associated with the existence of both fast and slow modes within the boundary layer. We find

$$\varepsilon^H_{++} E_+ = -i \frac{2X_S \varepsilon^2_S}{a_S \delta} D \int_{-\infty}^{+\infty} d\xi' \, F\left(\frac{|\xi - \xi'|(\xi + \xi')}{a_S}\right) \exp[i\bar{n}_S(\xi - \xi')] \, D'E_+(\xi').$$

The subscript S denotes the ion species at the second harmonic resonance.

The third problem can be expressed as a linear combination of the two previous cases studied above. The nonlocal operator in the field equation now involves two integral terms. We obtain

$$\varepsilon^H_{++} E_+ = -i \frac{2X_S \varepsilon^2_S}{a_S \delta} D \int_{-\infty}^{+\infty} d\xi' \, F\left(\frac{|\xi - \xi'|(\xi + \xi')}{a_S}\right) \exp[i\bar{n}_S(\xi - \xi')] \, D'E_+(\xi')$$

$$+ i \frac{2X_F}{a_F \delta} \int_{-\infty}^{+\infty} d\xi' \, F\left(\frac{|\xi - \xi'|(\xi + \xi')}{2a_F}\right) \exp[i\bar{n}_F(\xi - \xi')] \, E_+(\xi').$$

## III. DISCUSSION AND RESULTS

In the first problem, where the minority ions are at the fundamental resonance in the resonance surface and no other species are resonant, there is no slow mode within the layer. Using the field equation derived, we prove that the transmission coefficient is the same as that obtained for geometrical optics, and that there is no wave reflection for waves incident from the high field side. We solve numerically the field equation obtained for this case, and we show that in the case of low field side incidence the absorption significantly increases by up to 50% as the parameter $a$ approaches two. This increase is due to a decrease in wave reflection. Since there is no reflection for the high field side incident waves, poloidal effects are very small for such incident waves. Figure 1 illustrates the absorption contours for low field side incidence in the plane of $a$ vs. the ratio of $^3$He density to the electron density, with D as the majority ions in a CIT application: $B = 10T$, $k_\parallel = 5m^{-1}$, $T = 5keV$.

In the second problem the majority ion species is at second harmonic resonance within the singular layer. We prove that reciprocity relations of the type found in Ref.1 hold in this case for a pair of related systems, one with $+B_p$, the other with $-B_p$. We are unable to obtain a formula for the transmission coefficient or for the high field side reflection coefficient. For further information about the wave propagation we rely on the numerical investigation. We find that there is a reduction in the transmission rate, and an increase in the low field side incidence reflection rate. However, there is an increase in the energy flux absorption rate mainly due to a reduction in the mode conversion. Figure 2 compares the local and nonlocal absorption for $a = 1$, for the cases of high and low field side incidence, for a CIT example with $B = 10T$, $R = 1.75m$, $T = 10keV$, $n_e = 5 \times 10^{20} m^{-3}$. The resonant ion species is T, with nonresonant D with a 50-50% D to T ratio. We see that the absorption of wave energy flux incident from the high field side is substantially increased. This is particularly true for smaller values of parallel refractive index. The increase in the low field side absorption is somewhat smaller than in the high field side incidence case, because the reflection rate for the nonlocal case is larger than that for the local case. Variation of the propagation parameters as a function of $a$ is shown in Fig.3, for $k_{||} = 1$ m$^{-1}$, or $n_{||} = 0.469$. We do not present results for the case of Bernstein mode incidence since our numerical results are not accurate enough for this case.

In the third problem, where the singular layer contains two resonant species, the majority at the second harmonic and the minority at fundamental resonance. The reciprocity theorems proven before also hold in this case. Again we have found no other results in transmission or reflection coefficients. We rely heavily on numerical analysis for further information about wave propagation. We find that there is again a decrease in transmission coefficient. The gain in the absorption rate, however, depends strongly on the minority ion density which is at the fundamental resonance. There is a significant increase in the absorption rate for small values of minority density. However, this gain is quickly lost when this density becomes approximately 1% of the electron density. Figure 4-6 illustrate the wave propagation characteristics for high and low field incident fast mode for the same CIT example for the ion-ion hybrid resonance, and for $k_{||} = 0, 1, 5m^{-1}$, respectively. The resonant ion species are $^3$He and T, with 50-50% D to T ratio. We give in the same plots the corresponding values obtained for the local problem where the effect of the poloidal field is neglected. The results of the local problem are sketched by thin lines. We use solid line, dotted line, dashed line, and dash-dotted line to denote the transmission, mode conversion, reflection, and absorption, respectively. We see that the high field side reflection is zero within the accuracy of computation. The low field reflection is increased by a factor between 2 and 3, whereas the mode conversion is reduced by almost the same amount. The transmission coefficient is significantly smaller. Almost half of the energy flux of the high field side incident waves is absorbed. There is a more modest increase in the absorption for the low field incidence for small values of the minority density.

In conclusion, when the poloidal field is not tangent to $B = constant$ surfaces, there is an increase in the absorption of wave energy by plasma near the resonance layers for the case of the minority fundamental ICRH and also for the case of the pure second harmonic ICRH. However, if there are ion species within the same resonance layer which are in resonance at the fundamental and second harmonic, then the change in gain in energy absorption depends strongly on the density ratio of these species. We need to extend our calculations covering the parameter domain more extensively in order to reach a more complete understanding of this phenomenon.

## IV. REFERENCES

[1] K. Imre and H. Weitzner, (accepted for publication in Plasma Phys. and Cont. Fusion.) Also NYU Report MF-116 *Wave Propagation across Ion Cyclotron Resonance Harmonic Layers*, August 1987.

[2] K. Imre, H. Weitzner and D.B. Batchelor, *Wave Propagation at Ion Cyclotron Resonance in Large Toroids*. AIP Conference Proceedings No. 159, ed. S. Bernabei and R.W. Motley, p358 (1987). 7th APS Topical Conference on Applications of Radio-Frequency Power to Plasmas, Kissimmee, FL, May 4-6, 1986.

[3] E.F. Jaeger, D.B. Batchelor, K. Imre, H. Weitzner, D.C. Stevens, A. Bers and V. Fuchs, *Full Wave Modeling of Ion Cyclotron Heating in Tokamaks*, (submitted to 12th International Conference on Plasma Physics and Controlled Nuclear Fusion, IAEA November 13-20, 1988 Nice, France).

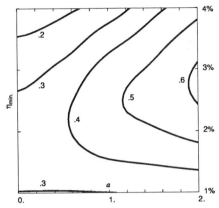

FIG.1 Absorption contours for LFI for minority fundamental ICRH in the plane of $a$ and $n_{He}/n_e$.

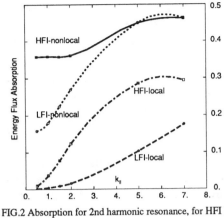

FIG.2 Absorption for 2nd harmonic resonance, for HFI and LFI , and for $a=0$ and 1.

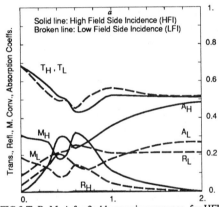

FIG.3 T, R, M, A for 2nd harmonic resonance, for HFI and LFI , for $k_{\parallel}=1m^{-1}$ against $a$.

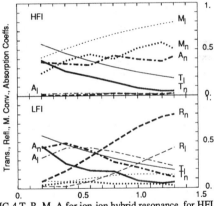

FIG.4 T, R, M, A for ion-ion hybrid resonance, for HFI and LFI: $a=1$, and $k_{\parallel}=0m^{-1}$ against $n_{He}/n_e$.

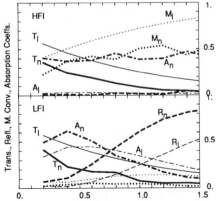

FIG.5 T, R, M, A for ion-ion hybrid resonance, for HFI and LFI: $a=1$, and $k_{\parallel}=1m^{-1}$ against $n_{He}/n_e$.

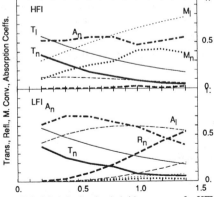

FIG.6 T, R, M, A for ion-ion hybrid resonance, for HFI and LFI: $a=1$, and $k_{\parallel}=5m^{-1}$ against $n_{He}/n_e$.

# COMPARISON BETWEEN THE DIELECTRIC-FILLED RECTANGULAR AND VACUUM FOLDED WAVEGUIDE AS FAST-WAVE LAUNCHERS IN THE ICRF

N. T. Lam, J. E. Scharer and O. C. Eldridge
University of Wisconsin, Madison

## ABSTRACT

In reactor situations, a waveguide may have certain advantages (such as power handling capability and structural rigidity) over loop antennas as fast wave launchers in tokamaks. We carry out a numerical analysis of a dielectric-filled rectangular waveguide, using a formalism that gives the wave reflection coefficient from the plasma (or the equivalent plasma impedance), once the plasma surface admittance tensor and the eigenmodes of the unloaded launcher are known. In the context of a slab model, we obtain the plasma surface admittance tensor by solving the wave differential equations in a cold plasma, assuming a radiation condition from inside the plasma. The eigenfunctions of the rectangular waveguide are readily written down. For the vacuum folded waveguide, semi-analytical expressions for the TE mode eigenfunctions have been obtained by a Ritz variational method. Some problems concerning the calculation of the reflection coefficient for a folded waveguide are discussed.

## INTRODUCTION

For coupling ICRF power into tokamak plasmas, coil antennas have succeeded in producing up to 1.5 MW/antenna and fluxes of 7 kW/cm². In a fusion reactor environment, a waveguide launcher may be more advantageous because of its structural rigidity, compactness and power handling capability. We present a numerical study of a dielectric-filled rectangular waveguide suitable for near-term coupling tests of ICRF heating. Some preliminary results for a vacuum folded waveguide will also be discussed.

## THEORY

Let $x$ = radial direction; $y$ = poloidal direction; $z$ = toroidal direction. In the waveguide, the transverse fields can be written as linear combinations of TE and TM modes, i. e.

$$\vec{E}_T^w = \sum_\ell \vec{E}_\ell(y, z) \left[ A_\ell \exp(i\beta_\ell x) + B_\ell \exp(-i\beta_\ell x) \right] \tag{1}$$

$$\vec{H}_T^w = \sum_\ell D_\ell^w \vec{H}_\ell(y,z) \left[ A_\ell \exp(i\beta_\ell x) - B_\ell \exp(-i\beta_\ell x) \right] \qquad (2)$$

where $\beta_\ell$ = axial propagation constant of the $l$ mode; $\vec{E}_\ell(y,z)$ and $\vec{H}_\ell(y,z)$ = mode transverse fields; $D_\ell^w$ = mode admittance. The TE$_{10}$ mode corresponds to $\ell = 1$, and it is the only mode which can propagate for the waveguide dimensions of interest.

In the slab model, the plasma fields are Fourier-analysed as, e. g.

$$\vec{E}_T^p = \frac{1}{4\pi^2} \int dk_y dk_z \vec{E}_T^p(k_y, k_z, x) \exp[i(k_y y + k_z z)] \qquad (3)$$

Continuities of $\vec{E}_T$ and $\vec{H}_T$ at the interface ($x = 0$) and mode orthogonality yield a formula for the reflection coefficient $\Gamma = B_1/A_1$ in terms of the surface admittance tensor $\overline{\overline{Y}}_p$ defined as

$$\vec{H}_T^p(k_y, k_z, 0) = \overline{\overline{Y}}_p \cdot \vec{E}_T^p(k_y, k_z, 0) \qquad (4)$$

Assume a cold plasma. To estimate $\overline{\overline{Y}}_p$, we use the method of Bers and Theilhaber[1,2] specialized to the case $E_z = 0$ (this approximation is good for the plasma parameters we are considering, since the slow wave is strongly evanescent). Also, the main tensor element turns out to be $Y_{zy} = H_z/E_y$. The differential equations for the transverse fields become a first-order system for $E_y$ and $H_z$. Assuming a radiation condition just before the lower hybrid resonance position ($S = 0$), the differential equations determining $\overline{\overline{Y}}_p$ have been solved numerically by a Runge-Kutta algorithm. Given the reflection coefficient $\Gamma$, one can define an equivalent surface plasma impedance by the formula $Z_p = Z_w(1+\Gamma)/(1-\Gamma)$ with $Z_w$ = waveguide impedance for the TE$_{10}$ mode.

## RESULTS FOR CIT AND TFTR

In table I, we list the power reflection coefficient $|\Gamma|^2$ and plasma impedance for a CIT-like plasma with the following parameters : edge density/center density = 1% ; major (minor) radius = 170 (55) cm ; $B_0 = 10.4$ T; 1:1 D-T plasma with $n_0 = 2.0 \times 10^{14}$ cm$^{-3}$ and 5% He$^3$. To simulate the H-mode, we take the plasma density to consist of a pedestal followed by a region of gaussian variation. The waveguide has a width of 30 cm and is filled with a dielectric of $\epsilon_r = 81$. The heating is at the second harmonic frequency of tritium $f = 95$ MHz ($Z_w = 51.7\ \Omega$). Table I lists the power reflection $|\Gamma|^2$ and plasma impedance for various waveguide heights and pedestal lengths. Up to 10 TE and 10 TE modes have been included included in the calculation. The numerical integration over $k$ space has been carried out from $n_y(n_z) = -40$ to $+40$. Also, the assumption of 1% density at the edge makes $S < 0$ everywhere in the integration region. This eliminates the lower hybrid resonance problem.

Table I $|\Gamma|^2$ and $Z_p$ in CIT for various waveguide heigths

|  | H(cm) | Power reflection coefficient | Plasma impedance ($\Omega$) |
|---|---|---|---|
| pedestal | 30.0 | 0.35 | $48.3 + j73.8$ |
| length | 20.0 | 0.45 | $53.0 + j95.5$ |
| =0.0 cm | 10.0 | 0.65 | $41.0 + j124.3$ |
| pedestal | 30.0 | 0.65 | $33.2 + j111.9$ |
| length | 20.0 | 0.77 | $25.0 + j130.3$ |
| =10.0 cm | 10.0 | 0.87 | $16.5 + j148.9$ |

We see that the $L$ to $H$ mode transition can significantly degrade launcher-plasma coupling. Note the general increase of the reactance and decrease of the resistance as the power reflection increases. Qualitatively, a plot of $\mathcal{RE}(Y_{yz})$ vs. $k_y$ and $k_z$ shows a contraction of the region of propagation in $k$ space upon increasing pedestal length. These results point out the need for a more careful launcher design, to take into account the loading variation in the $L$ to $H$ mode transition.

Next we present calculations for the supershot mode in TFTR. The density has both parabolic and gaussian components while the magnetic flux surfaces undergo a large Shafranov shift = 30 cm. Specifically, we consider a plasma with the following parameters : major (minor) radius = 262 (96) cm; $B_0$ = 5.2 T; $D(He^3)$ composition with $n_0 = 1.2 \times 10^{14}$ cm$^{-3}$ and $He^3$ = 5% electron density; heating at the first minority harmonic $f$ = 47 MHz;edge density/center density = 20%. The waveguide is placed 5 cm from the plasma "edge" and its width is set at 60 cm. The region of integration in $k$ space is again chosen to extend from $n_y(n_z)$ = - 40 to + 40. The starting point for the integration is set at x = 32 cm. Table II lists $|\Gamma|^2$ and $Z_p$ for various waveguide heights. Note the low power reflection coefficients in all cases. This is mostly due to the relattive high density at the plasma edge.

Table II $|\Gamma|^2$ and $Z_p$ in TFTR for various waveguide heigths

| H(cm) | Power reflection coefficient | Plasma impedance ($\Omega$) |
|---|---|---|
| 60.0 | 0.22 | $19.8 + j12.6$ |
| 30.0 | 0.18 | $23.2 + j14.8$ |
| 10.0 | 0.15 | $35.1 + j31.2$ |
| 5.0 | 0.22 | $57.9 + j58.6$ |

## FOLDED WAVEGUIDE

The vacuum folded waveguide has been proposed by Owens[3] as an alternative to ridged waveguides and coil antennas in the ICRF. A 30 cm × 70 cm folded
waveguide with 9 fins has a cut-off frequency of about 60 MHz and can operate
in a frequency range suitable for CIT. Using the Ritz variational method, we
have written a code[4] which gives accurate estimates of the eigenvalues of the TE
modes of such a waveguide. To achieve this accuracy, the eigenfunctions must be
expreesed in terms of a very large number of basis functions. This means that a
calculation of the reflection coefficient using a reasonably large admittance matrix would require a large amount of computer time. So far, this has precluded
a reliable estimate of $Z_p$ for the folded waveguide. Using a sparse admittance
matrix, we have obtained preliminary results which will be reported elsewhere.

## ACKNOWLEDGMENTS

This work is supported by DOE grant DE-FG02-86ER53218.

## REFERENCES

1. A. Bers and K. Theilhaber., Nucl. Fusion **23**, 41 (1983).

2. N. T. Lam, J. L. Lee, J. Scharer and R. J. Vernon, IEEE Trans. Plasma
Science **PS-14**, 271 (1986).

3. T. L. Owens, IEEE Trans. Plasma. Science **PS-14**, 934 (1986)

4. E. J. Sigal, M. S. thesis, University of Wisconsin, unpublished (1987)

GYROKINETIC THEORY OF PERPENDICULAR ION CYCLOTRON ABSORPTION OF
THE FAST WAVE AT THE SECOND HARMONIC IN A NON UNIFORM MAGNETIC
FIELD

C.N. Lashmore-Davies and R.O. Dendy
Culham Laboratory, Abingdon OX14 3DB, U.K.
(UKAEA/EURATOM Fusion Association)

ABSTRACT

The propagation of the fast wave perpendicular to a
non-uniform magnetic field and through the second harmonic
resonance is analyzed. Gyrokinetic theory enables the problem to
be treated self-consistently so that the plasma response
explicitly includes the effect of the non-uniform magnetic field.
A straight magnetic field with a perpendicular gradient in
strength, $\underline{B} = B \underline{e}_z(1 + x/L_B)$, is used. The inclusion of the
variation of the magnetic field across an ion Larmor orbit leads
to a new perpendicular cyclotron damping mechanism. Instead of a
wave resonance, leading to mode conversion to the ion Bernstein
wave, there is now a wave-particle resonance which results in
direct ion dissipation. The optical depth resulting from this
absorption is shown to be the same as the corresponding quantity
obtained from a mode conversion analysis. Another feature which
emerges naturally from the gyrokinetic calculation is the kinetic
power flow. This is a reversible power flow which enables the
variation of the electromagnetic fields across the resonance
region to be calculated.

INTRODUCTION

In this paper we employ the arbitrary frequency gyrokinetic
theory of Chen and Tsai[1],[2] to analyze the propagation of the
compressional Alfvén wave (the fast wave) across the second
harmonic resonance in a hot plasma containing only one ion
species. The absorption of the fast wave is due to the combined
effect of the finite Larmor radius of the ions and the gradient of
the non-uniform magnetic field. Gyrokinetic theory provides a
systematic method for obtaining a self-consistent solution for the
particle response in an arbitrary, non-uniform magnetic field. In
particular, it includes the particle response to the variation of
the equilibrium magnetic field across their Larmor orbits.

We shall restrict the analysis to the case of the fast wave
propagating exactly perpendicular to the equilibrium magnetic
field and into the gradient in field strength. For this
calculation we adopt a specific model for the magnetic field of
the form

$$\underline{B} = \underline{e}_z B(1 + x/L_B).$$

## GYROKINETIC FORMULATION

Following Chen and Tsai[1,2], we start with the linearized Vlasov equation in particle phase space $(x, \underline{v})$:

$$[\partial/\partial t + \underline{v} \cdot \underline{\nabla}_x + (q/mc)(\underline{v} \times \underline{B}) \cdot \underline{\nabla}_v] \delta f = -(q/mc)(c\delta\underline{E} + \underline{v} \times \delta\underline{B}) \cdot \underline{\nabla}_v F \qquad (1)$$

Here $\delta f$ and $F$ are the perturbed and equilibrium distribution functions; $\delta\underline{E}$, $\delta\underline{B}$ are the perturbed electric and magnetic fields. The arbitrary frequency gyrokinetic equation[1,2] is obtained by transforming Eq.(1) to guiding centre phase space $(\underline{X}, \underline{V})$ where $\underline{X} = \underline{x} + \underline{v} \times \underline{e}_{\parallel}/\Omega$ and $\underline{V} = (\epsilon, \mu, \alpha)$. Here $\epsilon = v^2/2$, $\mu = v_{\perp}^2/2B$, $\underline{e}_{\parallel} = \underline{B}/B$, $\Omega = eB/mc$, $\alpha$ is the gyrophase angle defined by $\underline{v}_{\perp} = v_{\perp}(\underline{e}_1 \cos \alpha + \underline{e}_2 \sin \alpha)$ and $\underline{e}_{\parallel}$, $\underline{e}_1$ and $\underline{e}_2$ are local orthonormal vectors. We refer to Refs.1 and 2 for the derivation of the arbitrary frequency gyrokinetic equation

$$<L_g>_{\ell} <\delta H_g>_{\ell} \simeq i(q/m)[\omega(\partial F_{go}/\partial \epsilon) + (\ell\Omega/B)\partial F_{go}/\partial \mu]<\delta\psi_g>_{\ell} \qquad (2)$$

Here $<\hat{L}_g>_{\ell} = (v_{\parallel}\underline{e}_{\parallel} + \underline{v}_d) \cdot \underline{\nabla}_X - i(\omega - \ell\Omega + \ell\omega_{\alpha})$ and the subscript g refers to guiding centre coordinates. Note that magnetic field inhomogeneity enters $<\hat{L}_g>_{\ell}$ explicitly through the spatial dependence of $\Omega$. The quantities $v_{\parallel}$ and $\omega_{\alpha}$ are defined in Refs. 1 and 2 but are not required here and $\underline{v}_d$ is the equilibrium drift due to the magnetic field inhomogeneity. In obtaining Eq. (2), the perturbed quantities have been expanded as Fourier series in $\alpha$, which enters through the guiding centre transformation. Thus, $<\delta H_g>_{\ell} = (2\pi)^{-1} \int_0^{2\pi} d\alpha \, \delta H_g(X, \mu, \epsilon, \alpha) \exp(i\ell\alpha)$, where $\delta H_g$ is proportional to the perturbed distribution function[1,2]. It will be shown that the gyrophase information is important in the cyclotron resonance region. The inclusion of this information is an important feature of gyrokinetic theory, in contrast to locally uniform models where a gyrophase average is carried out before cyclotron resonance is considered. The quantity $<\delta\psi_g>_{\ell} = <\delta\psi_g - \underline{v} \cdot \delta\underline{A}_g/c>_{\ell}$ where

$\delta\underline{B} = \underline{\nabla}_x \times \delta\underline{A}$, $\delta\underline{E} = -\underline{\nabla}_x \delta\psi - (\partial/\partial t) \delta\underline{A}/c$ and $\underline{\nabla}_x \cdot \delta\underline{A} = 0$. At high frequencies $(\omega \sim \ell\Omega)$, only the zero order solution of Eq.(2) is required[1,2]. The coupling to neighbouring harmonics has, accordingly, been neglected[1,2], as have terms on the right hand side of Eq.(2) that arise from diamagnetic drift corrections. However, the operator $<L_g>_{\ell}$ retains $O(\rho/L_B)$ terms, which are responsible for the broadening of the resonance.

We now represent all perturbed quantities by the single mode form $\delta f(x) = \delta f_k \exp(ikx)$ corresponding to propagation into the equilibrium field gradient. Since $k_y = 0$, the term $\underline{v}_d \cdot \underline{\nabla}_X$ in $<L_g>_{\ell}$ is zero. Noting that $<\delta\psi_g - \underline{v} \cdot \delta\underline{A}_g/c>_{\ell}$ is evaluated with $\underline{X}$ constant, Eq.(2) yields

$$<\delta H_g>_\ell = (q/m)<\delta\psi_g>_\ell\{\omega\partial F_{go}/\partial\epsilon+(\ell\Omega/B)\partial F_{go}/\partial\mu\}/\{\ell\Omega(X)-\omega\} \quad , \quad (3)$$

where

$$<\delta\psi_g>_\ell = \exp(ikX)\{J_\ell(kv_\perp/\Omega)[\delta\phi_k - v_\parallel \delta A_{\parallel k}/c]-(v_\perp/ck)J_\ell'(kv_\perp/\Omega)\delta B_{\parallel k}\}$$

In order to calculate the perturbed current density for use in Maxwell's equations we must transform the guiding centre distribution function given by Eq.(3) back into particle coordinates to obtain $\delta f_k$. We obtain, assuming a Maxwellian equilibrium velocity distribution function,

$$\delta f_k = -\frac{2n_0 q \exp(-V^2)}{m \pi^{3/2}v_T^5}[\delta\phi_k - <\delta\psi>_{0k} + \sum_{\ell\neq 0}\frac{\omega L_B<\delta\psi>_{\ell k}\exp(-i\ell\alpha)}{\ell v_T (V_y - \varsigma_\ell)}] \quad (4)$$

$$<\delta\psi>_{\ell k}=\exp(ik\rho V_y)[(\delta\phi_k-(v_\parallel/c)\delta A_{\parallel k})J_\ell(k\rho V_\perp)-(v_\perp/ck)\delta B_{\parallel k}J_\ell'(k\rho V_\perp)], \quad (5)$$

where $\varsigma_\ell \equiv L_B[\omega - \ell\Omega(x)]/\ell v_T$. Note that in Eqs.(4) and (5) the velocity coordinates will eventually be expressed in terms of the normalized Cartesian variables $(V_x,V_y,V_z)$, where $V_x = V_\perp\cos\alpha$, $V_y = V_\perp\sin\alpha$, $V_z = v_\parallel/v_T$ and $V_\perp = v_\perp/v_T$.

Equations (4) and (5) are the key results of the gyrokinetic analysis, since they enable the currents producing the self-consistent electromagnetic fields to be calculated. These equations are valid for large $k\rho$, subject only to $k\rho << L_B/\rho$.

## SECOND HARMONIC ABSORPTION

Let us now consider the propagation of the fast wave across the second harmonic resonance in a single ion species plasma. In order to obtain the dispersion relation we must use Eqs.(4) and (5) to calculate the perpendicular resonant currents. Assuming $k\rho << 1$ we can obtain analytic expressions for these currents.[3] We also need the perpendicular currents due to the non-resonant ions and the electrons. These are obtained in the usual way from the locally uniform model. Substituting into Poisson's equation, we obtain

$$\{1 - \frac{c^2}{3c_A^2} + \frac{c^2}{8c_A^2} kL_B G_x(\eta_2)\exp(-k^2\rho^2/4)\}k^2\delta\phi_k$$

$$= -\frac{4}{3}\frac{c}{c_A}\frac{\Omega}{c_A}\{1 - \frac{3}{16} kL_B G_x(\eta_2)\exp(-k^2\rho^2/4))\}\delta B_{\parallel k} \quad (6)$$

Similarly, we obtain from Maxwell's equations

$$\{k^2 - \frac{\omega^2}{c^2} + \frac{4\Omega^2}{c_A^2} - \frac{i}{2}\frac{\omega_p^2}{c^2} kL_B G_y(\eta_2)\exp(-k^2\rho^2/4)\}\delta B_{\parallel k}$$

$$= \frac{4}{3}\frac{c}{c_A}\frac{\Omega}{c_A}\{1 - \frac{3i}{16} kL_B G_y(\eta_2)\exp(-k^2\rho^2/4)\}k^2\delta\phi_k \quad (7)$$

where $\quad G_x(\eta_2) = -i[1 + \eta_2 Z(\eta_2)] + \frac{k\rho}{2} Z(\eta_2) \quad (8)$

$$G_y(\eta_2) = [-\eta_2^3 - i\frac{3}{2}k\rho\eta_2 + \frac{1}{2} + \frac{3}{4}k^2\rho^2][1 + \eta_2 Z(\eta_2)]$$

$$+ i\frac{k\rho}{4} Z(\eta_2) - \frac{1}{2} \qquad (9)$$

and $\eta_2 = \zeta_2 - ik\rho/2$. Equations (6) and (7) describe the propagation of the fast wave in the vicinity of the second harmonic resonance. They also contain the ion Bernstein wave. The dispersion relation resulting from Eqs.(6) and (7) is

$$c_A^2 k^2/\omega^2 =$$

$$\frac{[1 - \frac{i}{8}kL_B\{(-\eta_2^2 - i\frac{3}{2}k\rho_2 - \frac{1}{2} + \frac{3}{4}k^2\rho^2)[1 + \eta_2 Z(\eta_2)] - i\frac{k\rho}{4}Z(\eta_2) - \frac{1}{2}\}\exp(-k^2\rho^2/4)]}{[1 + i\frac{3}{8} kL_B\{1 + \eta_2 Z(\eta_2) + i\frac{k\rho}{2} Z(\eta_2)\}]} .$$

$$(10)$$

Far from the second harmonic resonance, where $\eta_2 \gg 1$, Eq.(10) reduces to the uniform plasma dispersion relation. In the vicinity of the second harmonic resonance, we seek a perturbation solution of Eq.(10), assuming $k = \omega/c_A + \delta k$ where $\delta k$ is a thermal perturbation to the cold plasma solution. The imaginary part is

$$\text{Im } \delta k = (\omega^2 L_B/32c_A^2)\{1 + (2\zeta_2^2 - 5)[1 + \zeta_2 Z_r(\zeta_2) - (\omega\rho/c_A)\zeta_2^2 Z_i(\zeta_2)]\} \quad (11)$$

where we have expanded $Z(\zeta_2 - ik\rho/2)$ assuming $\zeta_2 \gg k\rho/2$. The terms in Eq.(11), proportional to the real part of the plasma dispersion function $Z_r$ come from the kinetic power flow which is reversible and does not contribute to the total absorption. The optical depth for the second harmonic resonance is obtained from Eq.(11) by evaluating $\tau = 2\int_{-\infty}^{\infty} \text{Im } \delta k(x)dx$ where only the term proportional to the imaginary part of the plasma dispersion function, $Z_i$, contributes, giving $\tau = (\pi/4)(\omega L_B/c_A)(v_A^2/c_A^2)$ which is the standard[4] optical depth for the second harmonic resonance. However, in contrast to the previous result, where the power lost by the fast wave is due to mode conversion to the ion Bernstein wave, here we have obtained a direct ion dissipation mechanism due to the variation of the magnetic field across the ion Larmor radius. Equation (11) yields a smooth absorption profile which is influenced by the kinetic power terms since these terms effect the variation of the electromagnetic fields across the resonance region. In conclusion, we have shown how gyrokinetic theory of perpendicular ion cyclotron resonance leads to a perpendicular ion cyclotron dissipation mechanism.

## REFERENCES

1. L.Chen and S.-T.Tsai, Phys. Fluids 26, 141 (1983)
2. L.Chen and S.-T.Tsai, Plasma Phys. 25, 349 (1983)
3. X.S.Lee, J.R.Myra and P.J.Catto, Phys. Fluids 26, 223 (1983)
4. D.G.Swanson, Phys. Fluids, 28. 2645 (1985)

# ICRF EDGE MODELING STUDIES

Ira S. Lehrman
Grumman Corporate Research Center, Princeton, NJ 08540

Patrick L. Colestock
Princeton Plasma Physics Laboratory, Princeton, NJ 08543

## INTRODUCTION

Ion Cyclotron Range of Frequencies (ICRF) heating is one of the current technologies being investigated for heating plasmas to ignition conditions that has shown promise with its favorable experimental results and its ability to penetrate into the core of high density plasmas.[1,2] Although promising, ICRF heating is not without its own set of complications. Often accompanying the desired ion heating is a number of undesired side effects which, if not controlled, could limit the usefulness of ICRF heating in reactor-sized devices. The most notable effects are an increase in the impurity concentration, and an uncontrolled rise in the plasma density.[3]

In this paper, edge plasma models are presented that attempt to explain the behavior observed during ICRF heating experiments. Since experimental measurements have identified energized antennas as significant sources of impurities and deuterium,[4] the models presented calculate the particle transport in the vicinity of the antenna. In addition, kinetic modifications to the edge plasma and their implication on material sputtering from the Faraday shield are investigated. The electric fields produced by the antenna are computed using the ANDES[5] and ORION[6] codes. These fields are then used in conjunction with the models developed in this paper.

## STRUCTURE OF THE ANTENNA NEAR FIELD

The metallic boundary condition on the surface of the Faraday shield requires that the tangential electric field be identically zero. Since the Faraday shield is not expected to greatly reduce the power radiated from the antenna, the electric field must, on average, be unchanged with the addition of the shield. Therefore, the arrangement of the shield elements acts to geometrically magnify the electric field in the gap region while suppressing the field in the region of the blades. For the shields employed in the PLT experiments, the gaps between shield blades were 0.5 cm, while the shield blades were 1 cm in height (poloidal extent). Hence the electric field in the gap region is three times larger than the average electric field. The complete field structure for a 250 V/cm electric field incident on the PLT Faraday shield is shown in Fig. 1.

## FLUID MODEL OF THE EDGE PLASMA

The ripple in the antenna near field results in strong ponderomotive force ($F_p \propto \nabla |E|^2$) that acts mainly on the ions. In order to obtain a self-consistent model for the edge plasma the ponderomotive force is incorporated into the ion fluid equations. The equation for the ion velocity is:

$$V_i^{(2)} = \frac{q}{m_i B_0^2} \left[ E^{(2)} \times B_0 + S_i \times B_0 \right] - \frac{\nabla P_i \times B_0}{m_i n_0 B_0^2},$$

The non-linear source term $S_i$ is the ponderomotive force (multiplied by $q/m_i$). The electric field $E^{(2)}$ is evaluated by requiring that electrons follow Boltzmann's relation. From the solution of the velocity and continuity equations, the ion flux to the Faraday shield is evaluated:

$$\Gamma_x = - n_0 \, V_{ix}^{(2)} - D_\perp \nabla \left( n_i^{(2)} + n_0 \right) .$$

The Fluid model was applied to the edge conditions of PLT. Figure 2 shows the model results for the particle flux to the shield versus rf power. These results appear to be consistent with the $D_\alpha$ emission measurements which also exhibit a linear increase with applied power.[4] The fluid model predictions are about 60% lower than what is experimentally observed. Unfortunately, the mechanisms for the release of deuterium in the energy range of the edge ions is not well understood, therefore we conclude that the fluid model is not unrealistic for explaining the increased release of deuterium during ICRF heating. The fluid model does however fail to explain the increase in the metallic release during ICRF heating which is larger that the corresponding increase in deuterium.

## KINETIC MODEL OF THE IONS

The time evolution of the energy of ions which transit close to the gap region of the Faraday shield is found from the velocity squared:

$$V^2 = V_{\tilde{y}}^2 + V_{\tilde{x}}^2$$

$$= V_0^2 + \frac{2 \, q V_0 \, E_{yo}}{m_i \, (\omega_{ci}^2 - \omega^2)} \left[ \omega_{ci} \sin(\omega t + \phi) \sin(\omega_{ci} t) + \omega \cos(\omega t + \phi) \cos(\omega_{ci} t) \right]$$

$$+ \frac{q^2 \, E_{yo}^2}{m_i^2 \, (\omega_{ci}^2 - \omega^2)^2} \left[ \omega_{ci}^2 \sin^2(\omega t + \phi) + \omega^2 \cos^2(\omega t + \phi) \right] .$$

the first term, $V_0$, is the initial ion velocity. The second term contains oscillatory quantities which time average to zero, while the time average of the last term is non-zero. From the equation for $V^2$, the time-averaged energy is obtained:

$$\frac{1}{2} m_i \langle V^2 \rangle = \frac{1}{2\pi} \int_0^{2\pi} \frac{1}{2} m_i V^2 \, d\phi = \frac{1}{2} m_i V_0^2 + \frac{q^2 \, E_{yo}^2 \, (\omega_{ci}^2 + \omega^2)}{2m_i \, (\omega_{ci}^2 - \omega^2)^2} .$$

The average energy of the ion is now equal to its initial energy plus an additional term known as the quiver energy. The quiver energy represents stored (reactive) energy which is given back as wave energy when the rf is turned off, that is unless the particle suffers a collision or impacts the shield. For the shields employed in the PLT experiments, the factor of three enhancement in the near field results in a factor of nine increase in the quiver energy.

The rf acts to broaden the distribution of particles in velocity space with the net result being an increase in the average energy of the distribution. Figure 3a shows the distribution that results for an initial 10 eV Maxwellian distribution of ions in an rf field of 750 V/cm. The average energy of the rf distribution is found to be 21.8 eV, which is close to the initial energy (10 eV) plus the quiver energy (12.6 eV). The probability distribution function, $V_\perp f(V_\perp)$, is shown in Fig. 3b. This increase in ion energy results in a significant increase in the sputtering yield of ions which impact the shield.[7]

## KINETIC MODEL OF THE ELECTRONS

Experimental measurements indicated that in flux tubes that lie close to the surface of the Faraday shield, electron heating occurs.[4] To explain the observed heating, electron trajectories are followed past the antenna. The distribution of electrons is then computed. Since the perpendicular (poloidal and radial) components of the electric field are expected to negligibly affect the electron energy (quiver energy $\ll kT_e/e$), the effect on electrons from $E_{\parallel}$ is only investigated. The parallel electric field is determined from the ORION code (Fig 4). The toroidal variation of $E_{\parallel}$ could not be determined from this code, therefore it is modeled as a gap field of the form:

$$E_{\parallel}(z) = \frac{E_{\parallel 0}}{2}\left[ \tanh\left(\frac{z-l/2}{s}\right) - \tanh\left(\frac{z+l/2}{s}\right)\right],$$

where $s$ is a shape factor and $l$ is equal to the toroidal projection of the field line between the shield elements. Shape factors of 0.1, 1 and 2 cm were investigated.

To model the electron heating observed in the PLT experiments, the electron distributions were computed for a range of spatial positions. From the resulting distributions, the average parallel electron energies were computed. The measured change in $T_e$ is plotted in Fig. 5 along with the predicted change in $T_e$ for the three different shape factors. The agreement suggests that $E_{\parallel}$ induced at the shield surface was responsible for the observed electron heating.

## DISCUSSION

The shield structure of the ICRF antenna is found to magnify the field in the gap region between shield elements while inducing ripple in the near field. This ripple results in ponderomotive force which increases the particle transport to the shield. Ion kinetic modeling indicates that the inductive near field acts to increase the energy of ions which impact the shield. The electron kinetic modeling suggests that $E_{\parallel}$ induced as a result of fast was excitation is responsible for the observed electron heating.

[1]P. L. Colestock, IEEE Trans. Plasma Sci. **PS-12** (1984) 64.

[2]K. Steinmetz, "ICRF Heating in Fusion Plasmas," in Applications of Radio-Frequency Power to Plasmas, Kissimmee, FL, Editors: S. Bernabei and R. W. Motley, AIP: NY (1987) 211.

[3]P. L. Colestock, *et al.*, J. Vac. Sci. Technol. **A3** (1985) 1211.

[4]I. S. Lehrman, P. L. Colestock, D. H. McNeill, *et al.*, Princeton Plasma Physics Lab. report PPPL-2607 (April 1989), also submitted to Plasma Phys. Controll. Fusion.

[5]I. S. Lehrman and P. L. Colestock, IEEE Trans. Plasma Sci. **PS-15** (1987) 285.

[6]E. F. Jaeger, D. B. Batchelor, and H. Weitzner, "Global ICRF Wave Propagation in the Edge Plasma and Antenna Regions with Finite $E_{\parallel}$," to appear in Nucl. Fusion.

[7]I. S. Lehrman, "A Study of Coupling and Edge Processes for ICRF Antennas," Ph.D dissertation, Univ. of Wisconsin-Madison (1988).

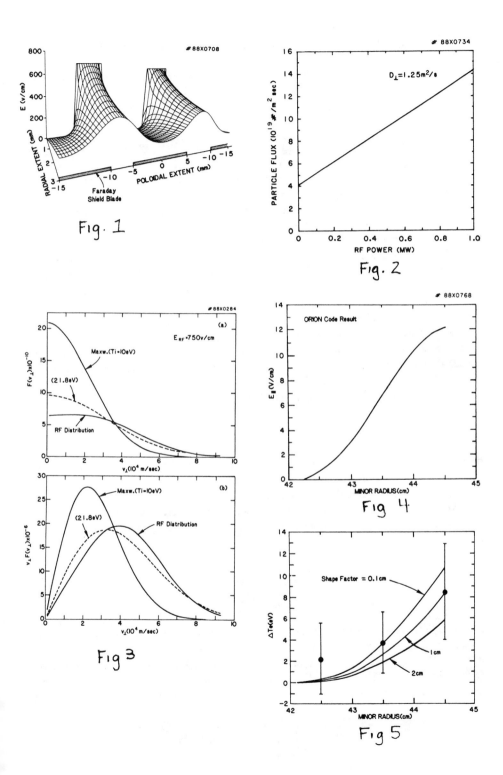

# 88X0708

E (v/cm)

Fig. 1

Faraday Shield Blade

RADIAL EXTENT (mm)

POLOIDAL EXTENT (mm)

# 88X0734

$D_\perp = 1.25 m^2/s$

PARTICLE FLUX ($10^{19}$ #/m² sec)

RF POWER (MW)

Fig. 2

# 88X0284

(a)

$E_{RF} = 750 v/cm$

Maxw.(Ti=10eV)

(21.8eV)

RF Distribution

$F(v_\perp) \times 10^{-10}$

$v_\perp (10^4 m/sec)$

(b)

Maxw.(Ti=10eV)

(21.8eV)

RF Distribution

$v_\perp F(v_\perp) \times 10^{-6}$

$v_\perp (10^4 m/sec)$

Fig 3

# 88X0768

ORION Code Result

$E_\parallel (V/cm)$

MINOR RADIUS (cm)

Fig 4

Shape Factor = 0.1cm

1cm

2cm

$\Delta Te(eV)$

MINOR RADIUS(cm)

Fig 5

# 60 MHz FAST WAVE CURRENT DRIVE EXPERIMENT FOR DIII-D*

M.J. Mayberry, S.C. Chiu, M. Porkolab,[†] V. Chan, R. Freeman, R. Harvey, R. Pinsker

General Atomics, San Diego, CA 92138

## INTRODUCTION

Non-inductive current drive is an essential element of the ITER program because it enhances high fluence nuclear testing during the technology phase of operations. By using fast waves in the ion cyclotron range of frequencies (ICRF), current drive efficiencies comparable to lower-hybrid current drive can be obtained with good penetration of wave power to the high temperature plasma core.[1] An additional advantage of the low frequency scheme is its technological simplicity due to the present availability of efficient, multi-megawatt rf sources in the ICRF.

The DIII-D facility provides an excellent opportunity to test the feasibility of the low frequency FWCD approach. By combining with high power (2 MW) ECH injection at 60 GHz, it should be possible to generate plasmas with central electron temperatures of $T_{e0} \simeq 4\,\text{keV}$, and by operating at a reduced toroidal field (B = 1 T) to increase the electron $\beta$, strong single-pass absorption ($\eta_{\text{abs}} \geq 0.3$) can be achieved. The availability of a wide port recess (1 m toroidal by 0.5 m poloidal) will enable a travelling wave spectrum to be launched with $N_{\|} \simeq 5\text{--}7$ at 60 MHz, which should be optimum for strong electron interaction. The resulting current drive efficiency should be sufficiently high to demonstrate FWCD at the $\sim 0.25 - 0.5\,\text{MA}$ level at moderate densities ($\bar{n} \simeq 1.3 \times 10^{19}\,\text{m}^{-3}$) using the existing 2 MW ICRF transmitter.

## FW ABSORPTION

In the present scheme, efficient FWCD is achieved by direct electron absorption due to Landau damping and transit-time magnetic pumping (TTMP). To avoid competing damping mechanisms such as ion absorption at cyclotron harmonics, mode mixing between the FW and the slow wave, or mode conversion to ion Bernstein waves at cyclotron harmonics,[2] we seek to maximize the single-pass absorption of the FW by electrons. For a Maxwellian electron distribution, the e-folding damping length at low frequencies $[(\omega/\omega_{pi})^4 (m_e c^2/T_e)^2 \ll 1]$ is given by:

$$\lambda_e^{-1} = 2k_{\perp\,Im} \simeq k_{\perp\,Re} \frac{\sqrt{\pi}}{2} \beta_e \xi_e e^{-\xi_e^2} \quad , \tag{1}$$

where $k_{\perp Re} \simeq \omega/v_A$, $v_A = c\omega_{ci}/\omega_{pi}$, $\beta_e = 2\mu_0 n_e T_e/B^2$, $\xi_e = \omega/k_{\|} v_{te}$, and $v_{te} = (2T_e/m_e)^{1/2}$. From the above expression it can be seen that the FW absorption increases directly with the electron beta, and decreases rapidly with magnetic field since $\lambda_e^{-1} \propto B^{-3}$. Equation (1) also shows that the fast wave absorption is maximized when $\xi_e \simeq 1$. For a $T_e = 4\,\text{keV}$ plasma, this condition corresponds to a parallel index of refraction, $N_{\|} = ck_{\|}/\omega = [256/T_e(\text{keV})]^{1/2} \simeq 8$. For the planned experiments with B = 1 T

---

* Work supported by U.S. DOE Contract DE-AC03-89ER52153.

† Massachusetts Institute of Technology, Cambridge, MA, 02139

and 60 MHz, the central ion resonance is $\omega = 8\omega_D$ for pure deuterium plasmas, and the ion damping is predicted to be relatively weak compared to the electron damping.

## FWCD MODELLING RESULTS

At the low phase velocities where the FW absorption is maximized, the absorption of rf power by trapped electrons can lead to a significant reduction in the FWCD efficiency. The FWCD efficiency including these trapping effects has been calculated numerically for a slab geometry using an adjoint technique with a momentum conserving collision operator.[3] The density and temperature profiles used in these calculations were: $n_e(r) = n_{e0}(1 - r^2/a^2)^{\alpha_n}$, $T_e(r) = T_{e0}(1 - r^2/a^2)^{\alpha_T}$, where $\alpha_n = 0.5$, and $\alpha_T = 5$. These are consistent with recent 60 GHz ECH results obtained with outside launch antennas and an O–mode polarization.[4] In those experiments, a central electron temperature $T_{e0} = 5\,keV$ was achieved with $\bar{n}_e = 0.9 \times 10^{13}\,cm^{-3}$, $B = 2.1\,T$, and $P = 1.0\,MW$. Recent analysis[5] of ECH results to date indicates that $\tau_E \propto B^{-0.75}$. Therefore, at $B = 1.0\,T$, we expect that $T_{e0} = 4\,keV$ should be obtainable with $4\,MW$ of combined ECH and FW heating power at a density of $\bar{n}_e = 1.3 \times 10^{13}\,cm^{-3}$ (below ECH cut-off) as long as the peaked profiles can be maintained.

Our calculations indicate that with a central electron temperature of $4\,keV$, complete sustainment of the plasma current with rf at the $0.5\,MA$ level should be possible if multiple-pass absorption by electrons occurs. The modelling results are shown in Figs. 1–4. In Figs. 1 and 2, the fractional single-pass absorption, $\eta_{abs}$ and driven current $I_{rf}$, are plotted as a function of the peak electron temperature, assuming a fixed value of $N_{\parallel} = 7$, and a line-averaged density of $\bar{n}_e = 1.3 \times 10^{13}\,cm^{-3}$. The calculated rf current depends on whether single-pass or multiple-pass absorption is assumed. Under the assumption of single-pass absorption, the current drive efficiency is a strong function of temperature; a $7\,keV$ plasma would be required for $0.5\,MA$ of current. However, multiple-pass absorption may be acceptable as long as we consider only a few reflections. In this case, $0.5\,MA$ of current should be achievable with a $4\,keV$ plasma.

A key factor in the design of the FWCD antenna is the $N_{\parallel}$ spectrum excited by the launching structure. Figures 3 and 4 show plots of $\eta_{abs}$ and $I_{rf}$ vs. $N_{\parallel}$ for a fixed central temperature of $4\,keV$ and the same parameters as earlier. The single-pass absorption approaches its maximum value at $N_{\parallel} \simeq 8$, which corresponds very closely to $\xi_e \simeq 1$. Under the assumption of single-pass damping, the maximum in $I_{rf}$ occurs at $N_{\parallel} \simeq 6$. At higher values of $N_{\parallel}$ the CD efficiency is reduced (even though absorption increases) because of the increased severity of trapping effects due to off-axis damping. At lower values of $N_{\parallel}$, the CD efficiency is reduced because of weak single-pass absorption. If multiple-pass damping occurs, however, then the optimum $N_{\parallel}$ value is reduced.

For the planned experiments on DIII–D, a four-loop antenna array will be installed in a $1\,m$ wide toroidal port recess on the outer midplane of DIII–D. The calculated spectrum for this antenna is shown in Fig. 5. This calculation[6] takes into account the plasma density profile (here assumed to be an L–mode profile) as well as realistic features of the antenna geometry including the shielding plates between adjacent current straps. For $90°$ antenna phasing, the primary peak in the antenna spectrum occurs at $N_{\parallel} \simeq 5$, which is very close to the optimum value; the secondary peak occurs at $N_{\parallel} \simeq -15$. The good antenna directionality (80%) is due in part to the presence of an evanescent region between the antenna and the FW cut-off layer which attenuates the higher-$N_\parallel$ spectral components. Further details of the electrical design of the FWCD antenna and the associated impedance matching circuitry are presented in a separate paper at this conference.[7]

## CONCLUSION

By using ECH to preheat the electrons, it should be possible to obtain strong single-pass ($\sim 30\%$) absorption of low frequency fast waves by electrons using a high-$N_\parallel$ phased-array of loops in DIII–D. Even when potentially deleterious trapping effects are taken into account, the FWCD efficiency at 60 MHz may be sufficiently high in DIII–D to demonstrate fully rf-driven plasmas at the 0.5 MA level, thus providing a convincing demonstration of the low frequency FWCD approach for ITER.

## REFERENCES

[1] D.A. Ehst, Argonne National Lab. Report ANL/FPP/TM-219, 1988.

[2] S.C. Chiu, et al., this conference.

[3] S.C. Chiu, et al., General Atomics Report GA–A1954, submitted to Nucl. Fusion.

[4] R. Prater, et al., in Proc. 12th Int. Conf. Plasma and Cont. Nucl. Fus. (Nice), 1988.

[5] B. Stallard, et al., General Atomics Report GA–A19349, to be published.

[6] S. Kinoshita, in Proc. IEEE 12th Symp. on Fusion Eng. (Monterey), Vol. II, p. 1374, 1987.

[7] F.W. Baity, et al., this conference.

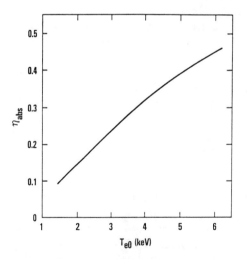

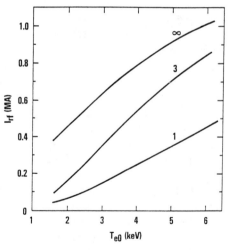

**Fig. 1.** Calculated single-pass absorption, $\eta_{abs} = P_{abs}/P_{in}$, as a function of the central electron temperature $T_{e0}$ for $f = 60\,\text{MHz}$, $B = 1\,\text{T}$, $\bar{n}_e = 1.3 \times 10^{13}\,\text{cm}^{-3}$, $N_\parallel = 7$.

**Fig. 2.** Calculated rf-driven current, $I_{rf}$, as a function of $T_{e0}$ for $P_{rf} = 2\,\text{MW}$; other parameters same as Fig. 1. Various curves correspond to single-pass and multiple-pass absorption, with the number of passes labeled.

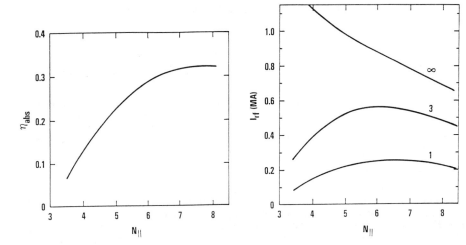

**Fig. 3.** Calculated single-pass absorption as a function of $N_\parallel$ for $T_{e0} = 4\,\text{keV}$; other parameters same as Fig. 1.

**Fig. 4.** Calculated rf-driven current as a function of $N_\parallel$ for $T_{e0} = 4\,\text{keV}$ and $P_{rf} = 2\,\text{MW}$; other parameters same as Fig. 1.

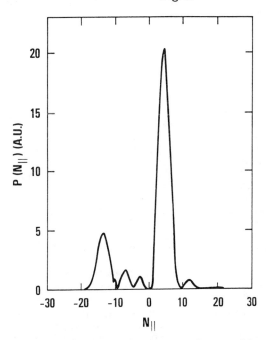

**Fig. 5.** Calculated power spectrum coupled to the plasma with a 1 m wide, four-loop antenna array with shielding plates between adjacent loops.

# Theoretical Studies of ICRH in Stellarators and Tokamaks*

P. E. Moroz
MIT Plasma Fusion Center, Cambridge, MA 02139 USA

## A. Effect of Magnetic Field Configuration on ICR Heating in Stellarators

ABSTRACT. All stellarators can be divided into two classes depending on the parameter $\zeta_s$, which can be easily expressed through the main parameters of the stellarator. Features of cyclotron absorption and mode conversion during fast wave ICR plasma heating are considered. The enhancement of absorption due to effects of non-zero $(\mathbf{B} \cdot \nabla)\mathbf{B}$ in the regime of majority fundamental cyclotron resonance is discussed.

INTRODUCTION. ICRH is now one of the most powerful and promising heating methods for tokamaks. In the last few years various regimes of ICRH well known for tokamaks were used in experiments in stellarators too. RF heating is more important for stellarators because the main regime of operation should be without ohmic heating. But the theoretical studies of ICRH in stellarators are lagging the experiments because of the complex structure of the magnetic field. Here we discuss some properties of the stellarator magnetic field of importance for ICRH. The regime of majority fundamental cyclotron heating that was successful in experiments on the L-2 stellarator is also considered.

ICRH IN STELLARATORS. Stellarators are toroidal devices where the magnetic field configuration is created by external currents. The magnetic field potential in the plasma can be expressed [1] in toroidal coordinates in terms of generalized Legendre functions $Q^m_{n-1/2}$ [2]

$$\Phi(\zeta, \theta, \phi) = aB_o \left[ \phi + \left( \frac{1 - \zeta \cos \theta}{\zeta} \right)^{1/2} \sum_{n,m} B_{nm} Q^m_{n-1/2} \left( \frac{1}{\zeta} \right) \sin(n\theta - m\phi) \right] , \quad (1)$$

where $a = (R^2 - r_o^2)^{1/2}$, $R$ and $r_o$ are the major and minor radii of the toroidal surface containing the windings. The coefficients $B_{nm}$ depend on the ratio of current in windings to magnetic field on the axis $I/B_o$ and on the shape of windings [1]. For ICRH the structure of $|\mathbf{B}|$ is very essential, because it defines the location of various resonant regions and surfaces in the plasma. The main feature of the stellarator magnetic field is the possible existence of saddle points of $|\mathbf{B}|$. In the straight stellarator approximation [3] the saddle point of $|\mathbf{B}|$ is placed on the magnetic axis, but in a real stellarator toroidal effects can easily move the saddle point even out of the plasma. For deriving an approximate condition on the parameters of the stellarator for separating these two classes of stellarators we introduce the parameter $\zeta_s$ such that $\zeta_s^{n-1} = 2^n n!/n^2 m^n \epsilon_n$, where $n, m$ are the numbers of the main helical harmonic in (1), which correspondingly equal the number of helical windings with the same direction of current and number of periods of magnetic field. The coefficient $\epsilon_n$ is given by $\epsilon_n = 8Im/(B_o cnR) \cdot \zeta_o K'_n(\zeta_o)$, $\zeta_o = mr_o/R$. Then in the case $\zeta_s > r_p/R$ (where $r_p$ is the mean plasma radius) the saddle point lies outside the plasma and the shape of the $|\mathbf{B}|$ field and the conditions for ICRH will be similar to those in a tokamak. In the opposite case $\zeta_s < r_p/R$ the saddle point lies in the plasma filled region. Then the tokamak-like $|\mathbf{B}|$ distribution is localized in the central part of the plasma $r < \zeta_s R$. In the region $r > \zeta_s R$ the structure of $|\mathbf{B}|$ has an essential helical character and ICRH may differ very much from that in a tokamak. In this case the regimes of fundamental cyclotron minority or harmonic cyclotron majority heating can be very efficient because of the enlargement of the cyclotron resonance region $|\omega - \ell\Omega_i| < k_\parallel V_i$

---

* Work supported by the MIT Nucl. Eng. Dept. and the US Dept. of Energy

if it includes the saddle point. Because the values of $|\mathbf{B}|$ for saddle points in different cross sections of the plasma column are not the same, ICRH will be more effective for a high temperature plasma where the resonance region can include all saddle points: $k_\| V_i/\omega > \ell n \epsilon_n I_n(m\zeta_s)$. For the ion-ion hybrid resonance regime of plasma heating a stellarator has many features, too. The multipole character of the $|\mathbf{B}|$ distribution in the region $r > \zeta_s R$ will result in the formation of $n$ regions within which the cyclotron or hybrid resonance occurs. Moreover, the preferred situation of fast wave incidence from the high field side can be easily reached in a stellarator with an outward antenna. In the regime of majority fundamental cyclotron heating the absorbed power can be much larger than calculated from usual "local" theory due to effect of non-zero $(\mathbf{B}\nabla)\mathbf{B}$ [4,5]. The coefficient of absorption enhancement is $\kappa = P/P_{loc} \simeq (\mathbf{B} \cdot \nabla)\mathbf{B}/k_\|^2 \rho_i B^2$, where $\rho_i$ is the ion Larmor radius. This expression can be used when $\kappa > 1$. In the central plasma region $r < \zeta_s R$ this effect is the same as for tokamaks because $\mathbf{B}$ should be averaged over fast oscillations but in the region $r > \zeta_s R$ the absorption in a stellarator is larger than in a tokamak by a big factor $P_s/P_t \simeq m_s q_t$.

## B. Importance of Alfven Resonance during ICRH

ABSTRACT. A theoretical analysis is presented for the effect of an Alfven resonance on propagation of the fast wave during majority fundamental ion cyclotron heating. Numerical calculations show a pronounced damping of the fast wave. The possibility of heating of the ions by a slow wave which is propagating out of the conversion region into the dense plasma is analyzed. Comparison with some published experimental results is carried out.

INTRODUCTION. In the experiments on majority fundamental cyclotron heating by the fast wave in tokamaks [6] and stellarators [7] very low quality factor of the eigenmodes and very short longitudinal damping length were observed. This behavior cannot be explained in terms of collisional absorption, Cherenkov absorption, or cyclotron absorption of fast waves, even if one takes account of the enhancement of cyclotron absorption due to the $(\mathbf{B}\nabla)\mathbf{B}$ effects (see part A). It was suggested [6] that the strong damping is associated with the presence of an Alfven resonance surface $\epsilon = 0$, where $\epsilon = \epsilon_1 - n_\|^2$. In Refs. [8,9] the effect of Alfven resonance on the fast wave was estimated (but only very approximately). In the present paper, we numerically solve a system of wave equations that allow us to do a comparison with the experiment.

ALFVEN RESONANCE. To study the effect of an Alfven resonance on the eigenmodes of a fast wave we must choose parameter values such that the condition $\epsilon = 0$ will be satisfied in the edge plasma where the dispersion relation for a "cold" plasma is a good approximation

$$n_\perp^4 \epsilon_1 + n_\perp^2 \left[ \epsilon_2^2 - \epsilon(\epsilon_3 + \epsilon_1) \right] + \epsilon_3(\epsilon^2 - \epsilon_2^2) = 0 \tag{2}$$

From this equation we can conclude that $Imn_\perp^2$ is zero except in a narrow "critical" region near the point of Alfven resonance. The corresponding system of wave equations in a cylindrical system of reference $(r, \phi, z)$ can be written in the form

$$\frac{1}{x}\frac{d}{dx}x E_\phi - \frac{m\epsilon_2}{x\epsilon}E_\phi = iB_z\left(1 - \frac{m^2}{\epsilon x^2}\right) - \frac{mn_z}{x\epsilon}\frac{dE_z}{dx}$$

$$\frac{dB_z}{dx} + \frac{m\epsilon_2}{x\epsilon}B_z = in_\perp^2 E_\phi + in_z\left(\frac{\epsilon_2}{\epsilon}\frac{dE_z}{dx} + \frac{m}{x}E_z\right) \tag{3}$$

and one more equation of second order for the $E_z$ component. Here $x = \omega r/c$. Equations (2) and (3) describe both fast and slow waves and are not singular. Using arguments similar to those for ion-ion hybrid resonance in Ref. [10] we can reduce (3) to a second order equation but only for the fast wave taking $E_z = 0$ and writing the factor $1/\epsilon$ as $n_\perp^2/(\epsilon^2 - \epsilon_2^2)$, where for $n_\perp$ we take only the fast wave root of (2). Two more ways of resolving the singularity $1/\epsilon$ [11] are either to introduce an artificial collision frequency $\nu \simeq \omega\omega_{pe}/2n_\parallel\Omega_e$ or make an analytical decision inside the critical region and use the relation $\ln(-\xi) = \ln(\xi) + \pi i$. All three methods give close results. Solving the wave equations (3) with appropriate boundary conditions and conditions at the antenna we calculated the dependence of leading resistance versus magnetic field. We found the quality factor $Q$ for resonances of different eigenmodes and calculated the longitudinal damping length $\lambda_D = 2QV_{gr}/\omega$ that was in good agreement with the experiment [7].

Antenna of width $d$ can effectively excite a wide spectrum of $k_\parallel < \pi/d$. To see what happens to waves with large values of $k_\parallel$: $\omega_{pi}(0)/c < k_\parallel < \pi/d$, we examined the dispersion relation in a hot plasma by numerical calculation. We found that for parameters of the stellarator L-2 the slow wave can penetrate into the central plasma closer to the ion cyclotron resonance and transfer energy to ions. This fact can explain the central heating of ions observed in experiment at L-2 stellarator. But this mechanism is possible only for rather hot electrons when parameter $\omega/k_\parallel V_e < 1$. This can explain the observed [12] critical electron temperature to obtain the high $T_i$ plasma.

## C. On the Splitting of Fast Wave by a Poloidal Magnetic Field

ABSTRACT. The splitting effect of fast wave eigenmodes with equal $|k_z|$ but opposite directions is discussed. An equation is found which determines the conditions for exciting various eigenmodes in a cylindrical plasma with arbitrary radial variation of plasma density and longitudinal and poloidal magnetic fields. Numerical solutions for parabolic density profile show substantial differences from the known approximate expressions for splitting of fast wave eigenmodes.

INTRODUCTION. Effect of splitting of fast wave eigenmodes by a poloidal magnetic field was predicted in Ref.[13] and experimentally detected in Refs.[14-16]. The following approximate expressions for the relative change in the mean plasma density due to the splitting were cited in [13,14] respectively

$$\frac{\delta\overline{n}}{\overline{n}} \simeq \frac{2}{3}\frac{mk_z a^2}{qR(1 + k_z^2 a^2/3)} \quad (4) \qquad \frac{\delta\overline{n}}{\overline{n}} \simeq 2.5\frac{mk_z c^2}{qR\omega_{pi}^2} \quad (5)$$

Here we derive and numerically solve a wave equation for the nonhomogeneous cylindrical plasma in a magnetic field $\mathbf{B} = B_z(r)\mathbf{e}_z + B_\phi(r)\mathbf{e}_\phi$ and compare results with (4) and (5).

SPLITTING OF FAST WAVE EIGENMODES. By writing the Maxwell equations in a cylindrical sytem of reference and considering that the tensor of dielectric permittivity $\bar{\epsilon}$ is given by the relation $\bar{\epsilon} = \hat{T}\hat{\epsilon}\hat{T}^{-1}$,

$$\hat{T} = \begin{pmatrix} 1 & 0 & 0 \\ 0 & \cos\gamma & \sin\gamma \\ 0 & -\sin\gamma & \cos\gamma \end{pmatrix} \qquad \hat{\epsilon} = \begin{pmatrix} \epsilon_1 & i\epsilon_2 & 0 \\ -i\epsilon_2 & \epsilon_1 & 0 \\ 0 & 0 & \epsilon_3 \end{pmatrix}$$

the following equation is found for the $E_\phi$ component of the wave electric field

$$(\epsilon_1 - n_3^2)E_\phi'' + E_\phi'\left\{\left(\frac{1}{x} + \frac{2\kappa\kappa'}{1+\kappa^2}\right)(\epsilon_1 - n_3^2) - n_2\left[\frac{n_2}{x} + \frac{2\kappa\kappa'}{1+\kappa^2}n_2 + \frac{n_2}{\epsilon - n_2^2 - n_3^2}\right.\right.$$

$$\cdot(\epsilon_1' + \frac{2m^2}{x^3}) + \frac{1}{x\sqrt{1+\kappa^2}}\left(\frac{m}{x} + 2\kappa n_z + n_z x \kappa'\right) + n_z(\kappa' - \frac{\kappa}{x})\bigg]\bigg\} + E_\phi\bigg\{\bigg[n_\perp^2 -$$

$$-n_2^2 - \frac{1}{x^2(1+\kappa^2)}\bigg](\epsilon_1 - n_3^2) - n_2\epsilon_2' - n_2\frac{\epsilon_1' + 2m/x^3}{\epsilon_1 - n_2^2 - n_3^2}\left[\frac{n_2}{x} - \epsilon_2 - n_z(\kappa' - \frac{\kappa}{x})\right] -$$

$$-\frac{1}{x\sqrt{1+\kappa^2}}\left[\frac{2mn_2}{x^2} - \epsilon_2\left(\frac{2m}{x} - n_z\kappa\right)\right] - n_2\left(\frac{n_z\kappa'}{x\sqrt{1+\kappa^2}} + \frac{\kappa\kappa'}{1+\kappa^2}\epsilon_2 + n_z\kappa''\right) -$$

$$-\frac{n_2^2\kappa^2}{x^2(1+\kappa^2)} - n_z(\kappa' - \frac{\kappa}{x})\left(\epsilon_2 - \frac{2m}{x^2\sqrt{1+\kappa^2}} + \frac{\kappa\kappa'}{1+\kappa^2}n_2\right)\bigg\} = 0 \qquad (6)$$

Here $x = \omega r/c$, $\kappa = B_\phi/B_z$, $\gamma = \text{arctg}\kappa$, $n_2 = (m/x - n_z\kappa)/\sqrt{1+\kappa^2}$, $n_3 = (n_z + m\kappa/x)/\sqrt{1+\kappa^2}$, $n_\perp^2 = \epsilon_1 - n_3^2 - \epsilon_2^2/(\epsilon_1 - n_3^2)$. The calculations show that (a) the symmetrical modes $m = 0$ can split; (b) with a change in sign of $k_z$ or $m$ the value of $\delta\bar{n}/\bar{n}$ not only changes sign but alters its value; (c) splitting is approximately proportional to $1/q$ only for large $k_z$ and in general has a more complex dependence. All these facts are contradictory to (4) and (5).

**Acknowledgments.** The author would like to thank Prof. D.J.Sigmar, Dr. P.T.Bonoli and Prof. M.Porkolab for stimulating and useful discussions.

### References

[1] L. M. Kovrizhnykh, P. E. Moroz, Sov. Phys. Techn. Phys. 9 (6), 735 (1983).

[2] E. W. Hobson, *The Theory of Spherical and Ellipsoidal Harmonics*, Chelsea, NY (1955).

[3] E. F. Jaeger, H. Weitzner, D. B. Batchelor, Report ORNL/TM-10223 (1987).

[4] L. M. Kovrizhnykh, P. E. Moroz, Proc. 10th IAEA Conf. on Plasma Phys. and Contr. Nucl. Fus. Res., London, 1984, Vol. I, p. 633.

[5] P. E. Moroz, Proc. 12th Eur. Conf. on Contr. Fus. and Plasma Phys., Budapest, 1985, Vol. 9F, Part II, p. 351.

[6] Equip. TFR. Proc. 6th IAEA Conf. on Plasma Phys. and Contr. Nucl. Fus. Res., Berchtesgaden, 1977, Vol. 3, p. 99.

[7] V. A. Batyuk et al, Proc. 11th IAEA Conf. on Plasma Phys. and Contr. Nucl. Fus. Res., Kyoto, 1986, Vol. 1, p. 481.

[8] F. W. Perkins, Nucl. Fus. 17, 1197 (1977).

[9] C. F. F. Karney, F. W. Perkins, Y.-C. Sun, Phys. Rev. Lett. 42, 1621 (1979).

[10] D. N. Smithe, P. L. Colestock et al, Nucl. Fus. 27, 1319 (1987).

[11] P. E. Moroz, Sov. J. Plasma Phys. 13, 317 (1987).

[12] T. Mutoh et al, Proc. AIP 7th Top. Conf. Appl. RF Power to Plasma, Kissimmee, 1987, p. 238.

[13] M. S. Chance et al, Bull. Am. Phys. Soc. 18, 1273, (1973).

[14] J. Adam et al, Proc. 5th IAEA Conf. on Plasma Phys. and Contr. Nucl. Fus. Res., Tokyo, 1974, Vol. 1, p. 65.

[15] J. C. Hosea, Symp. on Plasma Heating in Tor. Dev., Varenna, 1974, p. 61.

[16] Equip. TFR. Proc. 3rd Int. Symp. on Pl. Heat. in Tor. Dev., Varenna, 1976, p. 43.

# ION BERNSTEIN WAVE HEATING EXPERIMENTS ON PBX-M

M. Ono, G.J. Greene, and S. Bernabei
PPPL, Princeton University, Princeton, NJ 08542

## ABSTRACT

A multi-megawatt level IBWH experiment on PBX-M[1] is in preparation. The goal of the experiment is to contribute to the attainment of the high beta, second regime of stability. The high power IBWH will be used as an additional heating power source to supplement the existing 6 MW of NBI power to achieve higher $\beta$ values in PBX-M. Bulk ion heating via IBW excitation with localized, off-axis deposition can be used to modify the pressure profile for improved plasma stability at high $\beta$. The high power off-axis heating in principle can generate a significant bootstrap current [2]($\approx 30\ \%$) in the outer region of the PBX-M plasma complementing LHCD for broadening the current profiles. It is also interesting to note that the available rf power ($\approx 4$ MW) is comparable to the predicted power levels required for the rf ponderomotive stabilization of pressure driven modes (such as the high-n ballooning[3] and external kink modes[4]) for the closely fitted stabilizing shell configuration of PBX-M. There are, however, several experimental factors that require careful consideration in planning a high power experiment. Four important factors are discussed here in some detail: 1. Antenna location. 2. Effects of parallel electric fields. 3. Modification of launched wave spectrum due to antenna misalignment. 4. Possible interference of wave launching by protective limiters.

## PBX-M EXPERIMENTAL SET-UP

The PBX-M IBWH system utilizes two FMIT 40-80 MHz, 2MW, rf power supplies from the TFTR ICRF[5] system by means of rf switches. Four IBWH antennas are being fabricated to handle the available 4 MW of rf power. The IBWH antennas are placed in the PBX-M outer midplane region at R = 200 cm. The antenna is fabricated largely from silver plated stainless steel in order to minimize the rf resistivity. The Faraday shield is made out of 6-mm diameter TiC coated molybdenum rods in a double layered configuration. The separation between the adjacent rods is sufficiently small ($\approx 2$ mm) that it shields the antenna elements from particle and radiation bombardments. A bench test on a similar shield has shown that the presence of the shield causes negligible rf field attenuation. On both sides of the antenna, graphite protecting tiles are placed with about 3 mm protrusion from the Faraday shield surface. A voltage probe and Rogowski coil current probes are built into each antenna to provide direct rf antenna diagnostics.

## ANTENNA LOCATION

In order to heat the plasma effectively, the placement of the IBWH antenna plays an important role in obtaining good wave accessibility. Normally, the IBWH antennas are placed in the outer mid-plane region since it is the most accessible region of a tokamak. The wave launched from the mid-plane region propagates in an oscillatory trajectory toward the center until it reaches the cyclotron harmonic resonance layer.[6] However, if the antenna is placed significantly away from the mid-plane (poloidal angle $\geq 30\,^{\circ}$), the launched parallel wave number increases as the wave propagates toward the mid-plane. Ray-tracing calculation indicates that a significant up-shift of the parallel wave number occurs in this situation causing the wave to be absorbed via electron Landau damping near the plasma edge. Therefore, according to the ray tracing calculation, the placement of an IBW antenna is optimized at the low field mid-plane region.

## EFFECTS OF PARALLEL RF ELECTRIC FIELDS

Since the IBW antenna is oriented along the toroidal magnetic field direction, an rf electric field of as much as $\approx 300$V/cm along the magnetic field is expected at the antenna surface. The strong response of the electrons to the parallel electric field excites the electrostatic waves. This strong electron response also gives rise to a strong radial ponderomotive force for the low frequency IBWH since $F_{pond} \approx 1/16\pi$ $(\omega_{pe}^2/\omega^2) \, \partial E_{||}^2/\partial r$. This radially inward force on the electrons can induce a radial electric field causing the plasma to acquire a negative charge as observed in PLT.[7] The 300 V/cm parallel electric field also causes electrons to acquire large thrashing motion with energy $(U = e^2 E_{||}^2/m_e\omega^2)$ of $\approx 3$ keV at 30 MHz, a sufficient energy to ionize background neutrals. Such an ionization process in front of the antenna is not desirable since it can drain the rf power and make the wave launching physics unpredictable. In a typical tokamak discharge, the neutral pressure is usually sufficiently low ($\approx 10^{-6}$Torr) that the ionization is likely to have a negligible effect. However, it is also important to make sure that the effect of the locally enhanced recycling (such as a nearby limiter and/or the gas feed) is minimized at the antenna surface. In addition, on PLT, a titanium gettering has been used on the antenna surface to minimize the recycling. Also to reduce the electron secondary emission, TiC coated Faraday shield rods and graphite limiters are used for the PBX-M antenna.

If the plasma density between the Faraday shield gap is sufficiently high, $\omega_{pi}^2/\omega^2 > 1$, plasma sheath rectification can cause ion bombardment of the Faraday shield with energies similar to the gap potential ($E_i \approx 500$ eV )which can cause serious sputtering and impurity generation.[8] If the ionization problem is negligible, then the strong ponderomotive force may be able to push the plasma away from the surface to keep the density in the gap sufficiently low ( $n_e < 10^{10}$cm$^{-3}$). Another potentially serious Faraday shield effect involves the ripple field generated by the shield.[9] The ripple field has a very short wavelength characteristic with an effective parallel wavelength of the distance between the Faraday shield rods. For the PBX-M antenna, its effective phase velocity is about $3 \times 10^7$cm/ sec which is slow enough to interact strongly with the edge electrons (at a temperature of few eV) via electron Landau damping. The effect of the Faraday shield induced rf field ripple may be minimized to some extent by the double layered design[9] as in the case of the PLT and PBX/M experiments. Since the effects of parallel electric field tend to decrease with $\omega^{-2}$, they should be minimized by going toward higher frequency IBWH experiments.[10,11]

## CONSEQUENCE OF ANTENNA MISALIGNMENT

Another potential problem for IBWH in the low frequency regime is the effect of antenna misalignment caused, for example, by the magnetic field ripple. Since the IBW antenna position is chosen to be in the midplane region of low field side of the tokamak, the toroidal field ripple can be quite large ,approaching 1.5% for PBX-M and 0.8 % for PLT at the antenna position of R = 200 cm and 173 cm, respectively. In those devices, in order to launch desirable $k_{||}$-spectrum, the antenna typically spans a toroidal distance of twice the ripple distance of $L_r \approx 30$ cm. This field ripple causes the magnetic field line to move in major radius. For a typical tokamak experiment, this excursion distance $\Delta x_r$ can be few millimeters: for PBX-M, $\Delta x_r \approx 3$ mm and for PLT,

$\Delta x_r \approx 1.5$ mm.[12] The radial excursions are quite small in both PBX-M and PLT and are comparable to the fabrication and installation uncertainties.

For the second harmonic IBW launching case, to leading order in $(k_\perp \rho_i)^2$, the wave dispersion relation in the plasma edge can be expressed as

$$D = \alpha k_\perp^4 + \beta k_\perp^2 - \gamma k_\parallel^2 = 0 \qquad (1)$$

where $\alpha = 3 (T_i/m_i)[\omega_{pi}^2(4\Omega_i^2 - \omega^2)^{-1}(\omega^2-\Omega_i^2)^{-1}]$, $\beta = 1 - \omega_{pi}^2/(\omega^2-\Omega_i^2)$ and $\gamma = \omega_{pe}^2/\omega^2$. The above equation represents three types of wave dispersion relations depending on the plasma and wave properties. For the low density regime, $k_\perp^2 \approx \gamma k_\parallel^2/\beta$, it is an electron plasma wave. Near the lower hybrid resonance($\beta \approx 0$), $k_\perp^2 \approx (\gamma k_\parallel^2/\alpha)^{1/2}$, the wave is a warm-ion-mode. For higher density, $k_\perp^2 \approx -\beta/\alpha$, the wave becomes an ion Bernstein wave.

Misalignment of the antenna with respect to the magnetic field line can result in the launching of modes with shorter parallel wavelength than the value expected from the antenna geometry. This effect is particularly strong for the case of the direct launching of short perpendicular wavelength IBW. One can estimate the maximum parallel wavelength to be $\lambda_\parallel(\max) \approx \lambda_\perp L_r/\Delta x_r$. For launching the ion Bernstein wave directly (i.e. $\omega_{pi}^2/\omega^2 > 1$), one can show that $\lambda_\perp$ (cm) $\approx 4 \times 10^{-2}(T_i/\delta)^{0.5}$ where $\delta \approx (2\Omega_i-\omega)/\omega$. On the other hand, one would need to launch waves with $n_\parallel \leq 10$ to have a good accessibility to the plasma core which for the 30 MHz case would require $\lambda_\parallel \geq 100$ cm. This condition requires the launched waves to have $\lambda_\perp \geq 1.0$ cm for $\Delta x_r \approx 3$ mm (PBX-M) and 0.5 cm for $\Delta x_r \approx 1.5$ mm (PLT). This is equivalent to $T_i/\delta \geq 625$ for PBX-M and 144 for PLT. If $\delta \approx 0.1$, then the required edge $T_i$ is 63 eV for PBX-M and 14 eV for PLT. Since the required temperature would go up roughly as the square of the magnitude of the ripple and/or the misalignment, it is important to minimize the antenna misalignment. In early IBW excitation experiment,[13] careful antenna alignment was critical for successful wave launching. Because of the low frequency nature of the experiment, in order to avoid electron Landau damping, it was necessary to launch waves with very large values of $\lambda_\parallel$ (on the order of a few meters). The alignment problem was greatly reduced in the ACT-1 experiment using hydrogen[14] since going to higher wave frequency reduced the required $\lambda_\parallel$ to $\approx 30$ cm. Similarly, the alignment problem should decrease for future higher frequency waveguide IBWH experiments[10,11].

## LIMITER INTERFERENCE

If the antenna protective limiter protrudes significantly into the plasma compared to the Faraday shield surface, a different type of problem arises because the relatively "shallow" initial propagation angle of the launched wave packet can cause it to hit the side limiter. If the wave hits the graphite limiter, significant fractions of the power might be absorbed by the limiter and such a reflection may also introduce a change in the wave number spectrum. We define $\Delta x_{lim}$ to be the effective distance of the limiter protrusion with respect to the radiating surface.

To estimate the wave packet trajectory, the relevant quantity is the propagation angle which is a ratio of the radial and toroidal wave group velocities. In order to avoid the problem of hitting limiters, one would want to make the propagation angle $\theta$ to be greater than about 0.03 assuming that the overall net protrusion distance $\Delta x_{lim}$ is about 1 cm (making $\Delta x_{lim}/L_r \approx 0.03$) which can be a typical value for tokamak IBWH experiments. One can show from Eq.(1) that the propagation angle $\theta$ is given by

$$\theta \approx (4 \alpha k_\perp^3 + 2\beta k_\perp)/(2\gamma k_\parallel) = (k_\parallel/k_\perp)(\alpha k_\perp^4/\gamma k_\parallel^2 + 1) \qquad (2)$$

For the rf frequency of 30 MHz in the low density regime, one can launch the electron plasma wave whose propagation angle can be shown to be $\theta \approx 0.023$ ( $1.5 \times 10^{10} / n_e -1)^{0.5}$ where the plasma density $n_e$ is in cm$^{-3}$. In order for the angle to be larger than 0.03, the density must be lower than $5 \times 10^9$ cm$^{-3}$which is probably much lower than the expected values even in the scrape off region. As the density approaches the lower hybrid resonance (i.e. $\beta = 0$), one must use the dispersion relation of the warm-ion-mode giving $\theta \approx 2$ $k_{||}/k_\perp \approx 0.36$ $(V_{i}k_{||}/\omega)^{0.5}$ $\delta^{-0.25}$. One can see that to obtain larger values for the angle, one would require higher ion temperatures and slower waves. The values of $\omega/k_{||}$, however, is constrained by the accessibility condition $n_{||} \leq 10$. Therefore for $\delta = 0.1$, $n_{||} = 10$, the relation yields $\theta \approx 1.1 \times 10^{-2}$ $T_i^{0.25}$. This estimate suggests that in order to have $\theta \approx 0.03$, it is necessary to have an edge ion temperature above 55 eV which is not an unexpected value. However, since this effect has a relatively weak dependence on $T_i$, if $\Delta x_{lim}$ is larger than 1 cm, the required $T_i$ would quickly become prohibitively high.

For the ion Bernstein wave, the plasma density is sufficiently high that $\beta \approx -\omega_{pi}^2/(\omega^2-\Omega_i^2)$. Since $\alpha k_\perp^2 \approx -\beta$, $\theta \approx -\beta$ $k_\perp/\gamma k_{||} \approx 6 \times 10^{-4}$ $\delta^{0.5}$ $(\omega/k_{||}V_i)$. Here, the angle is larger for faster phase velocity, colder ion temperature, and for larger $\delta$. For $\delta = 0.1$ and $n_{||} = 10$, one obtains $\theta \approx 0.56$ $T_i^{-0.5}$ which indicates that as long as the ions are not too hot ($T_i < 350$ eV) and $\delta$ is not too small, the angle can be larger than 0.03 providing that one can launch the desired wave spectrum.

## ACKNOWLEDGEMENTS

The authors acknowledge useful discussions with Drs. M. Mayberry, H. Matsumoto, R. Pinsker, and M. Porkolab.

Work supported by DOE Contract No. DE-AC02-76-CHO-3073.

## REFERENCES

[1] M. Okabayashi, N. Asakura, R. Bell, et al., in Proceedings of the Twelfth International Conference on Plasma Physics and Controlled Nuclear Fusion Research, Nice, France October 1988, Paper IAEA-CN-50/A-2-2.

[2] F.L. Hinton and R.D. Hazeltine, Rev. Mod. Phys. 48, 239, (1976).

[3] D.A. D'Ippolito, J.R. Myra, and G.L. Francis, Phys. Rev. Lett. 58, 2216 (1987).

[4] D.A. D'Ippolito, Phys. Fluids 31, 340, 1988.

[5] J.R. Wilson, M.G. Bell, A. Cavallo, et al., in Proceedings of the Twelfth International Conference on Plasma Physics and Controlled Nuclear Fusion Research, Nice, France October 1988, Paper IAEA-CN-50/E-4-1.

[6] M. Ono, R. Horton, T.H. Stix, K.L. Wong, in Heating in Toroidal Plasmas, Commission of the European Communities, Brussels, 1981, Vol., I, p. 593-603.

[7] M. Ono, P. Beiersdorfer, R. Bell, et al., Phys. Rev. Lett. 60, 294 (1988).

[8] F.W. Perkins, Princeton Plasma Physics Laboratory Report PPPL-2452 (1987).

[9] Y. Sato, K. Sawaya, and S. Adachi, IEEE Transactions of Plasma Science 16, 574 (1988).

[10] A. Cardinali, R. Cesario, F. DeMarco, M. Ono, 16th European Conference on Controlled Fusion and Plasma Physics, Venice, Italy (March 1989).

[11] D.W. Ignat and M. Ono, Princeton Plasma Physics Laboratory Report PPPL-2583 (1989).

[12] M. Pelovitz, Private Communications.

[13] J.P.M. Schmitt, Phys. Rev. Lett. 31, 982 (1973).

[14] M. Ono, K.L. Wong, and G.A. Wurden, Phys. Fluids 26, 298 (1983).

310

# SAWTOOTH MIXING EFFECTS ON ICRF POWER DEPOSITION IN TOKAMAKS[*]

C.K. Phillips, A. Cavallo, P.L. Colestock, D.N. Smithe, G.J. Greene, G.W. Hammett,
J.C. Hosea, J.E. Stevens, J.R. Wilson
Princeton Plasma Physics Laboratory, P.O. Box 451, Princeton, NJ 08543

D.J. Hoffman, W.L. Gardner
Oak Ridge National Laboratory, P.O. Box 2009, Oak Ridge, TN 37831

## ABSTRACT

A series of experiments was performed on TFTR in which the toroidal field strength, and hence the location of the ICRF resonance layer, was varied.[1] Electrons were heated primarily via collisions with the fast tail ions produced by the rf heating. The net power deposited to electrons on axis can be directly inferred from ECE measurements[2,3] of the rate of rise of the central electron stored energy immediately following a sawtooth crash. Comparisons between these measurements and calculations of the net central electron heating obtained using the SNARF[4] tokamak simulation code as well as PPL's ray tracing/Fokker-Planck code[5] implied that the predicted power deposition was too localized. Results obtained by incorporating a sawtooth mixing model for the fast ions into the deposition codes are in good agreement with the experimental measurements.

## INTRODUCTION

Even though auxiliary plasma heating has become a major component of modern tokamak experiments, there are very few methods for directly measuring the applied power deposition profiles. One method which has been used with some success involves monitoring of the response of the electron temperature profile to the applied heating, either immediately following a sawtooth crash[2,3] or else in experiments when the applied power is modulated[6]. In recent ICRF heating experiments, comparisons of the electron heating rates, inferred from experimental measurements of the rate of rise of the electron temperature after a sawtooth, with the theoretical calculations from wave damping codes indicated that the codes were predicting central electron heating rates that were substantially above the experimental values[7]. Various mechanisms which can cause broadening of the calculated power deposition profiles have been suggested in the past, including finite banana width effects[4], loss of fast ions due to the sawtooth crash[7,9,10], radial transport of fast ions[7,8], and flattening of the density, temperature, fast ion distribution function, and power deposition profile due to the sawtooth mixing[7,9,11,12]. In this paper, theoretical calculations of the ICRF power deposition profiles to ions and electrons, which include a sawtooth mixing model, are presented which are in good agreement with experimental measurements of the electron heating rates following a sawtooth crash in ICRF-heated discharges in TFTR.

*Work supported by U.S.D.O.E. Contract No. DE-AC02-76-CHO-3073

## SAWTOOTH MIXING MODEL

Experimentally it has been observed that the plasma density, temperature and rotational transform profiles within the central region of the tokamak are flattened immediately following a sawtooth, or internal disruption. The effect extends out to the mixing radius, which is approximately $\sqrt{2}$ times the q=1 radius[1,2]. Charge exchange measurements of the fast ion flux near the mixing radius in TFTR[1] indicate that the fast ions are also affected by the sawtooth crash. It is therefore not unreasonable to assume that the fast ions are also redistributed within the mixing radius in a manner similar to the thermal ions.

The effect of sawtooth mixing on ICRF power deposition profiles has been investigated using two independent wave field and power deposition codes. One code uses a reduced order approximation to calculate full wave solutions for the wave fields[4]. The other code uses ray tracing[5] to generate the wave power deposition profiles. Both codes are coupled with an isotropic Fokker-Planck module which includes collisional thermalization as well as RF-induced quasilinear diffusion so that the thermalized power deposition profiles can be self-consistently determined. To simulate the effects of the sawtooth on the ICRF power deposition, the wave fields and thermalized power deposition profiles are first calculated using the magnetic fields and density and temperature profiles measured at the top of the sawtooth. The density and temperature profiles as well as the minority distribution function within the mixing radius are then replaced by their volume-averaged equivalents. The collisional transfer of power from the minority species to the background species is then recalculated. Calculations of the wave fields done with profiles measured at the bottom of the sawtooth differ insignificantly from those calculated using the pre-sawtooth profiles. However, the thermalized power deposition profiles are strongly affected by the sawtooth mixing, since the hot ions are spread out over the mixing volume on time scales short compared to the slowing down time. The collisional transfer of energy from the hot ions to the background species therefore occurs over a larger volume than the direct wave damping would predict.

## RESULTS AND CONCLUSIONS

Comparisons have been made between the code predictions and experimental measurements of the central electron heating rate for a series of TFTR discharges in which the value of the toroidal field, and hence the location of the ICRF minority resonance layer, was varied. The experiments were performed in He[4] majority- H minority plasmas with about 1.75 MW of ICRF power applied for about 0.5 sec at a frequency of 47 Mhz with a peak $k_{tor}$ of approximately 7 m$^{-1}$. The toroidal field was varied from 3.47 T (shot 37129), with the resonance layer located near the mixing radius of about 30 cm, to 3.22T (shot 37131), when the resonance layer was nearly on the magnetic axis. The deuterium and minority hydrogen concentrations were estimated to be 5%, while the carbon impurity concentration was estimated to be about 2%. The time evolution of the central electron density and the central electron temperature for these two discharges is shown in Fig. 1.

The calculated direct wave damping deposition profiles to the hydrogen minority are displayed in Fig. 2. For both shots, the wave damping deposition profile determined from the ray tracing code is more peaked than the corresponding profile from the full wave analysis. Some reasons for the discrepancy include the neglect of diffraction effects by ray tracing, neglect of rotational transform by the full wave analysis, use of cold plasma polarizations by the ray tracing code, and differences in the treatment of the equilibrium magnetic geometry between the two codes. ( The ray

312

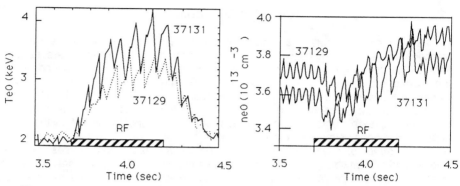

Fig. 1 Time evolution of the central electron density and temperature for on-axis and off-axis heating.

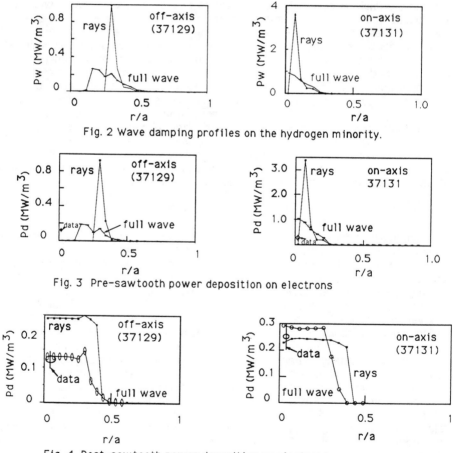

Fig. 2 Wave damping profiles on the hydrogen minority.

Fig. 3 Pre-sawtooth power deposition on electrons

Fig. 4 Post-sawtooth power deposition on electrons

tracing code uses concentric circular flux surfaces, while the full wave analysis includes shifts of the flux surfaces.) Code results for the thermalized electron power deposition profiles obtained without the sawtooth mixing model are shown in Fig. 3. Because of the relatively low minority density and significant power densities in these discharges, the minority ions are driven above the critical energy and hence slow down primarily on electrons. For comparison purposes, the experimental value for the central electron heating rate, as deduced from ECE measurements, is also indicated in the figures. Though both codes predict that nearly all of the input power is collisionally coupled to the electrons, the peak minority temperatures calculated with the ray tracing code are higher than those determined by the full wave approach by about the same ratio as the peak power densities predicted by the two codes. When no sawtooth mixing effects are included, both codes predict a far greater electron heating rate than was observed for the on-axis heating case. However, for the off-axis heating case, both codes predict no additional electron heating above the ohmic value whereas the data indicates a significant enhancement factor of about two. Corresponding results for the electron power deposition profiles immediately following the sawtooth crash are displayed in Fig.4. When the effects of the sawtooth mixing are included, the code predictions agree well with the experimental measurements. Results for the off-axis heating case are particularly significant, since the hot tail ions created off-axis are mixed into the center where they slow down on electrons, leading to an enhancement of the central electron heating rate above the ohmic value. In the full wave analysis, the sawtooth mixing reduces the average tail energy below the critical energy, so significant heating of the background ions is predicted. The sawtooth-averaged values from the ray code remain above the critical energy, so the ray code predicts that the tail energy will still be coupled primarily to the electrons. Verification of this difference will be explored using fine time-resolved ion temperature measurements during future experiments.

## REFERENCES

1. P.L. Colestock et.al., this conference.
2. E. Mazzucatto et.al., Proc. 10th Int. Conf. on Pl. Physics and Controlled Nuclear Fusion Res., (London), (IAEA, Vienna, Austria, 1985), vol.1, p. 433.
3. A. Cavallo et.al., Proc. 15th European Conf. on Controlled Fusion and Plasma Heating,(Dubrovnik), (EPS, 1988), Vol. I, 389.
4. D.N. Smithe et.al., this conference.
5. D.Q. Hwang, C.F.F. Karney, et.al., Princeton Plasma Physics Laboratory Report PPPL-1990, May 1983.
6. F. Tibone et.al., Proc. 15th European Conf. on Controlled Fusion and Plasma Heating,(Dubrovnik), (EPS, 1988), Vol.II, p. 709.
7. L.-G. Eriksson, T. Hellsten, et.al., Nuclear Fusion **29**,87 (1989).
8. L. Chen, V. Vaclavik and G.W. Hammett, Nuclear Fusion **28**, 389 (1988).
9. S.-I. Itoh, K. Itoh, A. Fukuyama, and T. Morishita, Proc. 14th European Conf. on Controlled Fusion and Plasma Heating,(Madrid), (EPS, 1987), Vol. II, 1204.
10. L.-G. Eriksson and T. Hellsten,Proc. 16th European Conf. on Controlled Fusion and Plasma Heating,(Venice), (EPS, 1989), Vol. III, 1077.
11. H. Romero and J. Scharer, Phys. Fluids B **1**,252 (1989).
12. B.B. Kadomtsev, Sov. J. Plasma Physics 1, 389 (1975).

# HIGH POWER ION BERNSTEIN WAVE EXPERIMENTS ON DIII–D*

R.I. Pinsker, M.J. Mayberry, M. Porkolab,† and R. Prater
General Atomics, San Diego, CA 92138

## INTRODUCTION

Previous tokamak experiments with Ion Bernstein Wave (IBW) heating,[1-3] have exhibited efficient central ion heating and associated improvement in particle confinement. The prospect of localized bulk ion heating, along with the prediction that coupling to the IBW should improve under edge conditions characteristic of H–mode plasmas,[4] motivated the present IBW program on the DIII–D tokamak. The 30–60 MHz tunability of the modified 2.25 MW FMIT rf power supply permits consideration of many different heating scenarios; however, experiments reported here have concentrated on reproducing the regime of Ref. 1, in which central ion heating at $3/2\Omega_H$ with a small $He^4$ minority ($3\Omega_{He^4}$) was observed. Therefore, the transmitter frequency was 38 MHz and the nominal central toroidal field was $B_T = 1.8$ T, which places the $3/2\Omega_H$ ($3\Omega_{He^4}$) layer a few centimeters outboard of the magnetic axis, and the $2\Omega_H$ layer just behind the radiating element in the antenna, as is required theoretically for efficient coupling to the IBW.[4]

## APPARATUS

The IBW coupler used in these experiments consists of a pair of cavity-type end fed loop antennas, oriented along the toroidal field, similar to the couplers used in all previous IBW heating experiments. Each current strap is 41 cm long and 16 cm wide. The loops are mounted colinearly in a single 1 m wide port at the outside midplane of DIII–D. The external impedance matching network includes a phase shifter between the two antenna feeds, which permits control of the launched $k_\parallel$ spectrum. In-phase operation yields a spectrum peaked at $k_\parallel = 0$, while operating with out-of-phase feeds produces a spectrum peaked at $k_\parallel \simeq \pm 7.4$ m$^{-1}$ ($n_\parallel \simeq 9.3$ at 38 MHz). Two different Faraday shields have been used: the first shield, used for most of the experiments reported here, consisted of a single row of 9.5 mm molybdenum rods, spaced on 19 mm centers, with 0.16 cm thick graphite tiles brazed onto the plasma-facing side. Recent experiments have employed an optically opaque Faraday shield consisting of two offset rows of 9.5 mm molybdenum rods coated with a thin layer of TiC spaced on 11 mm centers. With either shield, typical plasma loading yields a coupling efficiency of about 90%. Details of the antenna design are given in Ref. 5.

The total loading resistance $R_T$ is here defined as the sum of $2P_{net}/I_{max}^2$ for the two halves of the antenna, where $I_{max}$ is the maximum current in the transmission line. $R_T$ is measured with twelve voltage probes installed in the transmission lines near the antenna, and independently with dual directional couplers located at four points in the transmission lines. Edge electron density, temperature, and floating potential are measured with a moveable Langmuir probe located on the midplane near the IBW

---

* Work supported by U.S. DOE Contract DE-AC03-89ER51114.
† Massachusetts Institute of Technology, Cambridge, MA 02139

antenna. The same probe configured as an electrostatic rf probe is used to measure rf spectra in the edge plasma.

## RESULTS

In previous tokamak IBW experiments, the observed loading resistance has been found to be much larger than predicted theoretically. Furthermore, the theory predicts a sharp maximum in loading when an integral harmonic layer is located just behind the current strap;[4] this resonant behavior[6] has not been observed in most of the experiments. We have measured the loading in discharges in which the toroidal field was ramped down by 0.3 T during the 2 second rf pulse. With a fixed net rf power of 0.16 MW, a line averaged density of $\bar{n}_e = 3 \times 10^{13}\,\text{cm}^{-3}$, and $I_p = 1.0$ MA, the loading resistance with $k_\| = 0$ antenna phasing varied by less than 0.5 $\Omega$ from an average value of 4.5 $\Omega$ over the range of $B_T = 0.7 - 2.1$ T. The dependence of $R_T$ on density at the antenna was measured by varying the distance between the separatrix and the antenna in a lower single-null divertor configuration, with a fixed rf power of 0.25 MW and $I_p = 1.0$ MA, $\bar{n}_e = 3 \times 10^{13}\,\text{cm}^{-3}$, $B_T = 1.88$ T. The results are shown in Fig. 1. The dependence of the loading on the outer gap, and hence on density at the antenna is quite different for the two antenna phasings. The loading at high $k_\|$ drops as the edge density is lowered, while loading at low $k_\|$ is nearly independent of edge density until the gap is quite large, whereupon the loading tends to increase with further lowering of the edge density. In all cases, the absolute magnitude of the loading is about an order of magnitude larger than the theoretical prediction.[4]

Edge electron temperature and density measurements provide evidence of significant edge heating: upon injection of 0.5 MW of rf, the edge electron temperature rises from 7 eV to > 20 eV. The edge density does not change significantly. This edge heating may be correlated with the significant parametric decay observed, as well as with the substantial influx of metallic impurities that occurs upon rf injection.

An example of the parametric decay spectra observed at $B_T = 2.08$ T is shown in Fig. 2. In this case at least three pairs of decay waves are observed. The usual frequency selection rules are obeyed within experimental error for each pair of frequencies. Decay spectra were observed for magnetic fields in the range of $B_T = 0.7 - 2.1$ T. Our initial assessment indicates that depending upon magnetic field, decay into two (or more) Bernstein waves and/or decay into a Bernstein wave and an ion cyclotron quasimode may occur.[7,8] The multiple decay pairs owe their existence to the presence of the He$^4$ minority species in the hydrogen majority plasma. A detailed study of these nonlinear processes and their consequences is underway.

The evolution of a typical high power discharge in the regime studied here is displayed in Fig. 3, where an rf power of 0.50 MW is coupled to the plasma ($I_p = 1.4$ MA, $B_T = 1.81$ T, $\bar{n}_e = 3 \times 10^{13}\,\text{cm}^{-3}$) for 1 sec. The radiated power increases by 0.56 MW, evidently because of the influx of metallic impurities, primarily nickel. Note that there is no exposed surface containing Ni in the antenna or Faraday shield; the Ni almost certainly originates in the Inconel tiles lining the outboard wall of the vacuum vessel. The loop voltage rises slightly, the central $Z_{\text{eff}}$ rises by $\sim 0.2$, and the line-averaged density rises by 34%. Some evidence (ECE, spectroscopy) exists that the central electron temperature drops slightly. This must occur, as the stored energy decreases somewhat

316

during IBW injection. In experiments to date, no increase in stored energy has been detected under any conditions, with net injected IBW power of up to 0.6 MW.

One difference between the regime of our experiment and that of previous work is that under these conditions in DIII–D, the scaling of $\tau_E$ with $\bar{n}_e$ is saturated, i.e. increasing the density in the Ohmic phase of the discharge gives no increase in $\tau_E$. IBW heating and confinement improvement has been previously observed[2,3] when the target plasma is still in the regime of linear scaling of $\tau_E$ with $\bar{n}_e$. Our present experiments are aimed at reproducing this regime by operating at much lower target plasma densities.

### REFERENCES

[1] M. Ono, et al., Phys. Rev. Lett. **54**, 2339 (1985).

[2] M. Ono, et al., Phys. Rev. Lett. **60**, 294 (1988).

[3] J. Moody, et al., Phys. Rev. Lett. **60**, 298 (1988).

[4] S.C. Chiu, M.J. Mayberry, and W.D. Bard, "Theoretical Comparison of Coupling of a Recessed Cavity and a Conventional Loop Antenna for Fast Waves and Ion Bernstein Waves," General Atomics Report GA–A19600 (1989), submitted for publication in IEEE Trans. on Plasma Science.

[5] R.D. Phelps, M.J. Mayberry, and R.I. Pinsker, "Ion Bernstein Wave Antenna Design for DIII–D," presented at *15th Symp. on Fusion Tech.*, Utrecht, Netherlands, 1988.

[6] Y. Takase, et al., Phys. Rev. Lett. **59**, 1201 (1987).

[7] M. Porkolab and J. Moody, Bull. Am. Phys. Soc. **32**, 1939 (1987).

[8] R. Van Nieuwenhove, *et al.*, Nucl. Fusion **28**, 1603 (1988).

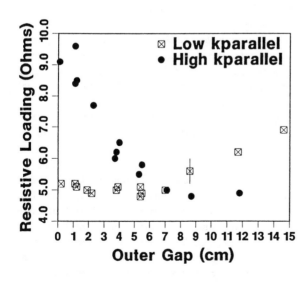

Fig. 1. Total resistive loading as a function of the gap between the separatrix and the outer limiter. The Faraday shield face was recessed 0.5 cm behind the surrounding graphite surface, which is 2 cm behind the limiter. Langmuir probe measurements indicate a density near the antenna radius of about $3 \times 10^{11}$ cm$^{-3}$ with a gap of 1 cm, and about $3 \times 10^{10}$ cm$^{-3}$ with a gap of 5 cm.

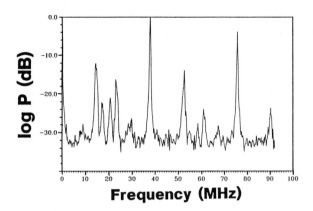

**Fig. 2.** Parametric decay spectrum obtained with $P_{rf} = 0.33$ MW, $B_T = 2.08$ T, outer gap = 5 cm. The power is measured relative to the pump power (38 MHz). Pairs of decay waves at $14.5 \pm 0.3$, 23.0 MHz, 17.2, 20.6 MHz, and 9, 29 MHz are evident, along with corresponding upper sidebands and the second harmonic of the pump.

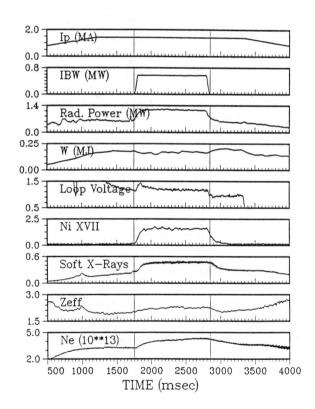

**Fig. 3.** Time evolution of a typical high power IBW discharge in DIII–D.

# PLASMA HEATING AND NONLOCAL ENERGY TRANSPORT
# DUE TO INVERSE MODE CONVERSION OF
# ION BERNSTEIN WAVES

Hugo A. Romero and G. J. Morales

Physics Department, UCLA

Los Angeles, CA 90024-1547

## ABSTRACT

An ICRF full-wave equation is used to investigate the process of inverse mode conversion of a short-wavelength ion Bernstein wave (IBW) into a long-wavelength fast magnetosonic wave. Appropriate boundary conditions are presented to set-up the numerical analysis. For conditions corresponding to fundamental minority heating in tokamaks (minority concentration above 2% and $k_\parallel < 9\mathrm{m}^{-1}$) most ($> 90\%$) of the incident IBW energy is mode converted into a fast wave that propagates towards the high field side of the device. This process can lead to an efficient transfer of thermally-generated IBW energy away from the tokamak chamber, or it may find applications as a diagnostic of ion behavior. Direct, high field side launch of an IBW at the second harmonic of the ion gyrofrequency can result in efficient ($> 90$ %) heating of the **majority** species if the ion temperature is above 1 keV. In all cases considered, the reciprocity of the fast wave to IBW and IBW to fast wave mode conversion coefficient has been observed.

## INTRODUCTION

The heating of tokamak plasmas using high power Ion Cyclotron Radio Frequency (ICRF) waves is achieved by launching a long-wavelength wave, the fast magnetosonic wave, from the low magnetic field side of the tokamak. Efficient localized heating at the ion cyclotron resonance layer results from cyclotron damping and linear mode conversion into a short-wavelength mode: the Ion Bernstein wave. We refer to this process as the direct problem. It is characterized by the resonant damping and transfer of energy of a large scale wave to a short scale wave in a narrow layer, and has been extensively investigated by several authors. In this paper, we present a full-wave investigation of the inverse process, namely the absorption and mode conversion physics of an ion Bernstein wave that is internally generated in the plasma, or externally coupled from the high magnetic field side of the tokamak. We have been motivated to study the physics related to the inverse mode conversion of an IBW because: i) the mode converted fast wave can be used as a diagnostic of ion behavior in tokamak experiments, ii) the conversion of thermally-generated IBW energy into fast wave energy can cause a nonlocal (and fast) transport of energy, and iii) efficient heating of the **majority** species can be achieved at the second (or higher) ion harmonics via direct launching of the IBW due to large $k_\perp \rho$ effects.

## BASIC EQUATIONS AND NUMERICAL ANALYSIS

A kinetic description of the fast and IBW interaction is required to take into account the mode conversion physics and the resonant absorption of the wave fields in the vicinity of the ion cyclotron resonances. The equation for the equilibrium distribution function contains first and second order corrections in $\rho/L$ ($\rho$ denotes the thermal gyroradius and $L$ the scale length of variation of equilibrium quantities) to obtain a self-adjoint formulation of the wave equation, which is written in the form

$$\frac{d}{dx}\left[A_{ij}\frac{dE_j}{dx}\right] + B_{ij}\frac{dE_j}{dx} + \frac{d}{dx}[B_{ij}E_j] + Q_{ij}\frac{dE_j}{dx} - \frac{d}{dx}[Q_{ij}E_j] + C_{ij}E_j = 0 \qquad (1)$$

where $E_j$ is the $j$-th cartesian component of the electromagnetic field. Expressions for the matrices of coefficients can be found in Ref. 1. The procedure for numerical solution is to obtain, in the vicinity of the resonance/mode conversion layer, four linearly independent solutions to Eq.(1). Outside of this region, the fields are written in their WKB form:

$$\mathbf{E}(x) = \begin{cases} \mathbf{E}_{\text{IBW}}{}^{(inc)} + R_{\text{IBW}}\,\mathbf{E}_{\text{IBW}}{}^{(ref)} + M_{\text{F}}\,\mathbf{E}_{\text{F}}{}^{(mc)} & x \leq x_l \\ \sum_{i=1}^{4} S_i\,\mathbf{E}_i(x) & x_l \leq x \leq x_r \qquad (2) \\ T_{\text{F}}\,\mathbf{E}_{\text{F}}{}^{(trn)} + E_{\text{IBW}}\,\mathbf{E}_{\text{IBW}}{}^{(ev)} & x \geq x_r \end{cases}$$

In Eq.(2), $\mathbf{E}_{\text{F}}$ and $\mathbf{E}_{\text{IBW}}$ denote the fast and IBW asymptotic forms of the wave fields. Requiring the continuity of $\mathbf{E}$ and its derivative at the points $x_l$ and $x_r$ results in a set of eight equations for the eight unknown coefficients, and the problem is formally solved.

## RESULTS

We first discuss results pertaining to the basic physics of the inverse mode conversion process. Figure 1 summarizes our principal observation in this regard, which is the reciprocity between the mode conversion coefficient of the processes fast wave $\rightarrow$ IBW and IBW $\rightarrow$ fast wave for the case of incidence from the high field side of the tokamak. The results in Fig. 1 pertain to an equilibrium configuration corresponding to fundamental minority heating of a (H)-D plasma in JET: major radius 3 m, minor radius 1.25 m, toroidal magnetic field 3T, minority concentration 5%, and electron density $3.5 \ 10^{13} \ cm^{-3}$. The majority and electron temperatures are 750 eV, while the minority (heated species) temperature is 2 keV. We plot the ratio of the absorption, transmission, and mode conversion coefficients that result from the incidence to the resonance layer, first, of a fast wave and then of an IBW. Note the great disparity between the absorption and transmission components, while the mode conversion coefficient ratio remains fixed near unity. We have observed the validity of this result in various ion concentrations and in different heating schemes.

We consider next in Figs.(2) and (3) the dependence of the inverse mode conversion co-efficients on the minority concentration and the $k_\parallel$ spectrum of the IBW for the equilibrium parameters used to obtain the results of Fig.(1). In Fig.(2), we indicate the variation of the absorption and mode conversion coefficients as the minority concentration is varied. Note that for minority concentrations above 2%, the IBW effectively transfers most of its energy ($> 90\%$) to a fast wave that propagates towards the high field side of the tokamak. This result points to the possibility that this process can cause an efficient nonlocal transport of energy since the mode converted fast wave can readily propagate, without appreciable attenuation, away from the resonance layer. In Fig.(3) we examine the sensitivity of the IBW inverse mode conversion coefficient to the $k_\parallel$ value of the IBW. It is found that unless the wave spectrum is centered at high $k_\parallel$ values ($> 8 \ m^{-1}$), the efficiency of the inverse mode conversion process remains high ($> 90\%$).

We present in Fig.(4) results corresponding to second harmonic heating of a 50-50 D-T plasma for the equilibrium parameters used to obtain the results of Fig.(1). The plasma temperature is 1 keV. The frequency of the wave is chosen to equal the second harmonic of tritium on the axis of the machine. The tritium absorption and IBW reflection coefficients are shown as a function of $k_\parallel$. We observe nearly complete absorption of the incident IBW energy for $k_\parallel > 3 \ m^{-1}$. Thus, if a compact IBW launcher could be accomodated on the high magnetic field side of the tokamak, this scheme has the advantage of heating the **majority** species directly. It might be expected that few tail particles should develop, and that the bulk plasma temperature increase at a faster rate than in the fundamental minority heating scheme.

## REFERENCES

1. H. Romero and J. Scharer, Nuc. Fusion **27**, 363 (1987); and H. Romero and G. J. Morales, UCLA Report PPG-1166.

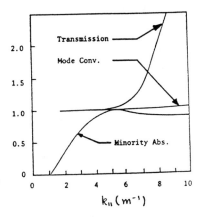

Fig. 1. Test of Reciprocity
Coefficient. Ratio of Abs.,
Mode Conv., and Transmission
Coefficients as a function
of $k_{\shortparallel}$.

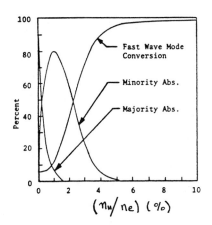

Fig. 2. Absorption and
Mode Conversion Coeff.
for high field side
incidence of IBW. (H)-D
fundamental minority
heating.

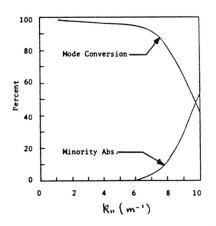

Fig. 3. Absorption and Mode
Conversion Coefficients for
high field side incidence of
IBW. (H)-D heating.

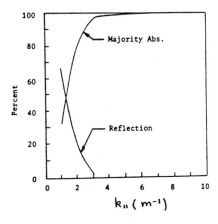

Fig. 4. Absorption and
Reflection for IBW high
field incidence. D-T
second harmonic heating.

# DETERMINATION OF FIELDS NEAR AN ICRH ANTENNA USING A 3D MAGNETOSTATIC LAPLACE FORMULATION*

P.M. Ryan, K.E. Rothe, J.H. Whealton, and D.W. Swain
Oak Ridge National Laboratory, Oak Ridge, TN 37831

## ABSTRACT

In the vicinity of an ICRH antenna strap, where there are no volume currents and a free-space wavelength is much longer than the dimensions of interest, Ampere's law reduces to a curl-free condition on the magnetic field, allowing a magnetic scalar potential to be defined. This scalar potential is a solution of the three-dimensional (3D) Laplace equation and satisfies the following boundary conditions on the magnetic field: (1) the line integral of the magnetic field around the current strap is equal to the current flowing in the strap and (2) the perpendicular component of the magnetic field vanishes at conductor surfaces (no flux penetration of perfect conductors). This formulation allows for the magnetic field solution of quite complex 3D geometries, such as poloidal current straps with asymmetric radial feeds or detailed Faraday shield geometries.

## INTRODUCTION

The design, evaluation, and optimization of ICRH antennas requires knowledge of the electromagnetic fields generated by the 3D launcher structure. Such knowledge permits evaluation of the antenna coupling and power density distribution, as well as the power transmission, reflection, and absorption of the Faraday shield. A 3D magnetostatic analysis has been developed to meet the following requirements:

(1) the ability to specify and modify the geometry of the structure quickly and easily, thus facilitating optimization studies;

(2) the reduction of computation time needed for a given degree of accuracy;

(3) the ability to treat a localized area of a large structure, such as the region of a Faraday shield element, in greater detail;

(4) the self-consistent calculation of current distributions, within the constraints of the model.

## THE MAGNETOSTATIC MODEL

The analysis uses Ampere's law, Fourier-analyzed in time and space:

$$\nabla \times \mathbf{B} = -j\,\mathbf{k} \times \mathbf{B} = \mu_0\mathbf{J} + j\omega\mu_0\varepsilon_0\mathbf{E} = j\left(k_0/c\right)\mathbf{E}, \tag{1}$$

where the current density $\mathbf{J}$ vanishes in the solution domain. The long-wavelength approximation may be used when $k_0/k \approx L/\lambda \ll 1$, where L is the characteristic scale length of the device. This approximation decouples the electric and magnetic fields and reduces Ampere's law to the curl-free static representation

$$\nabla \times \mathbf{B} = 0. \tag{2}$$

This allows $\mathbf{B}$ to be defined as the gradient of a scalar potential $\Psi_m$, which is a solution of Laplace's equation:

---

*Research sponsored by the Office of Fusion Energy, U.S. Department of Energy, under contract DE-AC05-84OR21400 with Martin Marietta Energy Systems, Inc.

$$\nabla \cdot \mathbf{B} = \nabla \cdot \left( \nabla \Psi_m \right) = \nabla^2 \Psi_m = 0. \tag{3}$$

Neumann boundary conditions are imposed on $\Psi_m$ on all conducting surfaces, which ensures that the normal component of $\mathbf{B}$ vanishes. The integral form of Ampere's law, which equates the total current flowing in the strap with the line integral of $\mathbf{B}$ along any contour that encloses the strap, is used to set the Dirichlet boundary conditions on the remaining surfaces:

$$\oint_c \mathbf{B} \cdot d\mathbf{l} = 2\int_{c_2} \nabla \Psi_m \cdot d\mathbf{l} = 2\left( \Psi_{m2} - \Psi_{m1} \right) = \mu_0 I , \tag{4}$$

where $\Psi_{m1} \equiv 0$ and $\Psi_{m2} \equiv \mu_0 I/2$.

## APPLICATIONS

Figure 1 shows the geometry of a representative current strap in a recessed cavity; a radial current feed is attached to the top of the strap, and the bottom of the strap is grounded to the cavity. Figure 2 shows a similar geometry, except that the strap has been divided in the poloidal direction into two loops which are isolated from one another by a ground plane. This is done to reduce the maximum voltage that appears along the strap and to minimize the current falloff by decreasing the ratio of strap length to wavelength. The long-wavelength approximation used in these calculations implies constant current along the length of the strap, which is a design ideal. Typical ICRH loop antennas are designed to have $k_0/k \approx L/\lambda < 0.1$; these calculations may be modified by $\cos(\beta y)$ effects for electrically long antennas.

Figures 3(a) and 3(b) show the radial and toroidal components of the magnetic field near the first wall of the tokamak for the geometry of Fig. 1; the asymmetry in the fields is due to the asymmetry in the current feed. Figures 4(a) and 4(b) show the same field components for the poloidally stacked array of Fig. 2; the ground plane has introduced higher poloidal wave numbers into the spectrum. The relative power handling of these two antennas may be estimated by integrating the square of the magnetic field. For the current-limited case, in which all straps carry the same maximum current, the power is reduced by 25% for the poloidally stacked array; the poloidal extent of the strap is reduced by 14%; the total length of each strap, and hence the maximum voltage, is reduced by 36%. For the more typical voltage-limited case, keeping the maximum voltage constant allows the loop current to be increased by 56% and the total power to be increased by 83%. The actual power handling increase for this example may be higher, depending on the frequency of operation, owing to the neglected finite-wavelength effects.

This analysis will also be used to study the transmission properties of Faraday shield geometries. The magnetostatic approach allows one to consider only one poloidal period of the structure, as shown in Fig. 5 for a two-tier shield. The upper and lower boundaries are now symmetry planes and Neumann boundary conditions are imposed; the scale parameter for the long-wavelength approximation, $L/\lambda$, is typically around $10^{-3}$ in such geometries. The transmission and shielding properties of the shield will be found from comparison with the fields in the absence of the shield. The magnetic field at the shield surface is equated with the surface current, which can be used to estimate the eddy current losses in good conductors.

324

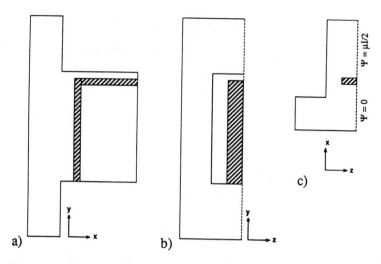

Fig. 1. Representative geometry for an asymmetrically fed current strap in a recessed cavity; (a) side view, (b) front view, (c) top view showing Dirichlet boundaries.

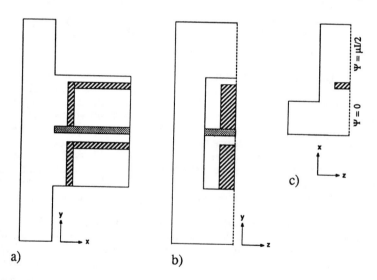

Fig. 2. Representative geometry for a poloidally stacked, isolated array.

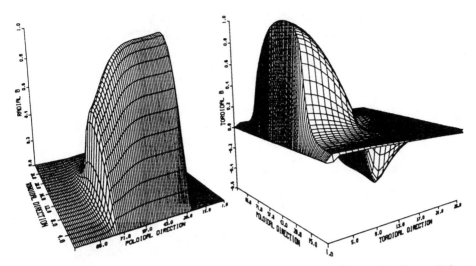

Fig. 3. (a) Radial and (b) toroidal magnetic field components close to the first wall for the antenna shown in Fig. 1.

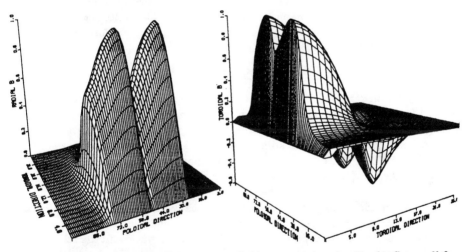

Fig. 4. (a) Radial and (b) toroidal magnetic field components close to the first wall for the antenna array shown in Fig. 2.

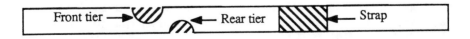

Fig. 5. Example of the poloidally periodic geometry used to model a two-tier Faraday shield (not drawn to scale).

326

# OBSERVATION OF PARAMETRIC DECAY WAVES DURING SECOND HARMONIC ICRF HEATING EXPERIMENT IN JT-60

M. Saigusa, H. Kimura, T. Fujii, N. Kobayashi, S. Moriyama,
T. Nishitani, K. Annoh, Y. Ogawa, S. Shinozaki, M. Terakado,
M. Ohta, T. Nagashima and JT-60 team
Naka Fusion Research Establishment
Japan Atomic Energy Research Institute,
Naka-machi, Naka-gun, Ibaraki-ken, 311-01, Japan

## ABSTRACT

Parametric decay waves were observed during second harmonic ICRF heating experiments in JT-60. The parametric decay process is probably that the pump wave decays into ion Bernstein waves (IBW) and ion cyclotron quasi-modes. Because the decay wave frequency at a toroidal magnetic field of 3.3-4.5 T agreed with the predicted frequency of IBW which was assumed to be excited parametrically near the ICRF antennas. The decay wave intensity decreased with increasing plasma current and with increasing toroidal magnetic field. Increase in radiation loss induced by ICRF heating at the plasma current of 1 MA was proportional to the intensity of the decay wave, but was not proportional to the pump wave power. Those results suggest that the parametric decay wave was one of the sources of the impurity production during the second harmonic ICRF heating experiments.

## INTRODUCTION

Ion cyclotron range of frequency (ICRF) heating is one of the most promising methods for a reactor plasma. Mechanisms of the impurity production during ICRF heating, however, have not yet been clarified thoroughly. Parametric instability during the second harmonic ICRF heating was observed in the several tokamaks (TFR[1], TEXTOR[2], ASDEX[3]). This process has been believed that the ion Bernstein waves (IBW) and the ion (or electron) cyclotron quasi-modes (IQM) were parametrically excited by the near electric field of the ICRF antennas[4].

The parametric decay process during ICRF heating might be one of the mechanisms of the direct energy deposition in edge plasmas and the impurity production. This paper reports the characteristics of parametric decay waves and the relations between the increase in the radiation loss and the decay wave intensity during the second harmonic ICRF heating in JT-60.

## OBSERVATION OF PARAMETRIC DECAY INSTABILITY

RF spectra were measured during the second harmonic ICRF heating (131 MHz $= 2\omega_{CH}$ at Bt $= 4.3$ T) of an ohmically heated plasma or a plasma preheated by NBI $(H^0, \sim 70keV)$ or LHRF (2 GHz or 1.74 GHz) in the limiter discharges. The main plasma parameters were Ip $= 0.7 \sim 2.7$ MA, Bt $= 3.3 \sim 4.8$ T, $\bar{n}_e = 0.7-8.7\times10^{19}$ cm$^{-3}$. Figure 1 shows the position of the RF probes in the poloidal cross section in JT-60. There are two pair of Langmuir probes attached on the top of the ICRF (P-11, R =

3.591 m) and the LHRF (P-18, R = 3.587 m) launchers. The measurement system which consists of two spectrum analyzers and two transient recorders can gather the RF spectra and the time evolution of fixed frequency signal at the frequency range from 10 MHz to 2.9 GHz (P-11) or to 22 GHz (P-18) during a shot[5].

A typical RF spectrum during ICRF heating which was measured by the probe placed under ICRF launcher is shown in Fig.2. Many decay waves were observed during ICRF heating, that is, the pump wave (131 MHz), its second harmonic wave (262 MHz), its third harmonic wave (393 MHz) and the parametric decay waves excited from the above three waves. The intensity of the harmonics waves of pump wave in Fig.2 were much more intense than that in vacuum condition. Therefore, those waves were not only due to the characteristics of RF generator, but also due to the nonlinear sheath effect at the ICRF antenna, which is similar to the observed waves in TEXTOR and ASDEX. The one of the decay wave frequencies, which seems to be the IBW decayed from the pump wave, versus the toroidal magnetic field is shown in Fig.3. The parametric decay process must obey the relation[4],

$$\omega_0 = \omega_1 + \omega_2, \ \bar{k}_0 = \bar{k}_1 + \bar{k}_2 \tag{1}$$

where $\omega_0$, $\omega_1$, $\omega_2$, $\bar{k}_0$, $\bar{k}_1$ and $\bar{k}_2$ are the angular frequencies of the pump wave, IQM and IBW and the wave number vectors of those waves, respectively. On the assumption that the one decay wave is the IQM ($\omega_{IQM} \simeq \omega_{CH}$), the frequency of the other decay wave (IBW) can be calculated with the above relations. The solid line and the broken line in Fig.3 show the IBW frequencies calculated by using Eq.(1), at the plasma surface and at the magnetic surface of 10 cm inside from the plasma surface in front of the center of ICRF antenna, respectively. The broken line agrees with the experimental data at the plasma current of 1.5 MA. The IQM signal, however, could not always be measured by the RF probes around the predicted frequency. The observation of IQM probably depended on the plasma condition around the RF probe. The excited layer of the measured IBW considering the band width and the predicted absorption layer ($\omega_{CH} = \omega_{IBW}$) are shown in Fig.1. The excited layer width considering the band width of IBW agrees with the launcher size.

Figure 4 shows the dependence of IBW signal intensity on the toroidal magnetic field. The signal intensity gradually decreased with increasing the toroidal magnetic field and abruptly decreased over 4.3T. The experimental data might be explained by the following reason. The predicted absorption layer ($1.5\omega_{CH} = \omega_{IBW}$) for IBW was located on the ICRF antenna at the toroidal magnetic field of 4.22 T, so that the excited IBW could be absorbed nearby the excited region.

## RELATION BETWEEN RADIATION LOSS AND DECAY WAVES

The intensity of the IBW signal decreased with increasing the plasma current. The similar dependence on the plasma current was seen in the radiation loss induced by the ICRF heating as shown in Fig.5. Figure 5 shows $\Delta P_{rad}/P_{IC}$ versus the averaged electron density in the limiter discharges. The maximum value of $\Delta P_{rad}/P_{IC}$ decreased with increasing a plasma current. The radiation loss gradually increased with the averaged electron density in same plasma current, except that the increase in radiation losses at a plasma current of 1.0 MA were particularly large and did not depend on the averaged electron density.

Figure 6 shows the relation between the radiation loss induced ICRF and IBW signal level at the toroidal magnetic field of 4.3 T. The increase in the radiation loss is proportional to the IBW signal at the plasma current of 1.0 MA, where the IBW signal level is not proportional to the pump wave power or the pump wave power over the averaged density. Therefore, IBW is probably one of the sources of impurity production during ICRF heating at Ip = 1.0 MA. This result may explain the reason why the $\Delta P_{rad}/P_{IC}$ at Ip = 1.0 MA was much larger than that at other plasma currents.

## CONCLUSION

The ion Bernstein wave which were parametrically excited by the near field of ICRF antenna were observed during the second harmonic ICRF heating experiment in JT-60. The predicted IBW frequency at Bt = 3.3 $\simeq$ 4.5 T agreed with the one of the measured decay wave frequencies. The IBW signal were dependent on the plasma current and the toroidal magnetic field. We found the relation the increase in the radiation loss during ICRF heating was proportional to the IBW signal at the plasma current of 1.0 MA and the toroidal magnetic field of 4.3 T. It suggests that IBW excited by the fast wave produced the impurity.

## REFERENCES

[1] V.K. Tripathi, C.S. LIU, Nucl. Fusion 26 (1986) 963.
[2] R. Van Nieuwenhove et al., in Proceedings of 15th Eur. Conf. on Controlled Fusion and Plasma Heating (Dubrovnic, Yugoslavia, 1988), Vol. 12B, Part II, p778.
[3] R. Van Nieuwenhove et al., Nucl. Fusion 28 (1988) 1603.
[4] F. Skiff, M. Ono, K.L. Wong, Phys. Fluids 27 (1984) 1051.
[5] JT-60 Team, Japan Atomic Energy Research Institute Report JAERI-M 88-063 (1988).

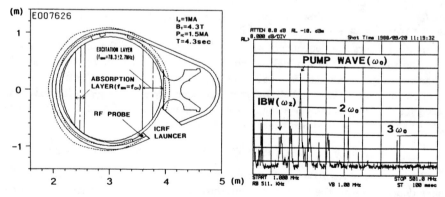

Fig. 1. Probe position in poloidal cross section of JT-60.

Fig. 2. Typical RF spectrum measured by RF probe during ICRF heating. Ip=1MA, B$_T$=4.3T, P$_{IC}$=1.5MW, $\bar{n}_e$=1.3×10$^{19}$cm$^{-3}$

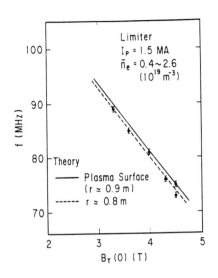

Fig. 3. IBW frequency versus toroidal magnetic field.

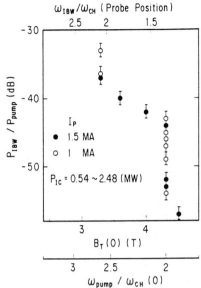

Fig. 4. IBW signal versus toroidal magnetic field.

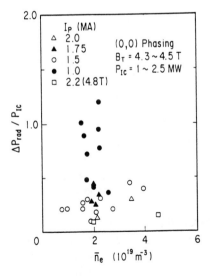

Fig. 5. $\Delta P_{rad}/P_{IC}$ versus averaged electron density.

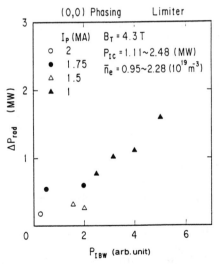

Fig. 6. $\Delta P_{rad}$ induced by ICRF heating versus IBW signal.

# ICRF FULL WAVE FIELD SOLUTIONS AND ABSORPTION FOR D-T AND D-$^3$He HEATING SCENARIOS*

J. Scharer and R. Sund
Electrical and Computer Engineering Department
University of Wisconsin, Madison, WI 53706-1687

## ABSTRACT

We consider a fundamental power conservation relation, full wave solutions for fields and power absorption in moderate and high density tokamaks to third order in the gyroradius expansion. The power absorption, conductivity tensor and kinetic flux associated with the conservation relation as well as the wave differential equation are obtained. Cases examined include D-T and D-$^3$He scenarios for TFTR, JET and CIT at the fundamental and second harmonic. Optimum single pass absorption cases for D-T operation in JET and CIT are considered as a function of the $k_\parallel$ spectrum of the antenna with and without a minority He$^3$ resonance. It is found that at elevated temperatures > 4 keV, minority (10%) fundamental deuterium absorption is very efficient for either fast wave low or high field incidence or high field Bernstein wave incidence. We consider the effects of a 10 keV bulk and 100 keV tail helium distribution on the second harmonic absorption in a deuterium plasma for JET parameters. In addition, scenarios with ICRF operation without attendant substantial tritium concentrations are found the fundamental (15%) and second harmonic helium (33%) heating in a the deuterium plasma. For high field operation at high density in CIT, we find a higher part of the $k_\parallel$ spectrum yields good single pass absorption with a 5% minority helium concentration in D-T.

I.    **INTRODUCTION.** We consider wave absorption in the ICRF for minority fundamental and second harmonic heating for D-T and D-$^3$He plasmas. Of primary concern is finding a single pass absorption of the ion species in the plasma core for $k_\parallel$ spectrum appropriate for a launching antenna.

To examine the efficiency of ICRF absorption we use a computer code corresponding to an appropriate definition power absorption and the associated conservation relation for inhomogeneous plasmas developed by McVey, Sund and Scharer [1]. The code correctly solves the propagation and coupling of incident fast magnetosonic waves from the low field side machine to ion Bernstein waves in the resonant core region.

**II.    D-T AND D-T-(³He) ICRF ABSORPTION IN CIT.** We first
examined the absorption for second harmonic tritium for high field, high
density CIT parameters ($B_0$ = 10 T, T = 10 keV, $n_e$ = $5 \times 10^{14}$/cm³, a = 0.55 m,
$R_o$ = 1.22 m and 50/50 D/T). The single pass absorption for tritium is quire
low (< 20%) for all of the $k_{\parallel}$ spectrum. As noted by Scharer et al. [2], the
tritium absorptivity always lies below that of the deuterium at the second
harmonic for this range of plasma parameters.

The 5% minority ³He case in a 45% - 45% deuterium-tritium plasma
as a function of the $k_{\parallel}$ spectrum for the above CIT parameters in the ± 20
cm central zone is illustrated in Fig. 1. One notes that the ³He single pass
absorption of 65% peaks at a high $k_{\parallel}$ of 18 m⁻¹. The associated electron
absorption is 12% and the tritium absorbs 4% of the incident fast wave
power from the low field side for a 20 cm absorption width near the core of
the machine. At elevated temperatures one has to be sure that the
electron absorption over the whole machine profile via Landau and
transit-time damping is not stronger than the ion heating in the core. One
also notes that a substantial fast wave reflection (<40%) from the helium
cyclotron resonance occurs for lower (<10 m⁻¹) values of $k_{\parallel}$.

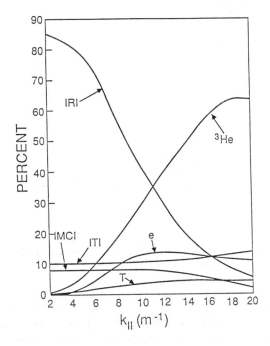

Fig. 1.   Low field fast wave scan for CIT

## III. SINGLE PASS FAST WAVE ABSORPTION FOR JET D-$^3$He

PLASMAS. We consider JET parameters with ne = 5 x 10$^{13}$/cm$^3$, B$_0$ = 3.45 T, R$_0$ = 3.0 m, a = 1.15 m, n$_{3He}$ = 5% n$_e$ and n$_D$ = 90% n$_e$ with T = 10 keV. We consider wave solutions in the zone R$_0$ ± 20 cm. The fast wave low field results show fundamental $^3$He absorption peaking at k$_{\parallel}$ = 5.5 m$^{-1}$ at a single pass absorption of 46% with < 4% electron heating and modest reflections. What was somewhat surprizing was the high field incident Bernstein wave case with 95% $^3$He core absorption over a wide range of the wave spectrum as shown in Fig. 2. The mode conversion process is substantial for lower values of k$_{\parallel}$.

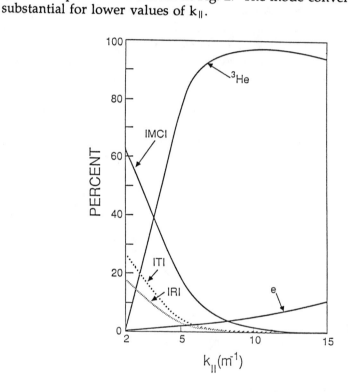

Fig. 2. High field Bernstein wave incidence for JET

Finally we consider a high power JET case with substantial $^3$He concentrations and bulk and tail $^3$He distributions at the second harmonic. The JET parameters are as in Fig. 2 except that T$_{3He}$ = 10 keV for the bulk distribution and T$_{3Het}$ = 100 keV for the tail temperature. The $^3$He bulk distribution is 25% n$_e$ with the tail density n$_{3Het}$ = 25% n$_{3Heb}$. The wave solutions and absorption are obtained in the inner ± 25 cm. In this case

$k_{\perp f} \, \rho_{tail} = 0.55$ so that this provides a test of the gyroradius expansion for the code. The wave frequency is set equal to $2 \, \omega_{C3He}$ on the axis at this higher helium concentration. Figure 3 illustrates the $k_\parallel$ scan. The tail absorption greatly dominates the absorption especially at lower values of $k_\parallel \leq 10 \text{ m}^{-1}$ At higher values of $k_\parallel \leq 15 \text{ m}^{-1}$ the transmission increases and the tail and bulk absorption are reduced.

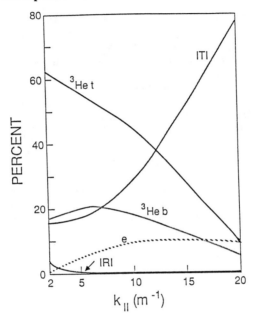

Fig. 3. Low field fast wave incidence for
JET with $^3$He tails and bulk distributions

IV. **SUMMARY.** These results indicate that good ICRF ion absorption can be obtained in hotter, moderate to high density tokamaks in both D-T-($^3$He) and D-$^3$He cases. However, particular attention to ICRF antenna design and coupling physics as well as core species concentration will have to be paid to avoid substantial reflections and dominant electron heating for higher temperature tokamaks.

**REFERENCES**

1. B. McVey, R. Sund and J. Scharer, Physical Review Letters <u>55</u>, 507 (1985).
2. J. Scharer, J. Jacquinot, P. Lallia and F. Sand, Nucl. Fusion <u>22</u>, 255 (1985).

*This research was supported in part by DOE Grant DE-FG02-86ER53218.

334

# LOADING, ABSORPTION, AND FOKKER-PLANCK
# CALCULATIONS FOR UPCOMING ICRF EXPERIMENTS ON ATF*

T. D. Shepard, M. D. Carter, R. H. Goulding, M. Kwon
ORNL Fusion Energy Division, Oak Ridge, TN  37831-8071

## ABSTRACT

ICRF experiments on ATF at the 100-kW level are planned for the current 1989 operating period. These plans include the $2\omega_{cH}$ regime at $f_{RF} = 28.88$ MHz, D(H) at 14.44 MHz, and $^4$He($^3$He) and D($^3$He) at 9.63 MHz. ECH target plasmas have $n_{e0} \lesssim 0.15 \times 10^{20}\,\mathrm{m}^{-3}$ and $B = 0.95$ T. The density and temperature profiles obtained are broader than those from 1988, owing to recent field error corrections. The values used for target-plasma parameters in the calculations were taken from initial 1989 ATF data. Loading and absorption calculations have been performed using the 3D RF heating code ORION with a helically symmetric equilibrium, and Fokker-Planck calculations were performed using the steady-state code RFTRANS with two velocity dimensions and one spatial dimension.

## INTRODUCTION

Medium-power ICRF heating experiments are currently in progress on the ATF torsatron. A maximum of 100 kW of RF power in the frequency range $5 \leq f_{RF} \leq 30$ MHz is currently available, and a prototype single-strap resonant-double-loop cavity antenna that can couple this power to the plasma is installed on ATF. The $2\omega_{cH}$, D(H), and D($^3$He) regimes all fall within this frequency range for second-harmonic X-mode ECH-produced plasmas at $B_0 = 0.95$ T, while only the minority regimes are within this range for fundamental O-mode ECH-produced plasmas at $B_0 = 1.9$ T. Experiments in all these regimes are planned; however, at this writing experimental data have been taken only in the $2\omega_{cH}$ regime, and ATF has run only at $B_0 = 0.95$ T.

Preliminary ICRF data were taken during the initial ATF operating period in 1988 by Goulding et al.,[1] and loading calculations similar to those presented herein were performed by Kwon et al.[2] In the current (1989) ATF operating period, broader density and temperature profiles are obtained due to recent corrections of field errors, and a few ICRF experimental shots have been produced, still only in the $2\omega_{cH}$ regime. The purposes of the work described herein are (1) to repeat Kwon's calculations to see if any differences in antenna loading are to be expected with the broader profiles, and (2) to perform Fokker-Planck calculations to see if it is reasonable to expect any nonthermal minority tail evolution when RF power is limited to 100 kW.

## LOADING AND ABSORPTION CALCULATIONS

Loading and absorption calculations were performed using the helically symmetric 3D RF heating code ORION.[3,4] This code performs a transform in the direction of symmetry and uses 2D finite differencing in the other directions. Finite-Larmor-radius hot plasma wave theory is used for the absorption model and mode conversion is handled using the Smithe-Colestock reduced-order dielectric kernel,[5] in which mode conversion appears as additional absorption. Although the antenna consists of only a single current strap, it was modeled using four straps to mock up the effects of current peaking on the sides of the current strap and image currents

Research sponsored by the Office of Fusion Energy, U.S. Department of Energy, under contract DE-AC05-84OR21400 with Martin Marietta Energy Systems, Inc.

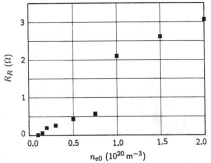

Figure 1: Antenna loading as a function of density for the $2\omega_{cH}$ regime.

---

in the side walls of the antenna cavity. This was found by Kwon[2] to be both necessary and sufficient to obtain agreement between the calculation and experimental measurements.

A repeat of the $2\omega_{cH}$ density-scan calculation by Kwon[2] is shown in Fig. 1. The parameters used in this calculation were typical for shots obtained after field error corrections and are essentially the same as those used by Kwon, except that the density profile shape was parabolic and the temperature profile was a parabola squared. (Kwon used $n(r) \sim [1 - (r/a)^2]^{1.8} + \text{const}$ and $T(r) \sim [1 - (r/a)^3]^5 + \text{const}$) Specifically, $f_{RF} = 28.88$ MHz, $B_0 = 0.95$ T, $T_{e0} = 700$ eV, $T_{i0} = 150$ eV, $n_{edge}/n_0 = 0.1$, $R_0 = 2.1$ m, $r = 22 \times 32$ cm, and the distance from plasma edge to current strap is 6.3 cm.

These results are very similar to those from Kwon's calculation except that Kwon's results exhibited strong loading peaks vs density that were not seen in the experiment. These peaks were eliminated in the present calculations by increasing the density in the region outside the plasma boundary, which suppresses edge modes. However, there is no evidence of a significant change in loading associated with the broader profiles in the interior of the plasma.†

For future high-power heating experiments, it is desired that the antenna be able to couple 1 MW of power to the plasma. In order to do this without breakdown, a loading of at least 2 $\Omega$ is required. Preliminary loading measurements indicate that we can expect this much loading at least at relatively high density in the $2\omega_{cH}$ regime. It may be possible to obtain improved loading by a simple modification to the antenna. The power spectrum and absorption profiles are shown in Fig. 2 for both low- and high-density cases. Each calculation is done using two different antenna spectra. The solid curves show the results for a "white" antenna current spectrum; i.e., the calculation was performed repeatedly at various values of $k_\parallel$ with the same antenna current at each $k_\parallel$. Such a spectrum would be produced by an idealized antenna in which the current strap consists of an infinitesimally thin filament and no return current flows in the cavity side walls. The dashed curves show the results for the actual antenna current spectrum. The antenna is evidently not well matched to the plasma, since the peaks in the actual spectrum are at high values of $k_\parallel$ out on the "tails" of the white-current-spectrum case. The lack of a $k_\parallel = 0$ component is due to the return currents flowing in the side walls of the antenna cavity. While a $k_\parallel = 0$ component is not desired for efficient heating, poor loading and broad absorption profiles result if too many nonzero but relatively small values of $k_\parallel$ are missing from the spectrum. It may be possible to obtain more low-$k_\parallel$ components by narrowing the antenna current strap (to reduce the distance between the strap edges and the cavity side walls) and/or by slotting the side walls. This should be investigated by performing more calculations and by performing tests on the ATF ICRF bench prototype antenna.

---

† The results presented here are too low by a constant scale factor because of a normalization error in the ORION code. The results are not sensitive to the variation in profile shapes described above. This bug did not appear in the version of ORION used by Kwon.

336

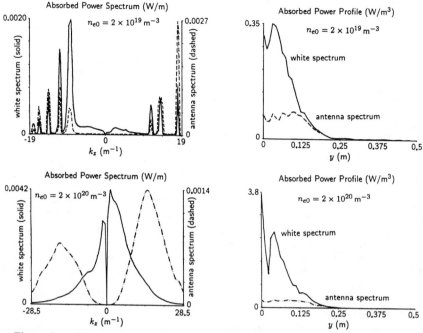

Figure 2: Power spectrum and absorption profile in low-density, weak-damping, eigenmode regime and in high-density, strong-damping regime. Profiles were not scaled to contain the same total RF power, but were all calculated using the same total antenna current.

## FOKKER-PLANCK CALCULATIONS

Minority tail generation and thermalization for a low-density D(H) case was studied using the 3D steady-state code RFTRANS[6]. This code solves the kinetic equation in two velocity space dimensions, speed and pitch angle, and in one real space dimension, radius, for the minority distribution function. Real space transport is modeled in this simulation by using a diffusion coefficient that is a constant in velocity and real space equal to 2.5 $m^2$/s. A cold uniform source is used to produce a parabolic density profile so that the global particle confinement time for ATF in the simulation is roughly 16 ms. For this diffusion coefficient, the equilibrium minority temperature is nearly equal to the equilibrium majority temperature for zero RF power with collisional heating given by electron and majority temperatures $T_{e0} = 700$ eV and $T_{D0} = 150$ eV. The profile shapes assumed in the calculation are the same as those used in the preceding $2\omega_{cH}$ loading calculations. Other parameters used were $B_0 = 0.95$ T, $f_{RF} = 14.44$ MHz, $n_{e0} = 0.2 \times 10^{20}\,m^{-3}$, and minority concentration $\eta_H = 5$ %.

The results of these calculations are shown in Fig. 3. With absorbed RF power $P_{RF} = 83$ kW, nonthermal minority distributions with average energy in excess of 1 keV are produced. Thermalization is reasonably efficient, with 75 kW of power deposited to the background plasma (primarily on deuterium). The collisional deposition profile is broader than the absorption profile owing to the transport effects included in the model. The anisotropy of the minority distribution function (at $r = 7$ cm) is also indicated by plotting $f_H(E, \theta)$ for two values of pitch angle $\theta$.

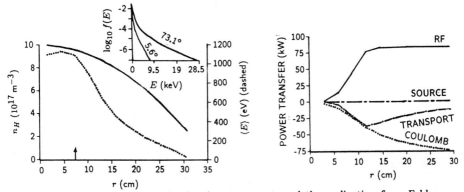

Figure 3: Hydrogen minority heating, transport, and thermalization from Fokker-Planck calculations.

## CONCLUSION

ICRF modeling calculations have been performed using plasma parameters and RF power levels that can reasonably be expected in near-term experiments on ATF. The results of loading and absorption calculations agree with existing experimental loading data and suggest that improved antenna loading may be possible if some simple modifications are made to the antenna design. Fokker-Planck calculations for minority heating indicate that experimentally significant tail production should occur, and that most of the absorbed power should be collisionally deposited on the majority ion species.

The question of whether the net deposited power of 75 kW is sufficient to produce significant bulk deuterium heating has not been addressed in detail. This power is approximately two times the electron-ion collisional exchange power, but the profile shape for the electron-ion exchange power is more peaked. Based on these simple considerations, one might hope for measurable bulk ion heating at these power levels, provided the electrons do not cool due to impurity influx. The results indicated by the Fokker-Planck calculations should be observable with the existing ATF diagnostic capability (the neutral particle analyzer, in particular) so it should be possible to obtain an initial evaluation of ICRF heating efficiency on ATF from these initial medium-power experiments.

## REFERENCES

1.  R. H. Goulding, F. W. Baity, M. Kwon, and D. J. Hoffman, Bull. Am. Phys. Soc. **33**(9), 1981 (1988) 5E10.

2.  M. Kwon, E. F. Jaeger, P. M. Ryan, and R. H. Goulding, Bull. Am. Phys. Soc. **33**(9), 2016 (1988) 6R1.

3.  E. F. Jaeger, D. B. Batchelor, H. Weitzner, and J. H. Whealton, Comput. Phys. Commun. **40**, 33 (1986).

4.  E. F. Jaeger, D. B. Batchelor, H. Weitzner, and P. L. Colestock, in Proc. 7th Topical Conf. Applications of RF Power to Plasmas (edited by S. Bernabei and R. W. Motley) AIP Conf. Proc. **159**, 394 (1987).

5.  D. N. Smithe, P. L. Colestock, R. J. Kashuba, and T. Kammash, Nucl. Fusion **27**(8), 1319 (1987).

6.  M. D. Carter, D. B. Batchelor, and E. F. Jaeger, Bull. Am. Phys. Soc. **33**(9), 1900 (1988) 2T3.

# SNARF ANALYSIS OF ICRF HEATING ON TFTR

D. N. Smithe, C. K. Phillips, G. W. Hammett, and P. L. Colestock
Princeton Plasma Physics Laboratory, Princeton, NJ 08543

## ABSTRACT

An ICRF heating model[1] which utilizes a poloidal mode expansion and contraction of the wave equation to second order, has been incorporated into the steady-state transport analysis code, SNAP[2]. A simple Stix-type isotropic Fokker-Planck calculation[3] models thermalization of the energy in the hot ion tails. Models of several other physical processes, such as enhanced RF damping from the hot tail and neutral beams, finite banana width effects, and sawtooth mixing have been incorporated in the code. This code has been used for data analysis of ICRF heating shots over TFTR's latest run. We will present analyses of these shots, including a toroidal field scan and simultaneous RF and beam heating.

## THE MODEL: SNARF=SNAP+RF

SNAP is a steady-state transport analysis software package which can take various TFTR measurement data such as magnetics, density and temperature profiles, and emission spectra, and combine them with models of plasma equilibrium and transport, beam deposition, and atomic physics, to provide estimates of internal processes which are not measured, or measurable. This paper describes the recent addition of an ion cyclotron RF model to SNAP, and uses the enhanced code to perform analysis of TFTR shots employing ICRH. The new RF software is comprised of two separate modules, a wave-propagation and deposition code, and an isotropic Fokker-Planck code.

The first module is the 2D (radial variation and poloidal harmonics) wave propagation and deposition code described in ref. [1], and commonly known as SHOOT. Several new features beyond the original software, ref. [1], have been introduced. Enhanced damping from the hot ion tails is modeled by increasing the Maxwellian temperature of the resonant ion species so that $^3/_2nT$ matches an estimate of the total stored energy from the isotropic Fokker-Planck model. Enhanced damping on the beam ions is modeled by increasing the Maxwellian temperature of the resonant beam species to match the estimation of beam stored energy from SNAP's beam physics model. Omission of these two effects would result in unphysically thin resonance layers, and therefore unphysically high estimates of power localization.

Mode-conversion is modeled via the contraction mechanism described in ref. [1], whereby the full 4th-order fast-wave root is employed in the finite-Larmor terms of the dielectric tensor. For this code, a simple separation of the mode-converted power from the ion damping is performed by separating the damping of the anti-Hermitian tensor from the damping associated with the imaginary part of the $k_\perp^2$. The mode-converted power is assumed to be deposited locally to the electrons. However, because of the enhanced damping from the hot tails, mode-conversion is rarely a dominant effect. Electron Landau and TTMP damping have also been added to the code. At heating levels and electron temperatures analyzed here, these are also small effects. However, with expected increases in available RF power, and with further beam+RF experiments planned, they could become noticeable.

The isotropic Fokker-Planck module follows Stix's calculation, ref. [3]. At each minor radius each ion tail is solved for individually, such that the RF quasilinear term is exactly balanced by the drag on background Maxwellian plasma species. The background temperatures are taken to be thermal temperatures as estimated by SNAP, or as measured directly. In the case of beam ions, the background temperature is raised by the estimated stored energy of the beam.

For the RF quasilinear term, a value of $k_\perp^2$ is required in the argument of the Bessel functions. This is taken to be the deposition-weighted average of the local fast-wave solution to the dispersion relation. Since the deposition model is truncated in Larmor radius, while the Fokker-Planck calculation uses untruncated Bessel functions, there is an inconsistency which can lead to power imbalance. To circumvent this, a multiplicative factor is introduced into the quasilinear RF term, and is iteratively selected so that the net power matches the deposition model. For off-axis heating, this correction is a few percent, or less. For highly focused deposition, small minority concentrations, or higher total heating power, the correction is more significant.

## BEAM vs OHMIC PLASMA HEATING

In this section, analysis and comparison of representative shots of RF heating in an Ohmic plasma (shot 37131) and in a neutral beam heated plasma (shot 37139) will be made. The RF power for both shots was around 1.8 MW. The Ohmic plasma was He-4 with H minority, $n_H/n_e\approx.05$ and $n_e=.36\times10^{20}$ m$^{-3}$, while the beam plasma was Deuterium with 4.7 MW D-beams and H minority, $n_H/n_e\approx.025$ and $n_e=.21\times10^{20}$ m$^{-3}$. Figure 1 shows the power deposition contours in the poloidal plane, as calculated by SNARF. In general, the beam plasma is predicted to see a much broader absorption region because of 2nd harmonic damping on the D-beam particles, and also because of the hotter ion thermal temperatures, 11.5 keV vs. only 1.5 keV for the Ohmic plasma.

In the Ohmic plasma, virtually all of the wave damping is on the H minority population. The Fokker-Planck analysis predicts an energetic H tail population with a peak effective temperature of around 210 keV. The total stored energy in the H tail is predicted to be about 120 kJ, or roughly 20% of the total kinetic energy of the plasma. Magnetic measurements indicated that the stored kinetic energy was strongly diamagnetic, consistent with a largely perpendicular H tail population. Because the tail is so energetic, most of the collisional drag is from electrons rather than ions, with $P_e$=1.4 MW, $P_{He-4}$=0.3 MW, and the remaining 0.1 MW going to impurities.

In the beam plasma, the wave damping was split with 1.4 MW going to the minority, and 0.4 MW going to D 2nd harmonic heating. The Fokker-Planck analysis predicted a peak effective H tail temperature of 650 keV (higher than Ohmic plasma because of the smaller H density) and a peak D effective temperature of 17 keV above the beam+thermal background. This may be compared to the peak beam effective temperature, which was predicted by SNAP to be roughly 30 keV above thermal (the beam injection energy was 100 keV). The stored energy in the ion tails was estimated to be 270 kJ in the Hydrogen and 100 kJ in the Deuterium, or roughly 32% of the total kinetic energy of the plasma. This may be compared to the estimated 560 kJ of kinetic energy in the D slowing down population, which comprised 48% of the kinetic energy. Magnetic measurements of the RF+beam heated plasma indicated no abnormally high diamagnetism, perhaps suggesting that the two co-beams, with largely parallel orientations were effectively balanced by the largely perpendicular ion tail populations. Because of the high H tail temperature, the drag on the tails was dominated by electrons, as was the case with the Ohmic plasma. Here, SNARF estimated $P_e$=1.25 MW, $P_D$=0.45 MW, and the remaining 0.1 MW going to impurities.

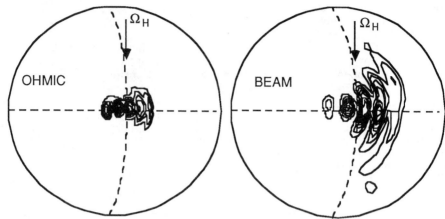

**Figure 1.** Local RF deposition in the poloidal plane for RF+Ohmic and RF+Beam heating. The lower ion temperatures of the Ohmic case result in better localization.

## TOROIDAL FIELD SCAN IN OHMIC PLASMAS

A series of four consecutive He-4 / minority H Ohmic shots (37129-37132) of nearly identical plasma parameters except toroidal field will be analyzed in this section. The most important consequence of varying the toroidal field is the varying location of the RF deposition vis-à-vis the magnetic axis and also the degree of focusing. The major radius locations of the exact H fundamental resonance for these four shots were $R_{res}$=2.93, 2.83, 2.73, and 2.62 m. The central magnetic axis was at $R_0$=2.73 m, thus shots 37129-31 were resonant on the outer side of the magnetic axis, shot 37131 was resonant on axis, and shot 37132 was resonant on the inside of the axis. The vessel major radius is at $R_V$=2.65 m; the code predicted the best focusing for shot 37132 and the worst for 37129.

**Table 1.** Toroidal Field Scan Data

| Shot # | $R_{res}-R_0$ [cm] | $P_e$ [W/cc] | $P_{He-4}$ [W/cc] | $T_{tail}$ [keV] | $W_{\perp SNARF}$ [MJ] | $W_{\perp EXP}$ [MJ] |
|---|---|---|---|---|---|---|
| 37129 | 20 | 1.36 | 0.63 | 71 | .453 | .467 |
| 37130 | 10 | 1.65 | 0.40 | 299 | .509 | .521 |
| 37131 | 0 | 1.80 | 0.28 | 378 | .531 | .544 |
| 37132 | -11 | 1.86 | 0.24 | 286 | .494 | .523 |

When the heating is on-axis and focused, the local deposition is a maximum. This results in the highest tail energies, and consequently the strongest collisional coupling to electrons. Table 1 shows the splitting of power between electrons and He-4, the peak tail temperature, SNARF's prediction of total kinetic energy of the plasma, and the diamagnetic measurement of stored energy. The modeling code predictions are consistent with the measured values of stored energy.

A measurement of the local on-axis electron heating can be derived from the observed rate of increase in $T_e$ after a sawtooth.[4] Since the temperature profile is flattened after a sawtooth, conduction and convection are small, and the observed temperature rise must be due to RF and Ohmic heating alone. It is observed that shot 37129 has a visible temperature rise on-axis, even though the RF deposition is expected to be completely off-axis. This rise is too great to be due to Ohmic heating. Conversely, shot 37131 is observed to have a temperature rise four times smaller than predicted by the RF model. Both these observations can be explained if one assumes that the sawtooth mixes the hot ion tails within the $q=1$ radius.[4]

A model of sawtooth mixing was added to the SNARF package, whereby the H velocity distributions, $f(r,v)$, calculated in the Fokker-Planck module are radially averaged within the $q=1$ radius, and partially averaged between the $q=1$ radius and mixing radius, $r_{mix}=\sqrt{2}r_{q=1}$. The collisional coupling of electrons to these mixed ion distributions is then recalculated to find a new estimate of central electron heating. These were found to be in much better agreement with the observed $T_e$ rises, as shown in Figure 2. It was further observed that the rate of $T_e$ rise was roughly flat across the central region, further supporting the hot ion mixing supposition. One interesting aspect of the mixing calculation is that the averaged ion velocity distributions can show a shape that has a lower effective tail temperature, which results in improved collisional coupling to the ions. Under certain circumstances, this improved ion coupling should be observable spectroscopically. Unfortunately the current data is insufficient for verifying this prediction.

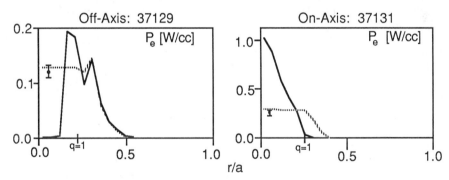

**Figure 2.** Radial profiles of predicted electron heating before a sawtooth (solid) and after sawtooth mixing (dotted). The center data points are measured from the $T_e$ rise.

A model of finite banana-width physics was also added to SNARF. The collisional coupling of the electrons to the tail distributions, $f(r,v)$, was spread out in radius over the banana width corresponding to an orbit of velocity $v$ with tips on the resonance layer. The net effect of this redistribution was to generally smooth the deposition pattern, however, no significant increase in the size of the deposition footprint occurred for the powers which were in use. This may be explained by noting that near the edges of the deposition pattern, the net local powers are small, resulting in low energy tails and hence thinner banana orbits.

## REFERENCES

1. D. N. Smithe, P. L. Colestock, R. J. Kashuba, and T. Kammash, *Nuclear Fusion* **27**, 1319 (1987).
2. H. H. Towner and R. J. Goldston, *Bull. of Amer. Phys. Soc.* **29**, 1305 (1984).
3. T. H. Stix, *Nuclear Fusion* **15**, 737 (1975).
4. C. K. Phillips, *et. al.*, this conference.

## METAL IMPURITY BEHAVIOR DURING ICRF HEATING ON TFTR[*]

J.Stevens, C.Bush, P.Colestock, G.J.Greene, K.W.Hill, J.Hosea,
C.K.Phillips, B.Stratton, S.vonGoeler, J.R.Wilson, W.Gardner[+],
D.Hoffman[+], and A.Lysojvan[#]
Plasma Physics Laboratory, Princeton University; [+]Oak Ridge National
Laboratory; [#]Kharkov Institute of Physics and Technology

### ABSTRACT
ICRF power levels of up to 2.8 MW were achieved during the 1988 experimental run on TFTR. The central metal impurity concentrations (Ti,Cr,Fe,Ni) and $Z_{eff}$ were monitored during ICRF heating by X-ray pulse height analysis and UV spectroscopy. Antenna phasing was the key variable affecting ICRF performance. No increase in metalic impurities was observed for $P \lesssim 2.8$ MW with the antenna straps phased 0-Π, while measureable increases in titanium (Faraday screen material) were observed for $\lesssim 1.0$ MW with 0-0 phasing.

### INTRODUCTION AND EXPERIMENTAL SETUP
A density rise and an influx of metal impurities are sometimes observed during ICRF experiments. It was shown on PLT that $Z_{eff}$ did not increase significantly with RF power when the tokamak was well conditioned and operated at sufficiently high plasma current. Nevertheless the impurity question is of increasing concern for TFTR and other large tokamaks as powers and pulse lengths are increased to reactor like values. Numerious papers have studied impurities associated with ICRF [1-2 for reviews]. The role of the antenna spectrum is now being studied experimentally on larger tokamaks[3], while theories [2,4-5] have not addressed the question of phasing.

ICRF experiments were begun on TFTR in 1988 with two antennas, each consisting of two loops which can be driven toroidally in-phase(0-0) or out-of-phase(0-Π). One antenna has TiC coated inconel Faraday screens with deep slots cut in the antenna casing. This antenna has been operated at powers up to 2.8MW, limited to date by the RF generator. The second antenna has carbon tiles bonded on water cooled inconel Faraday screens and has been operated at 0.5MW.

Large size plasmas were used during ICRF experiments on TFTR so that the antenna-plasma seperation is small enough for good loading. These plasmas had R=2.62m and a=0.96m versus a standard size of R=2.45m & a=0.82m. Loading is higher for in-phase operation than for out-of-phase operation and increases with smaller antenna-plasma seperation.

Two heating modes were used for ICRF experiments on TFTR: (1) H-minority operating at a central toroidal field of ≈3.3T, and (2) $^3$He-minority operating at ≈5.0T. The majority gas was either deuterium or helium. Helium was used for all experiments with ICRF alone (to avoid loading the limiter with deuterium) while experiments with ICRF in combination with neutral beams used low density deuterium discharges. The metal impurity concentration observed in TFTR depends mainly on the type of plasma discharge, with high density He plasmas [fig.1] having low $Z_{eff}$ and $Z_{metals}$ and low density, beam heated deuterium plasmas producing $Z_{metals} \approx 1$. This behavior with density is typical of all TFTR plasmas.

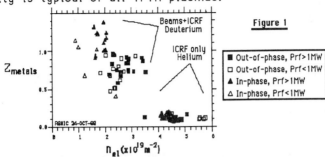

Figure 1

The titanium in the PPPL antenna Faraday screen provides a monitor of metal impurities generated directly from the antenna since it is the only significant source of titanium in TFTR. The concentration of titanium and other metals in the central part of the plasma is measured by monitoring the $K\alpha$ lines ($E \approx 4\text{-}8keV$) with a X-ray pulse height analysis system which views along a horizontal midplane chord. UV spectroscopy gave relative impurity levels similar to X-ray pulse height analysis.

RESULTS

The titanium concentration was negligable for standard size plasmas, even with 30MW of neutral beams. A measurable Ti signal was present for the large size plasmas with >1 MW of in-phase ICRF power in the H-minority regime [fig.2a] but was not present for up to 2.8 MW of out-of-phase power [fig.2b].

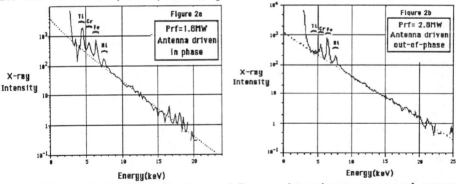

Figures 3a&b show the $Z_{Ti}$ and $Z_{metal}$ dependence versus rf power for in-phase and out-of-phase conditions in He plasmas. No change in the level of titanium or other metals versus rf power is observed for out-of-phase antenna operation with up to 2.8 MW, while in-phase operation shows an increase in $Z_{Ti}$ and $Z_{metal}$ above $\approx 1$ MW. A visible glow on the Faraday shields at the bottom of the antenna was also observed during in-phase operation. In-phase excitation (low $k_{\parallel}$) may produce higher local E-fields leading to higher rf sheath potentials [4] or to local acceleration of ions [5] such as 3rd harmonic $C^{+5}$ or 5th harmonic $C^{+3}$ and $He^{+1}$.

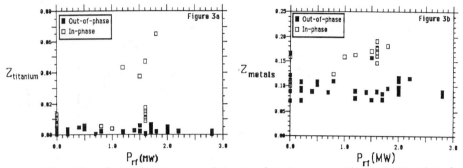

The high impurity level with the in-phase spectrum may also be related to poor first pass absorption of the in-phase spectrum. A toroidal field scan [fig.4], although somewhat limited, showed a minimum in $Z_{eff}$ due to metals and titanium around B0=3.25T which is the field where central heating is the best.

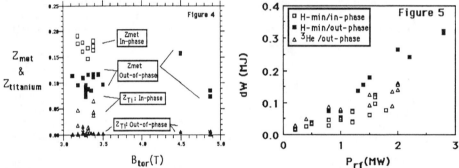

ICRF incremental confinement [fig.5] and heating efficiency varied with antenna phasing. Heating efficiency ($\epsilon = W/P_{rf}|_{trf}$) was ≈90% for out-of-phase operation and only ≈40% for in-phase operation. Thus, both impurity generation and poor heating efficiency are correlated with antenna spectra having small $k_{\parallel}$ (in-phase), while best results occur for the out-of-phase condition with a spectrum centered at $k_{\parallel} \approx 7m^{-1}$. The $^3$He-minority mode had heating results somewhat between the H-minority in-phase and out-of-phase modes and also showed metal impurity concentrations between the two modes. However, further optimizion of the $^3$He concentration is needed before a firm conclusion can be drawn for the $^3$He-minority mode.

Neutral beam heating and combined beams plus ICRF both produce measurable Ti levels in the large size TFTR plasmas [fig.6], suggesting that high power levels into the plasma, rather than ICRF power per se, can cause some of the impurity influx from the antenna. Comparing equal power shots with and without ICRF shows that 2 MW of ICRF does not produce any strikingly detrimental effects compared with the beams, especially when beam fueling with higher beam powers is taken into account. The $Z_{eff}$ due to titanium is still a small fraction of $Z_{metals}$ for $P_{beam} + P_{rf} + P_{OH} \lesssim 12MW$.

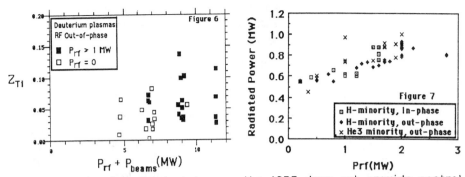

Figure 6

Figure 7

In contrast to neutral beams, the ICRF does not provide central fueling to dilute an initially high $Z_{eff}$ plasma. Nevertheless, during ICRF heating on TFTR the radiated power fraction decreased from $\approx 50\%$ to $\approx 30\%$ for up to 2.8MW of power, similar to the decrease typical of beam heating [fig.7]. Radiated power scales approximately linearly with stored energy.

## CONCLUSION

Driving the antenna out-of-phase improves the ICRF heating efficiency and reduces the metal impurity generation caused directly by the launched RF wave. However, metal structures near the hot plasma do contribute metal impurities approximately in proportion to total power in the plasma. These two conditions constrain the antenna design: the antennas should be far enough from the plasma to avoid impurity generation from the hotter scrapeoff flux while at the same time the antennas should be close enough to the plasma to provide sufficient RF loading for an out-of-phase (high $k_{\parallel}$) spectrum. These constraints have been satisfied so far on TFTR for plasmas with up to 12 MW of input power (2 MW of which is ICRF) for $\approx 0.5$ second pulses. Metal impurity levels coorelate approximately inversely with heating effectiveness. Higher power levels are planned for both antennas which should help resolve the trends reported here and also provide a comparison of impurities versus Faraday screen material.

*Supported by DOE Contract No. DE-AC02-76-CHO-3073.
[1] H.L.Manning, et.al., Nucl.Fusion, 26(1986)1665.
[2] I.S.Lehrman, Ph.D.dissertation, U.Wisconsin(1988).
[3] M.Bures, et.al., Plasma Physics and Controlled Fusion Research, 30(1988)149.
[4] F.W.Perkins, PPPL-2571, Nov.1988.
[5] S.Puri, Phys.Rev.Lett.,61(1988)959.

# ALCATOR C-MOD ICRF FAST WAVE ANTENNA DESIGN AND ANALYSIS AND EXPECTED PLASMA PERFORMANCE

Y. Takase, S. Golovato, M. Porkolab,

H. Becker, N. Diatchenko, S. Kochan, C. McCoy, N. Pierce

MIT Plasma Fusion Center, Cambridge, MA 02139

## ABSTRACT

The design and analysis of the fast wave antenna to be used for ICRF heating experiments on the Alcator C-Mod tokamak are described. A movable single-strap low-field-side-launch loop antenna will be used during the initial operation period. Structural analysis against disruption and thermal analysis are performed. Optimized two-strap antennae with twice the surface area are planned for high-power heating experiments. Antenna-plasma coupling studies indicate a poor ($< 20\%$) single-pass ion absorption for D($^3$He), and a radiation resistance of $\gtrsim 10\,\Omega$. Minority ions are not expected to be accelerated to very high energies under typical operating densities ($T_{eff} \lesssim 50\,\mathrm{keV}$ for $\bar{n}_e \gtrsim 3 \times 10^{20}\,\mathrm{m}^{-3}$), and energetic ions should mainly slow down on ions. The transport of minority ions broadens the energetic minority ion and slowing down profiles and reduces $T_{eff}$, but the effect is small for $\bar{n}_e \gtrsim 3 \times 10^{20}\,\mathrm{m}^{-3}$.

## ANTENNA DESIGN AND ANALYSIS

Alcator C-Mod is a compact, high-field, diverted tokamak which is scheduled to become operational in late 1990. Typical parameters are: $R = 0.665\,\mathrm{m}$, $a = 0.21\,\mathrm{m}$, $\kappa = 1.8$, $B \leq 9\,\mathrm{T}$, $I \leq 3\,\mathrm{MA}$. ICRF heating power of 4 MW at 80 MHz will be available. Minority heating utilizing either $^3$He (resonant at $B = 7.9\,\mathrm{T}$) or H (resonant at $B = 5.3\,\mathrm{T}$) in a D majority plasma is planned.

Initially only one antenna will be installed. The first antenna will be made movable to provide operational flexibility, but will be of a single-current-strap type because of restricted port width (0.20 m). During the initial operation period the condition for optimum coupling (plasma-antenna separation, in particular) will be determined. Preferred plasma shape and elongation will be established and disruption speed, scrape-off length, and heat flux will be measured. Based on information obtained during this period, optimized antennae for full power operation will be constructed.

The drawing of the first antenna is shown in Fig. 1. One of the major considerations in the design is mechanical strength against disruption-induced forces, which are very large in compact, high-field tokamaks. The antenna current strap is enclosed in a box which holds the Faraday shield rods and protection tiles. The back plate of the box provides mechanical strength. The current strap is grounded and supported mechanically at the center, and is fed from both ends through the feedthroughs which are attached directly to the back plate of the antenna box. The box sidewalls are slotted to allow the rf fields to escape from the side. The Faraday shield consists of two layers of triangular cross section rods. The front layer elements have the hypotenuse facing the plasma so that atoms generated by sputtering between the Faraday shield rods will not enter the plasma.[1] The second layer elements are oriented in the opposite direction to minimize the distance from the plasma to the current strap. The plasma-facing surfaces will be coated with TiC. Faraday shield rods are oriented 15° with respect to horizontal in order to align with the field line under typical operating conditions. Antenna box, Faraday shield, and current strap are made of Inconel (Nickel alloy) for mechanical strength and for reducing the disruption-induced eddy currents. All rf current carrying surfaces are plated with copper or silver. The antenna box is protected by Molybdenum tiles mounted on the sidewalls. The support points for the antenna are the top and bottom keyways mounted on the vacuum chamber and the bearing surfaces on the support tubes (which enclose the feedthroughs and the coaxial transmission lines) at the port extension flange. These support points guide the antenna for radial movement and provide support against disruption forces. The radial motion is controlled by a pair of drive screws driven by a single motor. Details of the transmission/matching system and the optimization studies of the current strap and box geometries are reported in a separate paper.[2]

The stress analyses of the antenna were performed for 1 MA/ms plasma disruptions at 3 MA and 9 T. Eddy currents and resultant forces are calculated by the SPARK code[3] and stresses are calculated by a commercial finite element code PAFEC.[4] A vertical disruption gives the largest forces and stresses. The largest components at the end of an upward-moving disruption are $M_x = 12\,\text{kNm}$ and $F_y = 18\,\text{kN}$, which combine to give a reaction of $F_y = -39\,\text{kN}$ at the bottom front keyway. Internal stresses are highest at the end of the sidewall slots, but they are below the allowable stress.

After 10 hours of 20-minute cycles of 3-second 6 MW shots and assuming 100 kJ of total rf dissipation per shot (mostly on the sidewalls), the antenna is expected to heat up to approximately 250°C. Momentarily after the shot, the tile temperature can climb up to 750°C whereas the sidewall temperature can rise to 500°C. Since this result depends on the value of interface conductance (which was taken to be $0.1\,\text{W/cm}^2\,{}°\text{C}$ in the analysis) and boundary conditions, the temperature will be monitored at several critical locations.

## ANTENNA-PLASMA COUPLING STUDIES

Antenna-plasma coupling was investigated by the slab-geometry ICRF coupling code developed by Brambilla,[5] both with radiating and reflecting boundary conditions. Single-pass ion absorption for D($^3$He) is predicted to be low (typically 10%, never much more than 20%) compared to the case of D(H) (typically $\gtrsim 50\%$). Optimum minority concentration for single-pass ion absorption for the D($^3$He) case is $\simeq 2\%$. Radiation resistance is calculated to be $\gtrsim 10\,\Omega$ for a single-strap antenna.

The field line pitch $B_p/B_t$ on the outside midplane is expected to be 10–20° under typical operating conditions. The radiation resistance is predicted to decrease by as much as 50% if the Faraday shield is misaligned from the field line by 15°. Most of this loss can be recovered if the Faraday shield can be aligned with the magnetic field. Significant ion Bernstein wave excitation is expected if the density at the antenna is too low ($\omega_{pi} \lesssim \omega$). It may be important to position the antenna close enough to the plasma so that such a condition is not realized during H-mode transitions, for example.

The possibility of $k_\parallel$ spectrum shaping by phasing two current straps was also investigated. Out-of-phase excitation of two current straps eliminates the low $k_\parallel$ component which may be responsible for impurity generation.[6,7] A two-strap antenna which fits in the same antenna box as the current one-strap design will generate a $k_\parallel$ spectrum which is peaked at very high $k_\parallel$ ($\simeq 25\,\text{m}^{-1}$) with a low radiation resistance. Phasing two single-strap antennas installed in adjacent ports will produce a $k_\parallel$ spectrum peaked at $k_\parallel \simeq 5\,\text{m}^{-1}$. Optimum $k_\parallel$ spectrum, peaked at $k_\parallel \simeq 10\,\text{m}^{-1}$ with acceptable radiation resistance and optimum single-pass ion absorption, is obtained when the spacing between the current straps is about half the distance between ports. The radiated $n_\parallel$

spectrum ($n_\parallel \equiv ck_\parallel/\omega = 6$ corrensponds to $k_\parallel = 10\,\mathrm{m}^{-1}$) of the "optimized" antenna is shown in Fig. 2(b), which can be compared with the single-strap antenna spectrum shown in Fig. 2(a). $P_F$, $P_I$, and $P_B$ represent the transmitted fast wave, ion absorption, and mode conversion powers, respectively. Such an antenna will have about twice the surface area of the single-strap antenna and will require internal assembly, but will allow operation at a lower power density of $\simeq 1\,\mathrm{kW/cm}^2$ for 2 MW of injected power. In addition, two double-strap antennae installed in adjacent ports can produce a traveling wave spectrum with $k_\parallel \simeq 5\,\mathrm{m}^{-1}$ for 90° phasing between adjacent straps.

## MINORITY ION TAIL FORMATION

The behavior of the minority ion distrubution function was studied with a bounce-averaged Fokker-Planck code (FPPRF)[8,9] coupled iteratively with a toroidal wave code (SHOOT).[10] Very energetic minority ion tail is not expected to be formed under typical operating conditions for Alcator C-Mod. For 2 MW of absorbed power and $n_{^3\mathrm{He}}/n_e = 2\%$, the "effective temperature" of the minority distribution function $T_{eff} \equiv 2\langle E\rangle/3$ is $\lesssim 50\,\mathrm{keV}$ for $\bar{n}_e \gtrsim 3 \times 10^{20}\,\mathrm{m}^{-3}$ and $T_{eff} \lesssim 10\,\mathrm{keV}$ for $\bar{n}_e \gtrsim 5 \times 10^{20}\,\mathrm{m}^{-3}$, and typically 70–95% of the power absorbed by minority ions is expected to slow down on ions. A highly peaked deposition profile is predicted in this density regime, as shown in Fig. 3. In the low density regime $\bar{n}_e \lesssim 1.5 \times 10^{20}\,\mathrm{m}^{-3}$, energetic minority distribution with $T_{eff} \gtrsim 200\,\mathrm{keV}$ may be formed. However, even in this regime more than half of the power absorbed by minority ions is expected to slow down on ions.

Spatial diffusion of minority ions broadens the spatial distribution of energetic ions and the heating profiles, but the effect is not very dramatic at higher densities. At $\bar{n}_e = 3 \times 10^{20}\,\mathrm{m}^{-3}$ the maximum $T_{eff}$ is reduced from 60 keV to 50 keV by including neoclassical diffusion, and to 25 keV by including 10 times neoclassical diffusion. This effect becomes more pronounced at lower densities. At $\bar{n}_e = 1.5 \times 10^{20}\,\mathrm{m}^{-3}$ the maximum $T_{eff}$ is reduced from 170 keV to 90 keV by including neoclassical diffusion, and to 50 keV by including 10 times neoclassical diffusion. Ripple diffusion is found to be unimportant.

## ACKNOWLEDGMENTS

We would like to thank Dr. D. Weissenburger for the use of the SPARK code, Dr. M. Brambilla for the use of the ICRF coupling code, Dr. G. Hammett for the use of the FPPRF/SHOOT code, and physicists and engineers at PPPL, ORNL, GA, JAERI, JET, and ASDEX for useful discussions. This work was supported by U.S. DOE Contract No. DE-AC02-78ET51013.

## REFERENCES

[1] F. W. Perkins, PPPL Report PPPL-2571 (1988).

[2] S. Golovato, et al., this conference.

[3] D. W. Weissenburger, PPPL Report PPPL-2040 (1983); PPPL-2494 (1988).

[4] PAFEC Ltd., Nottingham, UK.

[5] M. Brambilla, Nucl. Fusion **28**, 549 (1988).

[6] The JET Team, Plasma Phys. and Controlled Fusion **30**, 1467 (1988).

[7] J. R. Wilson, et al., IAEA-CN-50/E4-1 (IAEA, Nice, 1988).

[8] G. W. Hammett, Ph. D. Thesis, Princeton University (1986).

[9] G. W. Hammett, et al., Nucl. Fusion **28** 2027 (1988).

[10] D. N. Smithe, et al., Nucl. Fusion **27**, 1319 (1987).

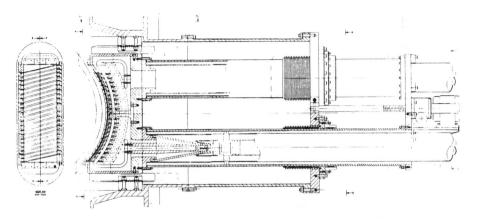

Fig. 1. The movable, single-strap antenna to be used on Alcator C-Mod initially.

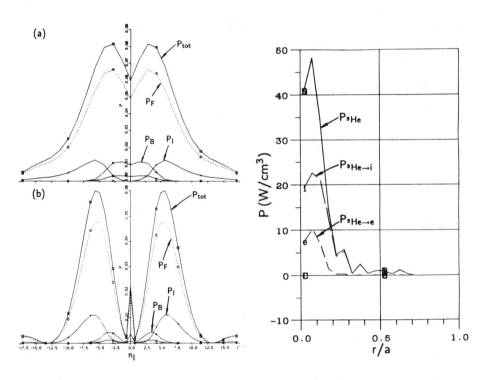

Fig. 2. The radiated $n_\parallel$ spectra for (a) single-strap and (b) "optimized" double-strap antennae.

Fig. 3. Power deposition profiles, $D(^3He)$, $n_{^3He}/n_e = 2\%$, $\overline{n}_e = 3 \times 10^{20}\,\mathrm{m}^{-3}$, $T_e = T_i = 4\,\mathrm{keV}$, $B = 7.9\,\mathrm{T}$, $P_{rf}^{abs} = 2\,\mathrm{MW}$.

# IBW EXPERIMENT ON JFT-2M TOKAMAK

H. Tamai, T. Ogawa, H. Matsumoto, K. Odajima, and JFT-2M Group
Japan Atomic Energy Research Institute, Naka, Ibaraki 319-11, JAPAN

## ABSTRACT

Ion Bernstein Wave(IBW) is launched to the JFT-2M tokamak plasma in order to investigate the coupling efficiency, the behavior of plasma heating, and the launching to H-mode plasma.

Plasma loading impedance increases as the separatrix approaches the IBW launcher, which indicates that the loading impedance increases proportionally to the edge electron density. Loading impedance with H-mode plasma falls to about 40% of that with L-mode plasma, mainly due to the drastic drop of the edge electron density.

Enhanced impurity radiation is observed during IBW heating. The maximum launched power is less than 150kW, due to the enhancement of radiation loss which exceeds the launched power by a factor of two or three. In spite of radiation enhancement, increase of stored energy is observed by IBW heating. This is mainly due to the density build-up, and the contribution from net plasma heating by IBW is small.

H-mode plasma produced by neutral beam heating is terminated by launching of IBW, which is considered to result from scrape-off plasma heating. The threshold IBW power for H to L re-transition is about 20kW.

## INTRODUCTION

Recent plasma condition in H-mode requires sufficient clearance from the limiter, and this means a decrease of the coupling with plasma for the conventional fast wave launcher. In contrast to these launcher, Ion Bernstein Wave(IBW) coupling is predicted not to decrease as the edge density decreases.[1] Therefore, in order to clarify the plasma coupling, and power absorption, IBW heating experiment is done on JFT-2M.[4] This report presents the experimental observation during IBW heating, which includes the investigation of launching to H-mode plasma.

## EXPERIMENTAL SETUP

The IBW launcher is $B_\theta$- $E_z$ type loop antenna which produces the RF current in the toroidal direction.[2] The launcher is installed at oblique port on JFT-2M, with angle of 43° to the horizontal plane. The heating regime is 3/2 harmonic ion cyclotron resonance of hydrogen($3/2\omega_H$) or third harmonic resonance of deuterium($3\omega_D$).[3] The wave frequency is 27MHz, and the resonance layer of hydrogen($3/2\omega_H$) or deuterium($3\omega_D$) is located at the magnetic axis.

## LOADING IMPEDANCE

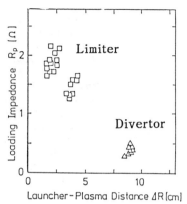

Fig.1 Loading impedance

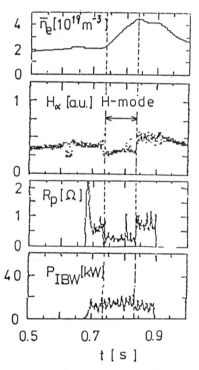

Fig.2 Loading impedance
with H-mode plasma.

Plasma loading impedance is investigated for plasma configurations of D-shaped limiter and upper single null divertor. Figure 1 shows the dependence on the distance between launcher and the separatrix($\Delta R$). The loading impedance becomes larger as plasma approaches the launcher. This result indicates that the plasma loading impedance increases as edge electron density increases, which is almost the same tendency as conventional fast wave launcher, and is the opposite result to the theoretical prediction.

Loading impedance with H-mode plasma is observed in upper single null divertor. Since the H-mode phase is terminated by launched power greater than 20kW as described below, the launched power is lower than that threshold. Figure 2 shows the temporal behavior of loading impedance. At the instant of transition, loading impedance falls rapidly from $0.65\Omega$ at L-phase to $0.25\Omega$ at H-phase. The distance between the separatrix and the launcher changes only slightly from 9.2cm to 9.6cm. However, the density in the scrape-off layer rapidly decreases at the transition to H-mode, and increases at the transition to L-mode as shown in Fig.4; which is similar behavior as that of the loading impedance. Thus, the decrease of loading impedance during H-mode phase is considered to be mainly due to the drop of scrape-off density.

## FEATURE OF PLASMA PARAMETERS

Typical plasma parameters during IBW are shown in Fig.3. After the initiation of the IBW pulse, the density, radiation loss, and loop voltage($V_l$) all increase. Electron temperature measured by electron cyclotron emission decreases. These behaviors show the increase of impurity contamination. In this case, the increase in radiation loss exceeds the launched IBW power by a factor of three. The ratio of total radiation loss to total input power increases from 40% at OH-phase to about 90% during IBW-phase.

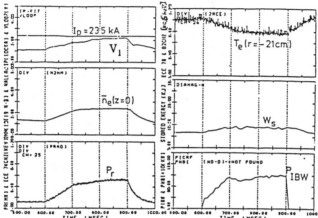

Fig.3 Temporal variation of typical plasma parameters.

Because of the remarkable radiation enhancement, launched IBW power is limited to less than 150kW.

In spite of the radiation enhancement, the stored energy($W_s$) increases slightly. The increase of stored energy is mainly due to the density build-up, and the contribution from net heating by IBW is quite small. Global energy confinement time($\tau_E$) is reduced from 40ms at OH-phase to 20ms at IBW-phase. This value is less than that obtained in mode conversion heating by high field side launcher on JFT-2M tokamak.[5]

## LAUNCHING TO H-MODE PLASMA

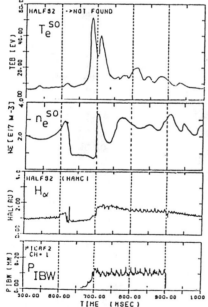

Fig.4 Temporal behavior during H-mode transition.

IBW power is launched to H-mode plasma produced by neutral beam heating in upper single null divertor. As shown in Fig.4, the H-mode transition occurs at 610ms with sudden drop in $H_\alpha/D_\alpha$ emission. At the onset of the IBW pulse at 700ms,

H-mode phase is terminated with sudden jump of $H_\alpha/D_\alpha$ emission. The termination of H-mode is observed with launched IBW power greater than 20kW.

Langmuir probe observes the electron temperature in the scrape-off plasma($T_e^{so}$). Before IBW pulse $T_e^{so}$ is the lower, constant value, and increases suddenly at the onset of the IBW pulse. The increase of $T_e^{so}$ shows that some amount of IBW power is deposited at the scrape-off layer.

If the scrape-off plasma heating by IBW is related to H-mode termination, two likely mechanisms are considered. One is that the increase of edge electron temperature might enhance the sputtering from the chamber wall, and through the impurity accumulation H-mode would be terminated. The other is that the edge plasma heating would destroy the temperature pedestal at the plasma edge which is required to keep the H-mode.[6]

## SUMMARY

The Ion Bernstein Wave(IBW) experiment on JFT-2M tokamak is summarized as follows:

Plasma loading impedance increases as the separatrix approaches the IBW launcher. Loading impedance with H-mode plasma is less than that with L-mode plasma mainly due to the lower edge density.

Enhanced impurity radiation exceeds the launched IBW power by a factor of two or three, and total radiation loss reaches 90% of the total input power. The enhanced radiation loss limits the launched IBW power to less than 150kW.

An increase of stored energy is observed during IBW, and is mainly due to the density build-up.

Launching to the H-mode plasma shows a rapid re-transition to L-mode at very low launched power($\sim$20kW). This is considered to be related to edge plasma heating.

## REFERENCES

1. M. Brambilla, Nuclear Fusion 28 549 (1988).
2. M. Ono, et al., in Proc. of 11th Int. Conf. on Plasma Phys. and Cont. Nucl. Fusion Research I-pp.477 (Kyoto, Nov. 1986 IAEA-CN-47/F-I-3).
3. Y. Ogawa et al., Nuclear Fusion 27 1379 (1987).
4. H. Tamai et al., JAERI-M 89-036 (1989).
5. H. Matsumoto, et al., Nuclear Fusion 27 1181 (1987).
6. K. Hoshino, et al., Physics Letters A 124 299 (1987).

# RADIATION FORCE EFFECTS ON MAGNETOHYDRODYNAMIC EQUILIBRIA

J. A. Tataronis
University of Wisconsin-Madison, Madison, WI 53706

## ABSTRACT

Quasi-static radiation forces, produced by ICRF waves, can modify tokamak equilibria by breaking the characteristic axisymmetry and producing magnetic island structures. The induced magnetic islands form about rational magnetic surfaces and magnetic axes of the unperturbed tokamak. In this paper, we present a perturbation analysis of a toroidal magnetohydrodynamic (MHD) plasma equilibrium that is under the influence of forces produced by low frequency waves. Conditions for magnetic island formation are derived.

## INTRODUCTION

Forces exerted by rf electric and magnetic fields can interact strongly with plasma, modifying properties and topology of equilibrium magnetic surfaces[1,2]. RF forces give rise to local magnetic field structures that cause magnetic surfaces to break into island chains centered about rational surfaces and magnetic axes of the unperturbed plasma configuration. In this paper, we present a perturbation analysis of the perturbed plasma configuration. We demonstrate that magnetic islands are governed by a nonlinear partial differential equation that has the form of the Grad-Shafranov equation for a helically symmetric plasma equilibrium, modified to account for the driving radiation force. An added effect of ICRF waves is current drive due helicity transfer to the plasma from the waves[3]. We speculate that our equations may have applications to this current drive process.

## PERTURBATION ANALYSIS

Our analysis is based on the governing equations of ideal MHD, modified to account for radiation force effects,

$$\mathbf{J} \times \mathbf{B} - \nabla P + \mathbf{F}_{rf} = 0 \tag{1}$$

$$\nabla \times \mathbf{B} = \mu_o \mathbf{J} \tag{2}$$

$$\nabla \cdot \mathbf{B} = 0 \tag{3}$$

where $\mathbf{F}_{rf}(\mathbf{r})$ designates the effective force density produced by the rf electric and magnetic fields. As shown elsewhere[4], $\mathbf{F}_{rf}(\mathbf{r})$ is derived by expanding the ion equations of motion to second order in the radiation field amplitudes, and averging over time.

We assume an axisymmetric equilibrium with magnetic field, $\mathbf{B}_o(\mathbf{r})$, and pressure, $P_o(\mathbf{r})$. The magnetic field forms nested magnetic surfaces, $\Psi_o(\mathbf{r}) = $ const., where $\Psi_o(\mathbf{r})$ satisfies, $(\mathbf{B}_o \cdot \nabla)\Psi_o = 0$. The radiation force perturbs the magnetic field, plasma pressure and magnetic surface structure of the axisymmetric state,

$$\mathbf{B}(\mathbf{r}) \; = \; \mathbf{B}_o(\mathbf{r}) \; + \; \mathbf{b}(\mathbf{r})$$

$$P(\mathbf{r}) \; = \; P_o(\psi_o) \; + \; p(\mathbf{r}) \tag{4}$$

$$\Psi(\mathbf{r}) \; = \; \Psi_o(\mathbf{r}) \; + \; \psi(\mathbf{r})$$

where lower case variables, $(\mathbf{b}, p, \psi)$, identify the induced perturbations, which we order as the applied force, $\mathbf{F}_{rf}(\mathbf{r})$. Linearization of Eqs. (1) - (3) produces governing equations for $\mathbf{b}$ and the total perturbed pressure, $\delta\pi \equiv p \; + \; \mathbf{B}_o \cdot \mathbf{b}/\mu_o$,

$$\frac{1}{\mu_o}(\mathbf{B}_o \cdot \nabla)\mathbf{b} \; + \frac{1}{\mu_o}(\mathbf{b} \cdot \nabla)\mathbf{B}_o \; - \; \nabla(\delta\pi) \; = \; \mathbf{F}_{rf} \tag{5}$$

$$\nabla \cdot \mathbf{b} \; = \; 0 \tag{6}$$

A more tractable form of Eqs. (5) and (6) results after introduction of contravariant components of $\mathbf{b}$, $(b^1, b^2, b^3)$, based on magnetic surface coordinates of the axisymmetric configuration, $b^i \equiv \mathbf{b} \cdot \nabla x^i$, where $x^1$ is a radial-like coordinate that labels the unperturbed magnetic surfaces, $\Psi_o(\mathbf{r}) \equiv$ const., while $x^2$ and $x^3$ are respectively poloidal and toroidal angular coordinates that span the constant $\psi_o(\mathbf{r})$ surfaces. The following scalar equations govern $(b^1, b^2, b^3)$ and $(\delta\pi)$,

$$\left.\begin{array}{l} \dfrac{\partial \delta\pi}{\partial x^1} + \gamma_1(\mathbf{B}_o \cdot \nabla)b^1 + \gamma_2 b^1 \; = \; \gamma_3 F_{rf}^1 + \gamma_4 b^2 + \gamma_5 b^3 \\[3mm] \dfrac{\partial \sqrt{g}\, b^1}{\partial x^1} \; = \; - \dfrac{\partial \sqrt{g}\, b^2}{\partial x^2} - \dfrac{\partial \sqrt{g}\, b^3}{\partial x^3} \end{array}\right\} \tag{7}$$

$$\{I\,(\mathbf{B}_o \cdot \nabla) + K\} \cdot \begin{bmatrix} b^2 \\ b^3 \end{bmatrix} = \begin{bmatrix} \alpha_1 F_{rf}^2 + \alpha_2 \dfrac{\partial \delta\pi}{\partial x^2} + \alpha_3 b^1 \\[3mm] \beta_1 F_{rf}^3 + \beta_2 \dfrac{\partial \delta\pi}{\partial x^3} + \beta_2 b^1 \end{bmatrix} \tag{8}$$

where $K$ is a 2 x 2 matrix of coefficients that depend on the axisymmetric pressure and magnetic field, $I$ represents a 2 x 2 identity matrix, and $(F_{rf}^1, F_{rf}^2, F_{rf}^3)$ designate contravariant components of $\mathbf{F}_{rf}(\mathbf{r})$. Note that the operator, $\{I(\mathbf{B}_o \cdot \nabla) + K\}$, contains only an ordinary derivative along force lines of $\mathbf{B}_o$. There are no derivatives with respect to the radial coordinate, $\Psi_o$, signifying that $\Psi_o$ is simply a parameter. Formally, solution of Eqs. (7) and (8) proceeds by inverting Eq. (8) on each unperturbed magnetic surface, $\Psi_o \equiv$ const., to express $(b^2, b^3)$ in terms of $(\delta\pi, b^1)$,

$$\begin{bmatrix} b^2 \\ b^3 \end{bmatrix} = \{I\,(\mathbf{B}_o \cdot \nabla) + K\}^{-1} \cdot \begin{bmatrix} \alpha_1 F_{rf}^2 + \alpha_2 \dfrac{\partial \delta\pi}{\partial x^2} + \alpha_3 b^1 \\[3mm] \beta_1 F_{rf}^3 + \beta_2 \dfrac{\partial \delta\pi}{\partial x^3} + \beta_2 b^1 \end{bmatrix} \tag{9}$$

and then substituting the result in Eq. (7), leaving coupled equations for $(\delta\pi, b^1)$. In Eq. (9), $\{I (B_0 \cdot \nabla) + K\}^{-1}$ represents an integral operator with a kernal specified by the Green's function of $\{I(B_0 \cdot \nabla) + K\}$. If $\{I(B_0 \cdot \nabla) + K\}^{-1}$ does not exist, spatial singularites in the variables, $(b^2, b^3)$, are implied. It is the existence of spatial singularities in the perturbed variables that imply magnetic island structures. The condition for non-existence of $\{I(B_0 \cdot \nabla) + K\}^{-1}$ is that the homogeneous equation, $\{I(B_0 \cdot \nabla) + K\}u = 0$, possess a nontrivial, single-valued, solution for u over the specific magnetic surface of the unperturbed tokamak. On rational magnetic surfaces, the domain of integration of this equation is the closed field line of finite length. It is on these lines that we expect to find solutions.

## INDUCED MAGNETIC ISLANDS

A convenient way to view the formation of magnetic islands is in terms of plasma currents that the perturbing magnetic fields produce about rational magnetic surfaces of the tokamak. Given $F_{rf}(r)$ and $B_0(r)$, and neglecting the pressure gradient [**low beta assumption**], we compute the perpendicular component of the induced plasma current from Eq. (1),

$$ J_\perp = \frac{B_0 \times F_{rf}}{B_0^2} \tag{10} $$

The parallel component of the current is computed from the condition, $\nabla \cdot J = 0$,

$$ J_\parallel = \sigma B_0, \text{ where, } B_0 \cdot \nabla \sigma = \nabla \cdot \left[ \frac{B_0 \times F_{rf}}{B_0^2} \right] \tag{11} $$

The magnetic field, $B$, induced by the resulting total current, $J_\perp + J_\parallel$, is governed by Ampère's law,

$$ \nabla \times B = \mu_0 \left[ - \frac{B_0 \times F_{rf}}{B_0^2} + \sigma B_0 \right] \tag{12} $$

which can be solved via Green function techniques. It is convenient to express $B$ in terms of a vector potential, $B = \nabla \times A$. Magnetic surfaces, $\psi(r) = $ const., spanned by $B(r)$, are then related to the covariant componet, $A_3(r)$,

$$ \psi(r) - \psi_0(r) \sim A_3(r) \tag{13} $$

Equation (13) leads to island chains that depend on harmonic fields contained in **B**. Cary and Kotschenreuther[5] adopted this iteration procedure to identify magnetic island structures in stellarators.

Near rational magnetic surfaces of the tokamak, $\psi(\mathbf{r})$ satisfies a nonlinear partial differential equation, similar in form to the Grad-Shafranov equation, but modified to account for the driving rf force. Closed helical magnetic force lines on rational surfaces induce a local helical symmetry in the plasma geometry, suggesting the use of helical coordinates to represent $\psi(\mathbf{r})$. Let $(x^1, x^2, x^3) \equiv (r, \varphi, z)$, where $\varphi$ is the helical angle, $\varphi \equiv m\theta + kz$, $(r, \theta, z)$ are cylindrical coordinates, and $(m, k)$ are azimuthal and axial winding numbers of the closed helical force line. To find $\psi(\mathbf{r})$ near rational surfaces of $\mathbf{B}_0(\mathbf{r})$, we invoke local helical symmetry, meaning that the equilibrium depends only on $r$ and $\varphi$, and express $\mathbf{F}_{rf}(\mathbf{r})$ in terms of an effective scalar potential, $\Phi(\mathbf{r})$, and a magnetization vector, $\mathbf{M}(\mathbf{r})$: $\mathbf{F}_{rf}(\mathbf{r}) = -n_0 \nabla\Phi + \mathbf{B} \times (\nabla \times \mathbf{M})$, where $n_0$ is the number density of the background ions[4,6]. For ICRF waves, **M** is parallel to **B**, $\mathbf{M} = \chi\mathbf{B}$, where $\chi(\mathbf{r})$ is an effective magnetic susceptibility that varies as the ICRF wave intensity. Litwin[6] has pointed out that this decomposition of $\mathbf{F}_{rf}(\mathbf{r})$ suggests treating the MHD plasma as a magnetized medium, characterized by a magnetic intensity vector, $\mathbf{H}(\mathbf{r}) \equiv \mu\mathbf{B}(\mathbf{r})$, where $\mu$ is an effective permeability, $\mu(\mathbf{r}) \equiv 1-\chi(\mathbf{r})$, and a magnetization current, $\nabla \times \mathbf{M}$. Use of the magnetization notation in Eq. (1) leads to the following governing equation for $\psi(\mathbf{r})$,

$$\frac{1}{r}\frac{\partial}{\partial r}\left(\frac{m^2 r\mu}{m^2 + k^2 r^2}\frac{\partial\psi}{\partial r}\right) + \frac{m^2}{r^2}\frac{\partial}{\partial\varphi}\left(\mu\frac{\partial\psi}{\partial\varphi}\right) =$$

$$- \pi'(\psi) - \frac{m^2}{\mu}\frac{I(\psi)I'(\psi)}{m^2 + k^2 r^2} - \frac{2m^3 kI(\psi)}{(m^2 + k^2 r^2)^2} - \frac{\Phi\rho'(\psi)}{\mu} \tag{14}$$

where $\rho(\psi)$ is the mass density expressed as a function of $\psi(\mathbf{r})$, $\pi(\psi)$ is an effective pressure, $\pi(\psi) \equiv P(\mathbf{r}) - \rho\Phi$, and $I(\psi)$ is a free functional of $\psi$. If the rf force is set equal to zero, Eq. (14) reduces to the familiar Grad-Shafranov equation in helical coordinates. Solutions of Eq. (14) reveal island chains that are driven by the rf force. We are considering the use of Eq. (14) to model current drive produced by circulary polarized ICRF waves[3]. The current drive results from helicity injection.

This work is supported by the U.S.DOE under contract DE-FG-88ER53264.

## REFERENCES

1. J. M. Myra, Phys. Fluids **31**, 1190 (1988).
2. J. A. Tataronis and C. Litwin, Bull. Am. Pys. Soc. **33**, 2014 (1988).
3. R. Mett and J. A. Tataronis, University of Wisconsin Phaedrus Program, Report No. PTMR89-2 [submitted to Phys. Rev. Letts.]--- also this conference.
4. M. L. Sawley, Ecole Polytechnique Fédérale de Lausanne Report No. LRP 260/85 (1985).
5. J. R. Cary and M. Kotschenreuther, Phys. Fluids **28**, 1392 (1985).
6. C. Litwin, private communication.

# PERFORMANCE OF INTERNALLY SEALED ICRF ANTENNAS IN CCT[*].

R.J. Taylor, M.L. Brown, K.F. Lai and J.R. Liberati.
University of California at Los Angeles, Los Angeles, Ca. 90024.

## ABSTRACT

The internally sealed ICRF antennas in CCT no longer show impurity problems in the steady state or in the pulsed mode. Experience over a 6 month period indicates a minimum of one order of magnitude reduction in impurity generation for comparable input power (50 kW) used in previous experiments. The antenna radiation resistances agree with computer calculation within experimental error. In the Fast Wave Current Drive regime cavity resonances are absent. The power transfer to the plasma is nearly ideal. It is now clear that the power goes to the ions at the edge, where confinement is poor. The absorption mechanism is not understood. Wave penetration to the center is only possible for extremely low ( $< 10^{12}/cm^3$ ) densities.

## INTRODUCTION

Fast wave performance is of interest in tokamaks for ignition and for current drive. This performance was perceived as limited by antenna physics and technology. At UCLA we have focused on testing the whole range of possible antenna configurations first in Microtor/Macrotor and now in CCT (Continuous Current Tokamak). The stress on antenna technology seemed important because of the rather mixed results in plasma heating and current drive achieved over the last 10 years world wide with ICRF, including impurity effects.

In this paper we will describe the latest changes that we thought were needed in order to leave the non-reactor technological problems behind us.

In order to achieve high reliability and low impurity generation, both in the pulsed and in the CW modes, the following innovations have been made. (1) All high voltage components have been placed outside of the vacuum area by the use of internal dielectric seals. (2) An external Faraday shield has been added on the atmospheric side. It reduces the electrostatic fields on the dielectric window which may come in contact with a tenuous plasma. (3) The internal Faraday is used only as a heat and particle shield with a fine segmentation to keep the intersegment voltages below 30 volts.

[*]Supported by USDOE Contract DE-FG03-86ER53225.

## ANTENNA CONFIGURATION

CCT now houses five antennas; two 450 degree fast wave helical
structures located 90 degrees toroidally for remaining experiments
to check phasing, one Bernstein wave on the outside midplane as a
good faith gesture, one 120 degree fast wave on the outside to make
contact with possible CIT experiments, and one 90 degree fast wave
on the inside to check possible fine points of propagation and
absorption just in case it will be revealing.

Figure 1a shows a typical antenna structure as viewed from the
plasma side. The radial distribution of the elements is shown in
Figure 1b where the vacuum and atmospheric sides of the dielectric
window are labeled for clarity.

The plasma sees a set of water cooled limiters and a slightly
recessed Faraday shield. The Faraday segmentation is fine enough to
reduce possible edge interaction energies to 30 volts. The segments
are conduction cooled at the edge limiters. Behind the Faraday
grill the plasma finds a teflon window. The plasma density at the
teflon window is below $10^9/cm^3$ and the temperature is below 5 eV.
We expect a mild metallization of this surface to take place from
the Faraday shield. Since the teflon will also ablate slightly, a
stable equilibrium is likely to exist. A true resolution of this
matter cannot be modeled and occasional maintenance may be
required.

In order for this interface to be stable we placed a Faraday shield
on the atmosphere side of the window to prevent the electric fields
from the current strap to go through the teflon and to agitate the
tenuous plasma on the vacuum side. This is a rough Faraday shield
since no ionized matter is likely to be in contact with it. The
cooling of the window is by forced gas or liquid dielectric.
Additional water cooling is provided for the current strap when gas
cooling is used in the high voltage cavity.

## LOADING RESISTANCE

Small signal loading resistance was measured over a wide range of
plasma parameters and checked by finite difference computation. The
experimental data is shown in Figure 2 for the ICRH second harmonic
mode (about 7 MHz). In CCT, due to its small B·a (B = 3kG, a = 0.4
m, n < 1e13/cm$^3$) this regime is near the magnitosonic cutoff and
cavity modes are dominant. However we can also enter the continuum.
Notice that the cavity modes are not damped due to the thin second
harmonic layer. Significant damping of the cavity modes however
can be seen in minority H and He$^3$ experiments (not shown).

The loading resistance at higher harmonics even in CCT can show
single pass behavior. This is also shown in Figure 2. At 70 MHz
(near where we use the FMIT power source) the radiation resistance
is high and devoid of cavity resonances. Here the group velocity is

along the field lines and anomalous absorbtion of the waves is the
likely reason for the lack of cavity effects (and not the resonance
overlap).

In any case, it must be made clear that the cold plasma theory
codes predict self interference even with a single antenna.
Whenever interference can be shown to be present, phasing of the
wave can be accomplished by a proper antenna design. Therefore,
experimentally we find that phasing is only possible near cutoffs.

In Figure 2 we also show the loading resistance of the fast wave
antenna in a low field (100 G) low density ($10^{11}$/cm$^3$) plasma used
in CCT for discharge cleaning and for transport driven current
studies. The lack of phasing here is clear and is believed to be
due to anomalous damping also (not collisional or Landau).

## POWER HANDLING

As it is mentioned above we have succeeded in operating the new
antennas in CCT without ambiguities due to impurities. These
experiments were performed using FWLH at about the 20th ion
harmonic. The power levels were modest (50 KW) nevertheless the
implications are clear.

Figure 3 shows the high power response of the plasma. The impact of
the FWLH energy on the various signal in a diminishing order are:
(1) fast particle generation (CX), (2) edge plasma density, (3)
floating potential at edge, (4) plasma current, (5) hydrogen
influx/recycling (not shown). Below measurability (in the noise)
are possible changes in: (1) loop voltage (not shown), (2) ECH
fundamental/second harmonic emission, (3) line average density, (4)
impurity content (see CX+UV trace where the fast particles 'shine
through' in the prompt response and no significant rise is seen in
the UV radiation just after the rf pulse).

The above list combined with the nature of the radiation resistance
and its dependence (or lack of) on density, frequency, magnetic
field, antenna geometry and the various tokamaks studied over the
years by us and others clearly indicate that (1) we cannot drive
satisfactory currents with FWLH, and (2) we have no theoretical
basis to develop a near term understanding of the perplexing
experimental results.

## CONCLUSIONS

We are satisfied with the range of technological capabilities in
dealing with the performance of ICRF antennas in CCT for both the
CW and the pulse modes of operation. The difficulty with ICRF and
particularly with high density heating and high frequency current
drive may not be technological. In particular, we need to
understand why these waves interact with the ions and why we cannot
control the internal k-spectrum well above the cutoff.

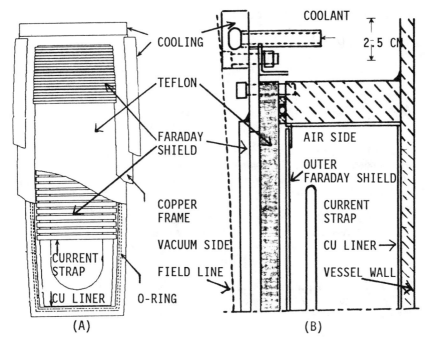

Fig.1 Sealed ICRF Antenna in CCT.

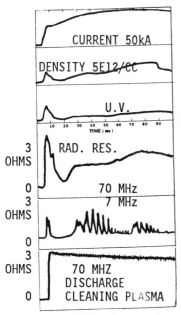

Fig.2 Loading Resistance.

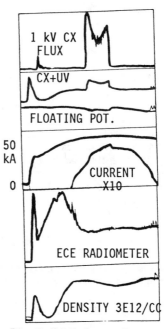

Fig.3 FWLH Power Response.

362

# PRELIMINARY RESULTS OF COMBINED ICRF-NBI HEATING IN TEXTOR

A.M. Messiaen[1], M. Gaigneaux, J. Ongena, P.E. Vandenplas, G. Van Wassenhove,
R.R. Weynants[1], J-M. Beuken, T. Delvigne, P. Descamps[2], F. Durodié, M. Jadoul,
R. Koch, D. Lebeau, X.M. Shen[3], S.S. Shinohara[4], D. Van Eester[2], R. Van Nieuwenhove,
G. Van Oost
Laboratoire de Physique des Plasmas - Laboratorium voor Plasmafysica
Association "Euratom-Etat belge" - Associatie "Euratom-Belgische Staat"
Ecole Royale Militaire - B 1040 Brussels - Koninklijke Militaire School

H. Conrads, H. Euringer, F. Hönen, M. Lochter, H. Kever, R. Uhlemann, G. Wang[5]
G. Bertschinger, P. Bogen, K-H. Dippel, H.G. Esser, K-H. Finken, E. Graffmann,
H. Hartwig, E. Hintz, K. Höthker, B. Kardon[6], L. Könen, M. Korten, Y.T. Lie, R. Moyer[7]
A. Pospieszczyk, D. Reiter, D. Rusbüldt, U. Samm, J. Schlüter, B. Schweer,
H. Soltwisch, F. Waelbroeck, G. Waidmann, P. Wienhold, J. Winter, G.H. Wolf
Institut für Plasmaphysik, Kernforschungsanlage Jülich, GmbH
Association "Euratom-KFA", D-5170 Jülich, FRG

## ABSTRACT

ICRH and NB, Co-injection, are compared at the same power of 1.4 MW and then combined. The diamagnetic energy increases linearly with the total power. At low plasma current, the combination of ICRH and NB, Co-injection, is possibly synergetic and leads to nearly zero loop voltage.

## INTRODUCTION

The paper describes the first results of the combined operation of ICRH and NBI on TEXTOR in order to compare them and look for possible synergetic effects. The newly installed NBI system of TEXTOR consists of one Co and one Counter beamline (each line rated for 1.5 MW at 50 kV) which are described in Ref. 1. For the present experiments only H injection into a D plasma has been used. One of the two pairs of low field side ICRH antennae[2] is utilized and fed in the out of phase configuration by one of the two upgraded 2 MW generators. This latter configuration does not lead to a drop in plasma loading with respect to the in phase feeding and its use together with carbonized walls heated at 350°C reduces significantly the interaction with the wall during ICRH[2]. All the generator power can then be easily applied to the antenna pair and about 85 % is coupled to the plasma. Minority heating in a (H)-D plasma is used at a frequency of 29 MHz with $B_t$ = 2T. The best ICRH results are obtained for a H/D concentration ratio of ~ 5 % (spectroscopically monitored by $H_\alpha/D_\alpha$ line intensity ratio). Short NBI pulses were used to minimize the H concentration increase during NBI + ICRH operation. It proved nevertheless difficult to maintain the optimal H concentration under all conditions and this influences the comparison when ICRH is involved.

### COMPARISON BETWEEN NBI,CO AND ICRH

Fig. 1 shows the results of a density scan at $I_p$ = 470 kA for a pure ohmic discharge and for discharges heated by equal amounts ($P_{NI} = P_{RF}$ = 1.4 MW) of NBI,Co or RF power

---

[1] NFSR, Belgium. [2] EEC grantee, IPP, KFA, Jülich. [3] A.S., Hefei, P.R. China. [4] University of Tokyo, Japan. [5] SIP, Leshan, P.R. China, [6] CRIP, Budapest, Hungary. [7] UCLA.

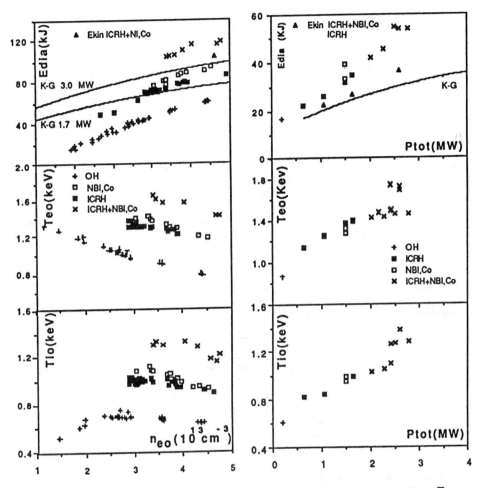

Fig. 1 $E_{dia}$, $T_{eo}$ and $T_{io}$ versus $\bar{n}_{eo}$ for OH, pure ICRH or NBI,Co and NBI,Co + ICRH discharges at $I_p$ = 470 kA ($P_{RF}$ = $P_{NI}$ = 1.4 MW). The corresponding predictions of the Kaye Goldston law (K-G curves) are shown.

Fig. 2 $E_{dia}$, $T_{eo}$, $T_{io}$ versus Ptot ($\bar{n}_{eo}$ = 2.1 $10^{13}$ cm$^{-3}$) for OH and heated discharges at Ip = 220 kA. The corresponding prediction of the Kaye Goldston law is also given.

or both combined. We see that the general trends previously observed[3,4] for mode conversion ICRH are retrieved whichever heating method being applied : the ohmic energy $E_{OH}$ has a neo-Alcator behaviour, slightly saturated for the highest density in the present experimental situation, whereas the auxiliary heated discharge exhibits a parallel evolution of the diamagnetic energy $E_{dia}$ which can be expressed by a nearly constant $\Delta E_{dia}$ for a given $\Delta P_{tot}$ ($P_{tot} = P_{RF} + P_{NI} + P_{OH}'$, where $P_{OH}'$ is the remaining ohmic power). Similar shifts with respect to the OH curve are observed for the central electron and ion temperatures $T_{eo}$ (from ECE, averaged on the sawtooth period) and $T_{io}$ (from neutrons). Comparing NBI and ICRH results, one sees that these two methods behave quite similarly.

Simulations by means of TRANSP for NBI[8] and our ICRH[9] codes confirm this relatively modest difference under the experimental situation pertaining to Fig. 1. Firstly, the power is shared between electrons (0.6) and ions (0.4) in roughly the same proportion[6]. Secondly, the power deposition profile for NBI is somewhat more peaked, with half the power deposited within a radius of 16 to 22 cm as opposed to 20 to 28 cm for ICRH. This might account for the small difference in the temperature and energy increments. Thirdly, the anisotropic perpendicular tail energies $E_{\perp a}$, causing a difference between the diamagnetic and isotropic kinetic energies according to the expression $E_{dia} = E_{kin} + 3/2\ E_{\perp a}$, are nearly the same in both cases. $E_{\perp a}$ equals 3 to 5.5 kJ for ICRH[7] and 3-5 kJ for NBI.

On Fig. 1 we have also indicated the prediction of the Kaye-Goldston scaling law[5] for $P_{tot} = 1.7$ MW (pure ICRH or NBI) and $P_{tot} = 3.0$ MW corresponding to the combined operation Also shown is a value of $E_{kin}$ for NBI+ICRH.

## COMBINED OPERATION RESULTS

Fig. 2 shows the results of a power scan, taken at constant central chord density $\bar{n}_{eo}$ of $2.1\ 10^{13}$ cm$^{-3}$ and for low current $I_p = 220$ kA. It is obtained by adding a variable amount of RF power to a constant beam power of 1.4 MW; also pure RF points are included. One sees again the linear offset behaviour expressed by the slope of $E_{dia}$ versus $P_{tot}$ which is defined by the diamagnetic confinement time $\tau_{inc,dia}$. Fig. 1 indicates that $\tau_{inc,dia}$ is only slightly density dependent and Fig. 2 shows that $\tau_{inc,dia}$ is not very different for ICRH, NBI,Co or the two heating methods combined. The energy increments due to NBI and ICRH are additive and there is no clear synergetic effect. A similar behaviour is also observed for $T_{eo}$ and $T_{io}$.

On Fig. 2 are also shown some values of $E_{kin}$ for pure ICRH and the combined operation. If, due to the presence of the contribution of $E_{\perp a}$, $\tau_{inc,dia}$ does not exhibit the $I_p$ scaling seen previously for mode conversion ICRH heating[4], $\tau_{inc,kin}$, which corresponds to the thermal part of the energy, scales for the three heating methods as previously stated[4]. Comparing the results with the Kaye-Goldston scaling the improvement factor with respect to the latter law increases with $\bar{n}_{eo}$, $P_{tot}$ and $1/I_p$ and reaches 1.8 for $E_{dia}$ at 220 kA. If $E_{kin}$ is considered, the improvement factor is only at most 1.25.

Operation with NBI, Counter has also been performed and is being evaluated. Combined results (with ICRH + NBI, Co + NBI, Counter) have been obtained up to 4 MW and, at the present time, the best combination is ICRH+NBI,Co.

## NON INDUCTIVELY DRIVEN CURRENT

The ICRH + NBI,Co operation allows the loop voltage $V_l$ to drop significantly. An example is shown in fig. 3 for a discharge at $I_p = 220$ kA, for which also $E_{dia}$, $\bar{n}_{eo}$ and the power stacking are shown. A simulation of $V_l$ by means of TRANSP of the ohmic

($Z_{eff}$ = 2) and combined ($Z_{eff}$ = 1.5) phases is also given. In the latter phase, the simulation predicts a bootstrap current of 56 kA and a beam driven current of 53 kA. Further studies are underway to ascertain whether the experimental achievement of zero loop voltage might be due to a possible synergetic ICRH-beam drive effect not yet incorporated in TRANSP.

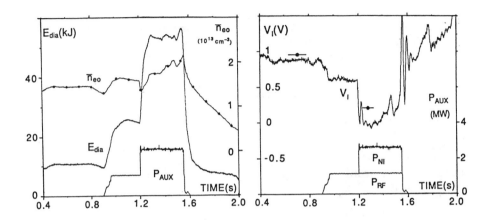

Fig. 3 $E_{dia}$, $\bar{n}_{eo}$, $V_l$ and $P_{aux} = P_{NI} + P_{RF}$ evolution versus time in a shot (# 35860) heated by pure RF or combined ICRH + NBI,Co. The value of $V_l$ predicted by TRANSP for OH and combined heating phases are indicated by —•— .

## REFERENCES

1. H. Conrads et al., Eur. Conf. Abstracts (Proc. 16th Eur. Conf. Venice, 1989) Vol.13B, part III, p. 1221.
2. J-M. Beuken et al., Eur. conf. Abstracts (Proc. 15th Eur. Conf. Dubrovnik, 1988) Vol. 12B, part II, p. 777 ; A.M. Messiaen et al. Plasma Phys. Contr. Fusion 31 (1989).
3. G.H. Wolf et al.,Plasma Phys. Contr. Fusion 28, 1413 (1986).
4. R.R. Weynants et al., Eur. Conf. Abstracts (Proc. 14th Eur. Conf. Madrid, 1987) Vol. 11D, part I, p. 197.
5. S.M. Kaye, R.I. Goldston, Nuclear Fusion, 25, 65 (1985).
6. R.R. Weynants et al., Plasma Physics and Controlled Fusion research 1988 (Proc. 12th Int. conf. Nice, 1988) paper IAEA-CN-50/E-2-1 (1989).
7. R.R. Weynants et al., Eur. Conf. Abstracts (Proc. 16th Eur. Conf. Venice, 1989) Vol. 13B, part I, p.7.
8. R.J. Hawryluck, in Physics of Plasmas close to Thermonuclear Conditions, edited by B. Coppi et al., Vol. 1, p. 19, ECE Brussels, 1980.
9. D. Van Eester, Proc. Varenna-Lausanne Int. workshop on Theory of Fusion Plasmas, Chexbres, Oct. 1988.

# ICRF FARADAY SHIELD PLASMA SHEATH PHYSICS: THE PERKINS PARADIGM

J. H. Whealton, P. M. Ryan, and R. J. Raridon,
Oak Ridge National Laboratory, Oak Ridge, TN 37831

## ABSTRACT

Using a 2-D nonlinear formulation which considers the plasma edge near a Faraday shield in a self consistent manner, progress is indicated in the modeling of the ion motion for a Perkins embodiment. Ambiguities in the formulation are also indicated, the resolution of which will provide significant insight into the impurities generation for ICRH antennas.

## INTRODUCTION

Ion Cyclotron Heating (ICH) at high power densities (5-10 kW/cm$^2$) offers several challenges—one of which is the anticipated high rates of heavy metal impurity generation and outflux. An understanding of the plasma edge near such antennas is an important part of eliminating or mitigating this problem. The present work reports progress toward this understanding.

The plasma edge problem presents formidable difficulties of treatment, particularly near the ICH antenna. One difficulty is the extreme nonlinearity of the equations describing the plasma sheath for an edge plasma on the order of $10^{12}/cm^3$. A second is the dimensionality involved; the complex geometry near an ICH antenna, with a Faraday shield and local limiters, demands the imposition of boundary conditions in all three spatial dimensions. A third is the time scales involved; an accurate model may need to account for electron motion in the magnetized plasma over short electron time scales, ion motion over many rf/ion cyclotron periods, and impurity distribution evolution over long, quasi-steady state periods. The model also needs to connect with the properties of the bulk plasma, preferably in an iterative, self-consistent manner.

## MATHEMATICAL FORMULATION

The Fokker-Planck equation for each species of charged particles, and Maxwell's equation for the potentials, in the Lorentz gauge, are considered:

$$\left\{ \frac{\partial}{\partial t} + v \cdot \nabla_r + \frac{q}{m} \left[ v \times (B_0 + \nabla \times A) - \nabla \phi - \frac{1}{c} \frac{\partial A}{\partial t} \right] \cdot \nabla_v \right\} f(r,v,t) = \beta g(r \cdot v) \tag{1}$$

$$\nabla^2 \phi(r,t) = -4\pi \left\{ N_{eo} \exp \left[ e(\phi(r,t) - \phi_p(r,t))/kT \right] - \int dv\, f(r,v,t) \right\}, \tag{2}$$

assuming a Boltzmann distribution for the electrons and neglecting $\partial^2 \phi/\partial t^2$ (Debye length much smaller than free space wavelength). The determination of $A$ in Eq. (1) in general can be taken from a solution of the

*Research sponsored by the Office of Fusion Energy, U.S. Department of Energy, under contract DE-AC05-84OR21400 with Martin Marietta Energy Systems, Inc.

homogeneous Helmholtz equation in 3-D [Ref. 1] or in at least 2-D [Ref. 2] or the 3-D scalar magnetostatic models, $\nabla^2 \Psi = 0$ [Ref. 3], where $\Psi$ is the magnetostatic potential.

The nonlinear formulation of Eqs. (1)-(2) constitutes a self-consistent description of: the sheath potentials, the onset of charge separation in the plasma, ponderomotive forces, ion Bernstein wave launching and damping, ion acoustic waves, edge plasma ion turbulence, sheath rectification, charged impurity ejection, and a whole host of near field phenomena, many of which are probably undiscovered at this point, but are expected to be important. This formulation is quite general and has been used numerically for long time scale sheath problems in the past [Ref. 4-5].

## PERKINS' PARADIGM

An example that will be considered is the "Perkins" Faraday shield [Ref. 6] and associated paradigm. By use of Faraday's law one can replace the induction term, $\partial A/\partial t$, in Eq. (1), by a boundary value problem on the scalar potential such as shown in Fig. 1. In the metallic elements E is assumed zero and thus (in Lorentz gauge) $\nabla \phi = (1/c) \partial A/\partial t$. We will extend Perkins' analysis to two dimensions.

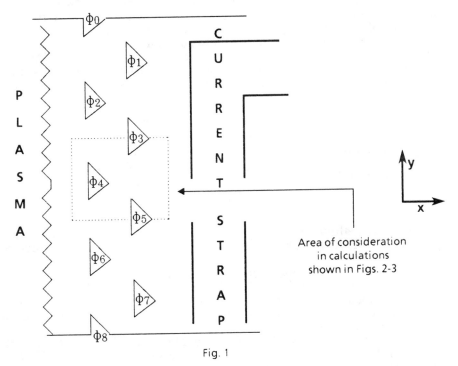

Fig. 1

As an illustrative calculation, we assume an immobile ion density in steady state for a case where the plasma potential, $\phi_p (r, t)$, is uniform and equal to $\phi_5$, which is the maximum potential considered (Fig. 2). The potential $\phi_4$ is halfway between $\phi_3$ and $\phi_5$. It is the intention of this configuration that there would be sheath fields only on the right-hand side of the $\phi_4$ electrode, so that sputtered material would go into the antenna or Faraday shield region as opposed to entering the confinement plasma. The presence of sheath fields on the left-hand side of the electrode is due to the imposition of the plasma potential $\phi_p(t)$, with respect to that electrode potential. Therefore, we can see the model and the results are very much dependent upon the local plasma potential. The local plasma potential is influenced by the potentials of the boundaries intersected by the magnetic field lines.

Time dependent ion trajectories from solution to Eqs. (1)-(2) are shown in Fig. 3. A uniform ion generation rate has been assumed. During this portion of rf cycle, ions are striking the $\phi_3$ electrode with relatively high energy. The edge ions that missed the $\phi_5$ electrode on the last half cycle are circulating due to the magnetic field and the negligible electric field. Near the $\phi_4$ electrode, the dominant motion during this time interval is in the E X B direction.

## CONCLUSIONS

The results in Fig. 3 are to be considered as preliminary in several respects: (1) There are ambiguities in the plasma potential which are an important feature of the model. This can be resolved by a full 3-D treatment of the boundary conditions; alternatively, plasma potentials may be imposed by geometrical consideration along a magnetic field line in conjunction with Faraday's law. (2) Numerical stability and variation of parameters' consistency have not yet been established. The status of ion acoustic-like waves routinely found in the solutions are not yet validated. Some space charge deposition issues have not yet been resolved.

## ACKNOWLEDGEMENTS

We wish to thank P. S. Meszaros, K. E. Rothe, W. R. Becraft, E. G. Elliott, and H. H. Haselton for their help and assistance. We wish to also thank F. W. Perkins (at PPPL) for his suggestions.

## REFERENCES

1. J. H. Whealton, G. L. Chen, R. J. Raridon, R. W. McGaffey, E. F. Jaeger, M. A. Bell, D. J. Hoffman, *J. Comput. Phys.*, **75**, 168-189 (1987).

2. E.F. Jaeger, D. B. Batchelor, H. Weitzner, J. H. Whealton, *Computer Physics Communications* **40**, 33 (1986).

3. P. M. Ryan, K. E. Rothe, J. H. Whealton, D. W. Swain. 8th Topical Conference on Radio Frequency Power in Plasmas, Irvine, California, May 1-3, 1989.

4. J. H. Whealton, R. W. McGaffey, P. S. Meszaros. *J. Comput. Phys.* **63**, 20 (1986).

5. J. H. Whealton, R. J. Raridon, M. A. Bell. NATO/ASI on High Brightness Accelerators, Pitlochry, Scotland, 1986.

6. F. W. Perkins, PPPL Report #2571 (1988).

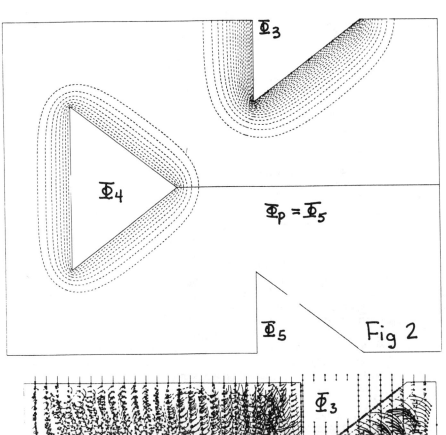

Fig 2

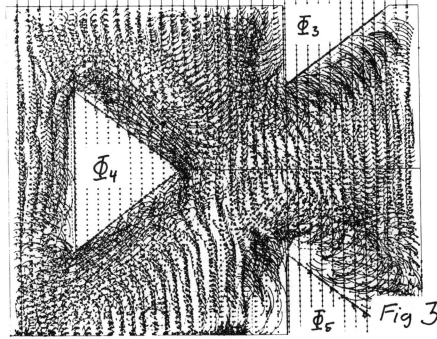

Fig 3

# THE TFTR ICRF ANTENNA - POWER FEED AND PHASE CONTROL OF TWO MUTUALLY COUPLED ANTENNA ELEMENTS[*]

J.R. Wilson, P.L. Colestock, G.J. Greene, J.C. Hosea, C.K. Phillips, J.E. Stevens
Princeton Plasma Physics Laboratory, Princeton University

D.J. Hoffman, W.L. Gardner
Oak Ridge National Laboratory

## ABSTRACT

A transmission line model for the TFTR ICRF antenna is presented. The antenna is considered as being composed of mutually coupled transmission lines. Expressions for the input impedance of such a structure are given. Details of the considerations in setting up the antenna feed structure are presented. The requirement of dual feed drive coupled with the desire for phase and amplitude control place restrictions on the operating circuit.

## INTRODUCTION

The ICRF heating system for TFTR includes two antenna boxes.[1,2] Each antenna box contains two poloidal radiating elements which are separated toroidally. The antenna box located at Bay M was designed to handle 6 MW of rf power at 47 MHz. Two rf generators are required to provide this power, therefore each radiating element is driven by an independant generator. An exploded view of the antenna box is shown in fig.1. The poloidal loops are driven from each end and are shorted in the midplane forcing the antenna current to a maximum there. Since there is only one generator for each loop, the top and bottom power feeds of each loop are connected together to form a resonant circuit with single power feed. By adjusting the length of transmission line joining the top and bottom feeds, the phase and amplitude relationship of the top and bottom antenna currents can be controlled. Since the septum between the two toroidally displaced loops is slotted to enhance the flux linkage between the loops, the mutual inductance between them is significant. The effect of this mutual coupling is to complicate the impedance presented to the two rf generators. An antenna circuit model is described in section 2 that allows an understanding of this impedance relationship. The process of tuning the transmission system to allow operation at 47 MHz will be described in section 3. The final section will discuss the combined operation including matching transformers, phase shifters, and tuning stubs.

## CIRCUIT MODEL

The circuit used to model the antenna box plus top to bottom resonant connections is shown in fig. 2. The four radiating current straps are modelled as transmission lines having a characteristic impedance of $Z_0 = 30\ \Omega$, a $1/\beta = v_{ph}/c = .65$, and a length of 0.4 m. The two top transmission lines are considered to have a mutual coupling as are the two bottom transmission lines. No coupling is assumed between top and bottom lines. Externally, the top and bottom transmission lines are connected by lengths of standard $Z_0 = 50\ \Omega$ coaxial transmission line that allows power to be fed at a simple coaxial tee.

Consider two transmission lines that are coupled through a mutual inductance per unit length M=kL, where L is the self inductance per unit length. The voltages and currents can then be described by the following matrix equations.[3]

$$V' + AI = 0$$
$$I' + BV = 0$$

where

$$A = \begin{bmatrix} -j\omega L - R & -j\omega kL \\ -j\omega kL & -j\omega L - R \end{bmatrix} \text{ and } B = \begin{bmatrix} -j\omega C & 0 \\ 0 & -j\omega C \end{bmatrix}$$

L,R,and C are the inductance,resistance and capacitance per unit length. Solution of these two coupled equations for voltages of the form $V = ce^{\gamma X}$ yields the following equation for $\gamma$.

$$\gamma^4 + 2a\gamma^2 + (a^2 - b^2) = 0$$

where $a = \omega^2 CL - j\omega CR$ and $b = \omega^2 kLC$. The solutions to this are $\gamma_{\pm}^2 = \gamma_0^2 \pm \omega^2 kLC$ with $\gamma_0^2 = -\omega^2 CL + j\omega CR$ the normal expression for a transmission line. The voltages and currents are then given by the following expressions

$$V_1 = a_1 \cosh(\gamma_- x) + a_2 \sinh(\gamma_- x) + a_3 \cosh(\gamma_+ x) + a_4 \sinh(\gamma_+ x)$$

$$V_2 = -a_1 \cosh(\gamma_- x) - a_2 \sinh(\gamma_- x) + a_3 \cosh(\gamma_+ x) + a_4 \sinh(\gamma_+ x)$$

$$I_1 = a_1/Z_- \sinh(\gamma_- x) + a_2/Z_- \cosh(\gamma_- x) + a_3/Z_+ \sinh(\gamma_+ x) + a_4/Z_+ \cosh(\gamma_+ x)$$

$$I_2 = -a_1/Z_- \sinh(\gamma_- x) - a_2/Z_- \cosh(\gamma_- x) + a_3/Z_+ \sinh(\gamma_+ x) + a_4/Z_+ \cosh(\gamma_+ x)$$

where $Z_{\pm} = Z_0(1 \pm k)^{1/2}$. At the center of the antenna, $x=0$, the loops are shorted and $V_{1,2} = 0$. This implies that $a_1 = a_3 = 0$. At $x = l_0$, the other end of the radiating elements, we then have

$$V_{1t} = a_2 \sinh(\gamma_- l_0) + a_4 \sinh(\gamma_+ l_0) \quad V_{2t} = -a_2 \sinh(\gamma_- l_0) + a_4 \sinh(\gamma_+ l_0)$$

$$V_{1b} = a_2 \sinh(\gamma_- l_0) + a_4 \sinh(\gamma_+ l_0) \quad V_{2b} = -a_2 \sinh(\gamma_- l_0) + a_4 \sinh(\gamma_+ l_0)$$

To solve for the input impedance $Z_{in}$ of fig. 2 we need three equations to eliminate three of the four unknowns. These equations are that the top and bottom voltages when transformed to the feed and load points are equal and that the transformed admittance at the output point $Y_2 = -Y_{load}$. Applying these three conditions yields the following expressions for the input admittance.

$$Y_{in} = \frac{y_1 + j\tan(\beta l_1)}{1 + jy_1\tan(\beta l_1)} + \frac{y_2 + j\tan(\beta l_2)}{1 + jy_2\tan(\beta l_2)}$$

$$y_1 = \frac{y_3 \xi_+ - 2}{2y_3 \xi_x \xi_+} \quad y_2 = \frac{y_4 \xi_+ - 2}{2y_4 \xi_x \xi_+}$$

$$y_4 = \frac{y_3(\alpha F + \delta G) - \beta F + \delta E}{y_3(\gamma G - \alpha H) + \beta H + \gamma E} \quad y_3 = \frac{-[(y E + G)(\gamma F + \delta H) - \beta + E(\delta F + \gamma H)]}{[(E + y G)(\gamma F + \delta H) + \alpha + G(\delta F + \gamma H)]}$$

where A=cos($\beta l_1$), B=cos($\beta l_2$), C=jsin($\beta l_1$), D=jsin($\beta l_2$), E=cos($\beta l_3$), F=cos($\beta l_4$), G=jsin($\beta l_3$),

$H = j\sin(\beta l_4)$, $\alpha = 2A\xi_x + C\xi_x$, $\beta = A\xi_+ + 2C$, $\gamma = 2B\xi_x + D\xi_+$, $\gamma = B\xi_+ + 2D$ and

$$\xi_+ = \sqrt{1-k}\tanh(\gamma l_0) + \sqrt{1+k}\tanh(\gamma_+ l_0) \quad \xi_x = \sqrt{1-k^2}\tanh(\gamma_+ l_0)\tanh(\gamma l_0)$$

## TUNING THE SYSTEM

The feed and load points are located near the voltage maximum of the resonant loops. Figure 3a shows the input impedance as a function of frequency if the output port is open circuited. This graph shows the characteristic double humped response of two coupled resonant circuits. If the output port is short circuited the input impedance is single humped (figure 3b). The circuit is tuned by succesively shorting one side while tuning the other side to exactly 47 MHz by varying the length of the resonant transmission circuit. The feed lines are then extended to quarter-wave matching sections (fig. 4). These are located at an electrical distance such that a low real impedance at 47 MHz is presented at their input. the purpose of the transformers is to lower the VSWR in the long feed lines between the antenna and the matching circuitry which is external to the TFTR test cell.

## DUAL ANTENNA FEED

The overall antenna feeed circuit is shown in figure 4. The antenna circuit discussed above can be described as a two port network. The input powers to each port of the network can be described by the following equations.

$$P_1 = \frac{1}{2}\mathrm{Re}\left(Z_{11}I^2 + Z_{12}I^2 e^{i\varphi}\right) \quad P_2 = \frac{1}{2}\mathrm{Re}\left(Z_{12}I^2 e^{-i\varphi} + Z_{22}I^2\right)$$

where $I_1 = I$, and $I_2 = Ie^{i\varphi}$ have been assumed. Since the network is symmetric $Z_{11} = Z_{22}$ and $Z_{12} = Z_{21}$ and by inspection for $P_1 = P_2$, only $\varphi = 0$ or $\pi$ is allowed. Therefore the toroidal phasing of the loops only allows in-phase or out-of-phase excitation for equal powers in each loop driven by equal power generators.

Considering the overall circuit of figure 4 as a two port network, the forward and reflected voltages at the inputs to the tuning structure can be related as follows,

$$\begin{bmatrix} V_1^- \\ V_2^- \end{bmatrix} = \begin{bmatrix} S_{11} & S_{12} \\ S_{21} & S_{22} \end{bmatrix} \begin{bmatrix} V_1^+ \\ V_2^+ \end{bmatrix}$$

where for a passive circuit $S_{12} = S_{21}$ and $S_{ii} = V_i^- / V_i^+$ with a matched termination on the other port . For the case of zero reflection $V_1^- = V_2^- = 0$ and $V_2^+ = V_1^+ e^{i\varphi}$ then

$$S_{11}S_{22} = S_{12}^2 \quad \text{and} \quad \varphi = \frac{1}{2i}\ln\left(\frac{S_{11}}{S_{22}}\right)$$

From this expression it can be seen that a totally symmetric antenna circuit can have its phase varied from 0 to $\pi$. If the circuit is not symmetric, then the generator phase need not be changed to change the antenna current phasing but $S_{11}$ will no longer equal $S_{22}$.

*Supported by DOE Contract No. DE-AC02-76-CHO-3073.

1. J.R. Wilson, et.al. Proc. Applications of RF Power to Plasmas 7th conf. Kissimee Fl 1987 p294.
2. D.J. Hoffman, et.al. Proc. Applications of RF Power to Plasmas 7th conf. Kissimee Fl 1987 p302.
3. R.A. Chipman, Transmission Lines (McGraw-Hill, N.Y.,1968)

373

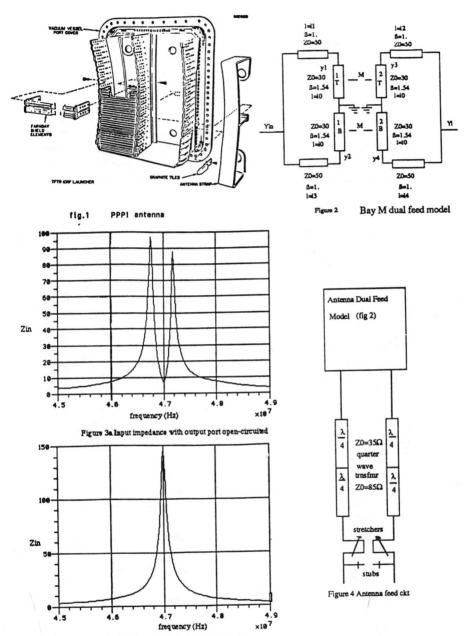

fig.1    PPPl antenna

Figure 2    Bay M dual feed model

Figure 3a Input impedance with output port open-circuited

Figure 3b Input impedance with output port shorted

Figure 4 Antenna feed ckt

# ANALYTIC THEORY OF ICRF MINORITY HEATING*

*Huanchun Ye and Allan N. Kaufman*
Physics Department and Lawrence Berkeley Laboratory
University of California, Berkeley, CA 94720

## ABSTRACT

We present a one-dimensional analytic theory of the ICRF gyroresonant absorption and mode-conversion, for the problem of minority fundamental resonance. Using the wave phase-space method, and the theory of linear mode conversion therein, we obtain explicit expressions for the coefficients of transmission $(T)$, reflection $(R)$, conversion $(C)$, absorption $(A)$.

Ion cyclotron resonant heating is one the two main methods in tokamak heating today. Both the majority (denoted $M$) second-harmonic and minority (denoted $m$) fundamental resonances are employed, but the minority heating appears to be predominant in present-day experiments. The majority second-harmonic heating problem has been solved analytically [1], using the phase-space method and the theory of linear mode conversion, for the one-dimensional slab model. In this paper we extend our method to treat the fundamental resonance of the minority ion species.

First we discuss briefly the phase-space concept and the theory of linear mode-conversion. Consider the wave equation: $D(x, k \to -i\partial)E(x) = 0$. To the lowest order in WKB approximation, it leads to the familiar Hamiltonian equations for rays: $dx/dt = \partial\omega/\partial k$, $dk/dt = -\partial\omega/\partial x$, where $\omega(\mathbf{x}, \mathbf{k}; t)$ is the solution of $D(\mathbf{x}, t; \mathbf{k}, \omega) = 0$. Conventionally the above wave equation is regarded as a differential equation in $(\mathbf{x}, t)$. However, in order to fully understand the wave dynamics, it is essential to treat $\mathbf{x}$ and $\mathbf{k}$ on an equal footing. There are many advantages to this phase-space point of view. For instance, one would never run into caustic singularities [2]. Another advantage, which is more important for our present work, is that we may have a $\mathbf{k}$-independent frequency function $\omega$, corresponding to a mode that travels only in $\mathbf{k}$-space [3]; thus the physics would be obscured if we insist on staying in $\mathbf{x}$-space. When two dispersion surfaces cross in phase-space, where the two modes have the same frequency and wave vector, and thus linear mode-coupling will occur. One mode, traveling on its own dispersion surface, transfers part of its energy to the other mode, which goes off on the other dispersion surface [Figure 1]. This is called *linear mode conversion*. When the coupling is localized in phase-space, the mode-conversion problem can be solved analytically [4].

The main idea in our analysis of ion gyroresonant absorption is to interpret it as mode conversions. We observe that the $N^{\text{th}}$ harmonic gyroresonance conditions $\omega = k_\parallel v_\parallel + N\Omega(\mathbf{x})$ ($N = 1$ for fundamental resonance) are in fact the dispersion relations of Case-van Kampen $(CVK)$ modes, one for each value of $v_\parallel$. Therefore

*Work supported by US DOE under contract No. DE-AC03-76SF00098

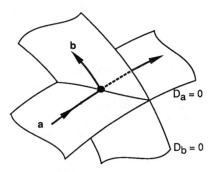

Figure 1. Linear mode conversion.

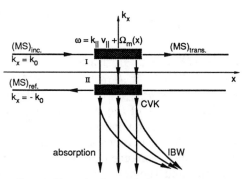

Figure 2. Dispersion diagram for ion gyroresonance.

ICRF heating can be viewed as mode conversions from the magnetosonic ($MS$) wave to a continuum of $CVK$ modes [1]. Note that these $CVK$ modes travel in k-space only, with velocity $dk/dt = -N\nabla\Omega$.

We shall study the problem in the one-dimensional slab model: $\mathbf{B}(\mathbf{x}) = \hat{z}B(x)$, where $B(x) = (1 + x/L_0)B_0$. We assume a uniform plasma density, and allow for small but finite $k_\parallel$. The dispersion relation for the $MS$ wave is $k_x^2 - N_{\perp0}^2\omega^2/c_A^2 = 0$ ($N_{\perp0}^2$ is defined below). The phase-space diagram of the dispersion relations of the $MS$ wave and three representative $CVK$ modes is shown in Figure 2. We see that the $CVK$ modes cross the two branches (incidence and reflection) of the $MS$ wave at two separate places. Thus we break the whole process into several steps, each of which can be analyzed explicitly: (1) the incident $MS$ wave crosses the resonance layer and excites the $CVK$ modes; (2) the $CVK$ modes propagate in $k_x$-space; (3) the $CVK$ modes cross the reflection branch and convert part of their energy to the reflected $MS$ wave; (4) the $CVK$ modes continue to travel in $k_x$-space. But because they are kinetic in $v_\parallel$, they contain collective modes, among them the weakly damped ion-Bernstein wave ($IBW$); it can leave the resonance layer, and be absorbed by electron Landau damping. The percentage of the energy that goes into the $IBW$ defines the conversion coefficient. What is left in the $CVK$ modes represents the direct absorption by the resonant ions.

We list our results here; a more detailed discussion follows.

$$T(\eta) = \exp(-2\eta), \qquad R(\eta, \kappa) = \left(\frac{4\eta}{2+\eta}\right)^2 |F(\eta, \kappa)|^2,$$

$$C(\eta, \kappa) = \frac{(2-\eta)^2}{8\eta} R(\eta, \kappa), \qquad A(\eta, \kappa) = 1 - T(\eta) - \frac{(2+\eta)^2}{8\eta} R(\eta, \kappa),$$

where

$$\eta \equiv \frac{\pi}{4}(k_0 L_0)\frac{\omega_m^2}{\omega_M^2}\frac{(\gamma_M - 1)^2 N_{\perp0}^2}{[1 + (\gamma_M - 1)N_\parallel^2]^2}\frac{v_z^2}{v_M^2}, \qquad \kappa \equiv (k_0 L_0)\frac{k_\parallel v_m}{\omega},$$

$$F(\eta, \kappa) \equiv \int_{-\infty}^{+\infty} d\xi \, G(\xi) \exp i \left( 2\kappa\xi - \frac{\eta}{\pi} \int_{-\infty}^{\xi} d\xi' \int_{-\infty}^{+\infty} d\xi'' \frac{G(\xi'')}{\xi'' - \xi' + i0^+} \right),$$

$\gamma_M \equiv \omega/\Omega_M$, $N_\parallel \equiv k_\parallel c_A/\omega$, $N_{\perp 0}^2 \equiv [1 - (\gamma_M + 1)N_\parallel^2][1 + (\gamma_M - 1)N_\parallel^2]/[1 + (\gamma_M^2 - 1)N_\parallel^2]$, $k_0 \equiv N_{\perp 0}\omega/c_A$, $v_m$ is the minority thermal speed, and $G(v_\parallel) \equiv g_m(v_\parallel) - \omega^{-1}k_\parallel v_m^2 g_m'(v_\parallel)$, ($g_m(v_\parallel)$ is the unperturbed minority $v_\parallel$-distribution).

The validity of the two-step mode-conversion approximation is that the $k_x$-width of each mode-conversion region, which is the inverse of the $x$-width of the resonance layer, is much smaller than the separation of the two branches. The width of the resonance layer comes from two contributions, one due to the mode coupling, and the other due to the Doppler shift caused by the thermal spread of $v_\parallel$. In the case of minority fundamental resonance, the coupling parameter $\delta$ is the ratio of minority to majority plasma frequency: $\delta = \omega_m/\omega_M$; hence the width of the resonance layer is given by $\omega - \Omega_m(x) - k_\parallel v_m \sim \omega\delta$, which leads to $\Delta x \sim L_0(\delta + k_\parallel v_m/\omega)$; thus the validity condition is $(\Delta x)^{-1} \ll 2k_0$.

The derivation of the equations can be outlined as follows. From Maxwell's equations we have $\nabla \times \nabla \times \mathbf{E}(\mathbf{x}) = 4\pi i\omega c^{-1}\mathbf{J}(\mathbf{x}; f)$, where the current $\mathbf{J}$ depends functionally on the perturbed particle distribution function $f^{(1)}$. The familiar procedure is to solve the linearized Vlasov equation for $f^{(1)}$, by integrating along the unperturbed orbit, and obtain the current $\mathbf{J} = \boldsymbol{\chi} \cdot \mathbf{E}$, where $\boldsymbol{\chi}$ is the linear susceptibility. But it is well known is that this introduces a (rapid varying) resonant denominator into $\boldsymbol{\chi}$. We can avoid this difficulty by omitting the resonant particles from $\boldsymbol{\chi}$; they remain as external current on the right-hand side of the wave equation. The motion of these particles is governed by their Vlasov equation. After some algebra, and using congruent reduction [5] to eliminate the other components of the electric field, we obtain two coupled equations:

$$D_J(x; v_\parallel) \, J(\,\cdot\,; v_\parallel) = \frac{i}{\omega} \, E(\,\cdot\,), \qquad D_E(k_x) \, E(\,\cdot\,) = -\frac{i}{\omega} \int dv_\parallel \, J(\,\cdot\,; v_\parallel),$$

where the dispersion functions are given by:

$$D_J(x; v_\parallel) = \frac{\omega - k_\parallel v_\parallel - \Omega_m(x)}{\omega\omega_m^2 \, G(v_\parallel)},$$

$$D_E(k_x) = \frac{[1 + (\gamma_M^2 - 1)N_\parallel^2](k_0^2 - k_x^2)}{(\gamma_M^2 - 1)(\frac{1}{2}k_x^2 + k_\parallel^2) - (\gamma_M - 1)\omega^2/c^2},$$

Here we keep terms $O(k_\perp^2 \rho_m^2)$. $J = J_x - iJ_y = e_m \int dv_x dv_y (v_x - iv_y)f^{(1)}$ is the resonant current, and $E = E_x - iE_y$ is the component of electric field that rotates in the ion sense. In the dispersion function $D_E$ we have used the cold plasma approximation, and ignored the $x$ dependence of Alfvén speed $c_A$. Choosing either the $x$- or $k_x$-representation, one of the equations is algebraic. We can derive the following conservation law for the wave-action flux:

$$\frac{\partial}{\partial x} \left[ (\dot{x})_E \frac{\partial D_E}{\partial \omega} \, E^2(x, k_x) \right] + \frac{\partial}{\partial k_x} \left[ (\dot{k}_x)_J \int dv_\parallel \frac{\partial D_J}{\partial \omega} \, J^2(x, k_x; v_\parallel) \right] = 0,$$

where $E^2(x, k_x)$ and $J^2(x, k_x; v_\parallel)$ are the Wigner functions [2] of $E$ and $J$ respectively. From this equation we can identify the wave-action flux associated with $CVK$ modes.

The solution of these equations closely follows that in [1], so we just outline it here. For the two mode conversions (steps 1 and 3), we linearize the dispersion function about $\mp k_0$: $D_E(k_x) \approx (k_x \mp k_0)V_E$, where $V_E = 4k_0[1 - (\gamma_M - 1)N_\parallel^2]^2/(\gamma_M - 1)^2 N_{\perp 0}^4$. Then the coupled equations become a first order ODE, and can be solved easily. In between the two mode-conversion regions (step 2), we ignore the coupling. Thus $CVK$ modes propagate according to a simple equation:

$$\left[x(v_\parallel) - i\frac{d}{dk_x}\right] J(k_x; v_\parallel) = 0,$$

where $x(v_\parallel) = -L_0 k_\parallel v_\parallel/\omega$. Below the second mode-conversion region, we cannot ignore the coupling. Eliminating $E$ we obtain the following $CVK$ equation:

$$\left[x(v_\parallel) - i\frac{d}{dk_x}\right] J(k_x; v_\parallel) = \frac{L_0 \omega_m^2}{\omega^2 D_E(k_x)} G(v_\parallel) \int dv_\parallel' \, J(k_x; v_\parallel'),$$

whose solution can be obtained by the method of [6]. However, in that calculation, the $IBW$ can only appear as a superposition of $CVK$ modes. In order to project out the $IBW$, we use the spectral deformation technique [7].

In conclusion let us highlight the main points involved in our analysis: (1) gyroresonant absorption as mode-conversion Case-van Kampen modes; (2) linear mode conversion in phase-space, and the closely related congruent reduction theory; (3) spectral deformation technique, which helps us to project out the ion-Bernstein wave. A more detailed account of this work is in preparation.

References:

[1] H. Ye and A. N. Kaufman, *Phys. Rev. Lett.* **61**, 2762 (1988).
[2] R. G. Littlejohn, *Phys. Rep.* **138**, 193 (1986); S. W. McDonald, *Phys. Rep.* **158**, 337 (1988).
[3] H. Ye and A. N. Kaufman, *Phys. Rev. Lett.* **60**, 1642 (1988).
[4] L. Friedland, G. Goldner and A. N. Kaufman, *Phys. Rev. Lett.* **58**, 1392 (1987); A. N. Kaufman and L. Friedland, *Phys. Lett.* **A**, 387 (1987).
[5] L. Friedland and A. N. Kaufman, *Phys. Fluids* **30**, 3050 (1987).
[6] G. Bateman and M. D. Kruskal, *Phys. Fluids* **15**, 277 (1972).
[7] J. D. Crawford and P. D. Hislop, *Ann. Phys.* (New York) (in press).

# A FAST-WAVE CURRENT DRIVE SYSTEM DESIGN FOR ITER[*]

J. J. Yugo[†]
Fusion Engineering Design Center, Oak Ridge, TN 37831-8218

D.B. Batchelor, P.L. Goranson, R.H. Goulding, E.F. Jaeger, P.M. Ryan
Oak Ridge National Laboratory, Oak Ridge, TN 37831

## ABSTRACT

The development of an efficient current drive system is an essential factor in a steady-state tokamak reactor. For the International Thermonuclear Experimental Reactor (ITER) one of the current drive systems being considered is fast wave current drive (FWCD) in the ion cyclotron range of frequencies (ICRF), supplemented by lower hybrid current drive (LHCD) in the low density edge plasma, although LHCD can also be utilized for current ramp-up. Three of the candidate current drive systems (neutral beams, electron cyclotron and ion cyclotron fast wave) presently have deficiencies either in the technology of sources and transmission systems or in the experimental physics basis. Engineering design and integration of each system will contribute to the basis on which the eventual system choice can be made. This FWCD system design begins to define and address the issues related to implementing such a system on a steady-state tokamak reactor. Plasma modeling by Batchelor, Carter, and Jaeger[1] at ORNL with the two-dimensional, full-wave code ORION[1] provides the basis on which this system has been designed.

## PHYSICS BASIS

Table I gives the basic ITER parameters that were used for the study of fast wave current drive. The current drive system is designed to operate at a frequency of 60-65 MHz. The current-carrying straps are 0.2 m wide with a toroidal separation of 0.45 m and a poloidal length of 0.6 m. It has been found, for the proposed FWCD launcher array, that the maximum current drive efficiency in ITER occurs for a relative phase between antennas of $\Delta\phi = -0.45\pi$, to yield a spectral peak at $n_\| = 2.45$. The launched spectrum of the 12 element antenna and the absorbed power spectrum in the ITER plasma are shown in Fig. 1. The current drive efficiency for the antenna array with this phasing was found to be $I/P_{total} = 0.131$ A/W for a figure of merit $\gamma = 0.33$. The absorbed power radial profile is found to be strongly peaked at the center of the plasma and is primarily due to electron transit time magnetic pumping (TTMP) with off-axis Tritium ion absorption on the order of 10%. The driven current profile closely follows the electron power absorption profile.

---

[†]On assignment from TRW, Inc., Redondo Beach, CA, USA

[*]Research sponsored by the Office of Fusion Energy, U.S. Department of Energy, under contract DE-AC05-84OR21400 with Martin Marietta Energy Systems, Inc.

Table I.   ITER Design Parameters

| $R_0$ | 5.5 | m |
|---|---|---|
| $a_{plasma}$ | 1.8 | m |
| $a_{wall}$ | 1.9 | m |
| $B_0$ | 5.3 | T |
| $\kappa$ (elongation) | 2.0 | |
| $Z_{eff}$ | 2.26 | |
| $\eta_\alpha$ | 1.0 | % |
| $T_{e0}$, $T_{i0}$ | 33 | keV |

dashed: unweighted $f = 60$ MHz
solid: weighted by 12 element ITER antenna spectrum

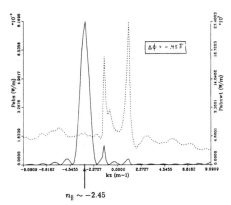

$n_{\parallel} \sim -2.45$

Fig. 1. Total absorbed power spectrum in ITER.

MECHANICAL DESIGN

A launcher array has been defined that meets the requirements for current drive in ITER[2].  The launcher is a 12 X 5 array of loop antennas (toroidal X poloidal).  Each of the antenna straps is fed by a coaxial line at one end, and is grounded at the other.  A double layer Faraday shield made of graphite-tiled Inconel tubes covers the front and sides of each resonant cavity.  The antenna straps are separated from each other poloidally by solid septa and toroidally by parially-slotted septa and Faraday shield elements.  The Faraday shield tubes, current straps, and the resonant cavity rear walls are water-cooled for long pulse or steady-state operation.  Overall plan and elevation views of the launcher, transmission and tuning systems installed in the ITER device are shown in Figs. 2 and 3.  Adjustable-length transmission line sections give phase adjustment capability and, along with a single stub tuner, allow impedance matching of each antenna.  The stub tuners, phase shifters, vacuum feedthroughs, and dc breaks are located outside of the vacuum vessel port as close to the antenna as practical.  The antenna array is assembled by radially inserting three main antenna modules through each of two adjacent midplane ports and then shifting them toroidally into their final position.  The antenna array is supported from blanket and shield modules which react the disruption loads into the vacuum vessel.

One alternative configuration being considered is to integrate the antenna array with the blanket modules and only bring the coaxial feed lines through the ports.  This arrangement would result in a superior spectrum due to the larger number of antenna loops in the toroidal direction but poses a more difficult integration problem.
A second alternative would launch fast waves from an array of folded waveguides, and would be particularly attractive for frequencies above about 100 MHz.

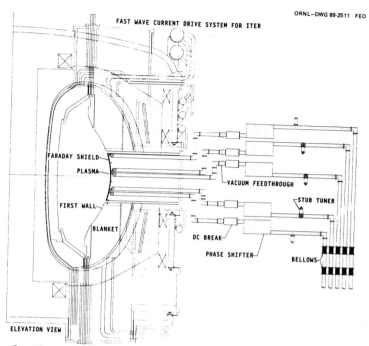

ORNL—DWG 89-2511 FED

FAST WAVE CURRENT DRIVE SYSTEM FOR ITER

FARADAY SHIELD

PLASMA

VACUUM FEEDTHROUGH

FIRST WALL

STUB TUNER

BLANKET

DC BREAK

PHASE SHIFTER

BELLOWS

ELEVATION VIEW

Fig. 2. Elevation view of FWCD launcher array, transmission line, and impedance matching system installed in ITER.

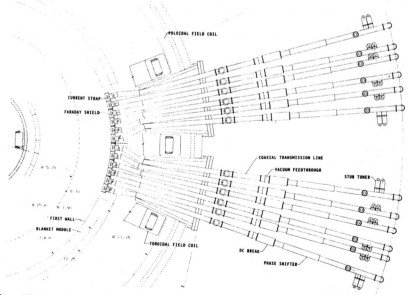

POLOIDAL FIELD COIL

CURRENT STRAP

FARADAY SHIELD

COAXIAL TRANSMISSION LINE

VACUUM FEEDTHROUGH

STUB TUNER

FIRST WALL

BLANKET MODULE

TOROIDAL FIELD COIL

DC BREAK

PHASE SHIFTER

Fig. 3. Plan view of the launcher array, transmission line and impedance matching system for a 120 MW FWCD system for ITER.

## EROSION

Analysis has estimated the flux of particles striking the first wall to be $1.0 \times 10^{16}$ cm$^{-3}$s$^{-1}$ with an energy between 10-200 eV. For this particle flux and energy the erosion of the Faraday shield surfaces is calculated to occur at a rate of $2.04 \times 10^{-8}$ mm/s for Beryllium and $2.18 \times 10^{-8}$ mm/s for Carbon. Thus, either coating would erode by 1 mm in approximately 10,000 hours of normal operation.

## ELECTRICAL CONFIGURATION

The antenna system described has been analyzed for its electrical performance at a fixed frequency of 65 MHz and relative phasing of $\Delta\phi = -0.45\pi$. Each antenna is a current-carrying strap fed at one end by a coaxial transmission line and grounded at the other end. A 100 MW FWCD system would require an input power of approximately 2.0 MW to each loop. The array, as designed, is capable of launching 100-140 MW depending on the antenna details and plasma loading. The peak voltage on the antenna straps is 54-64 kV for plasma loadings of 6.5-10.0 $\Omega$/m, while the peak antenna current is 650-850 A for the same range of plasma loading. For a factor of 1.7 increase in plasma loading and fixed tuning, the VSWR remains less than 1.6 for all but one strap. The mutual coupling between each pair of antennas was determined experimentally from a similar antenna array being tested at ORNL. Over the range of parameters considered it was possible to obtain very uniform currents in all of the loops. In all cases, the calculated parameters fall well within the technological and experimental limits of present day antennas. Improvements in voltage or current capability would allow reduction in the poloidal extent of the launcher array and hence a reduction in the number of components.

## SUMMARY

It appears from the physics modeling and engineering design that a FWCD system may be feasible and efficient in ITER. The technology needed in the system is essentially all available. If the physics basis for FWCD can be experimentally proven, it would provide a very attractive alternative to the other candidate current drive systems which require substantial improvements in technology to be considered viable.

## REFERENCES

1.   E. F. Jaeger, D. B. Batchelor, H. Weitzner, and J. H. Whealton, Computer Phys. Communications, **40**, 33 (1986).

2.   J. J. Yugo, D. B. Batchelor, R. H. Goulding, A Fast Wave Current Drive System Design For ITER, ITER Report IL-HD-7-9-U-1, February, 1989.

3.   D. Batchelor, E. Jaeger, M. D. Carter, Fast Wave Current Drive Modeling for ITER, ITER Report IL-PH-6-99-U-7, February, 1989.

CHAPTER 4

GENERAL RF/PLASMA INTERACTIONS

# ROTATING MAGNETIC FIELD CURRENT DRIVE - THEORY AND EXPERIMENT

I. J. Donnelly[*]

University of Sydney, NSW 2006, Australia

## ABSTRACT

Rotating magnetic fields have been used to drive plasma current and establish a range of compact torus configurations, named rotamaks. The current drive mechanism involves a ponderomotive force acting on the electron fluid. Recent extensions of the theory indicate that this method is most suitable for driving currents in directions perpendicular to the steady magnetic fields.

## INTRODUCTION

The use of a rotating magnetic field (RMF) to drive currents in plasmas has been well demonstrated over the past decade[1-5]. Currents of up to several kA have been driven in small plasma devices (major radii ~ 0.1 m), and used to generate compact torus configurations such as field reversed mirrors[1-3], FRCs[4] and compact tokamaks[5]. These devices have been named rotamaks. A schematic diagram of a spherical rotamak is shown in Fig. 1.

Theoretical analysis over the past few years has shown that it is convenient to separate RMF current drive into two categories. We shall call these: the standard model (Type-S), in which the oscillating plasma current $j$ is predominantly parallel to the steady magnetic field $B$; and the wave model (Type-W) in which a substantial component of $j$ is perpendicular to $B$. To date, theoretical work and experiments have been confined to low temperature plasmas in which plasma resistivity is the dominant dissipative mechanism. In this case the RMF drives an electron current by means of the time-averaged nonlinear force $<j \times b>$ exerted by the RMF, where $b$ is the oscillating magnetic field. This process, recently reviewed by Jones[6], drives all of the electrons rather than just a resonant group. Based on present understanding, we expect the $<j \times b>$ mechanism to also apply in hot plasmas for Type-S conditions, whereas resonant wave-particle interactions will probably be dominant under Type-W conditions.

[*]On attachment from the Australian Nuclear Science and Technology Organisation, Sydney, Australia.

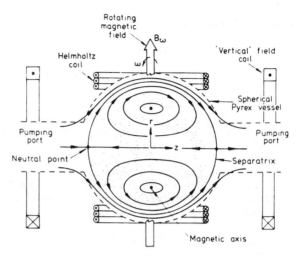

Fig. 1. A schematic diagram of the rotamak.

## THEORY

The principles underlying this current drive technique are best illustrated using $(r, \phi, z)$ geometry to describe the application of a rotating magnetic field

$$\mathbf{b}(\mathbf{r}, t) = (b_\omega \cos(\phi - \omega t), - b_\omega \sin(\phi - \omega t), 0) \tag{1}$$

to an infinite plasma cylinder of radius R and constant electron density n and temperature $T_e$. A magnetic field $\mathbf{B}(\mathbf{r}) = (0, B_\phi, B_z)$ can be applied to allow tokamak modelling and to satisfy equilibrium constraints. In theories developed to date the plasma response to the RMF has been derived from the MHD momentum and generalised ohm's law equations:

$$\rho d\mathbf{V}/dt = \mathbf{J} \times \mathbf{B} - \nabla P - \mathbf{M}_1 \tag{2}$$

$$\mathbf{E} + \mathbf{V} \times \mathbf{B} = \eta \mathbf{J} + (ne)^{-1}(\mathbf{J} \times \mathbf{B} - \nabla P_e) + (m_e/e)(d/dt)(\mathbf{J}/ne) - \mathbf{M}_2 \tag{3}$$

where the terms $\mathbf{M}_1$ and $\mathbf{M}_2$ arise from relaxation of the ion and electron fluid momentum due to particle loss and replacement, viscosity etc. (but not electron–ion collisions which are manifested in $\eta$). In the presence of oscillating fields, a time–average of Ohm's law gives

$$\eta \mathbf{J} = \mathbf{E} + \mathbf{V} \times \mathbf{B} + \langle \mathbf{v} \times \mathbf{b} \rangle - (ne)^{-1}(\mathbf{J} \times \mathbf{B} + \langle \mathbf{j} \times \mathbf{b} \rangle) + \ldots \tag{4}$$

This indicates that a steady current $\mathbf{J}$ may be driven by ponderomotive forces related to $\langle \mathbf{v} \times \mathbf{b} \rangle$, $\langle \mathbf{j} \times \mathbf{b} \rangle$ etc. In many cases this process is highly nonlinear because the wave response depends on the steady-state field configuration, which depends in turn on the current driven by the $\langle \mathbf{j} \times \mathbf{b} \rangle$ force.

Consider now the case (Type-S) in which the applied steady magnetic field is purely axial. The RMF induces an oscillating axial current $j_z(r)$ in the plasma edge, which tends to screen the RMF from the plasma (the skin effect). However, as the RMF amplitude is increased the $\langle j_z b_r \rangle$ interaction increases and drives an appreciable azimuthal current $J_\phi$ ($= - (\eta ne)^{-1} \langle j_z b_r \rangle$) once the condition $\omega_{ce}/\nu_{ei} \geq \lambda$ is satisfied, where $\omega_{ce} = eb_\omega/m_e$, $\nu_{ei}$ is the electron-ion collision frequency, $\lambda = R/\delta$ and $\delta$ is the resistive skin depth. Ion motion is negligible provided $\omega \gg \omega_{ci}$. A convenient measure of the driven current is $\alpha = I_\phi/I_\phi^{max}$ where $I_\phi^{max}$ corresponds to all electrons rotating synchronously with the RMF. The variation of $\alpha$ with $b_\omega$ depends on $\lambda$, as shown in Fig. 2(a). In as yet unpublished work, it has been found that these results are not strongly changed when more realistic density and temperature profiles are assumed.

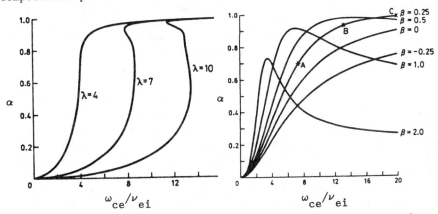

Fig. 2. Driven current vs. $b_\omega$ for (a) a plasma cylinder and (b) a plasma sphere. $\lambda = R/\delta$ and $\beta = B_z/b_\omega$.

The behaviour in Type-S cases can be summarised as follows. At low RMF amplitude the screening currents freeze the RMF out of the plasma and there is little current driven. However, when $\omega_{ce}/\nu_{ei}$ increases to a value around $\lambda$, the driven current density becomes appreciable, the screening is less effective and the RMF penetrates the plasma.

When a steady azimuthal field $B_\phi$ is added to the previous configuration the behaviour can change dramatically. In this case (Type-W) the $j_z B_\phi$ interaction leads to an oscillating force on the plasma, which generates plasma waves that allow penetration of the RMF, even at small RMF amplitude. This has been considered in detail[9,10] when $B_\phi \propto 1/r$, simulating the toroidal field in a compact tokamak. The results are complex, but the major points can be summarised as follows:

- if $\alpha \ll 1$ when $B_\phi = 0$, the addition of a moderate $B_\phi$ increases both the RMF penetration and the driven current, but the current decreases when $B_\phi$ becomes larger,
- if $\alpha \approx 1$ when $B_\phi = 0$, an increase in $B_\phi$ decreases the driven current, concentrating it near the surface, and has little effect on the near fully penetrated RMF except in some circumstances when it is enhanced because of the excitation of a resonant wave mode,
- steady diamagnetic axial currents are also generated.

These phenomena can be interpreted in terms of the excitation of a whistler wave (which propagates into the plasma) and a resistive mode (which is localised at the plasma edge). When $B_\phi$ is large the azimuthal current is driven by an interaction (mode-beating) between the $b_r$ of the whistler wave and the $j_z$ of the resistive mode. These results indicate that, except in the limits of low plasma temperature or very large $b_\omega$, current drive by the $\langle j \times b \rangle$ mechanism decreases in effectiveness if there is an appreciable level of propagating plasma waves excited by $j \times B$ forces.

Major advances have recently been made[11] using a spherical model of the plasma with no applied toroidal field. The solution technique involves expanding all fields and currents in vector spherical harmonics and numerically solving the differential equations in the radial co-ordinate. The solutions have been confined to values of $\lambda < 10$ because of computational difficulties, but the results obtained have been very useful in helping understand some of the experimental results. Much of the observed behaviour can be interpreted in terms of the following simple picture involving the screening currents that would be induced in a conducting sphere in the absence of any steady $B$. These flow predominantly in the axial (or poloidal) direction near the equator, and in the toroidal direction near the poles. In plasma, the interaction of the oscillating poloidal currents with the steady field will probably be minimised when the $B$ field, resulting from the driven current and the externally applied field, has a separatrix near the plasma edge as depicted in Fig.1.

Of course there will still remain an oscillating force on the plasma coming from the $j_\phi B_{pol}$ interaction near the poles, and this will excite whistler waves with possible adverse effects on the current drive efficiency in hot plasmas. Results for the driven current vs. $b_\omega$ are shown in Fig. 2(b) for a range of values of the applied axial field $B_z$. The curves are characterised by the parameter $\beta = B_z/b_\omega$. The peaks evident in the $\beta = 1.0$ and $\beta = 2.0$ cases correspond to configurations with the separatrix near the plasma boundary.

## EXPERIMENT

Rotating magnetic fields have been used to generate a wide range of rotamak configurations. For initial gas fill pressures in the mtorr range, the discharges have been reproducible and grossly stable, with time-averaged magnetic field structures as indicated in Fig. 1. Experiments [2, 12, 13] have ranged from low power, medium duration (5 kW, 20 ms) to high power, short duration (12 MW, 80 $\mu$s), with RMF frequencies of ~ 1 MHz. Hydrogen and argon plasmas have been used. In the past 5 years, efforts have concentrated on medium power, medium duration (100 kW, 20 ms) experiments in spherical [3, 5] and cylindrical [4] Pyrex vessels of radius 0.14 m and 0.07 m respectively. A large stainless steel vessel with internal RMF coils has also been used [13]. Typical parameters for these rotamaks are: $n_e = 10^{18} - 10^{19}$ m$^{-3}$, $T_e = 10 - 20$ eV, $T_i = 1 - 2$ eV, $I_\phi \sim 2$ kA, $B_{pol} \sim 5$ mT, $b_\omega \sim 2$ mT, and $\tau_E \sim 5$ $\mu$s. These modest values need to be improved before an assessment can be made of the potential of the rotamak as a fusion device. Nevertheless, many interesting results have already been obtained.

## CONFINEMENT AND SCALING

Measurements of the steady and oscillating magnetic fields allow derivation of the currents, and the steady $\mathbf{J} \times \mathbf{B}$ and ponderomotive $\langle \mathbf{j} \times \mathbf{b} \rangle$ confining forces. The latter is usually small (<20% of the total). For experiments in hydrogen, comparison of the $\mathbf{J} \times \mathbf{B}$ force with the pressure gradient derived from density and temperature measurements using Langmuir probes, indicates that the plasmas are in MHD equilibrium [3, 5]. Experimental results are compared with an MHD equilibrium model in Fig. 3.

Calculations of particle confinement times using classical diffusion theory [15] predict $\tau_p \sim 10$ $\mu$s, which is in reasonable agreement with experiment. However, we note that Bohm diffusion also predicts a similar $\tau_p$ in these low magnetic field cases.

In a spherical rotamak it has been found that the driven current scales almost linearly with the applied field [3]. However there is a limiting value of the applied field above which the plasma terminates. The reason for this termination is not known, but it may be due to power constraints since an increase in RMF amplitude (and available power) allows larger currents. An investigation [16] has been made of the relationship between the RMF amplitude and the driven current, which can be externally controlled, and the other plasma parameters. Although the RMF is responsible for creating and maintaining the rotamak configuration, its amplitude was found to have little effect on many of the plasma parameters. The temperature and density near the magnetic axis, and the particle confinement time were observed to scale mainly with the driven current, even though the same current can be obtained for a range of RMF amplitudes. However, the details of the equilibrium magnetic field configuration do depend on the rotating field. It is expected that the RMF will have even less effect in larger and hotter rotamaks when its amplitude should be much less than the steady fields.

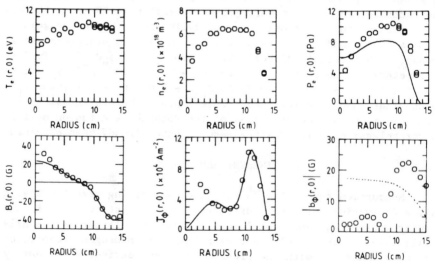

Fig. 3. Parameters in the equatorial plane of a spherical rotamak. The circles are experimental points and the lines are from an MHD equilibrium model.

## COMPACT TOKAMAKS

Compact tokamak configurations have been created by adding a toroidal field generated by an axial current. Plasmas with toroidal fields of up to 20 mT at R = 0.1 m were studied [5]. The equilibrium current density and magnetic field profiles were similar to those predicted for such devices [17]. The toroidal current was found to depend on the size of $B_\phi$ in a manner consistent with the RMF current drive theory. It was also observed that poloidal currents were generated. These were paramagnetic for low $B_\phi$ and diamagnetic for larger values. A possible explanation of this behaviour involves a combination of MHD equilibrium constraints and the current drive process.

## COMPARISON OF THEORY AND EXPERIMENT

Using the measured oscillating fields to derive the associated currents allows computation of the time-averaged force $\langle \mathbf{j} \times \mathbf{b} \rangle$. Collins et al. [18] have shown that $\eta J_\phi \simeq - (ne)^{-1} \langle \mathbf{j} \times \mathbf{b} \rangle_\phi$, with a resistivity 5 x the Spitzer value. They also found the following: in the presence of a toroidal field the $\langle \mathbf{j} \times \mathbf{b} \rangle_{pol}$ forces are in the appropriate directions to drive a diamagnetic current; for some values of $B_\phi$ there was enhanced penetration of the RMF; and a standing wave structure was sometimes evident in the plasma. These observations give strong support to the theory of RMF current drive described earlier.

It is convenient to discuss the efficiency of the RMF current drive process in terms of an ohmic heating figure of merit [19]. We write $P = (r\omega/v_d)\eta_{eff}J^2$ where $r\omega$ is the phase velocity and $v_d$ the effective electron drift velocity associated with the current. Typically, for experiments in hydrogen $r\omega/v_d \sim 5$ and $\eta_{eff} \sim 5$ x the Spitzer value [18].

## CONCLUSIONS

The extension of RMF current drive theory to situations in which the oscillating currents cross with steady magnetic fields has led to increased understanding of the $\langle \mathbf{j} \times \mathbf{b} \rangle$ current drive mechanism, which can be viewed as the process by which a wave deposits momentum in the electron fluid of a collisional plasma. Projecting from present knowledge it appears that the $\langle \mathbf{j} \times \mathbf{b} \rangle$ process should continue to work in hot plasmas, even for small $b_\omega$, provided that $\mathbf{j}$ is predominantly parallel to $\mathbf{B}$. This means that it is effective for driving steady current perpendicular to $\mathbf{B}$, as in the field-reversed mirror or FRC configurations. When applied to compact tokamaks with large $B_\phi$, RMFs will predominantly generate whistler waves (or compressional Alfven waves) which will drive current parallel to $\mathbf{B}$ by interaction with resonant particles; however, there still may be some energy loss and current drive in the plasma edge region because of the mode-beating effect.

Most RMF current drive theories have assumed conditions such that the ion motion is negligible. This is valid for existing experimental conditions, but it will not in general be true for Type-W current drive unless the wave frequency is well above the ion cyclotron frequency derived from the steady field ($eB/m_i$).

It should be noted that the RMF not only drives the current which establishes the confining magnetic field in the rotamak, but it also provides the power to ionise and heat the plasma. My expectation is that a significant increase in RMF power, applied to larger devices, will lead to hotter and less resistive plasmas, higher currents and magnetic fields, and better confinement. This scenario will soon be tested at Flinders University, South Australia where 800 kW of RMF power will be applied to plasma in a 200 litre Pyrex vessel.

I wish to acknowledge my colleagues at ANSTO and at Flinders University whose research has made this review possible.

## REFERENCES

1. W. N. Hugrass et al. Phys. Rev. Lett. <u>44</u>, 1676 (1980).
2. I. R. Jones. Int. Conf. Plasma Phys., Lausanne, Switzerland (1984) Vol.II, 473.
3. G. Durance el al. Plasma Phys. Contr. Fusion <u>29</u>, 227 (1987).
4. H. A. Kirolous et al. Proc 8th U.S. Compact Toroid Symp., University of Maryland (1987), 214.
5. G. A. Collins et al. Nuc. Fusion <u>28</u>, 255 (1988).
6. I. R. Jones. Comm. Plasma Phys. Contr. Fusion <u>10</u>, 115 (1986).
7. I. R. Jones and W. N. Hugrass. J. Plasma Phys. <u>26</u>, 441 (1981).
8. W. N. Hugrass. Aust. J. Phys. <u>38</u>, 157 (1985).
9. W. K. Bertram. J. Plasma Phys. <u>37</u>, 423 (1987).
10. P. A. Watterson. J. Plasma Phys. <u>40</u>, 109 (1988).
11. D. Brotherton-Ratcliffe and R. G. Storer. Plasma Phys. Contr. Fusion <u>30</u>, 967 (1988).
12. G. Durance and I. R. Jones. Phys. Fluids <u>29</u>, 1196 (1986).
13. M. Kuhnapfel et al. Plasma Phys. Contr. Fusion <u>29</u>,1501 (1987)
14. H. A. Kirolous et al. Plasma Phys. Contr. Fusion <u>31</u>, 79 (1989)
15. I. J. Donnelly et al. Aust. J. Phys. <u>40</u>, 393 (1987).
16. G. A. Collins et al. Plasma Phys. Contr. Fus. (1989) in press
17. Y.-K. M. Peng and D. J. Strickler. Nucl. Fus. <u>26</u>, 769 (1986).
18. G. A. Collins et al. J. Plasma Phys. <u>40</u>, 127 (1988).
19. W. N. Hugrass. Nucl. Fusion <u>22</u>, 1237 (1982).

# RF POWER APPLICATIONS FOR ITER

David A. Ehst
Argonne National Laboratory, Argonne, IL   60439

## INTRODUCTION -- WHAT IS ITER?

The International Thermonuclear Experimental Reactor (ITER) is envisioned to be a large tokamak with several missions.  In its initial phase of operation the DT plasma is driven inductively with enough confinement to provide an average neutron wall load   ~ 1 MW/m$^2$; this phase will allow a detailed assessment of the physics behavior of an ignited plasma.  The second, final stage of operation features certain structural modifications (breeding blanket) to test engineering performance of components under reactor-like conditions. For this latter stage, steady state operation is specified, and this requirement of noninductive current drive (CD) largely dictates the plasma parameters and technological needs for ITER.  In this paper we focus our attention on CD and heating in the steady state phase. For reference purposes, in this phase ITER is considered to have: major radius, R = 5.5 m; aspect ratio, A = 3.1; elongation, $\kappa$ = 2.0; field on axis, $B_o$ = 5.3 T; toroidal current, I = 18 MA; average electron density, $\bar{n}_e$ = 7.0 × 10$^{19}$ m$^{-3}$; average electron temperature, $\bar{T}_e$ = 18 KeV; and $Z_{eff}$ = 2.2.  Even with relatively flat density profiles we find   ~ 35% of I may be provided by the bootstrap effect,[1] if the current profile is appropriately controlled by the noninductive seed current (see Fig. 2).  Since the divertor plates may not survive more than ~ 50 full power disruptions, it is essential that this steady state CD provide quiescent plasma behavior. In the following we assess the possible roles to be played by various rf technologies for CD and heating.  Low cost equipment, flexible enough to provide multiple applications (heating, CD, preionization, etc.) appears highly desirable to enhance the attractiveness of ITER.

## PHYSICS OF CD AND HEATING -- EXPERIMENTAL VERIFICATION OF THEORY

Neutral beam CD experiments are in very good agreement with theory and have demonstrated CD at megampere levels.  Neutral beams (NB) could be almost guaranteed to provide steady operation and on this basis are a leading contender for auxiliary power to ITER.

The fast wave has shown excellent (ICRH) plasma heating, and, as predicted by theory, propagates readily into the high density central plasma.  At high harmonics of the ion cyclotron frequency $\left( \omega \gtrsim 10 \, \Omega_i \right)$ a small number of CD experiments have been done. Highlights of these CD experiments[2] include:  maximum current ~ 0.1 MA with $\gamma \equiv \bar{n}_e I R/P_{inj}$ = 0.04 (JFT-2M, ~ 1 keV); gas breakdown, current startup and rampup (Synchromak); and a rampup rate $\dot{I}$ = 300 kA/s (JIPP TIIU). Careful study shows these results in agreement with CD via the Landau resonance. Figure 1 suggests that CD in some experiments was accomplished at densities higher than would be possible with lower hybrid (LH) waves at those frequencies.

The best data base for CD with rf is with lower hybrid (LH) waves.[3] Despite some lingering issues (e.g., the spectral gap) the experiments are substantially in agreement with theory. On this basis, ironically, we cannot guarantee that LH waves will drive current in the interior of the ITER plasma, since the quasilinear damping process prevents wave penetration into dense plasma with $T_e \gtrsim 10$ keV.

Electron cyclotron (EC) waves provide excellent localized electron heating, but experiments yield CD efficiencies, $\gamma$, which are disappointingly small. The theory of ECCD leads to $\gamma$ predictions for ITER which are also relatively small. It appears that EC waves, which can provide various other ITER functions (e.g., preionization), are not destined to drive the bulk current.

## PROJECTED ELECTRIC POWER AND CAPITAL COST FOR CD

Both the physics parameter $\gamma$, the CD efficiency in the plasma, and the electric efficiency $\eta$ of the driver system figure in the economics of steady state operation. In the figure of merit $\gamma = \bar{n}_e R I/P_{inj}$, we define the denominator to be the power injected into the vacuum vessel to drive current (assuming there is no bootstrap effect); $P_{inj}$ generally includes significant launched power which is lost and does not contribute to CD. If a bootstrap current is present then less power is needed to provide the total I, in such a case the effective $\gamma^B$ is larger than $\gamma$. Finally, we define $\eta \equiv P_{inj}/P_{elec}$, where $P_{elec}$ is power from the grid.

Table I (compiled from INTOR[3] and ITER sources) displays calculations of $\gamma$ done for similar plasma conditions. The first four columns include $\gamma$ values obtained in a benchmark comparison done for INTOR. The best agreement appears for the NB results, reflecting the relative maturity of NBCD theory. The final column lists typical $\gamma^B$ values for ITER; in this case there is still uncertainty for the high frequency FW result, as trapped electron effects are incorporated in a very crude fashion. (For comparison, $\gamma \lesssim 0.25$ for CD with EC waves.[7])

The cost of CD systems is presently uncertain, in some cases (NB, EC waves) because the technology has not been developed, and in other cases (ICRF) because the performance requirements (e.g., frequency tunability) have not been specified. In terms of $P_{inj}$ we list rough estimates of the cost of power in the ITER time frame, assuming all technology development has taken place, for the various options, in Table II. To illustrate the cost variation of CD options consider first the NB: $\gamma^B = 0.61$ implies $P_{inj} = 114$ MW and a capital cost (@ $3.00) of $340M. In contrast, for the FW at low frequencies: if $\gamma^B \cong 0.45$ then $P_{inj}$ = 154 MW at a cost (@ \$1.25) of \$190M. Costs of this size are roughly 5% of the total ITER cost and are significant enough to warrant refinement of the cost estimates in the table.

Table I.
Benchmark Steady State $\gamma[10^{20}$ $A \cdot W^{-1} \cdot m^{-2}]$ Values for
$\bar{T}_e$ = 18-20 keV, $\bar{n}_e$ = 0.7 $\times$ $10^{20}$ $m^{-3}$, $Z_{eff} \cong 2$;
Reference in Brackets

| Driver | EC | Japan | USA | USSR | USA $\gamma^B$ |
|---|---|---|---|---|---|
| NBCD $0.5$ MeV $<$ $E_b$ $<$ $1.0$ MeV | 0.36 | 0.37 | 0.37 | 0.5 | .61 [6] |
| HFFW | -- | .59 [4] | 0.41 | 0.31 | .47 |
| f $\cong$ 300 MHz | | | | | |
| LFFW f = 17-70 MHz | .48 [5] | .36 [4] | 0.33 | 0.27 | -- |

Table II.   Cost and Efficiency of CW Power

| Technology | Cost/Watt, Injected | | Present | Near Future | Aggressive |
|---|---|---|---|---|---|
| NB (1 MeV, $D_o$) | \$3.00 | RFQ | -- | .34 | .68 |
| | | ESQ | -- | -- | .71 |
| EC (140 GHz) | --- | Gyrotron FEM | | .20 | .48 .60 |
| LH (8 GHz) | \$2.50 | | .17 (FTU) | .32 | .44 |
| LH (2.5 GHz) | \$2.00 | | | .35 | .48 |
| FW (250 MHz) | \$2.00 | | | .50 | .65 |
| FW ($<$ 70 MHz) | \$1.25 | | .43 (JET) [8] | .68 [8] | .84 |

When evaluating the electrical power required for CD we must consider the product of $\gamma^B \times \eta$. Predictions of $\eta$ are controversial for the same reasons as the cost estimates, but a plausible compilation of $\eta$ values is also given in Table II. The aggressive column assumes all technology is pushed to physical and material limits, but it may well turn out that such ambitious development programs would not be cost effective for ITER. Consider then the near future $\eta$ values promised with small modifications to existing technology. In this case the NB value of $\eta = 0.34$ leads to an electrical requirement $P_{elec} = P_{inj}/\eta = 335$ MW, while for the low frequency FW the value $\eta = 0.68$ results in $P_{elec} = 226$ MW. Hence, even though $\gamma^B$ is smaller for the FW, the electric power is less than for NBCD. Clearly, $\eta$ is crucially important in determining the relative attractiveness of CD options. To put $P_{elec}$ in perspective, we point out that the cryogenic power required for the superconducting coils of ITER is expected to be in excess of 100 MW.

## CURRENT DRIVE/HEATING WITH ICRF TECHNOLOGY

By ICRF we refer to a range $\Omega_T \lesssim \omega \lesssim 12\ \Omega_T$ or, for ITER, a frequency (f) in the range 20 MHz $\lesssim f \lesssim 250$ MHz. Except for an explicit discussion of the ion Bernstein wave (IBW), we consider only the fast (compressional Alfven) wave. This section includes composite descriptions of three ICRF systems drawn from several independent preliminary designs for ITER.

First we describe the "standard" CD technique which specifies a variable range, f = 54 MHz - 65 MHz. The lower f is used for $He^3$ minority heating during startup. Heating $He^3$ results in energetic tails and D-$He^3$ fusion, which amplifies the heating effect since all reaction products are charged. As the ion temperature increases the heating naturally evolves to harmonic, $2\ \Omega_T$, tritium heating. Some rf sources meanwhile are tuned to 65 MHz; this frequency displaces all ion resonances from the central plasma region and such waves are strongly damped on electrons. The ICRF antenna is phased to launch a travelling wave, resulting in CD. By a judicious choice of the power spectrum vs. the parallel index of refraction, $n_\parallel$, it is possible to accomplish efficient CD near the magnetic axis. In Fig. 2 the spectrum is assumed peaked at $n_\parallel = 1.5$ for the 65 MHz fast wave. Depending on the poloidal launch position a small ($\lesssim 10\%$) amount of power is lost to tritium heating, and over 70% of the power is absorbed by electrons in a single pass.

Several phased array antenna designs have been suggested for 65 MHz. Figure 3 shows a loop-antenna array with coax feeds integrated into blanket sectors.[9] Note the highly directional power spectrum radiated, which is typical of ITER designs featuring ten or more antenna elements in the toroidal direction. The coax feeds will require ceramic supports for the center conductor, located at low voltage points. Radiation effects on such insulators is not well characterized at present. An alternative all-metal design[10] would employ waveguide resonators in the blanket with anisotropic, diagonal radiating slots facing the plasma, a proposal which has been successfully tested on T 10. In all cases the radiating elements are covered with a Faraday screen, aligned with the total magnetic field. Parametric instabilities and impurity generation are minimized by various methods. First, the power intensity can be kept modest, ~ 5 MW/m$^2$, without claiming too much of the first wall area. Typically, ~ 5% of the first wall might be covered with antennas, but the effect on breeding in the blanket behind the antennas is even smaller. Second, the plasma-facing components of the antenna are coated with low Z material, such as Be. Third, avoidance of high power components with $n_{\parallel}$ = 0 has likewise been effective in reducing impurity generation; tests on TEXTOR[11] show that coupling to the plasma remains good if the antenna geometry is optimized. Antenna elements are connected to the rf generators with coax which is calculated to provide efficient power transmission for ITER, generally $\gtrsim$ 90% over 100 m.

The high power amplifier is a key element for the ICRF system. Recently developed tetrodes should provide generator units rated at ~ 2.5 MW output, and high efficiencies appear possible. For example, tests of the EIMAC X-2242 at Oak Ridge National Laboratory (ORNL) have produced an overall generator efficiency ~ 75%, including all power supply and driver losses. Generally, lower efficiencies are anticipated if large frequency tunability is demanded.

An additional feature of this standard ~ 65 MHz CD regime is the existence of two alternative techniques to generate current: minority heating and mode conversion to a slow wave. While these methods may not be as attractive as direct electron CD, they remain as viable options in this frequency range.

Let us turn next to a higher frequency CD regime, the FW near the tenth harmonic. Figure 2 illustrates the ray paths for rays at f = 250 MHz launched at poloidal locations off the midplane. Ion damping is insignificant at these frequencies, and the strong electron absorption permits a controlled deposition of current density.[1] The Japanese also studied this regime (f ~ 300 MHz) and found very high $\gamma$ values (see Table I). A sixteen-loop toroidal array was calculated to have excellent directivity (~ 100%).[4] Four-element loop arrays have already operated at 200 MHz on JFT2M. However, at these frequencies, all-metal folded waveguides may be preferable. In magnetic field measurements at ORNL, folded waveguides were found to provide nearly identical performance to loop couplers. If this same CD system had a variable frequency, it could heat ions (2 $\Omega_H$ or 4 $\Omega_\alpha$) at 160 MHz. Modelling for INTOR[12] showed excellent plasma heating is possible with the IBW at ~ 130 MHz. Otherwise, at f $\gtrsim$ 250 MHz this system would directly heat electrons. At these frequencies it appears that sources such as klystrodes would be appropriate.

One difficulty with CD at $\omega > \Omega_D$ is the possible loss of rf power damped on nonthermal alphas. The degree to which this is a problem has not been fully studied, but it is possible to completely avoid the issue by operating at $\omega < \Omega_D$. The Japanese[4] propose f = 17 MHz and predict $\gamma = 0.36$, while the Europeans[5] suggest f = 19 MHz and expect $\gamma = 0.48$; both groups find that the antenna spectrum can be highly directional. However, at these low frequencies single pass absorption is weak and radial standing waves appear, so control of the current density profile may become more difficult.

## LOWER HYBRID CURRENT DRIVE

The LHCD option is deemed useful to ITER when used in low $n_e$ plasma, during rampup and at the edge of the steady state discharge. For these purposes, $\lesssim$ 30 MW appears adequate (supplemented by larger NB or FW power for bulk CD). Spatial penetration is limited by the constraint on accessibility (minimum $n_\parallel$) and the effect of strong electron damping. While very narrow spectra, $\Delta n_\parallel \ll 1$, or high intensity power injection are speculative means of improving the penetration it is generally felt that LHCD is limited to the outer third of the steady state plasma (see Fig. 2).

Frequency selection for LHCD involves several considerations. Generally, higher f permits CD at larger density (see Fig. 1), roughly requiring f $\gtrsim$ 5 GHz for ITER. Also, f $\gtrsim$ 8 GHz tends to avoid perpendicular Landau damping on nonthermal alphas. In contrast, higher f generally results in lower $\eta$ (see Table II and the studies in Ref. 13). Changing to different technologies at higher frequencies (gyrotrons, overmoded waveguides, multijunction grills) restores $\eta$ somewhat. Likewise, the grill cooling arrangement varies with f; for f $\lesssim$ 5 GHz the thin septa must be internally cooled, whereas f $\gtrsim$ 8 GHz requires edge cooling of the relatively short septa.

## ELECTRON CYCLOTRON RF[7]

The main function of ECRF for ITER appears to be plasma startup and current profile control (modification of the local safety factor to avoid MHD instabilities). Power requirements are $P_{inj} \simeq$ 10-20 MW CW at f = 140 GHz. (Local current profile control will require frequency tunability.)

At high frequencies the rf intensity is large, estimates for ITER being in the 20-70 MW/m$^2$ range. Thus, the EC system occupies a small fraction of the first wall. Vacuum windows are specified in the transmission lines to isolate tritium from the outside environment, and these windows should transmit ~ 1 MW CW. The waveguides are designed to launch the ordinary wave from the low field side, and beam alignment appears important in determining the ray trajectories within the torus. If mirrors are utilized near the plasma they must be adequately protected from erosion and neutron damage.

Overall efficiency of the transmission line and rf generators is crucial to minimizing the cost of the EC system. A choice must be made between full waveguide or quasi-optical transmission, and careful design and/or alignment is needed in order to control mode polarization and minimize transmission losses. The gyrotron is identified as the best candidate rf source for ITER, considering its relative degree of development. Although there have been short pulse gyrotrons ($>$ 100 GHz) which have demonstrated high efficiency ($>$ 35%) at megawatt levels, there is very little experience with CW operation at high power. Varian expects to achieve $\geq$ 30% efficiency CW at 1 MW within the near future. The four design teams for ITER have foreseen overall values of $\eta \cong 0.17$-$0.25$, building on such technology.

## GENERAL PROBLEMS ASSOCIATED WITH RF APPLICATIONS

The alpha power during the ohmic, physics phase of ITER is roughly 200 MW, and a large fraction of this power is absorbed on the divertor plates, with peak heat loads of $\sim$ 12 MW/m$^2$ during H mode operation. During the technology phase the CD power adds to the alpha heating, greatly increasing the peak heat loads. The tremendous technical difficulty of cooling the divertor plates is exacerbated by CD, and as much as any other consideration, this is a strong incentive to minimize $P_{inj}$.

An additional concern during CD is the effect of $P_{inj}$ on confinement. A naive application of Goldston-like scaling laws would suggest that large $P_{inj}$ decreases energy confinement, but this issue deserves closer examination. The design of ITER would be aided if a data base were compiled to assess the differences in energy confinement between ohmic and noninductive CD discharges.

## CONCLUSION -- OPPORTUNITIES FOR RF APPLICATIONS

On the basis only of its proven CD capabilities, the NB option is currently the front runner for high power heating of ITER. On most every other count rf technologies appear superior for providing valuable services to ITER. The rf systems are reactor compatible, relatively inexpensive, well developed, and adequately flexible to provide a range of functions. Following is a partial catalog.

1. Vessel conditioning -- carbon tiles may require rf baking to 350°C to outgas after openings.
2. Plasma production -- has been done, for example, with IBW and whistlers at Nagoya, with radial ICRF eigenmodes on JET, with LHRF on PLT, as well as ubiquitously with ECRF.

3. <u>Current startup</u> -- this technique, already shown with LHCD on PLT, can eliminate the need for high loop voltages and permit the use of a thick vacuum vessel.

4. <u>Rampup</u> -- very slow current ramps ($\sim$ 20 min.) are desirable to minimize poloidal coil power demand, reduce eddy current heating of superconducting coils, and reduce thermal shock to the first wall/blanket, and should be possible with LHCD, for example.

5. <u>Extended ohmic burn length</u> -- combined ohmic/rf CD.

6. <u>Pressure, current profile control</u> -- eliminate sawteeth (LHCD on ASDEX); possibly enter second stability regime via safety factor modification (CD) or via kinetic effects (ICRH, ECRH); stabilize interchange mode with ICRF ponderomotive force (Phaedrus); possibly eliminate disruptions (CD).

7. <u>Burn control</u> -- subignited, driven plasma.

8. <u>Impurity control</u> -- ash removal by ICRF heating of banana tips on outer flux surfaces.

9. <u>Diagnostics</u> -- Alfven resonance to locate mode rational surfaces.

These and other more speculative applications will be ushered in if CD can be accomplished in rf experiments at high density and temperature.

## ACKNOWLEDGMENT

The author thanks W. Nevins, W. Becraft, and J. Yugo for informative discussions.

## REFERENCES

1. D.A. Ehst, Argonne National Laboratory Report, ANL/FPP/TM-238 (1989); also ITER-IL-PH-6-9-U-2.

2. D.A. Ehst, Argonne National Laboratory Report, ANL/FPP/TM-219 (1988).

3. INTOR Phase Two A, Part III, IAEA, Vienna (1988) Vol. 1.

4. H. Kimura et al., ITER-IL-PH-6-9-J-2.

5. J. Jacquinot et al., ITER-IL-HD-7-9-2.

6. R.S. Devoto et al., 16th EPS Conference, Venice (1989).

7. "Basic Design Engineering Meeting on the EC System for ITER," ITER-IL-HD-6-9-1.

8. C. Gormezano et al., "JET ICRF System," ITER Meeting on CD/Heating Technology, Garching, July 1988.

9. J. Yugo et al., "Fast Wave Current Drive Antenna Design Integrated with the ITER Blankets," ITER Session, Garching, March 1989.

10. V.V. Alikaev et al., ITER-IL-HD-6-9-S-2.

11. A. Messiaen et al., "Effect of Antenna Phasing and Wall Conditioning on ICRH in TEXTOR," ITER Meeting on CD/Heating Technology, Garching, July 1988.

12. U.S. Contribution to INTOR Phase 2A, Part 2 (1985), Vol. 1.

13. C. Gormezano et al., 14th SOFT, Avignon (1986); M. Sassi, ITER CH/Heating Technology Meeting, June 1988, Contr. No. - 40.

14. L.H. Sverdrup et al., 7th Top Conf. Appl. RF Power to Plasmas, Kissimmee (1987).

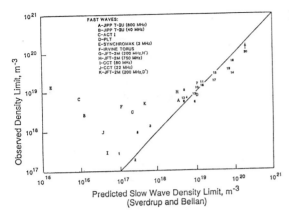

Fig. 1 Observed CD density limits for LH waves (numbers are citations in [1]) and FW, vs. slow wave theory [14].

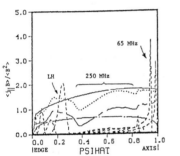

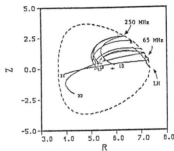

Fig 2. Multiple-wave current drive plus bootstrap current for ITER equilibrium: 30 MW of LH power; 74 MW of FW at 250 MHz; and 30 MW of ICRF fast wave at 65 MHz. Total current density is dotted line, whereas solid line is target current density of initial equilibrium; chain–dot line is bootstrap current density.

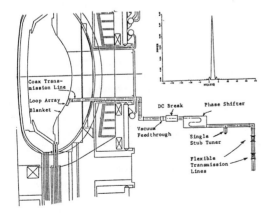

Fig. 3 Installed FWCD antenna, transmission line and impedance matching system. Inset shows spectrum for 12 element array.

# DISCRETE ALFVEN WAVES IN THE TORTUS TOKAMAK

Y. Amagishi
Shizuoka University, 836 Ohya, Shizuoka 422, Japan

M. J. Ballico, R. C. Cross and I. J. Donnelly[*]
University of Sydney, NSW 2006, Australia

## ABSTRACT

Discrete Alfven Waves (DAWs) have been observed as antenna resistance peaks and as enhanced edge fields in the TORTUS tokamak during Alfven wave heating experiments. A kinetic theory code has been used to calculate the antenna loading and the structure of the DAW fields for a range of plasma current and density profiles. There is fair agreement between the measured and predicted amplitude of the DAW fields in the plasma edge when both are normalised to the same antenna power.

## INTRODUCTION

A recent upgrade of the current in the TORTUS tokamak from 20 kA to 30 kA has led to the observation of DAW modes as peaks in the antenna resistance and as enhanced wave fields in the plasma edge [1]. TORTUS is a relatively small device of rectangular cross section with major radius R = 0.44 m and plasma minor radius ~ 0.1 m. The upgraded operating conditions are: $B_\phi$ = 1.0 T, $I_\phi$ = 30 kA, q(a) ~ 5 and $n_e$ ~ 1–4x$10^{19}$ m$^{-3}$. The experimental results reported here were obtained in a hydrogen plasma using 3 top–bottom pairs of poloidal antennas with appropriate phasing for the desired mode, and at a frequency of 3.2 MHz ( $\omega/\Omega_i$ = 0.21). DAWs are of interest as indicators of the imminent presence of an Alfven resonance layer in the plasma as the density rises [2], and measurement of their wavefields in the edge plasma gives the poloidal and toroidal wavenumbers (m,n) associated with the resonance layer. DAWs may also be used as a diagnostic for the q(r) profile [3].

In this paper we use the kinetic theory code ANTENNAS [4] to analyse the sensitivity of the DAW fields to the density and current profiles, and to relate wave amplitudes in the plasma edge to the central values. This shows that the central wave fields can be estimated within a factor of 2 using the fields measured at the plasma edge. When normalised to the same antenna power, we find fair agreement between the amplitude of the calculated and the measured DAW fields.

[*]Attached from the Australian Nuclear Science and Technology Organisation, Sydney, Australia.

## THEORY

A cylindrical model of TORTUS is used with a conducting wall at $r_w$ = 0.14 m. Wave fields of the form $b(r)\exp\{i(m\theta+n\phi-\omega t)\}$ are excited by antenna current distributions corresponding to Fourier components of poloidal or helical antennas [5] in the region $r_a \le r \le r_w$ ($r_a$ = 0.11 m). The plasma current density has the form

$$J_\phi(r) = [1.5I_\phi/\pi r_a^2 (1-\alpha/4)][1-\alpha(r/r_a)^2+(\alpha-1)(r/r_a)^4].$$

With this profile the safety factor is

$$q(r) = q(0)/[1-(\alpha/2)(r/r_a)^2+((\alpha-1)/3)(r/r_a)^4]$$

where

$$q(0) = 4\pi r_a^2 B_\phi(1-\alpha/4)/3\mu_0 RI_\phi \quad (= 3.1(1-\alpha/4) \text{ for TORTUS}).$$

The larger the parameter $\alpha$ the more peaked the current profile. Note that q is positive here, so the sign of n should be reversed when comparing with the results of Ballico et al.[1] Two density distributions have been considered: $n_1(r)$ which goes to zero at $r_a$; and $n_2(r)$ which goes to zero at $r_w$ and has $n_e(r_a)$ = $0.3n_{e0}$. In both cases the density for r < 0.05 m varies as $n_{e0}(1-\beta(r/r_a)^2)$ with $\beta$ = 0.87. Conditions for the existence of DAWs have been derived by several authors, e.g. Mahajan et al.[6] At frequencies $\omega$ < $0.5\Omega_i$, DAW modes can only exist in TORTUS when there is an appreciable toroidal current and when the Alfven resonance frequency profile

$$\omega_A(r) = v_A(r)|(n/r)(1+m/nq(r))|(1-(\omega_A/\Omega_i)^2)^{1/2}$$

is almost constant for an appreciable radial extent near the position of its minimum value. Near the origin

$$\omega_A(r) = \omega_A(0)[1+\gamma(r/r_a)^2], \quad \text{with } \gamma = 0.5[\beta-\alpha(1+(n/m)q(0))^{-1}].$$

We have only considered cases with $\gamma > 0$, for which the Alfven resonance enters the plasma at r = 0 and moves towards the edge as the plasma density increases. Positive m/n minimises $\gamma$, and negative m allows the optimum polarisation for DAWs when $\omega > 0.1\Omega_i$ so the $(-M,-N)$ DAW modes are dominant. For our model the $(-1,-1)$ mode has $\gamma(\alpha)$ values $\gamma(2)$ = 0.04 and $\gamma(1.4)$ = 0.20, while the $(-1,-2)$ mode has $\gamma(2)$ = 0.19 and $\gamma(1.4)$ = 0.30.

Fig. 1 shows the calculated DAW fields $|b_\theta(r)|$ for the indicated plasma models; results obtained were essentially identical for the poloidal and helical antenna models. A decrease of density in the edge region or a flattening of the current density profile both increase the ratio $|b_\theta(0)/b_\theta(r_a)|$, which lies within the range 8–25 for (−1,−1) DAWs. For (−1,−2) DAWs this ratio is about 40. The value of $b_\theta(r)$ varies only weakly with r in the edge region, so the central magnetic and density fields of the DAW can be estimated from edge measurements of $b_\theta$. In all cases, $b_\theta$ is significantly larger than $b_z$, as observed experimentally.

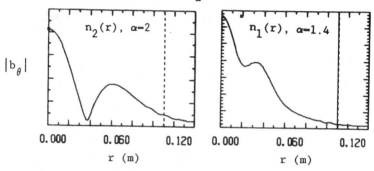

Fig. 1. DAW $b_\theta$ fields for the indicated plasma configurations.

### COMPARISON OF THEORY AND EXPERIMENT

Fig. 2 shows the time–dependence of the experimental electron density, the (−1,−1) DAW field $b_\theta$ at r = 0.11 m, and the resistive loading per antenna for the case in which two antenna pairs were phased to excite the (M=1,N=1) mode. Fig. 3 shows the resistance per antenna vs. $\omega$ calculated for the (−1,−1) mode and the plasma models indicated. As the current profile becomes less peaked the DAW becomes weaker; the same effect occurs if the density parameter $\beta$ is increased. Experimentally, the DAW does not give rise to a strong peak in the antenna loading as the density sweeps through the resonance value. This could be because the DAW excitation is marginal, cf the $\alpha$ = 1.4 curve in Fig. 3, but it is more likely to be because of the low Q value (~10) deduced for the DAW from the $b_\theta$ response. This value should be compared with the theoretical predictions of Q ~ 100. Other authors[2] have reported similar differences between Q values derived from experiment and theory.

In the case of two antenna pairs, each carrying 360 A current, the DAW calculated for the $n_1(r)$, $\alpha$ = 1.4 case has $b_\theta(r_a)$ = 15 $\mu$T. Provided that the DAW mode resistance peak is re-normalised to the plateau value associated with the continuum, the other density and current models predict values for $b_\theta(r_a)$ of up to 40 $\mu$T. The corresponding experimental value is 60 $\mu$T.

## REFERENCES

1.  M. J. Ballico et al., 16th EPS Conf. Contr. Fus. Plasma Phys., Venice (1989) 1203.
2.  G. A. Collins et al., Phys. Fluids 29 (1986) 2260.
3.  G. A. Collins et al., Plasma Phys. Contr. Fus. 29 (1987) 323.
4.  I. J. Donnelly et al., J. Plasma Phys. 35 (1986) 75.
5.  I. J. Donnelly et al., 13th EPS Conf. Contr. Fus. Plasma Heat. Schliersee (1986) I 431.
6.  S.M. Mahajan et al., Phys. Fluids 26 (1983) 2195.

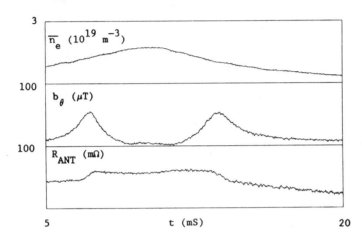

Fig. 2. $(-1,-1)$ DAW $|b_\theta|$ field at $r = 0.11$ m, and antenna resistance during a density rise and fall in TORTUS. $|b_\theta|$ shows the DAW clearly, whereas the increase in $R_{ANT}$ due to the DAW is similar to that caused by the resonance layer.

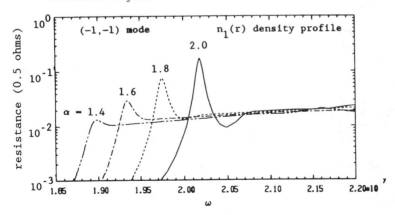

Fig. 3. Calculated antenna resistance vs. $\omega$ for the indicated current distributions.

# CURRENT DRIVE BY BEAT WAVES IN TOKAMAK PLASMAS

M. R. Amin and R. A. Cairns
Department of Mathematical Sciences, University of St Andrews
St Andrews, Fife KY16 9SS, U. K.

## ABSTRACT

Current drive in tokamak plasmas by a beat wave is considered in 2D geometry. The beat wave is excited by the nonlinear interaction of two intense microwave pulses (free-electron lasers) in the plasma. The three wave nonlinear interaction equations in steady state are solved numerically for this purpose. The 2D toroidal effect and the effect of finite spatial width of the pump microwave pulses are taken into account. To illustrate the principle, two types of tokamak are considered: one is small, typical of MTX, and the other one is larger, such as JET. In both cases, it is found that a good beat wave coupling exists for a Langmuir beat wave with a phase velocity of around 2.0 to 4.0 times the thermal velocity of the electrons. It has also been found that about 26-32% of the total input power of the pump microwaves is deposited in the beat wave. In these cases, there is almost complete pump depletion of the higher frequency pump microwave and the beat wave is completely damped on the plasma electrons within a very short distance.

## INTRODUCTION

Recently, intense pulsed radio frequency (RF) wave sources[1], known as free electron lasers (FEL's), have created new possibilities for plasma heating and current drive schemes in tokamaks.[2-5] The Beat Wave current drive scheme[2] uses the nonlinear beat of two high frequency intense microwave pulses such as are possible only with FEL's. Due to this nonlinear beat of the two microwave pulses, a longitudinal plasma wave is excited at the interaction region and this plasma wave is then damped on electrons leading to a current. One realization of this new idea of current generation is being examined in detail for the Microwave Tokamak Experiment (MTX) at Livermore.[5] The second source differs in frequency from the first by the plasma frequency and its intensity is perhaps as low as a few percent of the first source. Beat wave current drive in a tokamak offers some advantages over other RF current drive methods: this method provides excellent possibilities to localize the generation region and to control the current profile; there is no density limit in this method.

So far as we know all previous investigations on beat wave current drive[2,3,6] were restricted to 1D geometry. Here we study the possibility of beat wave current drive in tokamak plasmas taking into account the toroidal effect and the effect of finite spatial extent of the microwave pulses in 2D geometry which may be expected to be a reasonable approximation for waves incident in the median plane of the tokamak. We prefer opposed propagation of the pumps because of the strong nonlinear coupling and the large momentum flux deposited in the beat wave. We take the general three wave nonlinear interaction equations as our model equations in a steady state situation. We assume the electron plasma and cyclotron frequencies as:

$$\omega_p(z,x) = (\omega_e^2[1-(r-R_0)^2/r_0^2])^{1/2} \text{ and } \Omega_e(z,x) = \Omega_0(R_0/r) \text{ respectively. Where } \omega_e \text{ and}$$

$\Omega_0$ are respectively the electron plasma frequency and electron cyclotron frequency along the axis of the tokamak; $R_0$ and $r_0$ are respectively the major and minor radii of the torus, $r^2=(x_0-x)^2+z^2$ and $R_0+r_0=x_0$. We consider 2D geometry where we arrange the pump wave propagation directions such that $\theta_2=\pi-\theta_1$, $\theta_1$ and $\theta_2$ are the angles made by the wave vectors of the pumps with the +z axis (direction of the toroidal magnetic field). For this geometry, it has been found that the angle $\theta_B$, which is the angle between the wave vector $\vec{k}_B$ of the beat wave and +z axis, is very small ( $0^0 <$ $\theta_B < 4^0$ ); consequently, $\vec{k}_B$ is almost parallel to the toroidal magnetic field lines, so the damping of the beat wave is simply Landau damping.

The expression for beat wave current drive efficiency in 1D from a simple fluid model deduced earlier by Cohen[2] is slightly modified in our 2D geometry and takes the following form:

$$\eta_{BW} = [\ |e|q_e R_a \cos\theta_1\ /\ 2\pi q R_0 m_e v_r v_{llr}\ ], \tag{1}$$

where $|e|$ is the electron charge, $q$ is the safety factor, $v_r$ is the collisional slowing down rate of a fast electron, $v_{llr}$ is the resonant electron velocity; $q_e = \omega_B / \omega_1(1+\omega_2\rho/\omega_1)$ is known as the quantum efficiency and $R_a=\Delta J /J_1^{in}$ is known as the relative action transfer; $\rho$ is the input ratio defined by $\rho=J_2^{in} /J_1^{in}$. $J_1^{in}$ and $J_2^{in}$ are the input action flux densities of the two microwaves respectively. The price of the advantages of the beat wave current drive described earlier is the use of two intense microwave sources and the reduction of the current drive efficiency by the factor $q_e R_a$ in Eq.(1), where $q_e < 1$, $R_a < 1$.

## NUMERICAL SOLUTIONS

To solve the model equations numerically we apply an explicit finite difference scheme. We consider finite spatial extent pumps, with initially Gaussian profiles and with identical polarization for efficient coupling of the beat wave interaction. We take for example, two types of tokamak, one is smaller, typical of MTX at Livermore[3, 5] and the other one is larger, such as JET. The latter provides an interesting comparison, even though there are no plans, so far as we know, to use beat wave current drive in this or any comparable machine.

The FEL's microwave parameters are as follows:[5] $\omega_1$=240-280 (GHz), $\omega_2$=150-180 (GHz), peak power of the pumps of frequencies $\omega_1$ and $\omega_2$ are respectively as $P_1$=8.0 (GWatts) and $P_2$=5.0-0.0005 (GWatts), both with 3 cm × 8 cm illuminated cross section.

We present some graphics of the numerical solutions of our model equations in Figs. 1-4. We see from Figures 1 and 4 that the pump microwave pulses approach toward each other with the higher frequency pump from the left and the lower frequency pump from the right. After the interaction the pump microwave pulses are propagating with different shape of the profiles as can be seen from the graphics. We cannot see the profile of the beat wave amplitude in Figs. 1 and 4 because the excited beat wave is heavily Landau damped within a very short distance in the interaction region. Figure 2 shows the variation of the relative action transfer $R_a\cos\theta_1$ with x-distance for different propagation directions of the pump microwave pulses. We see that as the angle between the directions of propagation of the pumps decreases, the relative action transfer also decreases. Figure 3 shows the variation of the relative action transfer $R_a\cos\theta_1$ with x-distance for different input ratios $\rho$. We see that the relative action transfer increases with input ratio.

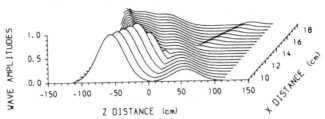

Fig. 1. MTX: Evolution of the wave amplitudes; $P_1$=8.0 GW, $P_2$=0.5 GW ($\rho$=0.1); $\theta_1$=5$^0$, $\theta_2$=175$^0$, $n_e$=0.8 × 10$^{14}$ (cm$^{-3}$), $T_e$=1.5 keV. The amplitudes are normalized by the initial peak amplitude of the higher frequency pump.

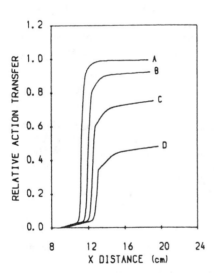

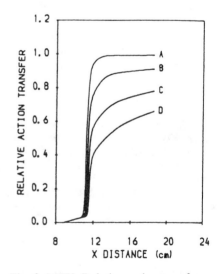

Fig. 2. MTX: Relative action transfer $R_a\cos\theta_1$ vs. x-distance for different $\theta_1,\theta_2=\pi-\theta_1$. The other parameters are same as in Fig. 1. The curves A,B, C and D correspond to $\theta_1$ = 5$^0$, 10$^0$, 15$^0$ and 20$^0$ respectively.

Fig. 3. MTX: Relative action transfer $R_a\cos\theta_1$ vs. x-distance for different $\rho$'s, $P_1$=8.0 GW. The other parameters are same as in Fig. 1. The curves A, B, C and D correspond to $\rho$ = 0.1, 0.01, 0.001 and 0.0001 respectively.

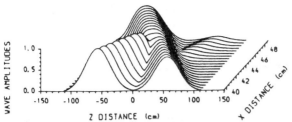

Fig. 4. JET: Evolution of the wave amplitudes; $P_1$=8.0 GW, $P_2$=5.0 GW ($\rho$=1.0); $\theta_1$=$5^0$, $\theta_2$=$175^0$, $n_e$=1.0 × $10^{14}$ (cm$^{-3}$), $T_e$=10.0 keV. The amplitudes are normalized by the initial peak amplitude of the higher frequency pump.

## RESULTS AND DISCUSSION

We define the fraction of energy deposited in the beat wave as $f_E$=$q_e R_a \cos\theta_1$. For a particular set of parameters in MTX the phase velocity of the beat wave is around $4.0v_{th}$, where $q_e$=0.32 for $\rho$=0.1, $R_a$=0.99 corresponds to $\theta_1$=$5^0$, therefore $f_E$ becomes ≈32%. Similarly for JET type parameters the phase velocity of the beat wave is around $2.0v_{th}$, $q_e$=0.41 for $\rho$=0.1, $R_a$=0.62 corresponds to $\theta_1$=$5^0$, therefore in this case $f_E$ becomes ≈26%. For MTX type parameters the scaling of the beat wave current drive efficiency is as $\eta_{BW}$(MTX) = $(2.46 q_e R_a \cos\theta_1 / Z_i^2 q)$ Amps / Watt. For $Z_i$=1, q=3 and 8 GWatts peak power, corresponding to 2 MWatts average power, and taking the optimum values of $q_e$, $R_a$ and $\cos\theta_1$ this leads to 520 kAmps of beat wave current in MTX. A similar calculation for JET parameters gives $\eta_{BW}$(JET) = $(0.30 q_e R_a \cos\theta_1 / Z_i^2 q)$ Amps / Watt. It has been found that the action transfer depends on the angle between the propagation direction of the pumps, it also depends on the input ratio $\rho$, the larger the input ratio the greater the action transfer. However, it is important to note here that, although the relative action transfer increases with the input ratio, the quantum efficiency decreases. As is mentioned earlier (Eq.(1)), the beat wave current drive efficiency is directly proportional to the product of these two quantities.

## REFERENCES

1. T. J. Orzechowski, B. R. Anderson, J. C. Clark, et al., Phys. Rev. Lett. 57, 2172 (1986).
2. B. I. Cohen, Comments Plasma Phys. Controll. Fusion 8, 197 (1984).
3. B. I. Cohen, R. H. Cohen, B. G. Logan, et al., Nucl. Fusion 28, 1519 (1988).
4. R. H. Cohen, B. I. Cohen, W. M. Nevins, et al., Current Drive by Intense Microwave Pulses, Rep. UCID-99583, Lawrence Livermore National Laboratory, Livermore (1988).
5. K. I. Thomassen, Plasma Phys. Controll. Fusion 30, 57 (1988).
6. J. A. Heikkinen, S. J. Karttunen, R. R. E. Salomaa, Nucl. Fusion 28, 1845 (1988).

410

# IONISATION LEVELS OF LOW DENSITY RF GENERATED NITROGEN PLASMAS

I. J. Donnelly[*]

University of Sydney, NSW 2006, Australia

E.K. Rose

ANSTO, PMB 1, Menai, NSW 2234, Australia

## ABSTRACT

Steady–state solutions for the densities of ionised states in nitrogen plasmas are found as a function of electron temperature. For plasmas with low confinement times the major ion species is $N_2^+$ unless the electron density is above $\sim 10^{11}$ cm$^{-3}$ when $N^+$ dominates.

## INTRODUCTION

Low temperature plasmas which are not in thermodynamic equilibrium are commonly applied to the processing of materials in fields such as plasma etching, sputter deposition, ion plating and ion nitriding, plasma–enhanced chemical vapour deposition and plasma immersion ion implantation. We have particular interest in the last application [1,2]. The ability to calculate the plasma density and temperature as a function of input power, filling pressure and confinement time is desirable because it will help in the optimisation of the plasmas used in these processes. Such calculations require knowledge of the electron ionisation cross–sections, which are available for the atomic species [3]. However, with the exception of $H_2$, there is a paucity of molecular cross–section data, and the dissociation and dissociative recombination cross–sections used here for nitrogen have been estimated from the hydrogen values [4]. Reaction rates were evaluated for electron temperatures in the range $1 \leq T_e \leq 50$ eV using the assumption that the electron energy distribution is Maxwellian. We restrict our analysis to plasmas with electron density $n_e$ (cm$^{-3}$), electron temperature $T_e$ (eV) and particle confinement time $\tau$ (s) which satisfy $n_e \tau \ll 10^9 T_e^{3/2}$ cm$^{-3}$s, in which case the dominant recombination mechanism occurs via ion–wall contact, so electron–ion recombination in the body of the plasma can be ignored.

[*]Attached from the Australian Nuclear Science and Technology Organisation.

The appropriate rate equations have been formulated and used to obtain the steady–state values, and the evolution in time, of the neutral and ionised population densities in low temperature ($T_e$ < 10 eV) nitrogen plasmas. Analytic and numerical solutions of the steady–state equations show that there is a threshold temperature below which there is no ionisation. For a small range of temperatures above this threshold there are two possible ionisation states, but at higher temperatures only the more highly ionised state exists. The threshold temperature decreases as the filling pressure or the confinement time increases. Ionisation levels have also been obtained as a function of input power density, assuming that the dominant energy loss comes from line radiation.

## THE RATE EQUATIONS

The following notation is used:

$x_1$ = number density of $N_2$; $\qquad\qquad$ $x_2$ = number density of $N_2^+$

$y_k$ = number density of $N^{(k-1)+}$ ($1 \le k \le 3$). $\qquad$ Reaction rates are:

$D_I$ = dissociative ionisation of $N_2$; $\qquad\qquad$ $I_M$ = ionisation of $N_2$

$D$ = dissociation of $N_2$; $\qquad\qquad$ $D^+$ = dissociation of $N_2^+$

$D_R$ = dissociative recombination of $N_2^+$; $\qquad$ $S_k$ = ionisation of $y_k$

Polynomial fits to the reaction rates as a function of $T_e$ are given elsewhere[4]. In the derivation of the time–dependent rate equations we have assumed the following: the density of each species is spatially uniform; the mass density of the system is time independent; the molecular ions have effective confinement time $\tau$ before reflecting from the vessel boundary as molecules; the atomic species have confinement times $\tau$; 90% of the atoms and ions reflect from the wall as atoms, 10% reflect as a component of a diatomic molecule[5]. When electron–ion recombination is dominated by the ion–wall interaction, the rate equations are:

$$\frac{dx_1}{dt} = -[(D_I + I_M + D)n_e)]x_1 + \frac{x_2}{\tau} + 0.05 \sum_{k=1}^{3} \frac{y_k}{\tau} \qquad (1)$$

$$\frac{dx_2}{dt} = -[(D^+ + D_R)n_e + \frac{1}{\tau}]x_2 + [I_M n_e]x_1 \qquad (2)$$

$$\frac{dy_1}{dt} = [(D_I + 2D)n_e]x_1 + [(D^+ + 2D_R)n_e]x_2 - [S_1 n_e + \frac{0.1}{\tau}]y_1 +$$

$$+ 0.9 \sum_{k=2}^{3} \frac{y_k}{\tau} \qquad (3)$$

$$\frac{dy_2}{dt} = [D_I n_e]x_1 + [D^+ n_e]x_2 + [S_1 n_e]y_1 - [S_2 n_e + \frac{1}{\tau}]y_2 \qquad (4)$$

$$\frac{dy_3}{dt} = [S_2 n_e]y_2 - \frac{y_3}{\tau} \qquad (5)$$

and

$$n_e = x_2 + y_2 + 2y_3 . \qquad (6)$$

## RESULTS

The complete set of rate equations (1)–(6) is too complica-ted for analytic treatment. However, a reduced set of steady–state equations, which model the most important ionisation processes for nitrogen when $n_e$ is low, can be obtained by setting $D^+ = D_R = S_2 = 0$. A quadratic equation of the form $an_e^2 + bn_e + c = 0$ is then obtained for $n_e$. This model shows that there are two critical temperatures, $T_c$ given by $c = 0$ (i.e. $(I_M + D_I)\tau = X^{-1}$) and $T_{cm}$ given by $b^2 = 4ac$. There is no ionisation when $T_e < T_{cm}$, there is a high and a low density solution for $n_e$ when $T_{cm} < T_e < T_c$, and one solution (high density) when $T_e > T_c$. This behaviour is also found numerically and is illustrated in Fig. 1 where the electron density is shown as a function of $T_e$ for the case of initial molecular density $X = 10^{14} cm^{-3}$ and confinement time $\tau = 20\mu s$. In this case $T_c = 4.20 eV$ and $T_{cm} = 3.39 eV$. For comparison the predictions of the analytic theory are also shown (here $T_{cm} = 3.45 eV$). It is apparent that the reduced equations give good agreement with the full model, except in the vicinity of $T_{cm}$. It is interesting that, on the lower branch of the solution locus, an increase in $n_e$ is associated with a decrease in $T_e$. Solving the time–dependent rate equations (1)–(6) at constant $T_e$ shows that the low density solution is unstable. Initial electron densities below (or above) this branch converge to zero (or to the upper branch), while all initial densities converge to the ionised steady state when $T_e > T_c$.

The input power density is a constraint which should be included in any realistic time–dependent calculation. Here we use an expression appropriate for atomic hydrogen[5] to estimate that there is an energy loss per ionisation of $\xi_i = 500/T_e$ eV due to line radiation. Thence the input power density is approximately $P = e\xi_i n_e/\tau$ Wcm$^{-3}$. In Fig. 2 the ionisation level is shown as a function of P for $X = 10^{14}$ cm$^{-3}$ and $\tau = 20$ $\mu s$. When $n_e < 10^{10}$ cm$^{-3}$ the number of atomic ions is less than 10% of the total ions. Therefore, low density nitrogen plasmas created by Maxwellian electrons will consist predominantly of $N_2^+$ ions. We find that the $N^+$ ions become more numerous than the $N_2^+$ ions when $n_e$ exceeds about $10^{11}$ cm$^{-3}$.

## REFERENCES

1. J. R. Conrad et al., J. Appl. Phys. 62, 4591 (1987).
2. J. Tendys et al., Appl. Phys. Lett. 53, 2143 (1988).
3. M. A. Lennon et al., Culham Report CLM–R175, (1986).
4. I. J. Donnelly and E. K. Rose, Aust. J. Phys. (1989) submitted
5. M. F. A. Harrison, Applied Atomic Collision Physics
   (Eds. H.S.W. Massey et al., North Holland 1984) Vol. 2.

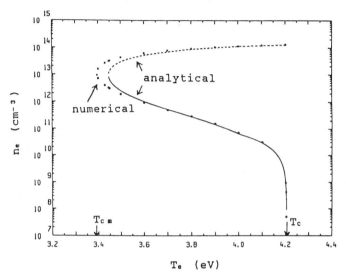

Fig. 1. Steady-state electron densities vs. electron temperature.

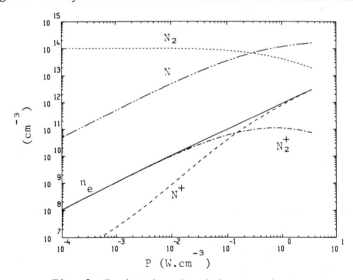

Fig. 2. Ionisation densities vs. input power.

PLASMA TOMOGRAPHY USING TRANSIENT SYNCHROTRON RADIATION

N. J. Fisch

Plasma Physics Laboratory, Princeton University, Princeton, NJ 08543

A. H. Kritz

Department of Physics, Hunter College, New York, NY 10021

ABSTRACT

We show by means of an example in a six-dimensional parameter space that induced transient radiation can inform on the dc electric field and other tokamak parameters.

INTRODUCTION

An examination of the transient, synchrotron radiation signal which arises from a deliberate, perturbation of hot tokamak electrons, can be quite informative.[1-2] The perturbation might be produced through brief heating of superthermal electrons by lower-hybrid waves. The plasma radiation response to this perturbation, in frequency-time space, forms a two-dimensional pattern that looks different under different plasma conditions. The parameters to which this radiation is sensitive include the dc electric field $E$, the ion charge state $Z_{\text{eff}}$, the angle of viewing with respect to the magnetic field $\theta$, the density $n$, and the precise velocity of the perurbed electrons.[3] Through a comparison of essentially all parameter sets that might possibly explain the transient signal, the relative probabilities of the competing parameter sets can be evaluated.

Details of the method have been worked out in the references; here, we explore the possibility of deducing an electric field at the plasma center. In a truly steady state, one could measure the loop voltage at the plasma edge to answer this question. An example for which the edge voltage may not be helpful is during nonohmic current-drive on axis. The plasma response at the periphery reflects the current-drive only after a magnetic relaxation time. Early experiments on current-drive by lower-hybrid waves, in fact, were ambiguous precisely because the experiments were too short. We illustrate here how from a transient radiation signal, one might detect a loop voltage on axis at the same time that the loop voltage vanishes at the plasma periphery.

SIMULATION MODEL

Suppose one detector is available, which observes along a vertical direction (at constant $B_{\text{tor}}$), so that the detector sums radiation arising from points all along the viewing direction. We are interested only in the *incremental* radiation arising perhaps from a brief surge of lower-hybrid power – the background radiation is ignored. We use here a very coarse-grained model in which radiation originates at just two points along the viewing direction, one point in the plasma center, and one peripheral point. We imagine that we are unsure as to how large a perturbation, $A$, was created at each of the two points. Likewise, we are unsure of the viewing angle with respect to the magnetic field at each of these points, since the $q$ profile is unknown. Given these unknowns, we wish to determine the dc parallel electric field at each point. We do imagine that the ion charge state is known and the same at each of the two points, and that the density at each point is known and different. Suppose further that the location in velocity space of the absorbed probing radiation is also known and the same at each point, possibly because of a resonance condition. Thus the detector sums

$$R(\omega, t) = R(\omega, t | n_c, Z_{\text{eff}}; E_c, A_c, \theta_c) + R(\omega, t | n_p, Z_{\text{eff}}; E_p, A_p, \theta_p), \tag{1}$$

where $c$ labels parameters at the plasma center and $p$ labels parameters at a peripheral point. Thus, the problem is to find the probability distribution over all possible parameter sets in the six-dimensional space $(E_c, A_c, \theta_c, E_p, A_p, \theta_p)$ given a very crude *a priori* probability distribution and the data $R(\omega, t)$.

Note that if the density were the same at the two radiating points, then there would be no way to distinguish radiation emanating from the plasma center from radiation emanating from the peripheral point. The density difference means that radiation from the denser place (generally but not necessarily the center) decays faster. Although only the sum of the radiation from the two points is measured, the different decay constants distinguish the contributions to this sum. The larger the density difference, the easier the distinguishing of the individual contributions.

We simulate the results of an experiment by calculating first the radiation response $R(\omega, t)$ in Eq. (1) for a specific set of plasma parameters. We assume, however, that in a realistic experiment this precise radiation response is not measured directly, *e.g.*, because of calibration errors, because of background radiation fluctuations, or because of imprecise assumptions concerning the governing physics. For the purposes of this simulation, we model all of these uncertainties by gaussian uncorrelated noise, *i.e.*, we imagine that we measure instead $R_x(\omega, t) = R(\omega, t) + \sigma(\omega, t)$, where $\sigma$ is an uncorrelated noise signal. Obviously, in the limit $\sigma \to \infty$, there is total degradation of our measurement, and we are left with the *a priori* probabilities as our best guess for the parameter set probabilities. For finite $\sigma$, our guess can be far more informed. Given the noise model, we can, of course, make precise statements concerning the probability distribution of parameters given the noisy data.

## MARGINAL PROBABILITY DISTRIBUTIONS

In Fig. 1a we show the marginal probability of deducing the electric fields given the data,

$$P_M(E_c, E_p) \equiv \sum_{A_c, A_p, \theta_c, \theta_p} P(E_c, E_p, A_c, A_p, \theta_c, \theta_p). \tag{2}$$

Here, the *a priori* probability is taken as flat over the range plotted, so that, clearly, the noisy data allows a significant refinement of the *a priori* probabilities. In this example, the true plasma parameters include a nondimensional dc electric field of 0.08 on axis, none off axis, and equal perturbation strengths $A$ at both locations. The noise level is 10% of the maximum incremental signal. The marginal probability distribution was calculated by comparing about $1.4 \times 10^6$ sets of plasma parameters to the noisy data.

Of particular interest is to compare the above result to the marginal probability distribution for the electric fields given the correct viewing angles, $P_M(E_c, E_p | \theta_c, \theta_p)$, shown in Fig. 1b. Apparently, the probability distribution for $(E_c, E_p)$ is not materially affected by our knowledge concerning the viewing angle; in either event, it is possible to discern the case at hand, where a loop voltage is induced in the plasma center but not at the plasma periphery.

Knowledge of the viewing angle is not critical in determining the electric fields not because the viewing angle is not an important factor in the radiation response, but rather because small changes in the viewing angles affect the radiation response in very different ways than do small changes in the electric fields. This can be seen by considering the marginal probabilities of the viewing angles, $P_M(\theta_c, \theta_p)$, which we depict in Fig. 1c. The viewing angles, from which we could deduce the $q$-profile, are each resolved on the order of $\pm 1°$. Clearly, in order to deduce so well the viewing angles from the radiation, the radiation must, in fact, be sensitive to these angles as well as to the electric field, but the deduction of these parameters from the radiation can proceed almost orthogonally.

## INFORMATION-THEORETIC ENTROPY

These ideas concerning the orthogonality of the parameters can be made more precise. Consider the information-theoretic uncertainty, or entropy, defined by $H \equiv -\sum_i P_i \log P_i$ which is a maximum for the *a priori* probability distribution, where the probability $P_i$ is flat over the allowable space. A measure of the utility of the data is the reduction in this uncertainty given the data. To this end, define $H_A$ as the maximal, *a priori*, uncertainty. A measure of the information content in the refined probability distribution, say, $P_M(E_c, E_p)$ in Eq. (1) is the relative uncertainty given the data,

$$S(E_c, E_p) \equiv -\frac{1}{H_0} \sum_{E_c, E_p} P_M(E_c, E_p) \log P_M(E_c, E_p). \tag{3}$$

A relative uncertainty $S = 1.0$ corresponds to the data being entirely useless, whereas the limit $S = 0.0$ corresponds to the data being entirely conclusive. The relative uncertainty $S(E_c, E_p)$ for the probability distribution in Fig. 2 is about 0.66, for the distribution in Fig. 2, $S(E_c, E_p|\theta_c, \theta_p) = 0.63$. This indicates that the relative *mutual information*, the uncertainty in $(E_c, E_p)$ resolved by knowing $(\theta_c, \theta_p)$, $I(E_c, E_p; \theta_c, \theta_p) = S(E_c, E_p) - S(E_c, E_p|\theta_c, \theta_p) = 0.03$, is rather small — indicating the relative orthogonality of the parameters to which we have alluded. The relative uncertainty for the probability distribution in Fig. 1c is 0.57.

The exact value of $S$ is dependent on the particular choice of the range of the plasma parameters — a larger range for these parameters appears to diminish further the relative uncertainty. Meaningful comparisons can be made only for the same parameter range. In Fig. 2 we show the relative uncertainty $S$ vs. noise $\sigma$ for peripheral densities at different fractions of the central density. Here, we note several important limits. At large noise levels, all data becomes useless as $S \to 1.0$. In the limit of vanishing noise, all data is conclusive, $S \to 0.0$. In the intermediate noise case, larger density differences allow finer distinctions to be made in the data.

In a more refined calculation, more points would be considered. In practice, however, rather than merely considering a finer mesh of independent radiating points, a functional parameterization of the spatial location of the radiation would be employed. This ansatz would significantly reduce the parameter space that need be considered.

## ACKNOWLEDGEMENTS

This work was supported by United States Department of Energy under contract numbers DE–AC02–76–CHO3073 and DE–FG02–84–ER53187.

## REFERENCES

[1] N. J. Fisch, *Plasma Phys. Control. Nucl. Fusion.* **30**, 1059 (1988).
[2] N. J. Fisch, and A. H. Kritz, to appear in *Phys. Rev. Lett.* (1989).
[3] N. J. Fisch, and A. H. Kritz, to appear in *Plasma Phys. Control. Nucl. Fusion* (1989).

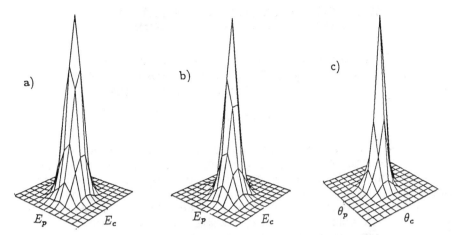

Figure 1. Joint marginal probability: a) $P_M(E_c, E_p)$; b) $P_M(E_c, E_p|\theta_c, \theta_p)$; c) $P_M(\theta_c, \theta_p)$. These marginal probability distributions are derived from the joint probability distribution $P(E_c, E_p, \theta_c, \theta_p, A_c, A_p)$. The parameter ranges plotted are $0.05 \leq E_c \leq 0.11, -0.03 \leq E_p \leq 0.03, -3.0° \leq \theta_c \leq 3.0°$, and $-3.0° \leq \theta_p \leq 3.0°$. The a priori probability distributions are uniform over these parameter ranges.

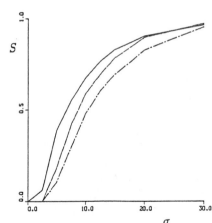

Figure 2. Relative entropy versus percent noise: Upper curve, $n_p/n_c = 0.9$; middle curve, $n_p/n_c = 0.7$; and lower curve, $n_p/n_c = 0.6$.

418

# LARGE SCALE DENSITY MODIFICATIONS :
## THEORY AND OBSERVATIONS

J. D. Hansen, G. J. Morales, L. M. Duncan[a], J. E. Maggs,
and G. Dimonte[b], Physics Department, University of California
at Los Angeles, Los Angeles, CA 90024-1547

## ABSTRACT

A 2-dimensional transport code is used to investigate the self-consistent formation of large scale ( several km) density depletions in the ionosphere due to HF heating under nighttime conditions. It is found that rotation of the absorption surface due to nonlinear refraction of ray trajectories near the reflection layer is an essential point in explaining the phenomena. Results obtained are in excellent agreement with observations made in the May 3-6, 1988 campaign at Arecibo.

## INTRODUCTION

Large scale density modifications can be produced in the ionosphere by localized temperature perturbations generated by an HF wave near its reflection layer. A schematic of the relevant geometry is shown in Fig. 1. Since the particle and heat transport is predominantly along the geomagnetic field, the process can be described by the 1- dimensional energy and continuity equations

$$\frac{3}{2}\left\{ n_e \frac{\partial T_e}{\partial t} + n_e v_s \frac{\partial T_e}{\partial s} \right\} + n_e T_e \frac{\partial v_s}{\partial s} = \frac{\partial}{\partial s}\left[ \kappa_e \frac{\partial T_e}{\partial s} \right] + Q - L_e , \qquad (1)$$

$$\frac{\partial n_e}{\partial t} = \frac{\partial}{\partial s}\left\{ D\left[ \frac{\partial}{\partial s}(n_e(T_e+T_i)) + n_e m_i g \sin\theta \right] - n_e u \cos\theta \right\} - \beta n_e + P, \qquad (2)$$

$$n_e v_s = - D\left[ \frac{\partial}{\partial s}(n_e(T_e + T_i)) + n_e m_i g \sin\theta \right] + n_e u \cos\theta , \qquad (3)$$

with s the coordinate along the magnetic field, $n_e$ the electron density, $T_e$ and $T_i$ the electron and ion temperatures, $v_s$ the field-aligned flow velocity, $m_i$ the ion mass, u the neutral wind velocity in north to south direction, $\theta$ the magnetic field dip angle, Q the HF source, P the local ionization source, and $\kappa$, $L_e$, D, $\beta$ the coefficients of thermal conduction, cooling, density diffusion and recombination, respectively[1]. The transport code solves these equations using a predictor-corrector tridiagonal method. Two dimensional effects can be modelled by solving on many adjacent field lines.

a) Clemson University. b) Lawrence Livermore National Laboratory.

## EXPERIMENTAL RESULTS

Experimental results from the heating campaign of May 3-6,1988 at the Arecibo Observatory show that large modifications $\delta(T_e/T_i)/(T_e/T_i)_0 \sim 2$, $|\delta n_e|/n_{eo} \sim 30\%$, occur in the nighttime ionosphere; these modifications proceed from broad symmetric heating patterns to narrow tubes shifted northward, and the process has a highly reproducible universal asymptotic state . The principal diagnostic is the received backscattered power of the 430MHz radar at Arecibo as a function of altitude. Since the received signal is proportional[2] to $n_e/(1 + T_e/T_i)$ both density depletions and temperature increases simultaneously cause a decrease in the signal. Figure 2 shows a typical asymptotic state achieved after 10-15 minutes of heating. Figure 3 displays the change in temperature assuming an unperturbed density (i.e. an upper bound). This exhibits a transition from an early broad and symmetric heating pattern to a narrow layer at an apparent higher altitude, thus implying a northward shift (see inset).

## INTERPRETATION AND COMPARISON WITH MODEL

Interpretation of the results requires an understanding of the role of the tilted magnetic field lines. Since transport is predominantly field-aligned, vertical positions along the diagnostic radar sample heating effects in different flux tubes. Thus the early time results of figure 3(a) show the mapping of the unperturbed horizontal pattern of the HF heater beam onto the vertical diagnostic. Figure 3(b) indicates a nonlinear modification of the heating pattern. We propose this results from nonlinear refraction due to large density depletions. To illustrate this point Fig. 4 shows ray trajectories[3] in a 30% field aligned density cavity such as could be generated by HF heating. It is seen that the reflection surface ($\omega_{pe}=\omega_{HF}$) aligns with the magnetic field. Since most of the energy is absorbed near the O-mode reflection layer, this causes intense localized heating in a narrow (~10 km) set of flux tubes.

To model nonlinear refraction effects, the absorption term Q in the transport code is modified. Density and temperature perturbations due to the HF source are solved on 51 adjacent field lines spaced in the meridian plane by 2 km. The effect of the density changes on the distribution of Q is computed as follows. At t=0 a horizontal Gaussian Q is assumed, with the power being absorbed in the neighborhood of the unperturbed reflection layer. As the density evolves, the reflection surface (actually the locus of the points $\omega_{pe}=\omega_{HF}$ across the computed field lines) is modified. Since O - mode rays near reflection have group velocity almost completely perpendicular to B, the original unperturbed Gaussian distributed HF rays are projected in this direction until they intercept the modified reflection surface between two field lines. The HF power contained in this ray is then assumed to be absorbed

420

proportionately on these two field lines, and the calculation proceeds with this modified Q. The density and temperature values along a vertical line cutting across the computed field lines is used to model the experimental diagnostic. The results are shown in Fig. 5. The early result (a) shows the original Gaussian spread of simulated HF beam pattern, followed by the narrow spike (b) of the late results. Despite the simplicity of the "ray tracing" algorithm used, the agreement between the model and experimental results (Fig. 3) is excellent.

## CONCLUSIONS

We have presented experimental and modelling results that demonstrate that large density and temperature perturbations can be produced in the ionosphere by HF waves. It should be stated that such results seem to depend upon low maximum density ionospheres during nighttime conditions. We have shown that these results can be accurately modelled using a transport code and a simple "ray tracing" algorithm. These results demonstrate that the key process is the nonlinear refraction of the HF beam across the geomagnetic field, not self-focussing of the beam. It is hoped that the generation of such large perturbations can be used to understand the maintenance of the nighttime ionosphere and features of the ionosphere-magnetosphere coupling.

## ACKNOWLEDGEMENTS

This work was supported by the Office of Naval Research. Arecibo Observatory is maintained by Cornell University under the sponsorship of the National Science Foundation.

1. Shoucri, M. M., G. J. Morales, J. E. Maggs, Ohmic heating of the polar F- region by HF Pulses, J. Geophys. Res., 89, 2907,1984.
2. Evans, J. V., Theory and Practice of Ionosphere Study by Thomson Scatter Radar, Proceedings of the IEEE, 57, 496,1969.
3. Jones, R. M., A three-dimensional ray-tracing computer program, ESSA Tech. Rept. IER-17/ITSA-17.

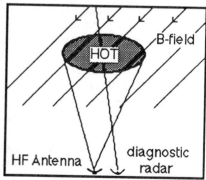

Figure 1. Schematic of experiment.

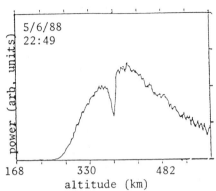

Figure 2. Backscattered 430 MHz power as a function of altitude showing asymptotic steady state. (HF on 10 min, $f_{HF}$=4.45 MHz, fof2 = 5 MHz, refl. ht. 380 km)

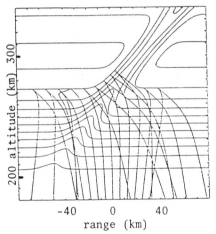

Figure 4. Ray trajectories in a 30% deep 10 km wide field-aligned Gaussian density perturbation.

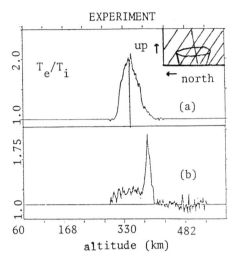

Figure 3. Change in $T_e/T_i$ using unperturbed $n_e$. HF on (a) 1 min and (b) 18 min. (Same parameters as Figure 2.)

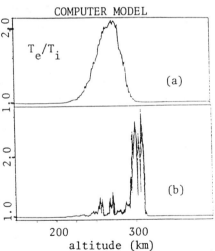

Figure 5. Early (1 min) (a) and late (10 min) (b) results of HF heating computer model. Compare with Figure 3.

# ANALYTIC EXPRESSION FOR MODE CONVERSION OF ELECTROSTATIC AND ELECTROMAGNETIC WAVES

D. E. Hinkel-Lipsker, B. D. Fried, and G. J. Morales

Dept. of Physics, University of California, Los Angeles, CA 90024-1547

## ABSTRACT

We have derived analytic expressions, in terms of Airy functions, for the fields and for the reflection and mode conversion coefficients of Langmuir and electromagnetic waves in an inhomogeneous, unmagnetized plasma. Our results, which are valid in the limit of nonrelativistic electron temperature, $T/mc^2 \ll 1$, agree with earlier numerical calculations.

## INTRODUCTION

When an electromagnetic wave of frequency $\omega$ is incident upon an unmagnetized plasma with wave vector $\underline{k}$ inclined at an angle $\Theta$ with the density gradient (assumed to be along the z axis), the wave is partially reflected at the electromagnetic cutoff, where $\omega_p(z) = \omega\cos\Theta$, and partially mode converted to an electrostatic Langmuir wave at the plasma resonance point, where $\omega_p(0) = \omega$. This "direct" problem is closely related to the "inverse" problem of an obliquely incident Langmuir wave, which is partially reflected at its cutoff and partially converted to an electromagnetic wave, that escapes the plasma.

Our analytic expressions for the mode conversion and reflection coefficients for both the direct and inverse problems in terms of Airy functions are valid to lowest order in an expansion in $T/mc^2$, where T is the electron temperature. The results agree closely with the numerical solutions obtained previously[1] for all values of $q \equiv (k_oL)^{2/3}\sin^2\Theta$, where $k_o = \omega/c$ and L is the density gradient scale length, with c the speed of light.

## ANALYSIS

We treat the electrons as a warm fluid whose unperturbed density near the reflection layer, $n_o = \overline{n}(1 - z/L)$, coincides with that of the static ions. We linearize the fluid equations, setting $n(\underline{r},t) = n_o + \{n_1(z)\exp[i(k_x x - \omega t)]\} + $ c.c.$\}$, etc., and assume $pn^{-\gamma} = $ constant, with $\gamma = 3$, as is appropriate for high phase velocity waves. From the continuity and momentum equations, together with Poisson's equation, $\nabla\cdot\underline{E}_1 = 4\pi q n_1 \equiv k_o\rho$, and Ampere's law we obtain a system of coupled second order equations for $E_z, \rho$ and $B = iB_y$:

$$d^2E_z/d\zeta^2 + (\zeta - n^2)E_z = bB(z) , \tag{1}$$

$$d^2\rho/d\zeta^2 + (\zeta - n^2)\rho = -E_z/\alpha , \tag{2}$$

$$d^2B/dZ^2 - (Z-N^2)^{-1} \, dB/dZ + (Z-N^2)B = S(z) = -iNQ \, (Z-N^2)^{-1} \, , \qquad (3)$$

where $Z \equiv (k_oL)^{-1/3} \, (k_oz) = \beta^{2/3}\zeta$, $N^2 = (k_oL)^{2/3}(k_xc/\omega)^2 = q = \beta^{-4/3}n^2$, $b \equiv i\alpha N(1-\beta^2) \, \beta^{-4/3}$, $Q =[(1-\beta^2) \, \rho - \alpha^{-1} \, dE_z/d\zeta]$ and $\alpha \equiv (k_oL\beta^2)^{1/3}$. To obtain an analytic solution of these equations, valid for small $\beta$, we exploit the great disparity (a factor of $\beta^{2/3}$) between the electromagnetic and electrostatic scales. Thus, we evaluate B on the right hand side of (1) at $Z = \zeta = 0$, which reduces (1) to an inhomogeneous Airy equation for $E_z$.

We consider first the inverse problem. For this case, $\rho$ and $E_z$ must be bounded for $\zeta \to -\infty$ and may have both incident and outgoing components for $\zeta \to +\infty$. A convenient solution of (1) which satisfies these boundary conditions is

$$E_z(\zeta) = \overline{Ai} + e_1 \, [\overline{Gi} + i\overline{Ai}] \, , \qquad (4)$$

where[3] $\overline{Ai} = \overline{Ai} \, (\zeta-n^2) \equiv Ai \, (n^2-\zeta)$, $\overline{Gi} = \overline{Gi} \, (\zeta-n^2) \equiv Gi \, (n^2-\zeta)$, and $e_1 = -\pi bB(0)$. (The amplitude of $E_z$ of the incident Langmuir wave in this case would be 1/2). From (4) and the asymptotic forms of Ai and Gi, it follows that the ratio of reflected to incident amplitudes is $R = 1 + 2ie_1$. With $E_z$ given by (4), we can solve (2) exactly:

$$\rho(\zeta) = \rho_o \, \overline{Ai} + \alpha^{-1} \, dE_z/d\zeta \, , \qquad (5)$$

where $\rho_o$ is a constant which turns out to be of order $\beta^{2/3}$ and hence negligible in the small $\beta$ limit.

To solve (3), we use the solutions of the homogeneous equation, which are derivatives of Airy functions, to construct a Green's function solution appropriate to the boundary conditions, namely that B is bounded for $Z \to -\infty$ and has no incoming wave component for $Z \to +\infty$. If we simplify the resulting expression for B(Z) by integration by parts and by again using the disparity of scale lengths, replacing, for example, $\int d\zeta' E_z \, (\zeta') \, \overline{Ai}\,{'} \, (Z'-N^2)$ by $\overline{Ai}\,{'} \, (-N^2) \int d\zeta' \, E_z \, (\zeta')$, we find, to lowest order in $\beta$,

$$B(Z) = i\pi N\beta^{4/3} \, \alpha^{-1} \, [B_1(0)B_2(Z) \int_{-\infty}^{\zeta} d\zeta' + B_2(0)B_1(Z) \int_{\zeta}^{\infty} d\zeta'] \, E_z \, (\zeta') \, , \quad (6)$$

and hence

$$B(0) = i\pi N\beta^{4/3} \, \alpha^{-1} \, \Gamma \, [1-i\pi^2 q\Gamma]^{-1} \, , \qquad (7)$$

where $\Gamma = Ai'(q) \, A_+{'}(q)$. (Details of the algebra involved in deriving (8) are given elsewhere.[4])

## RESULTS

From $R = 1 + 2ie_1 = 1 - 2i\pi bB(0)$, we have $R = (1+ i\pi^2 q\Gamma) (1 - i\pi^2 q\Gamma)^{-1}$. From the total energy flux, $\underline{S} = 2\text{Re} [(c/4\pi) \underline{E}_1 \times \underline{B}_1{}^* + p_1\underline{u}_1{}^*]$, we compute the mode conversion coefficient, $|\eta|^2$, i.e., the ratio of the outgoing electromagnetic energy flux $S_{em}^{out}$ and the incident electrostatic energy flux $S_{es}^{in}$. We find $|\eta|^2 = q|2\pi Ai'(q)|^2 \ |1-i\pi^2 q\Gamma|^{-2}$ and it is is readily verified that energy is conserved, $|R|^2 +|\eta|^2 = 1$.

The analysis for the direct problem is quite similar, but slightly simpler since there is no $p_o$ term in the expression for $p$. We find the reflection coefficients for the direct and inverse problem to be identical, consistent with the demonstration of Means et al.,[2] based on time reversal symmetry arguments. The absorption curve, i.e., $|\eta|^2$ vs. q, is presented in Fig. 1, together with the results obtained by previous authors. There is good agreement with the numerical solutions of Forslund et al.[1] for $\beta^2 = 0.015$ (the smallest value cited in that work), the deviations being of order $\beta^{2/3}$. Figure 2 is a plot of B(Z) vs. Z for the inverse problem, with $\beta^2 = .001$, $k_oL = 10$ and q = 0.5, approximately corresponding to the largest mode conversion coefficient in Fig. l. The field is "Airy-like" in nature, but due to the mode conversion, exhibits some structure on the electrostatic scale around Z = 0. This structure is less pronounced (and, of course, B is smaller in amplitude) for other q values, where the mode conversion is smaller. Plots of $E_z$ are not shown since $E_z$ is simply a superposition of Airy functions, as given by (4).

## ACKNOWLEDGEMENTS

This work was supported by the Office of Naval Research and the U.S.D.O.E.

## REFERENCES

1. D.W. Forslund, J.M. Kindel, K. Lee, E.L. Lindman, and R.L. Morse, Phys. Rev. A 11, 679 (1975).
2. R.W. Means, L. Muschietti, M.Q. Tran, and J. Vaclavik, Phys. Fluids 24, 2197 (1981).
3. H. A. Antosiewicz, Handbook of Mathematical Functions, (edited by M. Abramowitz and I.E. Stegun), 10th ed. (Dover, 1972), p. 446 ff.
4. D.E. Hinkel-Lipsker, B.D. Fried, and G.J. Morales, UCLA Plasma Physics Group Internal Report (PPG-1226, 1974).
5. A.D. Piliya, Ah. Eksp. Teor. Fiz. 36, 818 (1966) [Sov. Phys.-Tech. Phys., 11, 609 (1966).
6. T. Speziale and P.J. Catto, Phys. Fluids 20, 990 (1977).

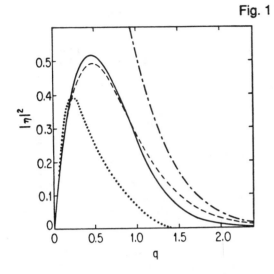

Fig. 1 Mode conversion coefficient $|\eta|^2$ as a function of q (solid curve). Also shown are the results of the numerical solution by Forslund, et al.[1] (- - - -), Piliya's[5] analytic result ($\bullet\bullet\bullet$), and the asymptotic approximation of Speziale and Catto[6] (— — — — — —).

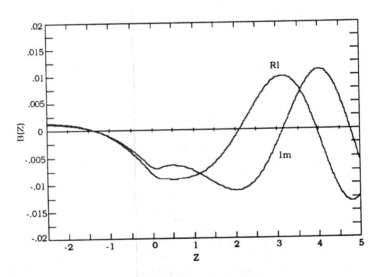

Fig. 2   Real and imaginary parts of B(Z) for q = 0.5, $\beta^2$=0.001, and $k_0L$ = 10. The mode conversion causes some deviation of B from the usual Airy function pattern near the resonance at Z = 0.

# Ponderomotive Forces, Three-Wave Coupling, and the Bispectrum[°]

T. Intrator and S. Meassick
Dep't. Nuclear Engineering & Engineering Physics
University of Wisconsin, Madison, Wisconsin, 53706

## ABSTRACT

In the Phaedrus-B Tandem Mirror plasma, antennas launch waves with high enough energy density $\delta B^2/8\pi nT \sim 3\times10^{-3}$ that ponderomotive forces can be significant. These effects are low frequency forces due to spatial gradients of high frequency (RF) wave fields, and have allowed us to experimentally verify the essential physical assumptions and conclusions of the theory and model[1]. The three-wave interaction is composed of applied RF, MHD and parametric coupled sideband waves. The Bispectrum, a Fourier transformed cubic correlation, is used to measure the local magnitude and phase of the three-wave interaction at each point in space, enabling a direct connection between experiment and theory. Wave-wave coupling of the type described by this approach and diagnosable with the Bispectrum may be important in auxiliary ICRF heated tokamaks.

## INTRODUCTION

In the Phaedrus-B Tandem Mirror plasma, loop antenna arrays launch sufficient wave fields in the Ion Cyclotron Range of Frequencies (ICRF) ($\delta B^2/8\pi nT \sim 7\times10^{-3}$) to generate significant ponderomotive forces. Ponderomotive effects are low frequency forces due to slowly varying spatial gradients of high frequency waves. Gradients of an applied wave envelope ($\nabla|Erf(r)|^2$) can interact "directly" with the plasma. But also any net plasma displacement (eg. instability) will also couple with an applied wave and induce sidebands (daughter waves) about the pump frequency.

## EXPERIMENT

In this experiment, the axially spaced loop antenna array[2] selected wavenumbers ($k_z$,m) and could be phased to either strongly excite fast magnetosonic waves and ponderomotively stabilize the central cell, or phased to weakly excite these fields. The phasing could be gated in less than 10μsec (short compared to MHD growth time), during a 20 msec discharge.

In the antenna far field, we measured the in-situ absolute magnitude and phase of vector fluctuating wave magnetic fields[3] using small orthogonal(4 mm diameter) loop probe sets to scan the r-z plane, during an "unstable" to "stable" transition. One can establish a direct correspondence between $B$ and $E$ from Faraday's law, $ik_\pm E_\pm - \partial_z E_\pm = \pm i\omega B_z/c$, neglecting electrostatic $E$ effects. Here $k_\pm$ is the axial wavenumber for each polarization component.

Experimentally, we have identified the coupling between this radiation pressure and the plasma response as a non-linear three wave interaction[4,5]. This involves the applied ICRF pump wave, the MHD plasma displacement, and sideband waves induced about the pump frequency. The explicit conservation of wavenumber and frequency demonstrate an essential feature of the theoretical model[1]. Furthermore, the major prediction seems to be correct; namely that the largest of the sideband forces tends to cancel the "direct" $|\nabla E_{rf}(r)|^2$ ponderomotive forces. We measure the local magnitude and sign of the wave-wave force with the Bispectrum[6].

## BISPECTRUM

The Bispectrum is a cubic correlation of fluctuating spectra,

$$\Gamma(\omega_1,\omega_2)=\langle F_1(\omega_1)F_2(\omega_2)F_3^*(\omega_3=\omega_1+\omega_2)\rangle \tag{1}$$

where the Fourier transformed constituents can be interpreted as as field components of three waves. At the point in space where the Bispectrum is calculated, the two parent waves at frequencies $\omega_1$, $\omega_2$ are presumed to have definite, measurable amplitudes and phases, of the form $F = |F|\exp(-i\omega t + i\mathbf{k}\cdot\mathbf{r} + i\phi)$. If the frequencies are constrained to sum resonantly, $\omega_1+\omega_2-\omega_3=0$, then the Re($\Gamma$) can be used to pick out the correlated coupling of these two parent waves with a non-linearly excited third daughter wave at $\omega_3=\omega_1+\omega_2$. Because of the sum rule, the complex $\Gamma$ has zero net frequency, and its phase corresponds to the phase of the daughter wave field with respect to the beating of the two parent waves. Only non random phase correlations survive the ensemble average $\langle\rangle$ (of 25 discharges for these data), so that the Bispectrum is a measure of the coherence, magnitude and sign of the three-wave interaction.

## CORRELATIONS of WAVE-WAVE FORCES

In the fluid limit ($kv_i \ll |\omega-\omega_{ci}|$) the expected form of the Fourier transformed ponderomotive radial force density[1] can be written in terms of a sum over polarizations ($\sigma$) of the measured fields

$$\delta F_p =-1/16\pi\Sigma\partial_r X_\sigma(r) \ [\xi\nabla_r|E_{RF}|^2+ E_U E_{RF}^* + E_L^* E_{RF}] \tag{2}$$

where the subscripts U, L denote the upper and lower sideband respectively, the susceptibility $X(n(r))$ changes radially, and the MHD displacement from equilibrium is denoted by $\xi=-\delta n/\nabla_r n$, which occurs at the MHD frequency $\omega_{MHD}$ in the lab frame. The local density $n(r,t)$ and change in density $\delta n(r,t)$ is measured with ion saturation probe arrays, and yields a magnitude and phase for $\xi$. The first term contains the gradient of the directly launched wave fields, and the dominant contribution is $\nabla_r|E_-|^2 \leq 0$ and points radially inward for all radii. The subsequent 3-wave terms contain the interaction forces of the RF fields with the sideband fields that coexist with the MHD density displacement ξ. Since all the terms are formally of the same order, one cannot a priori neglect any of the contributions. The magnitude and sign of the sideband forces relative to the $\nabla|E_{rf}|^2$ ponderomotive forces can be determined with a Bispectrum[5]

$$\Gamma(\omega_1, \omega_2) = \langle \xi(\omega_1) B(\omega_2) B^*(\omega_1 + \omega_2) \rangle \qquad (3)$$

of the resonant wave-wave interaction. These data were taken for small ξ and sideband $B_{\pm,SB}$, in the linear perturbation region of constant ratio $B_{SB}/\xi$.

In Figure 1, a surface plot of the real portion of the Bispectrum $-Re\Gamma(\omega_1, \omega_2)$ for the RHCP fields is shown so that the wave-wave couplings with cancelling phase relationship point up.

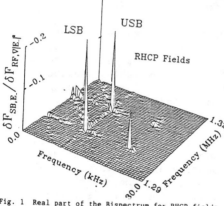

Fig. 1 Real part of the Bispectrum for RHCP fields, indicating cancelling wave-wave interaction forces.

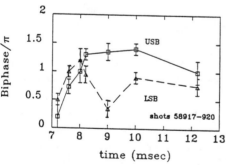

Fig. 2 Time history of the Biphase shows diagnoses the mutual interference of wave-wave forces.

Our Bispectrum shows an upper sideband peak at $\Gamma(\omega_1 = \omega_{MHD}, \omega_2 = \omega_{RF})$, and a lower sideband peak at $\Gamma(\omega_1 = \omega_{MHD}, \omega_2 = \omega_{RF} - \omega_{MHD})$. The vertical axis is in units of force density derivable from the Bispectrum, as a fraction of the first term in equation 2. Furthermore, the Biphase, $\beta(\omega_1, \omega_2) = \arctan(Im\Gamma^*/Re\Gamma^*)$ can be interpreted as the phase of the wave-wave interaction, so that $\beta = \pi(0)$ corresponds to an out of phase(in phase) cancellation(reinforcement) of the daughter wave sideband term relative to the parent RF and MHD waves in equation 2. The Biphases for the RHCP fields were $\beta_L(\omega_{MHD}, \omega_{RF} - \omega_{MHD}) = +0.9\pi \pm 0.12\pi$ and $\beta_U(\omega_{MHD}, \omega_{RF}) = -0.93\pi \pm 0.13\pi$, for the lower and upper sidebands respectively.

Moreover, if we follow the Biphase in time, we can trace the evolution of the wave-wave forces. Figure 2 shows the Biphase $\beta(t)$ as a function of time, where at $t \geq 7.5$ msec the antenna array was gated to suppress the launching of stabilizing ICRF waves. The Biphase shows that the character changes from non-interaction $(\beta \sim \pi/2)$ to cancelling interaction $(\beta \sim \pi)$ to a non-linearly saturated state $(\beta = ?)$ as time progresses.

## DISCUSSION

The theory and experiment concerning ponderomotive coupling of ICRF waves to plasma displacements are substantially consistent with each other. As in mirrors, the wave radiation pressure could significantly influence weakly unstable modes in tokamaks, The high wave field energy density in JET ($\delta B^2/8\pi nT \sim 4\times10^{-3}$) has prompted speculations[7] on the ponderomotive stabilization of internal kink modes.

The Bispectrum as a diagnostic has the capability of resolving the physics, or the action and reaction responsible for the wave-wave interaction. It is the mathematical entity corresponding to three-wave coupling, whether coherent or incoherent, because it is sensitive to the **magnitude** and **phase** of the participants. The ensemble average required for its calculation ensures that one can pick out the coherent energy channels, even in a grossly incoherent or turbulent system. The extension to higher order moments (4 or more waves) is straight forward. In fluids this type of non-linear coupling necessarily follows from the convective derivative in the Navier Stokes equation [$d\mathbf{v}/dt = \partial_t \mathbf{v} + (\mathbf{v}\nabla)\mathbf{v}$]. For example, Ritz et. al. have demonstrated[8] that three-wave coupling figures importantly in the edge turbulence and energy cascading in the TEXT tokamak. As another example, the scattering of Ion Bernstein Waves (IBW) by edge density turbulence in Alcator-C has been recognized as an impediment to wave penetration[9] into the plasma core. The physics of IBW-turbulence coupling could be easily investigated using the Bispectral correlations in a plasma edge region.

Work supported by USDOE grant #DE-FG02-88ER53264

## REFERENCES

1. J. Myra, D. D'Ippolito, G. Francis, Phys. Fluids **30**, 148 (1987).
2. R.P. Majeski, J.J. Browning, S. Meassick, N. Hershkowitz, T. Intrator, J.R. Ferron, Phys. Rev Lett. **59**, 206 (1987).
3. T. Intrator, S. Meassick, J.J. Browning, R. Majeski, N. Hershkowitz, Nucl Fusion, 4 (in press 1989).
4. T. Intrator, S. Meassick, J.J. Browning, R. Majeski, N. Hershkowitz, Phys. Fluids B **1**, 271, (1989).
5. S. Meassick, T. Intrator, J.J. Browning, R. Majeski, N. Hershkowitz, Phys. Fluids B **1**, 1049 (1989).
6. Y.C.Kim, E.J. Powers, Phys. Fluids **21**, 1452 (1978).
7. Ch. Ritz, E.J. Powers, R.D. Bengtson, Phys. Fluids B **1**, 153 (1989)
8. C. Litwin, Phys. Rev. Lett. **60**, 2375 (1988).
9. Y. Takase, D. Moody, C.L. Fiore, F.S. McDermott, M. Porkolab, J. Squire, Phys. Rev. Lett **59**, 1201 (1987).

# GREEN'S FUNCTION FOR RF-DRIVEN CURRENT IN A TOROIDAL PLASMA

Charles F. F. Karney, Nathaniel J. Fisch, and Allan H. Reiman

Plasma Physics Laboratory, Princeton University, Princeton, NJ 08543-0451

## ABSTRACT

The Green's function for rf-driven currents in a toroidal plasma is calculated in the long mean-free-path limit.

## INTRODUCTION

The "adjoint" technique provides an elegant method for computing the current driven in a plasma by externally injected radio-frequency waves. This technique involves using the self-adjoint property of the linearized collision operator to express the current in terms of a Green's function, where the Green's function is proportional to the perturbed distribution in the presence of an electric field. This technique was first applied by Hirshman[1] for the case of neutral-beam driven currents in a homogeneous magnetic field. The effect of the trapped particles found in toroidal confinement systems was included by Taguchi.[2] The application of this technique to rf-driven currents was made by Antonsen and Chu,[3] Taguchi,[4] and Antonsen and Hui.[5] Approximations to the Green's function have been given by a number of authors.[5-10]

In this paper, we solve the adjoint problem numerically in toroidal geometry using the collision operator for a relativistic plasma.[11-13] The pertinent approximations are: the device is assumed to be axisymmetric; the mean-free-path is assumed to be long compared to the device (the "banana limit"); drifts of the electrons away from the initial flux surface are neglected; in addition the expansion of the collision operator in Legendre harmonics is truncated after the $P_1(\cos\theta)$ term. By posing the problem in terms of a Green's function, we are, of course, also assuming that the plasma is close enough to equilibrium for the collision operator to be linearized, and that the wave-driven flux is known.

## BASIC EQUATIONS

In the long mean-free-path limit, the electron distribution on a particular flux surface is, to lowest order, a function of the collisionless constants of motion. We choose to express the distribution in terms of the "midplane" coordinates[14] $(u_0, \theta_0)$, the magnitude and direction with respect to the magnetic field of the momentum per unit mass (henceforth called just the momentum) at the position where the magnetic field is minimum. Measuring position on the flux surface by the length $l$ along the field line from this point, the momentum $(u, \theta)$ at an arbitrary position is

$$u = u_0, \quad \sin^2\theta = b\sin^2\theta_0,$$

where $b = B(l)/B(0)$. Particles with $\sin^2\theta_0 > \sin^2\theta_{\mathrm{tr}} = 1/b_{\max}$ are *trapped*; other particles are *passing*. Assuming that the rf is sufficiently weak, the distribution satisfies

$$f_{\mathrm{m}}\langle \widehat{C}(f/f_{\mathrm{m}}) \rangle = \frac{1}{\lambda}\frac{\partial}{\partial u_0}\cdot\lambda\mathbf{S}_0, \tag{1}$$

where $f_{\mathrm{m}}\widehat{C}(f/f_{\mathrm{m}})$ is the collision operator linearized about a Maxwellian $f_{\mathrm{m}}$,

$$\langle A \rangle = \frac{1}{\tau_b}\int\frac{dl}{v_\parallel}A$$

is the bounce-averaging operator, $\tau_b = \int dl/v_\parallel$ is the bounce time, $\lambda = \tau_b v_0\cos\theta_0/L$, and $L$ is total length of the field line (from one intersection with the midplane to the next). $\mathbf{S}_0$ is the rf-induced flux in momentum space expressed in midplane coordinates. This is related to the local rf-induced flux $\mathbf{S}$ via

$$S_{0u} = \langle S_u \rangle, \quad S_{0\theta} = \left\langle S_\theta\frac{\tan\theta_0}{\tan\theta} \right\rangle.$$

We should also include a term in eq. (1) which reflects the slow heating of the background electrons.[3] However, this term does not contribute to the current carried by $f$.

The power dissipated by the wave between two neighboring flux surfaces is

$$W = \frac{L\,dV}{\int dl/b}P_0, \tag{2}$$

where $dV$ is the elemental volume between the two surfaces and

$$P_0 = m \int d^3u_0 \, \lambda S_0 \cdot v_0. \qquad (3)$$

The current density at the midplane is

$$J_{0\|} = q \int d^3u_0 \, v_0 \cos\theta_0 f. \qquad (4)$$

At an arbitrary point the current density is $J_\| = bJ_{0\|}$. The total toroidal current flowing between two neighboring flux surfaces is

$$I = \frac{Q \, dV}{\int dl/b} J_{0\|}, \qquad (5)$$

where

$$Q = \int \frac{dl}{2\pi R} \frac{B_\zeta}{B}$$

is the safety factor, $R$ is the major radius, and $B_\zeta$ is the toroidal magnetic field.

Rather than determine $J_{0\|}$ directly by solving eq. (1), we consider the adjoint problem,

$$\langle \hat{C}(\chi) \rangle = -q \frac{v_0 \cos\theta_0}{\lambda} \Theta, \qquad (6)$$

where $\Theta = 1$ for passing particles and 0 for trapped particles. This is the equation for the perturbed electron distribution is the presence of a toroidal loop voltage $TL/Q$. The rf-driven current density is then given by[3]

$$J_{0\|} = \int d^3u_0 \, \lambda S_0 \cdot \frac{\partial\chi}{\partial u_0}. \qquad (7)$$

We will express the current drive efficiency by the ratio $\eta = J_{0\|}/P_0$. Another useful measure of efficiency is in terms of the macroscopic variables $I$ and $W$, namely

$$\frac{I}{W} = \frac{Q}{L}\eta.$$

## BOUNCE-AVERAGED COLLISION OPERATOR

The linearized collision operator is made up of three terms

$$\hat{C}(\chi) = \left( C^{e/e}(f_m\chi, f_m) + C^{e/e}(f_m, f_m\chi) \right.$$
$$\left. + C^{e/i}(f_m\chi, f_i) \right)/f_m, \qquad (8)$$

where $C^{a/b}(f_a, f_b)$ is the collision operator for distribution $f_a$ colliding off distribution $f_b$. The electron-ion term is computed in the Lorentz limit (with $m_i \to \infty$). It can be combined with the first term and bounce averaged to give

$$\frac{1}{u_0^2}\frac{\partial}{\partial u_0} u_0^2 D_{uu}\frac{\partial\chi}{\partial u_0} + F_u\frac{\partial\chi}{\partial u_0}$$
$$+ \frac{D_{\theta\theta}}{u_0^2}\frac{1}{\lambda\sin\theta_0}\frac{\partial}{\partial\theta_0}\sin\theta_0\lambda\left\langle\frac{\tan^2\theta_0}{\tan^2\theta}\right\rangle\frac{\partial\chi}{\partial\theta_0},$$

where $D_{uu}$ and $F_u$ are the coefficients of energy diffusion and drag due to electron-electron collisions, and $D_{\theta\theta}$ is the pitch-angle scattering coefficient due to collisions with both electrons and ions. These are given by one-dimensional integrals over a Maxwellian distribution.[13]

Let us now turn to the second term in eq. (8). Since $\chi$ is odd in $u_\|$, we can expand $\chi(u, \theta, l)$ in terms of spherical harmonics as follows:

$$\chi(u, \theta, l) = \sum_{k \text{ odd}} \chi_k(u, l)P_k(\cos\theta),$$

where $P_k$ is the Legendre polynomial of degree $k$ and $\chi_k(u, l) = (2k+1)\int_0^{\pi/2}\chi(u, \theta, l)P_k(\cos\theta) \times \sin\theta \, d\theta$. The linearized collision operator is a spherically symmetric, so that its angular eigenfunctions are spherical harmonics. This allows us to write the term $C^{e/e}(f_m, \chi_k(u, l)P_k(\cos\theta))/f_m$ as $\tilde{C}_k \equiv P_k(\cos\theta)I_k(\chi_k(u, l))$, where $I_k$ is a linear integral operator. Transforming to midplane coordinates, we find

$$\chi_k(u, l) = (2k+1)b\int_0^{\pi/2}\chi(u_0, \theta_0)\tilde{P}_k\sin\theta_0 \, d\theta_0,$$

where $\tilde{P}_k = P_k(\cos\theta)\cos\theta_0/\cos\theta$ and we have used the fact that $\chi$ is zero for trapped particles. The bounce-averaged collision term becomes

$$\langle\tilde{C}_k\rangle = \frac{1}{\lambda}\int\frac{dl}{L}\tilde{P}_kI_k(\chi_k(u, l)).$$

Evaluating these expressions is simplified by decomposing $\tilde{P}_k$ into midplane Legendre harmonics:

$$\tilde{P}_1 = P_{1,0},$$
$$\tilde{P}_3 = -(b-1)P_{1,0} + bP_{3,0},$$
$$\tilde{P}_k = \sum_{k'=1,3,\ldots}^k G_{k,k'}P_{k',0},$$

432

where $P_{k,0} = P_k(\cos\theta_0)$ and $G_{k,k'}$ is a polynomial in $b$. The collision term can now be written as

$$\langle \widetilde{C}_k \rangle = \frac{1}{\lambda} \sum_{k',k'' \text{ odd}}^{k} H_{k,k',k''} P_{k',0} I_k(\chi_{k'',0}) \Theta,$$

where

$$H_{k,k',k''} = \frac{2k+1}{2k''+1} \overline{bG_{k,k'} G_{k,k''}},$$

$\chi_{k,0} = \chi_k(l=0)$, and $\overline{A} = \int A\, dl/L$. In particular, we have

$$\langle \widetilde{C}_1 \rangle = \overline{b} \frac{\cos\theta_0}{\lambda} I_1(\chi_{1,0}) \Theta.$$

At present, we include only the $k=1$ term, ignoring all terms $\langle \widetilde{C}_{k\geq 3} \rangle$. We can estimate the error incurred by comparing the results we get for the electrical conductivity with those of Rosenbluth et al.[15] This indicates that the relative error in $\chi$ is on the order of $0.05\sqrt{\epsilon}$ where $\epsilon$ is the inverse aspect ratio.

## NUMERICAL SOLUTION

Putting all the terms in eq. (6) together, we obtain for the passing particles

$$\frac{1}{u_0^2} \frac{\partial}{\partial u_0} u_0^2 D_{uu} \frac{\partial\chi}{\partial u_0} + F_u \frac{\partial\chi}{\partial u_0}$$
$$+ \frac{D_{\theta\theta}}{u_0^2} \frac{1}{\lambda \sin\theta_0} \frac{\partial}{\partial\theta_0} \sin\theta_0 \lambda \left\langle \frac{\tan^2\theta_0}{\tan^2\theta} \right\rangle \frac{\partial\chi}{\partial\theta_0}$$
$$+ \overline{b} \frac{\cos\theta_0}{\lambda} I_1(\chi_{1,0}) + q \frac{v_0 \cos\theta_0}{\lambda} = 0. \quad (9)$$

We solve this integro-differential equation numerically in the domain $0 \leq \theta_0 \leq \theta_{\text{tr}}$, with boundary condition $\chi(\theta_0 = \theta_{\text{tr}}) = 0$.

A simple magnetic field configuration with circular flux surfaces is chosen. Designating the poloidal angle by $\phi$, we choose

$$R = R_0(1 + \epsilon \cos\phi),$$
$$B_\zeta = B_{\zeta 0}/(1 + \epsilon \cos\phi),$$
$$B_\phi = B_{\phi 0}/(1 + \epsilon \cos\phi),$$
$$b = (1+\epsilon)/(1 + \epsilon \cos\phi),$$

This gives $l/L = \phi/2\pi$, $Q = (B_{\zeta 0}/B_{\phi 0}) \times \epsilon/\sqrt{1-\epsilon^2}$, and $L = 2\pi R_0 Q \sqrt{1 + B_{\phi 0}^2/B_{\zeta 0}^2}$.

We normalize velocities and momenta to the thermal speed $u_t = \sqrt{T/m}$, times to the inverse collision frequency $\nu_t^{-1}$, where $\nu_t = nq^4 \times \log\Lambda/4\pi\epsilon_0^2 m^2 u_t^3$, $\chi$ to $qu_t/n u_t$, efficiency $\eta$ to $q/m u_t \nu_t$. The plasma is characterized three dimensionless parameters: $T/mc^2$, $\epsilon$, and the effective ion charge state $Z$.

Level curves for $\chi(u_0, \theta_0)$ for a typical case are shown in fig. 1. In computing the efficiency we specialize to waves which push the particles parallel to the magnetic field. First, we suppose that the wave is absorbed in a single location in momentum space, i.e., $\mathbf{S}_0 \propto \delta(\mathbf{u}_0 - \mathbf{u}_0')\hat{\mathbf{u}}_{0\parallel}$. The efficiency is given by $\eta = (\partial\chi/\partial u_{0\parallel})/v_{0\parallel}$ evaluated at $\mathbf{u}_0'$. This is shown as a function of $\mathbf{u}_0'$ in fig. 2. This shows where in momentum space we should try to have waves absorbed in order to maximize the efficiency.

We consider current drive by waves which are Landau damped. We assume that the rays pierce the flux surface at a single poloidal angle $\phi'$ where $b = b'$, and that the wave does not alter the slope of the electron distribution appreciably. In this case, we have $\mathbf{S} \propto f_m \delta(v_\parallel - v_{\text{ph}}) \times \delta(\phi - \phi')\hat{\mathbf{u}}_\parallel$ and the bounce-averaged flux is given by $\lambda \mathbf{S}_0 \propto f_m \delta(v_0 \cos\theta' - v_{\text{ph}})\hat{\mathbf{u}}_{0\parallel}$, where $\sin^2\theta' = b' \sin^2\theta_0$. The current drive efficiency can be calculated by inserting this form for $\mathbf{S}_0$ into eqs. (3) and (7) and performing the integrals numerically. The resulting efficiencies are given in fig. 3. This confirms that toroidal effects reduce the efficiency of current drive and that this reduction can be minimized if the waves are absorbed on the high-field side of the torus where there are fewest trapped particles.

## ACKNOWLEDGEMENTS

The authors would like to thank Dave Ehst, Steve Hirshman, and Dieter Sigmar for enlightening discussions. This work was supported by the U.S. Department of Energy under contract DE–AC02–76–CHO–3073.

## REFERENCES

[1]S. P. Hirshman, Phys. Fluids **23**, 1238 (1980).

[2]M. Taguchi, J. Phys. Soc. Jpn **51**, 1975 (1982).

[3]T. M. Antonsen, Jr. and K. R. Chu, Phys. Fluids **25**, 1295 (1982).

[4]M. Taguchi, J. Phys. Soc. Jpn **52**, 2035 (1983).

[5]T. M. Antonsen, Jr. and B. Hui, IEEE Trans. Plasma Sci. **PS-12**, 118 (1984).

[6]K. Yoshioka, T. M. Antonsen, Jr., and E. Ott, Nucl. Fusion **26**, 439 (1986).

[7]K. Yoskioka and T. M. Antonsen, Jr., Nucl. Fusion **26**, 839 (1986).

[8]V. S. Chan, Phys. Fluids **30**, 3526 (1987).

[9]G. Giruzzi, Nucl. Fusion **27**, 1934 (1987).

[10]R. H. Cohen, Phys. Fluids **30**, 2442 (1987).

[11]S. T. Beliaev and G. I. Budker, Sov. Phys.-Dokl. **1**, 218 (1956).

[12]B. J. Braams and C. F. F. Karney, Phys. Rev. Lett. **59**, 1817 (1987).

[13]B. J. Braams and C. F. F. Karney, Phys. Fluids **1B** (July 1989), PPPL–2598 (Mar. 1989).

[14]G. D. Kerbel and M. G. McCoy, Phys. Fluids **28**, 3629 (1985).

[15]M. N. Rosenbluth, R. D. Hazeltine, and F. L. Hinton, Phys. Fluids **15**, 116 (1972).

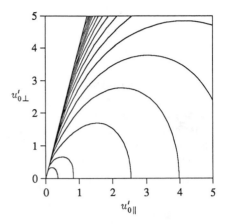

Figure 2: Contour plot of efficiency $\eta$ of current drive with point excitation of the rf, i.e., $S_0 \propto \delta(\mathbf{u}_0 - \mathbf{u}_0')\hat{\mathbf{u}}_{0\|}$. Here, $Z = 1$, $\epsilon = 0.03$, and $T/mc^2 = 0.05$. The levels of the contours are given by $\eta = 2j$ for integer $j \geq 1$ increasing from the origin outwards.

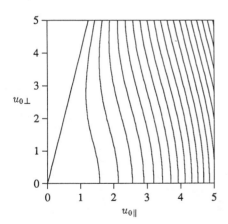

Figure 1: Contour plot of $\chi(u_0, \theta_0)$ for $Z = 1$, $\epsilon = 0.03$, and $T/mc^2 = 0.05$. The levels of the contours are given by $\chi = 5j$ for integer $j \geq 0$ increasing from left to right.

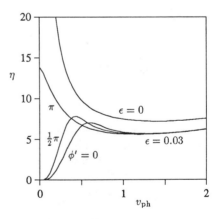

Figure 3: Efficiencies for current drive by Landau-damped waves for $Z = 1$, and $T/mc^2 = 0.05$. The top curve gives the efficiency for the case of a uniform magnetic field $\epsilon = 0.0$. The other curves are for $\epsilon = 0.03$ and three different poloidal angles $\phi'$ at which the wave is absorbed.

# RF INDUCED TRANSPORT IN TOKAMAK PLASMAS

**K. Kupfer, A. Bers, and A. Ram**
*Plasma Fusion Center, Massachusetts Institute of Technology,*
*Cambridge, Massachusetts 02139*

## ABSTRACT
We investigate the effects of coherent RF wave-fields on the transport of particles in a tokamak by coupling the chaos induced in velocity space to particles' radial guiding center drifts. In the following calculation, it is shown that localized, phase coherent structures can induce transport fluxes which are outside the scope of conventional random-phase theories.

## MODEL OF RF INTERACTION

We develop a model of circulating super-thermal electrons interacting periodically with a localized RF parallel electric field. Two types of localizations are considered: case I, the RF field intensity is localized poloidally and is uniform toroidally; and case II, the intensity is localized toroidally and is uniform poloidally. In both cases, on each transit along the magnetic field line from $z = -L/2$ to $+L/2$, an electron's interaction with the RF field is taken to be

$$\frac{dz}{dt} = v \tag{1}$$

$$\frac{dv}{dt} = \alpha e^{-z^2/d^2} \cos(z - t), \tag{2}$$

where $\alpha = (eE_{\parallel}k/m\omega^2)$ is the normalized field strength, $d$ specifies the width of the interaction region, and $v$ is the parallel velocity of the particle normalized to $\omega/k$; $z$ is normalized to the wave-number $k$ and $t$ to the frequency $\omega$. Since we have neglected the magnetic well in (2), the electron's perpendicular velocity, $v_{\perp}$, is taken to be a constant. The radial guiding center drift, ignoring the perpendicular RF fields, is simply

$$\frac{dr}{dt} = \rho_0(v^2 + v_{\perp}^2/2)\frac{2\pi}{L_p} \sin\left(\frac{2\pi z}{L_p} + \theta_n\right), \tag{3}$$

where $r$ is the instantaneous radial position, $\rho_0 = q\omega/k\omega_{ce}$, $L_p = 2\pi R_0 kq$, $q$ is the safety factor, $R_0$ is the major radius, and $\theta_n$ is the poloidal position of the particle as it transits the RF field at $z = 0$ for the $n^{th}$ time. It is here that the crucial difference between case I and II occurs. In case I, the periodicity length of the interaction, $L$, is equal to $L_p$, which means that $\theta_n$ is a constant equal to $\theta_0$, the poloidal angle about which the RF field intensity is localized. In case II, $L = 2\pi R_0 k$ and $\theta_n = \theta_1 + (n-1)2\pi/q$, so that the interactions occur at a multiplicity of poloidal angles.

If after an electrons $n^{th}$ transit through the wave-packet the time is $t_n$ and the parallel velocity is $v_n$, then integrating equations (1) and (2) in $z$ from $-L/2$ to $L/2$ gives the values of $t_{n+1}$ and $v_{n+1}$. This mapping can be iterated for a given initial condition $(t_0, v_0)$ and is area preserving when $v_n^2$ is plotted against $t_n$ (modulo $2\pi$). When $\sqrt{\pi}L\alpha d > 4$, the first order fixed point at the $v = 1$ resonance is unstable and the map contains a connected region of stochasticity, bounded in parallel velocity by two KAM surfaces on either side of this resonance; the lower one near $v_-$ and the upper near $v_+$. In the present work we will assume $\alpha d \ll 1$, so that the orbits can be expanded about the unperturbed values during a single pass through the RF fields, and $L \gg d$, so that the fields are indeed localized. Note that for present day lower-hybrid current drive experiments typical values of these parameters are $L/d \approx 10^2$, $\alpha \approx 10^{-2}$, and $d \approx 2\pi$ corresponding to a parallel wavelength of 5 to 10cm. In this limit the KAM surfaces near $v_\pm$ are located by the two solutions of $K(v_\pm) = 1$,

$$K(v) \equiv \left(\frac{L\alpha d\sqrt{\pi}}{v^3}\right) e^{-(1-\frac{1}{v})^2 d^2/4} \tag{4}$$

where $K(v)$ is the local standard map parameter [1]. Large primary islands are embedded in the upper portion of the stochastic phase space where $K$ dips below 4 so that the first order fixed points in this region are elliptic.

## RADIAL TRANSPORT

The radial motion of a single electron is obtained by integrating (3) as the electron transits the wave-packet and its parallel velocity fluctuates. In general, as an electron diffuses in the stochastic region of parallel phase space it also diffuses radially. The radial diffusion is found to be sensitive to the poloidal localization of the fields; field structures extending over a broad range of poloidal angles diffuse particles across flux surfaces faster than fields extending over a narrow range. This effect can be visualized by allowing the field envelope on the right-hand side of (2) to become a delta-function. In this case the radial equation is easy to integrate and one finds that

$$\Delta r_N = \rho_0 \sum_{n=1}^{N} \Delta v_n \cos \theta_n \ . \tag{5}$$

$\Delta r_N$ is the change in average radial position resulting from $N$ instantaneous jumps in parallel velocity, where the $n^{th}$ jump, $\Delta v_n$, occurs at the poloidal angle $\theta_n$. Note that we ignored $v_\perp^2/2$ compared to $v^2$ when integrating (3) to obtain (5) in this particularly simple form; retaining this term does not affect the following argument. In case I, all the impulsive kicks occur at the same poloidal angle, thus $\cos \theta_n$ can be pulled out of the sum in (5). The remaining sum is simply $v_N - v_0$, which is bounded as $N \to \infty$ because the stochastic orbits cannot diffuse past the KAM surfaces at $v = v_\pm$ in parallel phase space. Thus an electron being scattered in an infinitely narrow range of poloidal angle remains within a scale length $\rho_0$ of its initial radial position so that there is no radial diffusion. On the other hand if the impulsive kicks occur at two or more values of $\theta_n$, as in case II, then the sum in (5) is not bounded and $(\Delta r_N)^2/N = \rho_0^2 \langle \Delta v^2 \rangle/2$ as $N \to \infty$, where $\langle \Delta v^2 \rangle$ is the mean square jump in parallel velocity after one transit through the fields. The averaging is done over an ensemble of initial conditions spread uniformly throughout the stochastic phase space. This result is obtainable from a random-phase, quasi-linear theory. In case II, the relationship between radial diffusion and $\langle \Delta v^2 \rangle$ remains valid for a finite width field structure.

Finally, we must look back at case I when the interaction is no longer confined to a delta-function. The appropriate result replacing (5) is

$$\Delta r_N = -[\sum_{n=1}^{N}(1-\frac{1}{v_n})v_n^2 K(v_n)\sin\xi_n]\frac{\pi d^2 \rho_0 \sin\theta_0}{L_p^2} \quad , \tag{6}$$

where $\xi_n = t_n + L_p/2v_n$ and $\Delta r_N$ is the change in radial position at the poloidal angle $\theta_0$ after $N$ interactions with a wave-packet localized about $\theta_0$. The delta-function limit can be obtained by setting $\alpha = \text{constant}/d$ and letting $d$ go to zero, in which case (6) shows us that $\Delta r_N$ goes to zero like $(d/L_p)^2$ so that the radial diffusion is significantly reduced over that in case II. But more interesting than this is the fact that a dynamical, or ensemble average of $\Delta r_N$ does not vanish for particles diffusing through the portion of parallel phase space where $K(v)$ is between 4 and 1. This is entirely due to the existence of large primary islands surrounding the stable fixed points at $\xi = 3\pi/2$. As an electron diffuses in the phase space around an island its $\sin\xi_n$ is on the average positive, unless it gets stuck oscillating around the edge of the island (figure 1). The net result is a radial convection which scales as $(d/L_p)^2\rho_0$ per transit for the electrons populating this region of phase space. Also, we see that this convection can be inward or outward depending on $\sin\theta_0$.

## CONCLUSIONS

In considering the effect of phase-coherent, localized RF structures on plasma transport we have developed a simplified model for circulating super-thermal electrons interacting with an RF parallel electric field. As a particle is scattered by the RF in parallel phase space its radial guiding center drift is computed. Since the curvature and $\nabla B$ drifts are independent of toroidal angle, the toroidal localization of RF intensity (case II) has no special effect on the guiding center motion and the radial diffusion coefficient is equivalent to that predicted by a random-phase, quasi-linear theory. On the other hand, when the field intensity is localized poloidally (case I) radial diffusion is significantly reduced and a convective flow appears instead. These physical effects are not accounted for in conventional quasi-linear treatments [2,3].

## ACKNOWLEDGMENTS

This work was supported in part by DOE contract No. DE-AC02-78ET-51013 and by NSF grant No. ECS-8515032.

## REFERENCES

[1] A. J. Lichtenberg and M. A. Lieberman, University of California at Berkeley, Report UCB/ERL M88/5. Berkeley, California, 1988.

[2] T. M. Antonsen, Jr. and K. Yoshioka, Phys. Fluids 29, 2235 (1986).

[3] K. C. Shaing, Phys. Fluids 31, 2249 (1988).

FIGURE 1

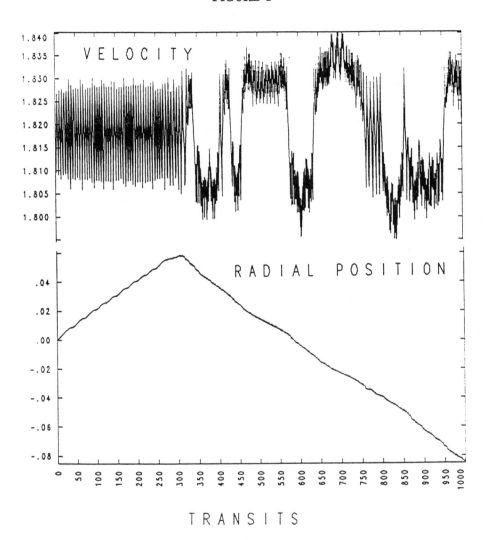

Parallel velocity (top figure) and the radial position (bottom) as a function of the number of transits through the wave packet. The radial position has a positive slope when the particle is in the vicinity of an island and a negative slope when it is diffusing between islands. ($\alpha = 10^{-2}$, $L_p/d = 10^2$, $d = 2\pi$, $\sin\theta_0 = 1$, and radial position is plotted in units of $\rho_0$.)

POWER DEPOSITION AND FIELD PENETRATION
IN A FIELD-REVERSED CONFIGURATION
GENERATED BY ROTATING MAGNETIC FIELDS

A. Kuthi, H. Zwi, B. Wells, and A.Y. Wong
University of California, Los Angeles, CA 90024-1547

## ABSTRACT

Power deposition profiles derived from measured equilibrium and field-penetration profiles in the RACETRACK rotating magnetic field driven FRC are presented. It is found, that significantly higher RF power can be deposited in the plasma than what is necessary to maintain the diamagnetic current. The Klima relations are reconciled with the higher power input because the excess power is delivered by waves posessing zero net angular momentum. Only the right-hand rotating component of the RF field penetrates fully, and this results in correct circular polarization of the fields on axis regardless of the imposed polarization by the antennas.

## INTRODUCTION

Stable high-beta plasma equilibria are routinely generated by the Rotating Magnetic Field (RMF) method[1]. The best known of these configurations is the Rotamak[2]. Of significant interest is the question of power necessary to drive a given total current. In order to gain better understanding of the current drive process and the scaling relations which allow predictions of operating parameters for high temperature collisionless plasmas, details of power deposition and field penetration were measured on an elongated RMF driven FRC.

## APPARATUS

The elongated FRC is generated in one straight section of the toroidal device RACETRACK[3]. The length of the section is 3 m, at the end of which the toroidal magnetic field increases from 2 mT to 6 mT as it turns around and connects to the other straight section. The field lines in RACETRACK are closed, no endwalls intersect the plasma. A squirrel cage antenna, 1 m long, 12.5 cm radius, at the midplane of the straight section drives the plasma currents by the phased vertical and horizontal oscillating magnetic fields it produces (Fig. 1). The antenna is excited by two individully adjustable 25 kW RF amplifiers. All components of the magnetic field are measured by 1 cm diameter magnetic loop probes, single turn for the oscillating fields (485 KHz) and 150 turns for the slow varying part of the $B_z$ field component. The density and electron temperature profiles are measured separately by double probes. The probes are radially scanned at the antenna midplane. Antenna terminal voltages and circulating currents are recorded synchronously with the probe signals. Operation at 5 x 10-5 torr

Argon pressure is made possible by 1 A of preionizing electrons from
a hot tungsten filament biased to -50 V.

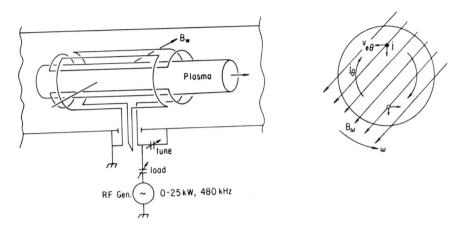

Fig.1. Rotating field antenna and induced plasma currents.

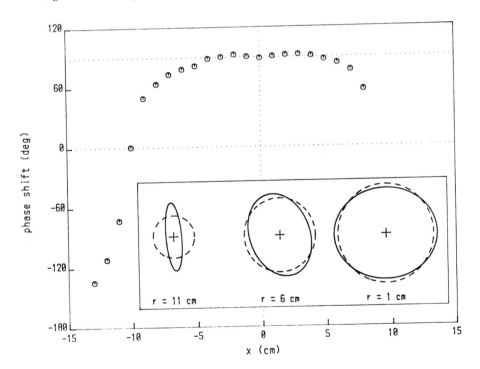

Fig.2.  Phase and amplitude profile of the rotating field.

OBSERVATIONS

When the RF power and thus the rotating field is above a
threshold (0.2 mT), complete field penetration of the right-hand
rotating component takes place. This can be seen from figure 2, the
radial distribution of the relative phase and magnitude between the
radial and azimuthal components of the measured magnetic field. The
field inside the plasma is always right-hand circularly polarized,
independent of the antenna polarization. The magnitude of the right
rotating field on axis corresponding to the radius of the Lissajous
figure is 0.4 mT.

The power deposition profile derived from the probe signals,
figure 3, is consistent with the total power leaving the antenna,
and with the fact that the left-hand rotating field is excluded,
thus depositing power mainly in the outer plasma regions by the
screening currents. The outer plasma layers are somewhat slower
rotating than the field, and the resultant increase of power input
by the right-handed wave compensates the momentum input of the left-
hand wave. The total power input is about ten times larger than the
power dissipated by the diamagnetic current alone. As a consequence
of preferential power input by the left-rotating wave and momentum
input by the right-handed wave the antenna loading is very sensitive
to the antenna polarization.

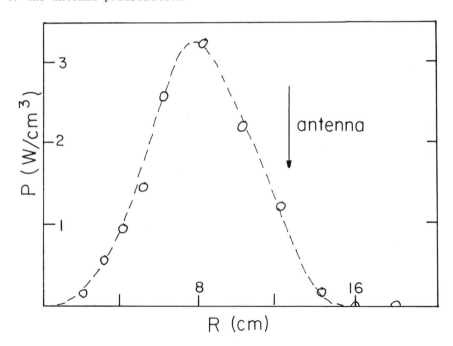

Fig.3. Distribution of RF power deposition by the screening
plasma currents.

The Klima relations connect the rate of angular momentum input with the power dissipated,

$$(w-w_0)<T>_z = P_{rf} \qquad (1)$$

where w is the angular frequency of the field, $w_0$ is the angular velocity of the electron fluid, $T_z$ is the axial component of the torque, and <> denotes time average over a period. These realations for the left and right rotating waves are satisfied, because the excess power input is by left-hand polarized wave, which in the rotating plasma frame has negligible momentum content per unit power compared with the right-hand wave.

The cost of this extra power input, however, is the distortion of the circularly symmetric equilibrium by an m = 2 elliptical component stationary in the laboratory frame. The importance of this distortion can be seen from the amplitude of the odd-harmonic mixing products appearing on the spectrum of the probe signals. In our case, this corresponds to a very slight ellipticity of the equilibrium, the third harmonic of the antenna frequency appearing on the magnetic probes at 15 dB below the fundamental. The consequences for transport can be similar to that caused by the quadrupole stabilizing fields of a theta pinch generated FRC.

Acknowledgement:

Work supported by US Department of Energy, contract #:
DOE IV DE FG03-86-ER53225.

References:

1.  W.N. Hugrass, I.R. Jones, M.G.R. Phillips, Nucl. Fusion 19,
    1546 (1979).
2.  W.N. Hugrass, I.R. Jones, K.F. McKenna, M.G.R. Phillips, R.G.
    Storer, H. Tuczek, Phys. Rev. Lett. 44, 1676 (1980).
3.  A. Kuthi, H.Zwi, L. Schmitz, D. Chelf, Rev. Sci. Instrum. 57,
    2720 (1986).

# MICROINSTABILITIES EFFECT ON MHD MODES

C. Litwin

University of Wisconsin, Madison, Wi. 53706

## ABSTRACT

The ponderomotive force due to high-frequency microinstabilities can affect the low-frequency macrostability. Two examples of this phenomenon are discussed. It is shown that an $\alpha$ particle loss-cone instability stabilizes the flute mode of an ignited, axisymmetric mirror plasma. In tokamaks, the ion-whistler instability, driven by an anisotropic population of energetic particles, stabilizes the internal kink mode for JET range of parameters.

## INTRODUCTION

Microinstabilities can occur under a variety of different circumstances. They are driven by a nonthermal character of the velocity-space distribution of plasma constituents and occur at high frequencies. Their effect in general is an enhanced fluctuation level which can lead to an anomalous transport. Consequently, microinstabilities are considered detrimental for the plasma confinement. In the present paper we shall show that microinstabilities can stabilize low-frequency fluid modes thus playing a beneficial role.

A link between the slow and fast time scales is provided by the ponderomotive force of the high-frequency unstable waves. Its effect on the low-frequency modes can be evaluated by employing the energy principle which, in addition to the usual MHD potential energy[1] $\delta W_f$, includes the contribution of the ponderomotive force of unstable waves $\delta W_w$.[2,3] To lowest order in the magnetic field curvature it can be represented in form

$$\delta W_w = \sum_\sigma \int d\tau (\boldsymbol{\xi} \cdot \nabla \epsilon_\sigma) E_\sigma^* \delta E_\sigma^{(plasma)} / 32\pi. \qquad (1)$$

Here $\delta E_\sigma^{(plasma)} = \boldsymbol{\xi} \cdot \nabla E_\sigma + \delta E_\sigma^{(lab)}$ is the wave field perturbation in the plasma rest frame, $d\tau$ is the volume element and $\boldsymbol{\xi}$ denotes the fluid displacement. The summation is performed over positive, negative and zero helicity components ($\sigma = +, -, \parallel$) and $\epsilon_\sigma$ denote the diagonal elements[4] of the dielectric tensor. We shall discuss, on two important example the effect of a microinstability on an MHD mode.

## ALPHA PARTICLE LOSS CONE INSTABILITY EFFECT ON THE $l = 1$ FLUTE MODE

In a mirror reactor $\alpha$ particles are characterized by a loss cone distribution which is unstable[5] to Alfvenic excitations. In the local theory the most unstable modes are left-circularly polarized waves which propagate parallel to the magnetic field. In a bounded plasma the radial wavenumber $k_r$ is finite and determined by boundary conditions. We shall model the plasma by a homogeneous column of radius $a$ surrounded by a vacuum region and a conducting wall at the distance $d$ from the plasma surface. A square density profile can be expected to describe adequately lowest radial harmonics which turn out to be most unstable. The magnetic field is assumed axial and uniform: $\mathbf{B} = B_0\hat{z}$. The eigenmode magnetic field then has form $B_\parallel = AJ_m(k_r r)\exp[i(kz + m\phi - \omega t)]$ where $k_r$ is expressed in terms of dielectric tensor elements. For a nearly left-circularly polarized such that $k \gg k_r$ it can be seen[6] that $k_r a \approx j_{m+1,s}$, a zero of Bessel function. Since the damping increases with $k_r$, it can be therefore expected that the most unstable eigenmode will be the lowest radial excitation $(s = 0)$ of an $m = -1$ wave. Indeed, this fact is confirmed by a detailed numerical solution of the dispersion relation.[7] The ponderomotive force of a left-circularly polarized lowest radial harmonic of an $m = -1$ wave below the ion cyclotron frequency is directed radially inward thus exerting a stabilizing influence on the $l = 1$ flute mode. However, the contribution of the wave field perturbation also has to be taken into account.

The perturbation of wave fields caused by a rigid displacement $\xi = \xi\hat{x}$ of the plasma column can be conveniently found by calculating wave fields in the plasma rest frame subject to perturbed, nonaxisymmetric boundary condition on the conducting wall. The latter, together with the continuity conditions on the plasma-vacuum boundary determines the perturbed wave fields in terms of the unperturbed ones. After a straightforward albeit lengthy algebra this leads to the expression

$$\delta W_w = \frac{\tanh kd}{\cosh^2 kd}ka(\epsilon_+ - 1)\langle|E_+|^2\rangle\xi^2/16 \tag{2}$$

where $\langle\rangle$ denotes the space average. Thus the microinstability has a stabilizing effect on the flute mode. This effect diminishes with growing distance between the plasma surface and the conducting wall. It also appears to vanish for $d = 0$ but this is only an artefact of our approximations. In this limit the wave is linearly polarized on the plasma surface and other terms have to be included. Comparing now $\delta W_w$ with $\delta W_f$ for a square pressure profile one finds the stability criterion:

$$\frac{\langle|B_+|^2\rangle}{B_0^2} > \frac{2\beta}{kR_c}\frac{\cosh^2 kd}{\tanh kd} \tag{3}$$

where $R_c$ is the curvature radius and $\beta$ is the ratio of thermal and magnetic pressures.

To model an ignited mirror plasma we assume that it is a maxwellian deuterium-tritium mixture while $\alpha$ particles are characterized by a slowing-down distribution with a loss-cone: $f_\alpha(v,\mu) = S_0\tau_s/(v^3 + v_c^3)$ if $v < v_0$ and $\mu < \mu_0$ and vanishing otherwise. Here $v$ and $\mu$ are the velocity and the cosine of the pitch angle, respectively, $\tau_s$ is the collisional slowing-down time, $v_c$ is the characteristic slowing-down velocity, $S_0$ is the rate of $\alpha$ particle production and $v_0$ is the velocity with which $\alpha$ particles are born. This model, while much simpler than the ones considered in Refs. [5] and [7] and therefore allowing for an analytic solution, leads nonetheless to parameters of the unstable spectrum within 15% of the values obtained in the above quoted references. The previously mentioned infinite medium dispersion relation is solved for the MARS mirror reactor range of parameters[8]: electron density $n_e = 3.3 \times 10^{14}$ cm$^{-3}$, electron temperature $T_e = 24$ keV, ion temperature $T_i = 28$ keV, the magnetic field at the midplane $B_0 = 4.7$ T and at the mirror throats $B_m = 24$ T, plasma radius $a = 0.5$ m, plasma-wall distance $d = 0.1$ m and plasma length $L = 130$ m. For the most unstable mode, $\omega/\omega_{c\alpha} = 0.24$, $ka = 7.5$, $k_r a = 2.1$ and $E_-(a)/E_+(a) = 0.1$. This demonstrates the validity of previously made approximations.

In order to determine the amplitudes of unstable waves we consider two different saturation mechanism: the quasilinear relaxation of the distribution and the resonance broadening. Hanson and Ott[5] argued that the instability saturates because the wave-particle scattering magnifies the size of the loss cone and thus increases the real frequency of the fastest growing mode. This in turn increases the ion cyclotron damping (predominantly on tritium). When the loss cone reaches a certain critical size $\mu_1$, the growth rate vanishes. The energy density of saturated waves can be estimated from the energy conservation. If the spectrum of unstable waves is sufficiently narrow so that $|\omega'| \ll \omega_{c\alpha}$ where $\omega'$ is the frequency in the reference frame of the fastest growing mode, the relaxation process happens through resonant particles diffusion on constant energy surfaces in this frame. To lowest order in $v_A/v_0$ where $v_A$ is the Alfven velocity, a particle with the initial velocity and pitch angle $(v, \mu)$ loses in the laboratory frame energy $\Delta\mathcal{E} = m_\alpha v_A v(\mu_0 - \mu)$. The total energy loss is obtained by an integration of the right-hand side of the above equation, convoluted with the $\alpha$−particle distribution, over the resonant region and over pitch angles with $\mu \in [\mu_1, \mu_0]$. The particle energy loss is equal the fluctuation energy. This leads to an estimate of the fluctuation level

$$|B_+|^2 = \frac{3}{4}\beta_\alpha \frac{1 - x_c^3}{\log(1 + 1/x_c^3)} \frac{(\mu_0 - \mu_1)^2(1 - \mu_0^2)}{\mu_0 + (1 - \mu_0^2)v_0/v_A} B_0^2 \qquad (4)$$

where $\beta_\alpha = 4\pi n_\alpha m_\alpha v_0^2/B_0^2$ and $x_c = v_c/v_0$. For previously quoted MARS parameters, $\mu_0 = 0.9$, $\mu_1 = 0.42$ and $|B_+|^2/B_0^2 = 1.2 \times 10^{-3}$. The wave energy density, while large, constitutes less than a percent of the plasma thermal energy under reactor conditions. We have also considered the resonance broadening as the saturation mechanism. It leads to a higher fluctuation level than the one found from the quasilinear relaxation, indicating that the latter is is the relevant saturation mechanism.

For the found fluctuation level Eq. (3) now shows that the flute mode is stable if the curvature radius is greater than 200 m for MARS range of parameters for which $kd = 1.5$. The actual curvature radius can be estimated from a simple formula[9] $R_c = L^2/8a(1 - \sqrt{1 - \mu_0^2})$ which yields $R_c = 8$ km for the MARS reactor. Therefore the flute mode is stabilized by the alpha particle instability. Because of this plasma self-stabilization it may be possible to design a thermonuclear mirror reactor without stabilizing MHD anchor cells which are a source of large radial losses.

## ION WHISTLER INSTABILITY EFFECT ON THE INTERNAL KINK

In experiments with the ion-cyclotron frequency (ICRF) heating in Joint European Torus (JET) an energetic minority ion population, with energies up to several hundred keV has been observed.[10] Fokker-Planck calculations[11] suggest that the distribution function of these energetic particles should exhibit a temperature anisotropy. Preliminary measurements[12] indicate that $T_\perp/T_\parallel > 3 - 5$. Such a distribution can be unstable to ion whistler modes which in turn may affect plasma MHD stability properties. In particular, the mode which is most susceptible to this effect is the internal kink mode. In the recent paper[14] it has been suggested that the internal kink mode can be stabilized by an application of externally excited ICRF waves. A similar influence can be exerted by internally excited ion whistler waves. In fact, the latter may be more efficient in this respect since they occur at a frequency much smaller than the ion cyclotron frequency and consequently are less strongly damped than the ICRF waves applied for the heating purposes. An analysis, similar to the one in the preceding section can be performed in order to determine the ion whistler instability effect on the internal kink mode. We shall only summarize the main results. Assuming that whistler waves are excited in the region occupied by energetic particles which are confined within radius $d$ such that $a \gg d > r_1$ where $r_1$ is the inversion radius, the sideband effect on stability can be neglected and only the first term on the rhs of Eq. (1) needs to be retained. Similarly as for the loss cone instability of $\alpha$ particles one can argue that the most unstable wave is the lowest radial harmonic of the $m = -1$ wave of a predominantly positive helicity. Such a wave has a stabilizing influence on the internal kink mode below the ion-cyclotron frequency.[14] For the ion-whistler instability in the JET range of parameters the relevant saturation mechanism is the magnetic trapping which takes place when $\gamma \sim \omega_b$ with $\omega_b$ being the magnetic bounce frequency and the proportionality factor estimated on the basis of particle simulation of the electron whistler instability.[15] Solving numerically the dispersion relation in order to find the growth rate $\gamma$, wave amplitudes are determined. For JET parameters and $T_\perp / T_\parallel \sim 3-6$, $B_+ \sim 20-90$ G. Comparing $\delta W_w$ with $\delta W_f$ for the ideal internal kink mode[16] it is then found that, for poloidal beta $\beta_p = 0.37$, the latter becomes stable for the safety factor on the axis, $q_0$, sufficiently close to unity: $|1 - q_0| \sim 0.015 - 0.24$. These parameters fall within the region observed during so called monster sawtooth periods. It is therefore plausible that the effect discussed here may be responsible for the monster sawtooth stability. This hypothesis is additionally supported by the fact that the monster sawtooth crash, when the ICRF heating is turned off occurs after a delay time comparable to the isotropization time of the energetic ions, i.e., the time of disappearance of ion whistler waves.

Supported by U.S. DOE contract DE-FG02-88ER53264

## REFERENCES

1. I.B. Bernstein et al. Proc. Roy. Soc. (London) A224, 348, 1954
2. J.R. Myra and D.A. D'Ippolito, Phys. Rev. Lett. 53, 914 (1984)
3. P.L. Similon and A.N. Kaufman, Phys. Rev. Lett. 53, 1061 (1984)
4. T.H. Stix, Theory of Plasma Waves, McGraw Hill, New York (1962)
5. J.D. Hanson and E. Ott, Phys. Fluids 27, 150 (1984)
6. C. Litwin and N. Hershkowitz, Phys. Fluids 30, 1323 (1987)
7. S.K. Ho et al. . Phys. Fluids 31, 1656 (1988)
8. B.G. Logan et al. Proc. Tenth Int. Conf. Plasma Phys. Contr. Fus. Res. 1984 (IAEA, Vienna, 1985), Vol. 3, p. 335
9. B. G. Logan, private communication
10. D. Campbell et al. Phys. Rev. Lett. 60, 2148 (1988)
11. J. Scharer et al. Nucl. Fusion 25, 435 (1985)
12. J. Jacquinot, private communication.
13. C.S. Wu and R.C. Davidson, J. Geophys. Res. 77, 5399
14. C. Litwin, Phys. Rev. Lett. 60, 2375 (1988)
15. S.L. Ossakow, E. Ott and I. Haber, Phys. Fluids 15, 2314 (1988)
16. M.N. Bussac et al. . Phys. Rev. Lett. 35, 1638 (1975)

# RF CURRENT DRIVE AND HELICITY INJECTION

R. R. Mett and J. A. Tataronis
University of Wisconsin-Madison, Madison, WI 53706

## ABSTRACT

Recent studies suggest that low frequency electromagnetic waves possessing helicity may drive currents in plasmas with greater efficiency than schemes based on linear momentum transfer to charged particles.[1,2] A key issue is the transfer of wave helicity into the helicity of the equilibrium magnetic field. Previously we examined the conversion mechanism by following the dynamics of a packet of arbitrarily polarized Alfvén waves propagating in an infinite uniformly magnetized plasma.[3,4] Analysis of the equations of magnetohydrodynamics with resistivity and viscosity yielded a set of nonlinear equations governing the Alfvén wave amplitude and an induced magnetic field. Solutions for the initial value problem[4] related the induced plasma current to the helicity of the wave and the difference between the magnetic diffusivity and viscosity. In the present investigation we examine solutions for the boundary value problem.

## INTRODUCTION

We have introduced the following wave scenario.[4] A strong, arbitrarily polarized Alfvén wave propagates in a plasma medium with resistive and viscous dissipation processes. In the course of time, the Alfvén wave decays and produces plasma perturbations that are caused by nonlinear forces exerted by the wave on the plasma. In an analysis based on the equations of incompressible MHD with a scalar conductivity $\sigma$ and a scalar kinematic viscosity $v$, we isolated from the induced perturbations components that evolve slowly in time. These slowly evolving components are associated with quasi-steady plasma currents. Our analysis reveals the following important features of low frequency current drive:[4] (1) the level of driven current scales as the net helicity of the Alfvén wave; (2) a current appears only if the Alfvén wave is spatially localized; (3) the current drive process requires dissipation. A curious result is that resistive and viscous dissipation produce opposing forces that can cancel the current drive process under special circumstances. A further important feature of our analysis is that it treats the dynamics of wave helicity injection with the problem posed either as an initial or boundary value problem.

## GOVERNING NONLINEAR EQUATIONS

An expansion about a dynamical equilibrium consisting of the nonlinear shear Alfvén wave was accomplished in terms of a small parameter $\varepsilon$ that introduces slow spatial and temporal variation in the strong Alfvén wave amplitudes $b(\varepsilon x, \varepsilon t)$ and $v(\varepsilon x, \varepsilon t)$. The amplitudes possess harmonic space and time dependence given by $e^{ik \cdot x - \omega t}$, where $k$ and $\omega$ are real constants obeying the Alfvén wave dispersion relation $\omega = \pm aB_0 \cdot k$, $a$ designating $(\mu_0 \rho)^{-1/2}$ where $\rho$ is the mass density and $\mu_0$ is the vacuum permeability. The magnetic diffusivity $\eta$ and viscosity $v$ were ordered according to $\eta = \varepsilon \eta'$ and $v = \varepsilon v'$. The following coupled nonlinear equations result.[4] The equation for the driving Alfvén wave amplitude $b$ that is zero order in $\varepsilon$ was found to be coupled to the induced zero harmonic fields $\tilde{b}^{(0)}$ and $\tilde{v}^{(0)}$ that are first order in $\varepsilon$,

$$[\partial_{t'} \pm v_A \partial_{z'} + ik \cdot (\tilde{v}^{(0)} \pm a\tilde{b}^{(0)}) + \tfrac{1}{2}k^2(v' + \eta')] \, b = 0, \tag{1}$$

while the induced zero harmonic fields are driven by the Alfvén wave amplitude,

$$\partial_{t'}\tilde{v}^{(0)} - a^2 B_0 \partial_{z'}\tilde{b}^{(0)} = 0, \qquad \partial_{t'}\tilde{b}^{(0)} - B_0 \partial_{z'}\tilde{v}^{(0)} = \nabla' \times (i\Lambda b \times b^*). \tag{2a,b}$$

In Eqs. (1) - (2), the prime associated with the differential operators denotes differentiation with respect to the appropriate primed variable $x' = \varepsilon x$, $t' = \varepsilon t$, while $\Lambda = k^2(v'-\eta')/4B_0 \cdot k$, and $v_A$ is the Alfvén speed $aB_0$. The source term in Eq. (2b) is the curl of an effective electric field that arises from the time average of $V \times B$ in Ohm's law. The physical mechanism represents a form of the dynamo or $\alpha$-effect. Observe that viscosity and resistivity are opposing effects, a result that has been anticipated by J. B. Taylor[5] on the basis of a linearized plane wave treatment (unpublished).

For parallel propagation, Eqs. (2a) and (2b) may be combined and written as

$$(\partial_{t'}^2 - v_A^2 \partial_{z'}^2) \, \tilde{b}^{(0)} = \partial_{t'} H, \tag{3}$$

where, in cylindrical coordinates,

$$H = (H_r, H_\phi, H_z) \equiv k(v'-\eta')(-\partial_\phi q/r', \partial_{r'}q, 0)/4B_0. \tag{4}$$

In addition, $q(x',t') = -2ebb^*/(1+e^2)$, $e = (1-e_p^2)^{1/2}$ where $e_p(x',t')$ is the eccentricity of the polarization ellipse, $r' = \varepsilon r$, $\partial_\phi = \partial/\partial\phi$, and $\partial_{r'} = \partial/\partial r'$. Assuming zero initial induced fields, the term in Eq. (1) containing $\tilde{b}^{(0)}$ and $\tilde{v}^{(0)}$ vanishes leaving a linear equation for $b$. Solving this linear equation as a boundary value problem with $b(r', \phi, 0, t')$ specified we obtain

$$b(r', \phi, z', t') = b(r', \phi, 0, t' \mp z'/v_A) \, e^{\mp k^2(v'+\eta')z'/2v_A}. \tag{5}$$

The solution for a cylindrical component of the induced magnetic field $\tilde{b}_r^{(0)}$ or $\tilde{b}_\phi^{(0)}$, designated by $u$, has the form[4]

$$u(r', \phi, z', t') = (2v_A)^{-1} \int_{z'-v_A t'}^{z'+v_A t'} dz'' \, H(r', \phi, z'', t' - |z' - z''|/v_A). \tag{6}$$

## SOLUTION FOR A PARTICULAR BOUNDARY VALUE

We evaluate the solutions given by Eqs. (5) and (6) for a slowly varying driving Alfvén wave amplitude at $z' = 0$ with propagation in $+z'$ direction given by

$$b(r, 0, t) = b\hat{\pi} \{1 + e^{\varepsilon(t - t_{on})/t_o} + e^{-\varepsilon(t - t_{off})/t_o}\}^{-1} e^{-\varepsilon(r/r_o)^2}, \tag{7}$$

where $b$ is a real constant, $\hat{\pi}$ is a constant polarization unit vector perpendicular to the z-axis, and $r_o$ and $t_o$ are scale lengths of order unity. Choosing normalized time, distance, and damping scales $\tau = t/t_o$, $\rho = r/r_o$, $\zeta = z/v_A t_o$, $\beta = k^2 v t_o$, $\gamma = k^2 \eta t_o$, the solutions for the driving wave amplitude and induced field become, respectively,

$$b(\rho,\zeta,\tau) = \hat{b}\hat{\kappa}U(\zeta)e^{-\varepsilon\rho^2-(\beta+\gamma)\zeta/2}\left\{1 + e^{\varepsilon(\tau-\zeta-\tau_{on})} + e^{-\varepsilon(\tau-\zeta-\tau_{off})}\right\}^{-1}, \quad (8)$$

$$\vec{b}^{(0)} = C(kr_o)^{-1}\rho e^{-2\varepsilon\rho^2} I_{BV}\,\hat{\phi}, \quad (9)$$

where

$$I_{BV} = (\beta-\gamma)\int_{\zeta-\tau}^{\zeta+\tau} d\zeta' U(\zeta')U(-\kappa)e^{-(\beta+\gamma)\zeta'}\left\{1 + e^{\varepsilon(\kappa+\tau_{on})} + e^{-\varepsilon(\kappa+\tau_{off})}\right\}^{-2}, \quad (10)$$

while U is the unit step function, $C = e\hat{b}^2/[B_0(1+e^2)]$, and $\kappa = \zeta'+|\zeta-\zeta'|-\tau$. These solutions are shown in Fig. 1. We point out that the solutions break down for sufficiently large time $\tau_b \sim (\beta+\gamma)^{-1}$ due to the expansion. In Fig. 1, $\tau_b \sim 100$.

The induced current density from Ampere's law consists of a radial and axial component. The axial component is given by

$$J_z^{(0)} = 2(\mu_0 r_o \rho)^{-1}(1 - 2\varepsilon\rho^2)\,\vec{b}_\phi^{(0)}, \quad (11)$$

which is proportional to $I_{BV}$. Therefore the effect of the driving wave on the axial current is cumulative, the rate of growth being proportional to the square of the driving wave field. No dissipation of the induced fields is present to this order in the expansion. In the formal limit $\tau \to \infty$, $I_{BVmax} \to f$ where the helicity transfer fraction $f = (\beta-\gamma)/(\beta+\gamma)$ and the axial current

$$J_{z\,max}^{(0)} \sim fC(2k\mu_0 r_o^2)^{-1}(1 - 2\varepsilon\rho^2)\,e^{-2\varepsilon\rho^2}. \quad (12)$$

The radial current profile is peaked on axis and reverses at a value $\rho \sim 2.1$ for $\varepsilon = 0.1$. There is a net axial current within a flux surface. When integrated over $\rho$ there is zero net current because solutions require $\nabla\cdot\mathbf{J} = 0$ with no currents extending to infinity.

The evolution of the magnetic helicity $h = \mathbf{A}\cdot\mathbf{B}$ was found for a uniform axial Gaussian beam initial value problem[4]. After appropriately distinguishing between the helicity of the wave and that of the induced field and after integrating over radius to produce the total helicity H per unit axial length, we find

$$H_w = C_H\,e^{-(\beta+\gamma)\tau}, \quad H_i = fC_H[1 - e^{-(\beta+\gamma)\tau}], \quad (13)$$

where $C_H = \pi r_o^2 e\hat{b}^2/[2\varepsilon k(1+e^2)]$. Helicity is conserved only for the special case of zero resistivity. This is because helicity dissipation is proportional to the resistivity alone. With zero resistivity, nonzero viscosity is necessary for the transfer of the wave helicity to the induced magnetic field. When the viscosity is zero with nonzero resistivity, a complete sign reversal of the total helicity takes place in the course of time. Yet the sense of spatial spiral of driving wave magnetic field and induced magnetic field is the same when resistivity dominates. The current is thus associated with induced magnetic helicity whose magnitude can range from +1 to -1 times the initial wave helicity depending on the relative magnitudes of the viscosity and magnetic diffusivity. A similar conclusion was reached by J. B. Taylor in another analysis.[5] In general, there is a complicated relationship between the helicity density of the driving wave and the induced current profile. We anticipate that current drive is enhanced near the Alfvén resonance.

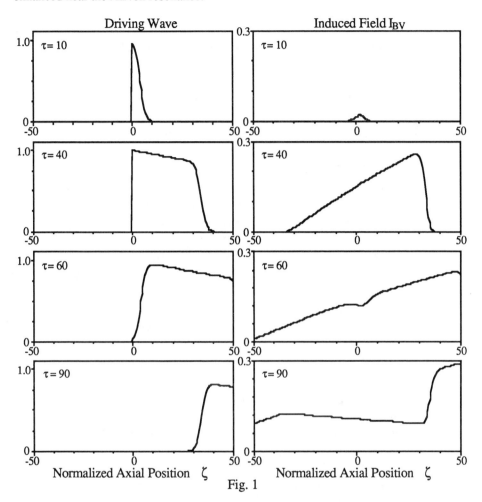

Fig. 1

This work is supported by U.S. DOE contract DE-FG02-88ER53264 and is part of the Phaedrus program at the University of Wisconsin. We gratefully acknowledge comments from J. B. Taylor that pointed out the importance of viscosity in this analysis. We also thank Noah Hershkowitz for his advice and support.

## REFERENCES

1. T. Ohkawa, General Atomics Report GA-A19379 (1988), to be published in Comments on Plasma Physics and Controlled Fusion.
2. V. S. Chan, R. L. Miller, and T. Ohkawa, General Atomics Report GA-A1919585 (1989), submitted to Phys. Fluids B.
3. R. R. Mett and J. A. Tataronis, 1989 Sherwood Conference.
4. R. R. Mett and J. A. Tataronis, submitted to Physical Review Letters.
5. J. B. Taylor, private communication.

# CENTRAL CURRENT DRIVE BY SYNCHROTRON RADIATION IN A TOKAMAK REACTOR

R.L. MEYER, I. FIDONE, G.GIRUZZI, LPMI, U.A. CNRS 835,

Université de Nancy I, FRANCE  and DRFC - CADARACHE, FRANCE

## Abstract

Current drive by synchrotron radiation is considered. The general formula for computing the generated current for a given asymmetric spectral distribution is presented. Preliminary numerical results on the current drive efficiency and radial profile are also shown.

## I.  INTRODUCTION

In a hot plasma (Te = 50 Kev) tokamak reactor, it is very tempting to use a small fraction of the synchrotron radiation with a non-zero parallel momentum for steady-state current drive[1]. Conceptual devices in which the centrally located synchrotron driven current acts as a seed for the bootstrap effect were discussed recently[2,3], using crude estimates of the radiated power and generated current. In order to assess the potential of synchrotron radiation as a current driver, we have undertaken an extensive and accurate study of the problem using the current drive efficiency of Fidone at al[4] and Trubnikov's[5] theory of synchrotron radiation. While the difficulty of achieving a wall configuration capable to create an asymmetric radiation spectrum with net nonzero momentum is recognized, here the intention is to identify the role of the plasma and radiation parameters which determine the driven current.

## II.  CURRENT DRIVE EFFICIENCY

A wave of frequency $\omega$ and parallel refractive index $N_{||}$ generates a toroidal current in a Maxwellian target plasma through momentum transfer and asymmetric resistivity[5] (the latter is in general predominant). The figure of merit is given by[4]

$$\Delta J/\Delta P = \mu\ G(N_{||},\alpha) = \epsilon\ \mu\ N_{||}\frac{1 + \alpha y\ [1 + \gamma_o\Gamma(\gamma_o)]}{(1 + \alpha y)\ \gamma_o\ \nu(\gamma_o)}, \tag{1}$$

where $\mu = mc^2/T_e$, $\alpha = \omega/\omega_c\mu$, $\Delta J$ and $\Delta P$ are in units $n_e e(T_e/m)^{1/2}$ and $n_e\ T_e$ $(4\pi e^4\ n_e\ \Lambda/m^{1/2}\ T_e^{3/2})$, respectively, y is the solution of the equation $shy - y = 1/\alpha\ (1 - N_{||}^2)$,

$\gamma_o = (1 + \alpha Y)/[1 + 2\ \alpha Y + \alpha^2\ (1 - N_{||}^2)\ (2 - chy + y^2)]^{1/2}$,

$\nu(\gamma) = (\gamma - 1)^{3/2}/(\gamma + 1)^{1/2}\ (\gamma^2 - 2\gamma\ln\gamma - 1)$

$\Gamma(\gamma) = [2 \, \gamma^2(\gamma + 2) \, \ln\gamma - (4\gamma - 1) \, (\gamma^2 - 1)]/\gamma(\gamma^2 - 1) \, (\gamma^2 - 2\gamma\ln\gamma - 1),$

and we assume the ion charge Z = 1. Equation (1) with $\epsilon$ = 1 is valid for sufficiently high values of $\omega/\omega_c$ and $\mu$. In fact, $\epsilon$ is a slowly varying function of $\omega/\omega_c$ and $\mu$ and a rigorous numerical computation of Eq.(1) shows that for Te = 50 KeV and $\omega/\omega_c$ > 5, $\epsilon$ = 1.2. In order to compute the spatial profil and the total current generated by synchrotron radiation, it is necessary to evaluate $\Delta P$ for a given asymmetric wall configuration and given values of $N_{||}$, $\omega$ and plasma parameters. Note that $\Delta P$ is the fraction of the reflected radiation with a nonzero average value of $N_{||}$ and $\Delta P$ = 0 for < $N_{||}$ > = 0. From Eq.(1)

$\Delta J = (mc^2/4\pi e^3 n_e \Lambda) \; G \; (N_{||}, \alpha) \; \Delta S_o \; 2k''(\ell) \; \exp \, (-2 \int_0^\ell K'' d\ell'),$ where

$K''(\ell) = \vec{K}'' \vec{V} g/Vg$, $\vec{K}''$ is the imaginary part of the wave propagation vector, $\ell$ is the abscissa along the ray path, Vg is the group velocity, $\Delta S_o$ is the energy flux, and the ordinary units are restored. $K''$ is obtained from the relevant dispersion relation. Using Eq. (1) we obtain

$\Delta I/\Delta W = (mc^2/4\pi e^3 \; n_e \Lambda) \; 2\pi \int_0^a r dr \; G \; \Delta P \; /(2\pi R) \; 2\pi \int_0^a \; r d\dot{r} \; \Delta P =$

$(mc^2/4\pi e^3 \; n_e \Lambda) \quad \int_0^a dr \; G(N_{||}, \alpha) \; p(r) \; /2\pi R \int_0^a \; dr \; p(r),$

where a and R are the minor and major radii of the torus of circular cross section, p(r) = p ($\ell$(r)), and p($\ell$) = 2 K''($\ell$) exp $(-\int_0^\ell 2K'' d\ell')$. For a homogeneous plasma

$\Delta I/\Delta W = (mc^2/4\pi e^3 \; n_e \Lambda) \; G \; (N_{||}, \alpha) = (\frac{20}{\Lambda}) \frac{1.56 \; G}{n_e \; R}$ (A/W), where R and n are expressed in meters and $10^{20} \; m^{-3}$, respectively. For $\omega/\omega_c$ = 9, $N_{||}$ = 0.5-0.7, $T_e$ = 50 Kev ($\Lambda$ = 20), $\Delta I/\Delta W$ = (0.3/$n_e R$)(A/W). We now compare this result with the corresponding for realistic profiles of $n_e$, $T_e$, and B.

III. NUMERICAL RESULTS FOR AN INHOMOGENOUS PLASMA

We now present some preliminary results on the radial profile of the generated current $\Delta J$ and the global efficiency $\Delta I/\Delta W$ for given values of $\theta$ and $\omega$. These results show that the generated current is located in the plasma core and characterized by relatively high values of the current drive efficiency. We consider a Tokamak device with a = 1.5 m, R = 4 m and B(0) = 60 KG. The density and temperature profiles are given by $n_e$ (r) = $n_e$ (0) (1 - $r^2/a^2$), $T_e$ (r) = $T_e$ (0) (1 - $r^2/a^2)^{3/2}$, where $n_e$(0) = $10^{14}$ cm$^{-3}$

and $T_e(0) = 50$ KeV. Values of $p(r)$ versus $r$ for the X - mode in the equatorial plane for $\omega/\omega_c(o) = 7$ and $\Theta = 60°$, $70°$ are shown in Fig.1. It appears that the main part of the power deposition as well as of the generated current lie within $r \approx 50$ cm. Similar results are obtained for different values of $\omega/\omega_c(o) > 5$. Note that for $\Theta = 60°$, the ray crosses the plasma axis twice. In Fig.2, we present the total wave power absorbed in a single transit $\eta = \int_o^a p(r)dr$ versus $\omega/\omega_c(o)$ for the parameters of Fig.1 In general, strong absorption occurs for $\omega/\omega_c(o) < 8$ and $\Theta > 50°$. Note however that weakly absorbed waves at $\omega/\omega_c(o) > 8$ are relatively more important for current generation since the maximum of the intensity occurs at high values of $\omega/\omega_c(o)$. Figure 3 shows $\Delta I/\Delta W$ versus $\omega/\omega_c(o)$ for the parameters of Fig.2 for $\Theta = 40°$, $60°$, and $70°$. It appears that the values of $\Delta I/\Delta W$ are in general significantly smaller than the homogeneous case ($\approx 0.07$ A/W) except for $\Theta \approx 60°$ and $\omega/\omega_c(o) > 9$. The temperature dependence of $\Delta I/\Delta W$ for $\Theta = 60°$ is presented in Fig.4. It is found that the current drive efficiency is a slowly increating function of $T_e(o)$.

REFERENCES

1 - J.M. DAWSON and P.K KAW, Phys. Rev. Lett 48,1730 (1982)

2 - J. JOHNER and I. FIDONE, in 12th Int. Conf. on Plasma Physics and Contr. Nucl. Fusion Research, Nice (1988), paper IAEA - CN - 50/G-3-5.

3 - K. YOSHIKAWA et al, in 12th Int. Conf. on Plasma Physics and Contr. Nucl. Fusion Research, Nice (1988), paper IAEA - CN - 50/G-3-4

4 - I. FIDONE, G. GRANATA, and J. JOHNER, Phys.Fluids 32, 2300 (1988)

5 - B.A Trubnikov, in Reviews of Plasma Physics, edited by M.A Leontovich, Consultants Bureau, New-York, Vol 7, 345 (1979)

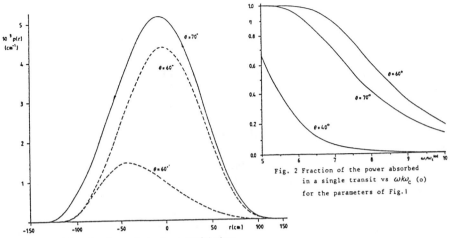

Fig.1 p(r) vs for the x-mode in the equatorial plane

Fig. 2 Fraction of the power absorbed in a single transit vs $\omega/\omega_c$ (o) for the parameters of Fig.1

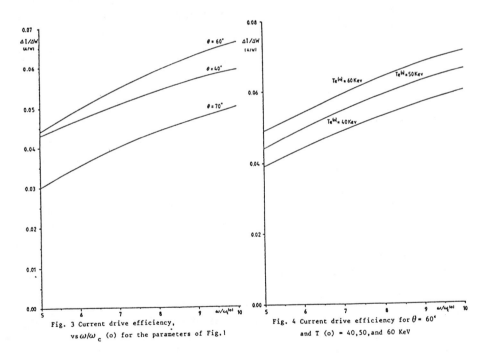

Fig. 3 Current drive efficiency, vs $\omega/\omega_c$ (o) for the parameters of Fig.1

Fig. 4 Current drive efficiency for $\theta = 60°$ and T (o) = 40,50, and 60 KeV

# RF-DRIVEN SOURCE OF SUPERTHERMAL OXYGEN ATOMS

R.W. Motley, J.W. Cuthbertson,
and W.D. Langer
Plasma Physics Laboratory
Princeton University
Princeton, NJ 08543

## ABSTRACT

An intense (4 ampere) low energy oxygen beam has been developed by neutralizing an oxygen plasma on a biased molybdenum plate. The plasma was exicted by a 1 kW, 2.45 GHz coaxial source in a 4 kG field.

## INTRODUCTION

It is estimated that the space station and telescope, to be lifted into orbit by NASA, will be bombarded by $\sim 10^{14}/cm^2sec$ oxygen atoms at 5-10 eV. Materials exposed to this chemically-active bombardment must resist oxidation. Some materials, such as silver and kapton insulators, will be quickly eroded unless they are protected by coatings. Reliable, intense laboratory source of $\sim 10$ eV oxygen atoms are needed for timely evaluation of space materials.

We have developed a facility in which oxygen ions from an RF discharge are accelerated into and neutralized on a biased metal plate to produce a 3-4 ampere beam of atoms in the 5-25 eV energy range. RF excitation avoids cathode poisoning that limits the oxygen content of simple arc and glow discharges. The facility is based on the lower hybrid plasma source developed by Motley et al.[1] and the plate neutralization demonstrated on the ACT-1 beam facility.[2]

The plasma source (Fig. 1) consists of an RF coaxial exciter at one end of a stainless steel vacuum chamber, 22 magnetic coils yielding a 4 kG solenoidal field, and a pulsed gas valve. A magnetron supplies up to 1 kW of RF power to the center pin of the coax. Up to 99% of the power can be matched from the waveguide to the plasma load by a triple stub tuner.

The coaxial exciter consists of a type N connector with a ceramic seal and a 4 cm long, 3 mm diameter stainless steel center electrode (Fig. 2). Just before RF power is applied to the coax, gas at a pressure of 1-2 mtorr is admitted to the chamber by a piezoelectric valve. Discharges consisting of 100% oxygen at duty cycles up to 10-15% are routinely initiated. The upper limit to the duty cycle is set by heating of the uncooled center pin. Floating the pin yields a higher plasma density, but produces more sputtering, especially in a neon discharge.

The plasma flows from the exciter a distance of 20 cm to a molybdenum plate cooled from the rear by a water-cooled copper plate. The plate, biased at 2-40 volts, is tilted at an angle of $\sim 60^{\circ}$ to the magnetic field. The ions are accelerated into the plate by the potential across the sheath. According to the TRIM code[3] approximately 80% of the ions should emerge from the plate as neutrals with a $\cos\theta$ beam spread at an energy of ~2/3 the ion energy.

According to lower hybrid wave theory[5] the maximum density for wave penetration, in the limit where $\omega_{pe}^2/\omega_{ce}^2 \gg 1$, and $\omega^2 > \omega_{ci}\omega_{ce}$, is

$$\omega_{pe}^2 < \omega_{ce}^2 \, n_{\parallel}^2 \, (1 - 1/n_{\parallel}^2)^2/4$$

i.e., the maximum plasma density scales as the magnetic field squared. As shown in Fig. 3, the measured ion current scales somewhat more slowly than $B^2$, whereas the center density drops sharply below 2.5 kG. Radial scans show that at low field an initially peaked density profile (1.4 cm FWHM) collapses after 1/2 msec into a hollow, broad profile (Fig. 4). The center density at 4 kG is $3\times10^{13} \text{cm}^{-3}$, the center temperature ranges from 7-16 eV and the maximum ion current measured is 4 amperes.

We have bombarded kapton, teflon and carbon films placed 9.3 cm from the neutralizer with beams of atomic oxygen and find that the rate of erosion of kapton is $\approx$0.9 microns/hr, ($10^{-23} \text{cm}^3/0$ atom). TFE teflon exposed to the beam shows no erosion. We believe that the erosion arises from chemical interaction between the samples and the oxygen beam, because samples of kapton exposed to non-oxygenic beams, such as neon, are unaffected.

## ACKNOWLEDGMENTS
We thank J. Taylor, W. Kineyko, D. Cylinder, R. Yager, and E. Thorsland for valuable help in constructing the apparatus. This work was supported by a research grant No. H83097B from the National Aeronautics and Space Administration.

## REFERENCES
[1]R.W. Motley, S. Bernabei and W. Hooke, Rev. Sci. Instrum. 50, 1586 (1979).
[2] W.D. Langer, S.A. Cohen, D.M. Manos, R.W. Motley, M. Ono, S.F. Paul, D. Roberts, and H. Selberg, Geophys. Res. Lett. 13, 377 (1986).
[3]J.P. Biersack and L.G. Haggmark, Nucl. Instrum. Methods 174, 257 (1980).
[4]J.W. Cuthbertson, W.D. Langer, and R.W. Motley, Proc. Am. Phys. Soc. 33, 1993 (1988).
[5]R. Troyon and R.W. Perkins, Proc. of Conf. on RF Heating in Plasmas (Texas Tech. U., Lubbok, TX, 1974) paper B4.

456

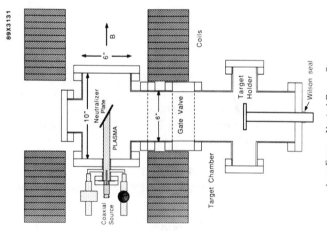

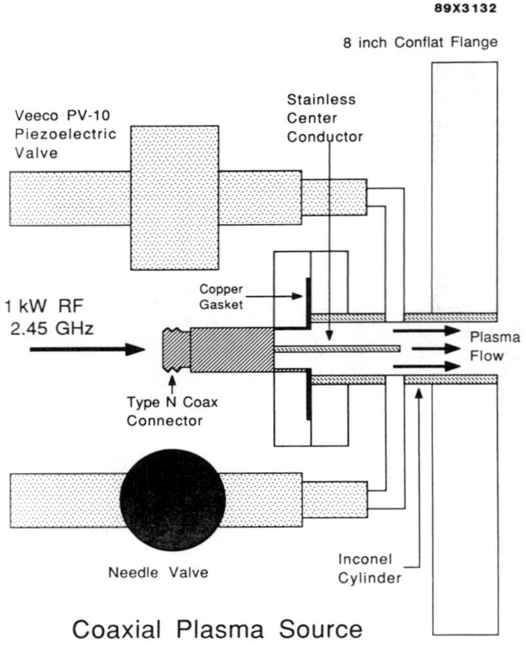

89X3132

8 inch Conflat Flange

Stainless Center Conductor

Copper Gasket

Plasma Flow

Inconel Cylinder

Veeco PV-10 Piezoelectric Valve

Type N Coax Connector

1 kW RF 2.45 GHz

Needle Valve

Coaxial Plasma Source

FIG. 2

89X3131

B

6"

10"

Neutralizer Plate

PLASMA

Coaxial Source

Coils

Target Holder

Wilson seal

6"

Gate Valve

Target Chamber

Low Energy Neutral Beam Source

FIG. 1

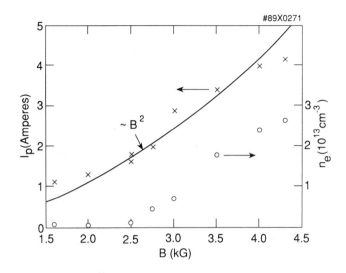

FIG. 3   Total ion current and electron density ( 4mm from  r = 0) versus
magnetic field.

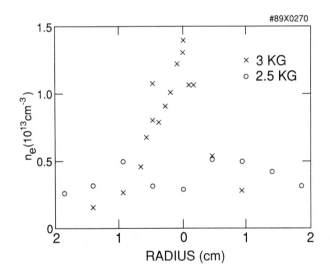

FIG. 4   Density profiles at 2.5 and 3 kG.

# KINETIC-ALFVEN-WAVE CURRENT DRIVE
# IN ELONGATED-CROSS-SECTION PLASMAS

S. Puri and R. Wilhelm

Max-Planck Institut für Plasmaphysik, EURATOM Association,
Garching bei München, Federal Republic of Germany

*Efficient kinetic-Alfven-wave current drive is possible in plasmas with elongated cross sections ($l > 1.5$) due to the reduction in the relative (compared with the antenna radius) penetration distance from the antenna to the plasma core. Localization of current drive close to the plasma axis alleviates the problems due to the trapped-particle effects. The net fraction $(1-\varepsilon^{1/2})/(1+0.5Z)$ of the wave momentum contributing to the subthermal current drive far exceeds the existing projections. For the bootstrap-current contribution in the range $0.9 \geq I_b/I_p \geq 0.1$, the current-drive efficiency $\gamma = Rn_{20}I/P$ assumes the values $22 \geq \gamma \geq 1.1$.*

## INTRODUCTION

Current drive using the subthermal kinetic-Alfven-wave (KIN) is studied. Subthermal (parallel phase velocity $v_p$ less than the electron thermal velocity $v_t$) waves possess the advantage of transferring high parallel momentum content to the electrons[1]. The method, however, lacks credibility because the wave momentum is primarily imparted to the trapped electrons which do not contribute to current drive.[2] We show that, notwithstanding the trapped particle effects, high-efficiency current drive is possible in elongated-cross-section plasmas using optimized antenna parameters.

## MOMENTUM-TRANSFER EFFICIENCY

Conservation of canonical angular momentum[3] for a trapped particle in a radio-frequency-driven, steady-state Tokamak implies

$$\frac{d}{dt}\left[mRv_\varphi + \frac{ieRE_\varphi}{\omega - k_\varphi v_\varphi}\exp\{i(\omega - k_\varphi v_\varphi)t - i\varphi_0\} + \frac{e}{2\pi}\Phi\right] = 0 , \qquad (1)$$

where the last two terms in the square brackets are contributed by $eRA_\varphi$, $(R, \varphi, z)$ are the cylindrical coordinates, $\Phi = \int_0^R 2\pi RB_z \, dR$, and $B_z$ is the $z$ component of the magnetic field due to the plasma current. Under resonant interaction of duration $\Delta t$, the particle gains (assuming a favorable phase angle $\varphi_0$ relative to the electric field) an angular momentum

$$R\Delta p_\varphi = mR\Delta v_\varphi = -eRE_\varphi \cos\varphi_0 \Delta t = \frac{e}{2\pi}(\Phi_0 - \Phi) = \frac{e}{2\pi}\Delta\Phi , \qquad (2)$$

and is pinched[4] inwards from flux surface $\Phi_0$ to $\Phi$ (Fig. 1). In steady state, collisions restore the particle to its original orbit with the starting velocity $v_{\varphi 0}$ and the flux surface $\Phi_0$, in the process releasing[3] a mechanical momentum $\Delta p_\varphi$ to the background ions and electrons *in proportion* to their respective collision frequencies. Thus the subthermal wave acts as a continuous pump delivering angular momentum to the trapped electron population resonant with the wave; whereas collisions redistribute the momentum to the bulk plasma over a period lasting $\varepsilon(\nu_{ee} + \nu_{ei})^{-1}$, where $\varepsilon = r/R$. Alternatively, the collisions may be modeled by an appropriate electric-field-spectrum term in Eq.(1). Summing Eq.(1) over the ensemble of accelerated population and requiring $\sum \Delta p_\varphi = 0$ (since $\sum \Delta v_{\varphi 0} \equiv 0$ and $\sum \Delta \Phi \equiv 0$ in the steady state) shows that the mechanical momentum imparted by the pump wave to the accelerated particle population is redistributed to the background plasma in the manner already described. Assuming that $\nu_{ei}$ and $\nu_{ee}$ are not materially affected by the truncated ($v_\varphi = v_p$, $v_\perp$ Maxwellian) distribution from their corresponding Spitzer values, the fraction of the original wave momentum transferred to the bulk-plasma electrons is given by $\sigma(Z) = (1 + \nu_{ei}/\nu_{ee})^{-1} = (1 + 0.5Z)^{-1}$. The untrapped fraction $\xi(\varepsilon) = 1 - \varepsilon^{1/2}$ of these bulk-plasma electrons participates in the current drive in a manner analogous to the ohmic current. *The fraction of the wave momentum contributing to the KIN current drive is $\sigma(Z)\xi(\varepsilon)$ and not $\Re = 1 - \exp\left[-(v_p/v_t)^2/2\varepsilon\right] \approx (v_p/v_t)^2/2\varepsilon$ as currently believed.*

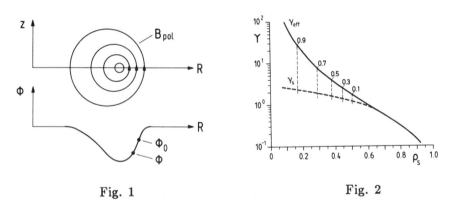

Fig. 1                Fig. 2

## CURRENT-DRIVE EFFICIENCY

The current density induced in the plasma is related to the power density by (energy input $U = mv_A \Delta v$ produces a current $j = e\sigma\xi\Delta v$ of duration $\tilde{\nu}_{ei}^{-1}$)

$$j(\rho) \approx \frac{e}{m} \frac{\sigma(Z)\xi(\varepsilon)\eta_A P(\rho)}{\tilde{\nu}_{ei}(\rho, \varepsilon) v_A(\rho)} , \qquad (3)$$

where $a$ is the plasma radius, $\rho = r/a$, $A = R/a$, $v_A(\rho) = v_p$ is the local Alfven speed, $\tilde{\nu}_{ei} = \nu_{ei}/\xi^2$ is the Spitzer collision frequency enhanced by the trapped-particle effects[5], and $\eta_A$ is the antenna coupling efficiency. The current-drive efficiency becomes

$$\gamma = \frac{<n_{20}> RI}{P_P} = \frac{<n_{20}>}{2\pi} \frac{e}{m} \frac{\sigma(Z)\eta_A \int_0^1 j(\rho)\rho\, d\rho}{\int_0^1 \xi^{-1}(\varepsilon)\tilde{\nu}_{ei}(\rho, \varepsilon) v_A(\rho) j(\rho)\rho\, d\rho} , \qquad (4)$$

where $< n_{20} > \times 10^{20} m^{-3}$ is the volume-averaged plasma density.

In the presence of the bootstrap current $(I_b)$, the demands on the current drive in the unfavorable outer region are significantly diminished.[2] Maximum current-drive efficiency is obtained if the wave driven seed current $I_s = I_p - I_b$ is spread with a constant density $j_s$ in the region $0 \leq \rho \leq \rho_s \approx \sqrt{I_s/q_a I_p}$, where $q_a$ is the safety factor at the plasma boundary and $j_s = (2/R)(B_t/\mu_0)$ is the maximum current density consistent with the MHD stability requirement $q \geq 1$. Assuming $n_e(\rho) = n_{e0}(1 - \rho^2)^{\chi_n}$ and $T_e(\rho) = T_{e0}(1 - \rho^2)^{\chi_T}$ yields the seed-current-drive efficiency ($\eta_A \approx 1$ for optimized antenna parameters[6])

$$\gamma_s = \frac{13.5 \mu^{1/2} \sigma(Z) < \beta >^{1/2} < T_{keV} > \int_0^{\rho_s} \rho \, d\rho}{Z < \ln \Lambda > \int_0^{\rho_s} \xi^{-3}(1 - \rho^2)^{-\alpha} \rho \, d\rho} \frac{\chi_T + 1}{\chi_n + 1}(\chi_n + \chi_T + 1)^{1/2} , \quad (5)$$

where $\mu$ is the atomic mass number, $< \beta >$ is the volume-averaged toroidal $\beta$, $< T_{keV} >$ is the volume-averaged electron temperature in $keV$, $< \ln \Lambda >$ is the weighted Coulomb logarithm, and $\alpha = 1.5\chi_T - 0.5\chi_n$. The pertinent figure of merit is contained in the effective current-drive efficiency given by

$$\gamma_{eff} = \frac{I_p}{I_s} \gamma_s . \quad (6)$$

Figure 2 is a plot of $\gamma_s$ and $\gamma_{eff}$ as a function of $\rho_s$ for $A = 4$, $\chi_n = 1$, $\chi_T = 1.5$, $\mu = 2.5$, $< \beta >= 0.05$, $< T_{keV} >= 15$, $Z = 1.5$, $q_a = 3.5$ and $< \ln \Lambda >= 17$. For $0.90 \geq I_b/I_p \geq 0.10$, one obtains $22 \geq \gamma_{eff} \geq 1.1$. This compares favorably with the alternative approaches such as lower-hybrid-resonance and transit-time-magnetic-pumping current drives. The ability to drive current close to the plasma axis makes the KIN current drive particularly suited for taking optimal advantage of the bootstrap current.

## KINETIC-ALFVEN-WAVE ACCESSIBILITY

$\gamma_s$ in Fig.2 falls off rapidly for $\rho_s > 0.5$ because of (i) enhanced collisionality and (ii) trapped-particle effects with a $\xi^3$ dependence; thus a central deposition of the wave energy is imperative for efficient current drive. Efficient wave conversion from the compressional mode to the KIN requires[6] $m = 1$ and $n = 5 - 8$. Fundamental Bessel function properties restrict accessibility[6] of the evanescent compressional mode launched by the antenna to $\rho \gtrsim 0.67$. Reduction in the relative (compared to the antenna radius) penetration distance from the antenna to the plasma core for elongated-cross-section plasmas, together with the shift in the plasma axis due to finite $\beta$ effects in reactor-grade plasmas suffices to make the entire plasma volume accessible to KIN.[7] Further improvement in central energy deposition would occur via the quasilinear enhancement of the radial damping length of KIN. Single pass KIN absorption, however, would not be endangered (i) due to the extremely short[8] ($\sim 1 \, cm$) linear absorption length for KIN, and (ii) because the magnetosonic cutoff surrounding the propagating KIN region would effectually thwart the escape of the radially trapped KIN.

## EDGE AND OTHER SPURIOUS EFFECTS

The low-frequency ($f \sim 1MHz$) operation (i) is undemanding on the antenna voltage, (ii) cuts down on direct coupling to the torsional surface wave[9], (iii) is not susceptible to cyclotron or $\alpha$-particle acceleration, and (iv) precludes the possibility of direct ion-Bernstein-wave excitation[10]; however, since at low $f$, the electrons are able to sweep over large distances in the toroidal direction during one half-cycle, antenna shielding with a Faraday screen assumes particular importance in order to avoid plasma production and density rise problems encountered in the TCA[11] experiment.

## INVERSE WARE PINCH AND THE BOOTSTRAP CURRENT

Inverse Ware[4] pinch effect responsible for the recovery of the mechanical momentum from trapped particles is closely allied to the concept of bootstrap current[2]. Disregarding the electric-field term, Eq.(1) shows that in steady state

$$\frac{d}{dt}\sum mRv_\varphi = -\frac{d}{dt}\sum \frac{e}{2\pi}\Phi = \frac{e}{2\pi}\Gamma\frac{\partial\Phi}{\partial R} , \tag{7}$$

where

$$\Gamma = \frac{\nu_{ee}}{\varepsilon}\left(\rho_{ce}\frac{q}{\varepsilon^{1/2}}\right)^2\left[\varepsilon^{1/2}\frac{dn_e}{dr}\right] = \nu_{ee}\rho_{ce}^2 q^2\varepsilon^{-3/2}\frac{dn_e}{dr} . \tag{8}$$

Since $\partial\Phi/\partial R = 2\pi RB_\theta$ ($B_z \equiv B_\theta$ in the equatorial plane), it follows that

$$\sum \frac{dv_\varphi}{dt} = \frac{\nu_{ee}\varepsilon^{1/2}T_e}{eB_\theta}\frac{dn_e}{dr} , \tag{9}$$

and

$$j_b \approx -\frac{\nu_{ee}}{\nu_{ei}}\frac{\varepsilon^{1/2}T_e}{B_\theta}\frac{dn_e}{dr} \approx -\frac{2}{Z}\frac{\varepsilon^{1/2}T_e}{B_\theta}\frac{dn_e}{dr} . \tag{10}$$

It is our pleasure to thank Prof. F. Engelmann and Prof. D. Pfirsch for helpful discussions during the course of this work.

## REFERENCES

[1] D. J. H. Wort, Plasma Phys. **13**, 258 (1971).

[2] R. J. Bickerton, Comments Plasma Phys. Controlled Fusion **1**, 95 (1972).

[3] W. K. H. Panofsky and M. Phillips, in *Classical Electricity and Magnetism*, Addison-Wesley, London(1962), p.431.

[4] A. A. Ware, Phys. Rev. Lett. **25**, 15 (1970).

[5] J. Wesson, in *Tokamaks*, Clarendon Press, Oxford (1987), p.95.

[6] S. Puri, Nucl. Fusion **27**, 229 (1987); Nucl. Fusion **27**, 1091 (1987).

[7] S. Puri and R. Wilhelm, in *Proc. 16th EPS*, Venice, 1989, p.1315.

[8] S. Puri, in *Proc. 8th IAEA Conf.*, Brussels, 1980 (IAEA, Vienna **2**, 51, 1981).

[9] A. B. Murphy, Plasma Phys. Contr. Fusion, **31**, 21 (1989).

[10] S. Puri, in *Proc. 3rd Topical Conf. on Radio-Frequency Plasma Heating*, Caltech, Pasadena, 1978; Phys. Fluids **22**, 1716 (1979).

[11] G. A. Collins et al., Phys. Fluids **29**, 2260 (1986).

# M=-1 ALFVÉN WAVE BEACH HEATING OF TWO-ION MIRROR PLASMAS

D.R. Roberts, N. Hershkowitz, R.P. Majeski, and D.H. Edgell

University of Wisconsin, Madison, WI 53706

## ABSTRACT

RF heating experiments in the central cell of the Phaedrus-B tandem mirror have utilized a rotating field antenna set to excite $m = -1$ cylindrical eigenmodes for hydrogen cyclotron resonance heating at a magnetic beach. Recent investigations have examined modifications to this heating imposed by the addition of deuterium or helium ion species. In addition to nearly 50% increases in $T_i$, we have observed evolution of the radial heating profiles with $n_M/n_H$ ($M = D$ or $He$). In particular, we discuss the correspondence of our observations with the predicted variations with $n_M/n_H$ in the antenna-wave coupling to the slow ion-cyclotron global and fast magnetosonic surface wave eigenmodes.

## INTRODUCTION

A successful method for heating magnetized plasmas utilizes the resonant coupling of electromagnetic wave energy to ions at the fundamental cyclotron frequency. The effectiveness of this process depends critically on the strength of the wave $E^+$ component, which is polarized in the direction of ion gyration. Among the several classes of waves supported by plasma, the slow mode generally carries this component most strongly. Consequently, the most dramatic heating has been achieved with antennas that couple either directly or indirectly, via mode conversion processes in the plasma gradients, to this resonant wave.

In tokamaks, multiple layers of wave mode conversion and resonance may be localized in the magnetic and plasma density gradients which lie normal to the confining magnetic field. Since the strongest wave damping processes have been associated with the magnetically placed ion-ion hybrid and ion-cyclotron resonances, electromagnetic energy traversing the column from a localized antenna at the edge may thus be strongly absorbed during a single radial pass. In contrast, a mirror geometry is characterized by a lack of significant transverse magnetic gradients and a corresponding absence of strong single-pass absorption mechanisms. The interference of reflected wave energy from the plasma and vacuum chamber boundaries subsequently results in the formation of wave eigenmodes characteristic of the overall system composition and geometry. In mirror devices, the excitation of such eigenmodes has been further encouraged by antenna structures which azimuthally encompass the plasma. Once established, these eigenmodes propagate axially from the antenna until eventually absorbed via damping processes in the graded mirror plasma column.

In the present work, we explore the consequences for mirror ICRF heating which result from the addition of a second (heavy) ion species. Most fundamentally, ion-ion hybrid and heavy ion-cyclotron resonances are introduced into the mirror magnetic gradient and may enhance eigenmode absorption. In addition, a second ion species modifies the dispersion and subsequent antenna coupling efficiency to the various system eigenmodes. As discussed in the remainder of this paper, we have been able to correlate this latter effect with the radial power deposition profiles observed in our experiments.

## CYLINDRICAL PLASMA EIGENMODES

The axial dispersion and radial field structure of cylindrically bounded wave eigenmodes in single ion-species plasma have been examined by several authors[1,2]. Typically, the models employ a cold, vacuum immersed, uniform plasma column of radius, $a$, which is homogeneous and infinite in axial extent. Tangential components of $\vec{E}$ and $\vec{H}$ are required to be continuous at the plasma/vacuum interface, simulating a free boundary. The simultaneous solution of Maxwell's equations then yields the parallel dispersion, $k_\parallel(\omega)$, and radial structure, characterized by $k_\perp$, of the system eigenmodes. For $\omega < \Omega_h$, two classes of Alfvén wave eigenmodes are found, termed fast magnetosonic and slow ion-cyclotron modes, whose local propagation characteristics are in general dominated by the fast and slow waves, respectively, of infinite plasma theory. Notable features of the slow modes include the relative independence of their parallel dispersion on the azimuthal mode number, $m$, as well as their resonant behavior for $\omega \sim \Omega_h$. In contrast, the fast mode dispersion depends critically on $m$, with the lowest radial order modes for $m < 0$ converting at the Alfvén resonance into the slow mode spectrum.

Recently, Akiyama *et.al.*[3] have shown that the addition of a second ion species imposes additional eigenmode cutoffs and resonances at the ion-ion hybrid and heavy ion species cyclotron resonance layers. Figure 1 shows the $m = -1$ eigenmode dispersion diagram for a homogeneous H-He plasma with $a = 15cm, n_e = 5 \times 10^{12} cm^{-3}$, and $n_{He}/n_e = 0.5$. For $\omega < \Omega_H$, a fast magnetosonic (F) as well as discrete sets of slow ion-cyclotron (S) modes are associated with each ion species.

In addition to their parallel phase velocities, the magnetosonic and ion-cyclotron modes are characterized by their radial e-m field distributions. For the magnetosonic mode, the radial profile of the ion heating field component, $E^+$, is described by the Bessel function, $I_0(|k_\perp|r)$ (peaked on edge), while for the slow ion-cyclotron modes, $E^+ \propto J_0(|k_\perp|r)$ (peaked on axis). In these expressions we have used $k_\perp = \sqrt{A^2(\gamma^2 - 1)}$ as a good approximation away from the Alfvén resonance, with $A^2 = k^2 - k_0^2 S$, $\gamma = k_0^2 D/A^2$ and $k_0 = \omega/c$, where $S$ and $D$ are components of the cold plasma dielectric tensor. The consequences of these contrasting eigenmode characteristics are demonstrated in our experimental results presented below.

## EXPERIMENTAL RESULTS

When operated as a simple mirror, the Phaedrus-B plasma is solely sustained and stabilized by RF in the central cell. Previous experiments in a flat central cell magnetic field with $\omega/\Omega_h \sim 0.75$ (far from cyclotron resonance) have demonstrated sufficient electron heating from the antenna near fields to sustain the plasma ($n_e \sim 2 \times 10^{12} cm^{-3}, T_e \sim T_i \sim 40eV$). However, present experiments in a magnetic beach configuration take advantage of wave $E^+$ cyclotron resonance heating while maintaining efficient wave coupling at the antenna where $\omega/\Omega_h < 1$. This scheme has resulted in dramatically enhanced parameters, $n_e \sim 1 \times 10^{13}$ and $T_{i\perp} \sim 500eV$, for the case of pure hydrogen plasmas.

464

Figure 2 illustrates the static magnetic field and antenna geometry in the central cell for the experiments. Appropriate phasing of currents through the two antenna elements selectively excites an $m = -1$ near field structure. In plasma, the near fields extend $\pm 25cm$ axially beyond the antenna array. The near field radial structure in this region determines the local power deposition (mostly into electrons for $\omega/\Omega_h \sim 0.8$) and subsequent radial distribution of coupled wave fields propagating axially into the cyclotron resonance beach. Fuelling for these experiments is accomplished by puffing a controlled mixture of H-He gas on the outboard, antenna side of the central cell.

Despite the broad $k_\parallel$ spectrum ($k_{max} \sim 0.03cm^{-1}$, $k_{-3dB} \sim 0.08cm^{-1}$) of the rotating field antenna, the evolution of the dispersion diagram of figure 1 with $n_{He}/n_e$ suggests a corresponding variation in the relative excitation of eigenmodes supported in the region $\omega/\Omega_h \sim 0.8$ under the antenna. In particular, the axial placement of the Alfvén and ion-ion hybrid resonances with respect to this region limits the range of fast magnetosonic mode support to $0.40 < n_{He}/n_e < 0.75$, while no $m = -1$ eigenmodes are supported for $n_{He}/n_e > 0.85$.

Figure 3 shows the evolution in plasma $T_{i\parallel}$, $T_{e\parallel}$, and $n_e$ radial profiles, as determined from endloss and microwave interferometry, with increasing $n_{He}/n_e$. The evolution of the bulk $T_{i\perp}$ as evaluated from the diamagnetic signal in the beach region is also shown. For the pure hydrogen case, the dominant antenna coupling is to the slow ion-cyclotron modes, albeit at large $k_\parallel$. As the helium concentration is increased, the degradation in bulk plasma and shift in radial profiles from core to edge peaked until $n_{He}/n_e \sim 0.35$ is associated with enhanced surface mode coupling. Such coupling may occur at lower $n_{He}/n_e$ than predicted by the eigenmode dispersion model due to radial density profile effects. Further increasing the helium fraction, the strongest heating and peaking of profiles is observed for $n_{He}/n_e \sim 0.45$. In particular, $T_i$ is enhanced nearly 50% above its pure hydrogen value. We attribute this to improved coupling to low $k_\parallel$ ion-cyclotron modes at this fractional concentration, perhaps associated with surface wave mode conversion through the Alfvén resonance. For still higher concentrations, the heating efficiency degrades until the ion-ion hybrid layer moves inboard of the antenna and the plasma is no longer sustained. We have verified our interpretation of this last feature from the corresponding heating variations in an H-D plasma.

## CONCLUSION AND ACKNOWLEDGEMENTS

The addition of a second ion species in the Phaedrus-B ICRF beach heating configuration has substantially enhanced central cell plasma parameters. Although we acknowledge that the presence of the steep magnetic gradient may further complicate a detailed description of the heating process, this improvement is consistent with the predicted enhanced coupling to slow wave eigenmodes for the range of experimental H-He and H-D ion concentrations.

This work has been supported by U.S. Department of Energy Grant No. DE-FG02-88ER53264. In addition, the primary author is grateful for support by the DOE Magnetic Fusion Energy Technology Fellowship program.

## REFERENCES

[1]F. J. Paolini, Phys. Fluids, **18**, 640-644(1975).
[2]K. Appert, J. Vaclavik, and L. Villard, Phys. Fluids, **27**, 432-437(1984).
[3]H. Akiyama, M.O. Hagler, and M. Kristiansen, IEEE Trans. Plasma Sci., **13**, 64(1984).

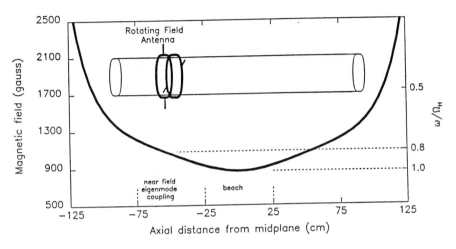

Figure **2**: Central cell beach heating configuration

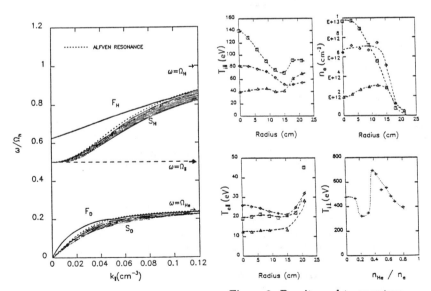

Figure 1: Eigenmode parallel dispersion for $n_{He}/n_e = 0.5$

Figure 3: Density and temperature profiles for $n_{He}/n_e = 0(\diamond)$, 0.25($\triangle$), and 0.45($\square$).

## CAPACITIVE PROBING OF PLASMA AND SURFACE
## POTENTIALS IN A MAGNETICALLY ENHANCED rf DISCHARGE

S. E. Savas and K. G. Donohoe
Applied Materials, Santa Clara, CA 95054

### ABSTRACT

A capacitive probe is used in a magnetically enhanced rf discharge to measure spatial and parametric dependences of the amplitude of the rf fundamental of the plasma and electrode potentials. The probe has a metal "paddle" tip that can be either immersed in the plasma or positioned flat on the exposed surface. The magnetically enhanced discharge uses capacitive coupled power at 13.56 MHz in the range of 1 $W/cm^2$ to 3 $W/cm^2$; $N_2$ pressure in the range between 10 mTorr and 125 mTorr, and magnetic field up to 100 Gauss. The rf amplitude of the plasma potential is found to have a large gradient within the electrode sheath (>1000 V/cm) and small variation (< 10 V/cm) in the glow region. The variation along B in the glow is found to be much smaller than that perpendicular to B. The magnitude of the gradient perpendicular to B increases with both gas pressure and magnetic field strength, consistent with our simple plasma current model.

### INTRODUCTION

The potential drops at the electrode and walls in rf plasmas used for semiconductor processing are very important since they determine damage to and contamination of devices[1]. These potentials have been hard to measure, however, in halogenated gas process plasmas. Furthermore, previous probes for measuring the plasma potential have not been capable of measuring electrode surface potentials or had sufficient resolution to measure spatial dependencies in thin sheaths[2].

rf magnetron discharges are known to have higher plasma densities and lower discharge voltages than unmagnetized capacitive rf discharges[3]. In order to verify the reduced sheath potentials and understand the rf potentials and currents in the plasma we have used these probes to measure the spatial and parametric dependences of the plasma and electrode potentials in our magnetically enhanced rf discharge.

### THE PROBE DESIGN AND DISCHARGE CHARACTERISTICS

The capacitive probes (See Fig. 1) were specifically designed for use in this process plasma[4]. They have quartz bodies, 9 mm in diameter, within which are standard ~50 coaxial cable (RG 59), with stainless steel paddle tips spot welded to stainless steel studs which are potted into recessed area at the tip. They are used with an rf spectrum analyzer, but have no need of auxiliary circuitry as in previous designs such as Ref 3. They have little problem with rf noise at 13.56 MHz.

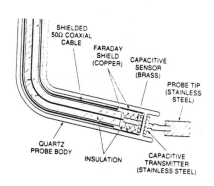

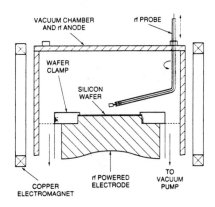

Fig. 1. Capacitive Probe.     Fig. 2. The Plasma Chamber

The plasma process chamber is shown in Fig. 2. It has a 33 cm inner diameter with the electrode having a diameter of 25 cm. The gap between the electrode and the top lid of the chamber is approximately 8 cm. There are four electromagnets at the side of the chamber which have phased currents so that the direction of the up to 100 Gauss magnetic field rotates in the plane of the electrode surface. The discharge operates with gas pressures from 10 mTorr to 125 mTorr, magnetic field from 0 Gauss to 100 Gauss and rf power at 13.56 MHz from 200 W to 800 W. Plasma densities have been measured in this and in similar systems[3] to range from $10^{10} \text{cm}^{-3}$ to $4.10^{10} \text{cm}^{-3}$.

## ELECTRICAL POTENTIALS IN CAPACITIVE rf PLASMAS

The rf potentials of probes, electrodes and plasmas in capacitive rf discharges have been the subject of experimental and theoretical work since the 1950's. Work by Garscadden[5], Koenig[6] and Coburn[7] has shown that for rf frequency of order 10 MHz or higher the dominant impedances in the discharge circuit are at the sheaths and these behave nearly as capacitors. In this case, simple models of rf potentials based on capacitive division of the discharge voltage may not be too inaccurate. If it is applied to our plasma a plot can be made of the time dependence of the various potentials (see Fig. 3). When immersed in this plasma, the tip of the rf probe floats at a nearly constant offset with respect to the plasma potential so long as the impedance to ground through the sensor is much greater than the sheath impedance of the tip. Experiments and simple calculations assure us this is the case.

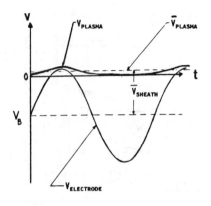

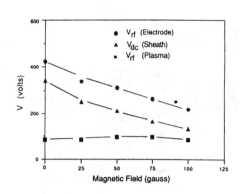

Fig.3.Electrode and plasma
potentials versus time.

Fig.4. Plasma and electrode rf
potentials and d.c.sheath drop.

## MEASUREMENTS OF PLASMA AND SHEATHS POTENTIALS

In Figure 4 the values of plasma , $V_p$, and electrode potentials, $V_e$, are shown along with the approximate d.c. sheath potential, $V_s = V_e - V_p$, as functions of the magnetic field strength. These data were taken at 400 Watts and 100 mTorr. They show the effect of the magnetic field is primarily to reduce the electrode rf voltage rather than to change the plasma potential. In Figure 5 the dependence of the sheath potential on the gas pressure is shown. Note how much more dependent the sheath drop is on gas pressure at zero magnetic field than at a field of 50 Gauss. This is not inconsistent with the reduction in the cross field electron mobility at these pressures and magnetic fields. the gyrofrequency at 100 Gauss is about $1.8,10^9 s^{-1}$, and the collision frequency, $_m$, for momentum transfer is $6.10^8 s^{-1}$ at 100 mTorr.

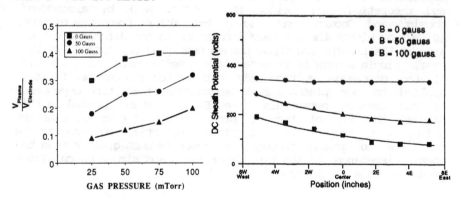

Fig.5. Sheath drop vs pressure.

Fig.6. Sheath drop vs position
perpendicular to B.

The spatial dependence of the sheath potential along a diameter perpendicular to the magnetic field is shown in Fig. 6. The increase in gradient with increasing magnetic field and pressure is clear, through the average potential decreases as a function of either.

Finally, we show in Fig. 7 the dependence of the rf potential amplitude at 13.56 MHz on the distance from the electrode surface. The notable thing in this data is the very clear edge of the sheath about 5 mm from the electrode surface. The electric field in the sheath region may be approximately inferred by taking twice the slope of the curve. Thus, E ranges from about a hundred volts per centimeter to as much as 5000 $V/_{cm}$ at the electrode surface. In the glow region E is of order $5V/_{cm}$.

In order to see if the above estimated electric field in the glow region is consistent with the classical electron cross-field mobility we have calculated the rf current density as a function of E and B. The expression uses the approximation that the polarization drift is negligible which is equivalent to $\omega_e \gg \omega_{rf}$.

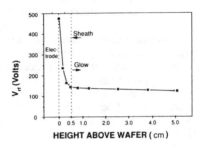

Fig.7. rf amplitude versus distance from the electrode.

One can easily derive the following relation between fields and currents:

$$(1) \quad j_{rf} = e n_e E_{rf} (\mu_\perp^2 + 1/B^2)^{1/2}$$

where $\mu_\perp$ is the cross field mobility, $\mu_0 (1 + (\Omega_e/\nu_m)^2)^{-1/2}$

REFERENCES

(1) J.E.Greene,C.R.C.Crit.Rev.Solid State Mafer.Sci.II,47(1983).
(2) for example, Neil Benjamin, Rev.Sci.Instrum.53, 1541 (1982).
(3) Ch. Steinbruchel et al, IEEE Trans.Plasma Sci. PS14, 137 (1986).
(4) S.E.Savas and K.G.Donohoe, submitted to Rev.Sci.Instrum.
(5) A.Garscadden and K.G.Emeleus,Proc.Phys.Soc.London, 79, 535 (1962).
(6) H.R. Koenig and L.I. Maissel, I.B.M.J. Res. Devel. 14, 168 (1970).
(7) J.W. Coburn and E. Kay, J. Appl. Phys. 43, 4965 (1972) and K. Kohler, J.W. Coburn et al J,. Appl Phys. 58, 3350 (1985).

470

# ELECTRODE SURFACE rf HARMONICS GENERATED BY THE NONLINEAR SHEATH IN A COAXIAL CAPACITIVE rf DISCHARGE

Stephen E. Savas
Applied Materials, Santa Clara, CA 95054

## ABSTRACT

rf harmonics of the 13.56 MHz excitation signal have been measured on the electrode surface in a large coaxial capacitive discharge. These are seen to have from 10% of the fundamental amplitude for the second harmonic to between 1% and 4% for the third and fourth harmonics. There is evidence that these modes propagate as TEM surface waves (Gould-Trivelpiece modes) along the length of the electrode. The Telegrapher's equations can be written for the system with non-constant shunt capacitance and admittance. The resulting nonlinear equation for the sheath voltage is solved for the harmonics to yield approximate agreement with their observed magnitudes.

## INTRODUCTION

Measurements have shown the presence of rf harmonics (up to 10th) on the power input lead to the electrode in capacitive rf plasmas used for semiconductor processing[1-3]. These harmonics only appear when the plasma is turned on. In small parallel plate systems used for etching it has been impractical to determine the source of the harmonics. In our system, however, the large – 65 cm tall, 32 cm wide – hexagonal electrode is coaxial with a 72 cm diam. cylindrical metal vacuum vessel, (see Fig.1) this permits good access for our capacitive, high impedance probes[4] to touch the electrode surface. These probes (see Fig. 2) allow the amplitudes of the fundamental and harmonics to be measured at three points on the electrode surface, nearly equally spaced along its length.

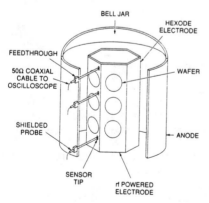

Fig. 1. Cylindrical Plasma Chamber.

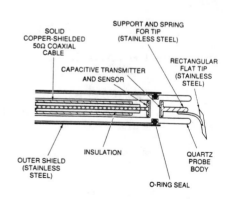

Fig.2. Shielded Capacitive Probes.

## EXPERIMENTAL RESULTS

The results of these measurements are shown in Figure 3 for plasma with and without a large capacity termination. They show consistent patterns in the maxima and minima of the amplitude of each mode as determined with an rf spectrum analyzer. The second harmonic is always about an order of magnitude smaller than the fundamental, while the third and fourth harmonics are several times smaller than the second. The magnitude of the variation of the amplitudes along the electrode's length increases with increasing frequency. The fourth harmonic seems to have its maximum at the top of the electrode and a minimum, which is several times smaller, between the middle and bottom points. This contrasts with the fundamental which decreases by about 10% from the electrode top to its bottom. This is what one would expect if the wavelengths of the modes are in nearly the same proportion, but about three times smaller than their free space values. Finally, it is notable that while the fundamental and fourth harmonic have their maximum amplitudes at the top of the electrode, the second and third harmonics' peak magnitudes are at the bottom of the electrode where the rf power is fed in.

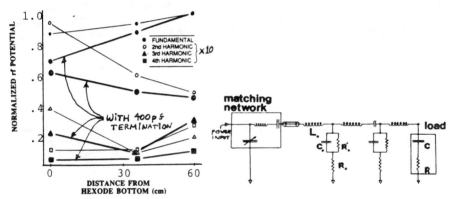

Fig.3. Spatial dependence of amplitudes of Fundamental and harmonics.

Fig.4. Electrode-plasma equivalent circuit, transmission line model.

## CIRCUIT ANALOG OF THE ELECTRODE SURFACE MODES

In order to understand the structure of the modes' amplitudes and the generation of the harmonics we could model the system as a coaxial transmission line or an antenna immersed in plasma[5]. The surface modes are TEM in which the radial electric field is essentially confined to the sheath with a thickness d while the azimuthal magnetic field penetrates well into the glow region with a resistive skin depth,        . These are Gould-Trivelpiece modes[6] with phase velocities less than c by a factor of        , d in these plasmas is about 1.5 cm and        varies from about 10 to 15 cm.

Using the transmission line analogy (see Fig.4 for the circuit) the impedance along the electrode is assumed to be a pure constant inductance, while the predominantly capacitive shunt impedance (across the sheath) is modeled as voltage dependent. The modified Telegrapher's equations can be written, and then combined to yield a non-linear PDE for the sheath voltage, $V_1$. The modified Telegrapher's equations are:

(1) $\quad \dfrac{\partial V_1}{\partial x} = - \dfrac{\partial I}{\partial t} \cdot L - \dfrac{\partial^2 I}{\partial x} \cdot R$

(2) $\quad \dfrac{\partial I}{\partial x} = - \dfrac{\partial V_1}{\partial t} \cdot C(V_1) - e \, j_e \exp(e V_1 / k T_e)$

Where the sheath voltage, $V_1$ is found from:

(3) $\quad V_1 \equiv V - \dfrac{\partial I}{\partial x} \cdot R$

We denote electrode voltage as V, the current as I, electrode inductance per unit length as L, shunt capacitance per unit length across the sheath as C and glow resistance per unit length as R. $T_e$ is the electron temperature and $j_e$ the electron flux, $(1/4) n_e v_e$ where $n_e$ is the density and $v_e$ the thermal speed. The equation which results for $V_1$, is:

(4) $\quad \dfrac{\partial^2 V_1}{\partial x^2} = L \cdot \left\{ C(V_1) \dfrac{\partial^2 V_1}{\partial t^2} + \dfrac{\partial C(V_1)}{\partial V_1} \cdot \left( \dfrac{\partial V_1}{\partial t} \right)^2 - \dfrac{e^2}{k T_e} j_e \exp\left(e V_1 / k T_e\right) \right\}$

$\qquad - R \dfrac{\partial^3 I}{\partial x^3}$

In order to solve equation (4) the sheath voltage is assumed to be a linear combination of the fundamental and harmonics, each with components propagating in both directions on the electrode.

(5) $\quad V_1(x,t) = \displaystyle\sum_{i=1}^{4} V_{1i}^{+} e^{i(\omega_i t - k_i x)} + V_{1i}^{-} e^{i(\omega_i t + k_i x)}$

Where the coefficients of each mode are related by the reflection coefficients which are calculated from the impedance of the termination at the top of the electrode, $Z_1$, and the characteristic impedance of the electrode-plasma system, $Z_p$.

(6) $\quad \left( V_{1i}^{-} / V_{1i}^{+} \right)_{x_T} = \left( Z_{Li} - Z_{oi} \right) / \left( Z_{Li} + Z_{oi} \right)$

where $x_T$ is the position of the end of the electrode. This is reasonable since the non-linear coupling terms on the right of equation (4) are negligible for much of the rf cycle since $C(V_1)$ is only weakly dependent on $V_1$ unless $V_1$ is less than $10 k T_e / e$. The admittance, which is proportional to the $\exp(e V_1 / k T_e)$, is essentially zero unless $e V_1 / k T_e$ is more than $-5$. When $e V_1 / k T_e$ is greater than $-10$ the right hand side of equation (4) becomes non-negligible. In order to calculate this we have assumed a specific form for $C(V_1)$:

$$(7) \quad C(V_1) = V_1 \cdot C_1 + C_0$$

where s is constant of order 10 which yields a maximum of $C(V_1)$ about an order of magnitude greater than $C_0$. Using this form for C equation (4) is first solved for the second harmonic by using only the fundamental for $V_1(t)$ in the terms on the right hand side. These terms, including all terms with first power time or space derivatives of $V_1$ contribute to the power flow to the harmonic when the sheath voltage is small. The right hand side of equation (4) thus reduces to:

$$(8) \quad L C_0 \frac{\partial^2 V_1}{\partial t^2} - L C_1 \frac{\partial^2 V_1}{\partial t^2} - \frac{e^2}{kT_e} j_e \, exp(eV_1/kT_e) - R \left\{ C_0 \frac{\partial^3 V_1}{\partial^2 x \partial t} + \right.$$

$$C_1 \frac{\partial V_1}{\partial t} \frac{\partial^2 V_1}{\partial x^2} + 2C_1 \frac{\partial^2 V_1}{\partial x \partial t} \cdot \frac{\partial V_1}{\partial x} + \frac{e^2}{kT_e} j_e \, exp(eV_1/kT_e) \cdot \frac{\partial^2 V_1}{\partial x^2} \left. \right\}$$

The resulting magnitudes for second, third and fourth harmonics in the solution of equation (4) are of the correct magnitude for physically reasonable values of s (i.e. about 5), but the spatial variations of the modes based on the reflection coefficients in equation (6) are not as observed. A numerical treatment of equation (4) will be necessary, we believe, to account for spatial profiles but we are encouraged that the basic model incorporates the essential physics of the phenomenon.

### REFERENCES

1. W. G. M. van den Hoek, C. A. M. De Vries and M. G. J. Heijman, J.V.S.T., B5, 647 (1987).
2. K. R. Stalder, private communication.
3. S. E. Savas and R. W. Plavidal, J.V.S.T. A6, 1775 (1988).
4. S. E. Savas and K. G. Donohoe, submitted to Review of Scientific Instruments.
5. S-H. Lin and K. K. Mei, IEEE Trans. on Antennas and Propag. Vol AP-18, 672 (1970).
6. A.W.Trivelpiece and R.W.Gould, J.Appl.Phys, 30, 1784 (1959).

# Emission of Ion and Electron Cyclotron Harmonic Radiation from Mode Conversion Layers

D. G. Swanson and Suwon Cho

Department of Physics, Auburn University, Alabama 36849

## Abstract

The asymmetry of cyclotron radiation from a mode conversion layer is presented for harmonics of the ion cyclotron frequency and the second harmonic of the electron cyclotron frequency for weakly relativistic electrons. The same form of Kirchhoff's law is found for all cases, relating the emission along each branch to the absorption of an incident wave along the corresponding branch. Results show that the fast wave radiation is more strongly asymmetric at the third harmonic than at the second harmonic of the ion cyclotron frequency, while the slow wave radiation ratio is about same. At the second cyclotron harmonic of weakly relativistic electrons, the asymmetry of radiation is found to be small at high temperature. The effect of equilibrium Bernstein wave radiation is also discussed.

## I.   Introduction

Whenever the external magnetic field is inhomogeneous so that mode conversion and cyclotron harmonic absorption may occur, there exist asymmetries between emission of radiation along the various plasma wave branches and those emission ratios are related to absorption ratios of corresponding driven wave problems via a form of Kirchhoff's law.[1] The primary results at the second harmonic of the ion cyclotron frequency include: (i) The slow wave radiates most of the energy when the tunneling layer is thin. (ii) More of the radiation is carried out by the fast wave on the low magnetic field side of the cyclotron radiation layer than by the fast wave on the high field side regardless of the tunneling thickness. The net asymmetry of the fast wave radiation may be less than that estimated from cyclotron radiation alone in actual experiments, however, since other radiation mechanisms exist. The slow Bernstein wave, which radiates away from the cyclotron harmonic layer, is usually absorbed before it reaches the plasma boundary. This means that it is also emitted by a thermal source which may be remote from the cyclotron layer. The thermally emitted Bernstein wave then is partially converted into fast waves through mode conversion. This effect decreases the asymmetry since the high field side fast wave conversion coefficient is usually larger than the low field side coefficient.

We have examined a broad class of mode conversion problems to show that the form of Kirchhoff's law is a general result. This includes the analysis of the second and third ion cyclotron harmonic and the electron cyclotron harmonic, which employs the same basic equation but with $\gamma < -1$ (which modifies the coupling of solutions in a nontrivial way). Since ion cyclotron harmonic emission has been proposed as a diagnostic for alpha particles which are born with high energy as fusion products, the third ion cyclotron harmonic emission may be important for alpha particle diagnostics. The electron cyclotron harmonic emission from weakly relativistic electrons is also important, because ECE is commonly used as a plasma temperature diagnostic.[2,3]

## II.   Kirchhoff's Law for Mode Conversion Layers

Using the second ion cyclotron harmonic as an example of the method, we begin with the mode conversion-tunneling equation and its adjoint equation in normalized variables[4]

$$f^{iv} + \lambda^2 z f'' + (\lambda^2 z + \gamma)f = 0, \tag{1}$$

$$F^{iv} + \lambda^2 z F'' + 2\lambda^2 F' + (\lambda^2 z + \gamma)F = 0. \tag{2}$$

The solutions are labeled such that $f_1$ is the solution with an incoming fast wave from $+\infty$, $f_2$ has an incident fast wave from $-\infty$, and $f_3$ has an incident slow wave. Both emission and absorption are labeled according to these branches. The asymptotic behavior of these solutions is summarized in Table I, where $\eta = \frac{\pi}{2}\frac{1+\gamma}{\lambda^2}$ and analytic expressions for the coefficients are available for the case with no absorption.[4] The adjoint solutions have exactly the same asymptotic relations as $f_k$ so that the lower-case letters can be replaced by capitals in Table I.

| $-\infty \longleftarrow z$ | $\psi$ | $z \longrightarrow +\infty$ |
|---|---|---|
| $T_1 f_- + D_1 \sigma_-$ | $e^{-\eta}\psi_1$ | $f_- + R_1 f_+ + C_1 s_-$ |
| $f_+ + R_2 f_- + D_2 \sigma_-$ | $-\psi_2$ | $T_2 f_+ + C_2 s_-$ |
| $C_3^- f_- + D_3 \sigma_-$ | $-e^{-\eta}\psi_3$ | $s_+ + R_3 s_- + C_3^+ f_+$ |
| $\sigma_+ + C_4^- f_- + D_4 \sigma_-$ | $-e^{\eta}\psi_4$ | $T_4 s_- + C_4^+ f_+$ |

Table 1: Asymptotic behavior of solutions of Eq.(7).

With absorption effects included for $\omega \sim 2\omega_{ci}$, Eq. (1) changes to[4]

$$\psi^{iv} + \lambda^2 z \psi'' + (\lambda^2 z + \gamma)\psi = h(z)(\psi'' + \psi) \qquad (3)$$

with $h(z) = \lambda^2(z - z_a)[1 + 1/\zeta Z(\zeta)]$, where $Z$ is the plasma dispersion function, $\zeta = (z_a - z)/\kappa$, $z_a$ is the resonant absorption point, and $\kappa$ is the normalized $k_\parallel$. The solution of Eq. (3) is given as $\psi_k(z) = \int_{-\infty}^{+\infty} G_k(z,y)h(y)\,dy$, where $G_k(z,y)$ is the Green function for solution $\psi_k$, given by[5]

$$2\pi i \lambda^2 \varepsilon G_k(z,y) = \mu_k F_k(y) f_k(z) + \begin{cases} F_1(y)f_2(z) + \varepsilon F_0(y)f_4(z) & \text{for } y < z \\ F_2(y)f_1(z) + \varepsilon F_4(y)f_0(z) & \text{for } z < y \end{cases} \qquad (4)$$

where $\mu_k$ is a normalizing constant which will be chosen such that $\psi_k \to f_k$ as $h(z) \to 0$, and $f_0 \equiv f_3 - f_1$, $F_0 \equiv F_3 - F_1$, and $\varepsilon = 1 - e^{-2\eta}$. For small $\kappa$, $h(z)$ may be approximated by a delta function so that Eq. (3) reduces to

$$\psi^{iv} + \lambda^2 z \psi'' + (\lambda^2 z + \gamma)\psi = 2\pi i \varepsilon \lambda^2 \alpha (\psi'' + \psi)\delta(z - z_a) \qquad (5)$$

where the constant multiplier $\alpha$ is arbitrarily chosen. Solutions are

$$\psi_k = f_k + \alpha(\psi_{ka} + \psi_{ka}'') \begin{cases} F_{1a}f_2(z) + \varepsilon F_{0a}f_4(z) & \text{for } z > z_a \\ F_{2a}f_1(z) + \varepsilon F_{4a}f_0(z) & \text{for } z < z_a, \end{cases} \qquad (6)$$

with $F_{ja} \equiv F_j(z_a)$. Differentiating Eq. (6) twice and adding it to itself, we find $\psi_{ka} + \psi_{ka}'' = F_{ka}/[1 - \alpha(F_{1a}F_{2a} + \varepsilon F_{0a}F_{4a})]$ as well as

$$\psi_k = f_k + \begin{cases} H_{k1}f_2(z) + \varepsilon H_{k0}f_4(z) & \text{for } z > z_a \\ H_{k2}f_1(z) + \varepsilon H_{k4}f_0(z) & \text{for } z < z_a \end{cases} \qquad (7)$$

where $H_{kj} = \alpha F_{ka}F_{ja}/[1 - \alpha(F_{1a}F_{2a} + \varepsilon F_{0a}F_{4a})]$. The asymptotic behavior of the solutions is summarized in Table I.

From the conservation of energy, the absorbed fraction for each incident wave solution is

$$\begin{aligned} A_1 &= 1 - |T_1|^2 - |R_1|^2 - |C_1|^2/\varepsilon = e^{-2\eta}|F_1|^2 \chi \\ A_2 &= 1 - |T_2|^2 - |R_2|^2 - |C_2|^2/\varepsilon = |F_2|^2 \chi \\ A_3 &= 1 - |R_3|^2 - \varepsilon|C_3^+|^2 - \varepsilon|C_3^-|^2 = \varepsilon e^{-2\eta}|F_3|^2 \chi \end{aligned} \qquad (8)$$

where $\chi = 2\beta_r - |\beta|^2(e^{-2\eta}|F_1|^2 + |F_2|^2 + \varepsilon e^{-2\eta}|F_3|^2)$ and $\beta = \beta_r + i\beta_i = [1 - \alpha(F_{1a}F_{2a} + \varepsilon F_{0a}F_{4a})]^{-1}$. In deriving Eq. (8), we have used $F_1^* = -F_2 + \varepsilon F_3$, $F_2^* = -e^{-2\eta}F_1 - \varepsilon F_2 - \varepsilon e^{-2\eta}F_3$, and $F_3^* = F_1 - F_2 - e^{-2\eta}F_3$ which can be obtained by comparing asymptotic relations.

For a point source located at $z = z_e$, the emission solution can be obtained by the same method as for the absorption problem except that incoming waves are not permitted. The result is obtained from the discontinuity terms (excluding $f_k$) in Eq. (4);

$$\psi_e = \begin{cases} F_2(z_e)f_2(z) + \varepsilon F_0(z_e)f_4(z) & \text{for } z > z_e \\ F_1(z_e)f_1(z) + \varepsilon F_4(z_e)f_0(z) & \text{for } z < z_e. \end{cases} \tag{9}$$

In terms of the asymptotic representations,

$$\psi_e \rightarrow \begin{cases} F_2(z_e)f_-(z) + \varepsilon F_4(z_e)e^\eta \sigma_-(z) & \text{as} & z \rightarrow -\infty \\ -F_1(z_e)e^{-\eta}f_+(z) + \varepsilon F_3(z_e)e^{-\eta}s_-(z) & \text{as} & z \rightarrow \infty. \end{cases} \tag{10}$$

Then ratios of emitted power along various branches are

$$\frac{E_1}{E_2} = e^{-2\eta}\left|\frac{F_1(z_e)}{F_2(z_e)}\right|^2, \qquad \frac{E_3}{E_2} = e^{-2\eta}\varepsilon\left|\frac{F_3(z_e)}{F_2(z_e)}\right|^2, \tag{11}$$

where $E_1$ and $E_2$ are the emitted powers along the fast wave branches on the high field side ($z > 0$) and the low field side respectively, and $E_3$ is the power along the slow wave branch on the high field side and it is multiplied by the relative power flux ratio $1/\varepsilon$ of the slow wave to the fast wave. Comparing Eq. (8) with Eq. (11), we discover the relations, $A_1/A_2 = E_1/E_2$ and $A_3/A_2 = E_3/E_2$, which are identities as long as $z_e = z_a$ regardless of the value of $\alpha$. This leads to a form of Kirchhoff's law for a mode conversion layer

$$\frac{A_1}{E_1} = \frac{A_2}{E_2} = \frac{A_3}{E_3}, \tag{12}$$

so the ratio of the emitted power to the fraction of power absorbed is the same for all branches of waves whenever absorption and emission occur at the same point. The same relations have been proved for the third ion cyclotron harmonic and the second electron cyclotron harmonic.[6]

## III.  Results and Conclusions

$E_1/E_2$ for the second harmonic is plotted in Fig. 1a which shows strong asymmetry of emission the two fast wave branches. Similar plots of $E_1/E_2$ are shown for the third ion cyclotron harmonic in Fig. 1b and for the second electron cyclotron harmonic in Fig. 1c. In each ion case, the delta function result is the solid line, and the other case has $k_\parallel \neq 0$ and is calculated *assuming that the coherent absorption model is valid for distributed sources.* Compared with the result at the second harmonic, $E_1/E_2$ is smaller at the third ion harmonic, i.e., the fast wave radiation is more highly asymmetric, while $E_3/E_2$ remains about the same as in the second harmonic case.[6] In fact, the coherent absorption model may not represent accurately the incoherent radiation from a distributed source, but we show it as a first approximation. In each case, the slow wave radiates most of the energy when the tunneling layer is thin, and more of the radiation is carried out by the fast wave on the low field side of the magnetic field than by the fast wave on the high field side regardless of the tunneling thickness.

It is likely that in actual experiments, the observed asymmetry will be less than that shown, since one never sees only cyclotron emission. When the slow Bernstein wave emitted from the cyclotron layer is absorbed at a remote location in the plasma (presumably Landau damping), we assume that in equilibrium an equivalent amount of energy is radiated back along the Bernstein branch toward the mode conversion layer. If the remote absorption/emission region is at a different temperature from the cyclotron harmonic layer, or there is a nonthermal source of Bernstein waves, then the outgoing and incoming slow wave energy may not balance.

When the Bernstein wave is assumed to be in equilibrium so that the incident power is to equal to the net emitted power, we set up the relation

$$I_3 = I_3 R_3 + E_3 = I_3 (R_3 + \xi A_3) \tag{13}$$

to get $\xi = (1 - R_3)/A_3$ where $\xi = E_j/A_j$, and $I_3$ is the incident power flux of the Bernstein wave. Then the emission ratio $E_1/E_2$ is given by

$$\frac{E_1}{E_2} = \frac{C_3^+ A_3 + (1 - R_3) A_1}{C_3^- A_3 + (1 - R_3) A_2} \tag{14}$$

which is shown in Fig. 1a for the second harmonic and in Fig. 1b for the third harmonic, where the two upper curves relate to $E_1/E_2$ with the incident equilibrium Bernstein wave and the two lower curves relate to the case without any incident slow wave for comparison. Fig. 1c also shows the effects of the equilibrium Bernstein wave in the upper two curves for two different temperatures. The asymmetry of the fast wave radiation is reduced in each case, and is completely eliminated as $\eta \to 0$ as it should, since in the zero tunneling thickness limit, the uniform plasma radiation model is recovered.

From the results which indicate a remote source of Bernstein waves, it appears feasible to distinguish between nonthermal cyclotron emission and nonthermal Bernstein wave emission. This is due to the fact that cyclotron emission and mode converted Bernstein wave emission are asymmetric in opposite senses. By measuring both the magnitude and the asymmetry ratio, it may be possible to unravel the contributions from the thermal and nonthermal emissions and identify whether the nonthermal source is related to cyclotron emission (fast wave process) or to Bernstein wave emission (an electrostatic wave process).

This work has been supported by the U. S. DOE under contract DE–FG05–85–ER–53206D.

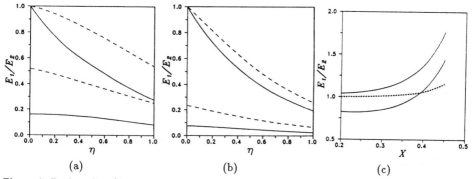

(a)  (b)  (c)

Figure 1: Ratios of $E_1/E_2$ for a) $\omega \sim 2\omega_{ci}$, b) $\omega \sim 3\omega_{ci}$, c) $\omega \sim 2\omega_{ce}$. In each case, lower curves include cyclotron emission only and upper curves include Bernstein emission. For a) and b), — is the delta function case and - - - is with $\kappa = 1$. For c), — is .5 keV, - - - is 1.5 keV.

## References

[1] D. G. Swanson and Suwon Cho, Phys. Rev. Lett. (to be published).

[2] John Wesson, **Tokamaks**, (Clarendon Press, Oxford, 1987), Chapter 10.

[3] M. Bornatici, R. Cano, O. De Barbieri, and F. Engelmann, Nucl. Fusion **23**, 1153(1983).

[4] D. G. Swanson, **Plasma Waves**, (Academic Press, Inc., Boston, 1989), Chapter 6.

[5] D. G. Swanson, Phys. Fluids **21**, 926 (1978).

[6] Suwon Cho and D. G. Swanson, Submitted to Phys. Fluids.

On the deterioration and Improvement of on Plasma Confinement
During Additional Heating in Magnetic Fusion Traps

K.Uehara,O.Naito,M.Seki and K.Hoshino

Naka Fusion Establishment,Japan Atomic Energy Research Institute,
Naka,Naka,Ibaraki,Japan

A simple physical picture indicates that the confinement time decreasing with the heating power should be deduced to the Bohm like diffusion and improving with plasma current in tokamaks should be deduced to the characteristics of Z pinch. The travelling type wave heating has a possibility to improve the confinement by the inward $\vec{E}_{rf} \times \vec{B}$ drift due to resonant electrons against the diffusive bulk electrons.

1 Introduction

A new trouble is appearing in a magnetic fusion research when the additional heating power increases ,that is,the more additional heating power $P_{IN}$ is injected,the less energy confinement time $\tau_E$ is obtained such as Kaye-Goldston scaling.[1] If we obey the two body collisional theory,$\tau_E$ must be improved with the plasma temperature and must decrease with plasma density n and in the result $\tau_E$ must be improved with $P_{IN}$ except for including trapped particle effect.

In this paper,we present a simple physical picture of the deterioration and improvement of the plasma confinement during the additonal heating in magnetic fusion traps,whereas we abandan the explanation due to the wave turbulence and the trapped particle instability.[2]

2 Experimental Survey

The empirical scaling of $\tau_E$ has a power dependence of $\tau_E \propto P_{IN}^{-0.58}$ One of direct ways to overcome this problem is to enhance the plasma current $I_p$ in torus and the minor radius. Figure 1 shows $\tau_E$ against $I_p$ in JT-60 case,in which the additional heating power is NBI.[3] Flowing plasma current in torus may at least improve the confinement rather than deteriorating confinement due to enhancing fluctuations. On the other hand,$\tau_E$ is often improved near to the Joule heated plasma in the lower hybrid current drive (LHCD) at relatively lower density region even when the additional heating[4] power is injected into torus. In JT-60,[5] the improved confinement time is observed during the combination experiment of LHCD and neutral beam injection (NBI) as shown in Fig.2.

3 Simple Physical picture
(1) Power balance and Bohm diffusion

To get a simple physical picture on the confinement charac- teristics the zeroth order approximation must be initially performed. An achieved plasma temperature during the additional heating is obtained by the power balance of the input and loss power, which is expressed by

$$\frac{d}{dt}\int \frac{3}{2} n(T_e + T_i) dV = P_{IN} - P_{RX} - \frac{1}{\tau_E}\int \frac{3}{2} n(T_e + T_i) dV , \quad (1)$$

, where $P_{RX}$ is the sum of the radiation loss and the charge excahnge loss. In the stationary state $(d/dt=0)$, we get

$$\int \frac{3}{2} n(T_e + T_i) dV = (P_{IN} - P_{RX})\tau_E . \quad (2)$$

Here we asuume that $\tau_E$ is replaced by the particle confinement time as $\tau_E = a_p^2/D$. When we assume $T_e = T_i$ and $\int dV = \pi a^2 2\pi R = 2\pi^2 a^2 R$, where R is major radius in torus, we can get $\tau_E$ corresponding to various diffu- sion process if we substitute each type diffusion coefficient into D. We can easily find that $\tau_E$ in Bohm type diffusion ( $D = D_B = kT_e/16\delta eB$ , where k is Boltsman constant, e is electronic charge, B is magnetic field and $\delta$ is anumerical factor) shows a scaling as $\tau_E \propto P_{IN}^{-0.5}$ .

(2) Pinch effect

Here, let's consider simply the reason why $\tau_E$ is improved by $I_p$ in tokamaks is the pinch effect due to $E_{\emptyset} \times B_{\theta}$ drift, where $E_{\emptyset}$ is an inductive electric field and $B_{\theta}$ is poloidal magnetic field. In tokamak discharge, $E_{\emptyset}$ is generated in plasma and all electrons feel the force $F = eE_{\emptyset}$ which is caused as as anti-reaction of the external momentum transfer to flow the plasma current. The resulting $\vec{F} \times \vec{B}/eB^2$ drift confines higher temparature plasma to reduce the anomolous cross field diffusion in the following

$$\Gamma_{out} - \Gamma_{pinch} = -D_B \frac{\partial n_1}{\partial r} - \frac{n E_{\emptyset} B_{\theta}}{B^2} = -D_{OH}\frac{\partial n_2}{\partial r} \quad (3)$$

where $B_t$ is toroidal magnetic fieldi and $B = \sqrt{B_t^2 + B_{\theta}^2}$. The confine- ment time is obtained substituting eq.(3) into D, that is , $\tau_E \propto n^{0.5} R^{0.5} a_p^2 P_{IN}^{-0.5}$ ( $1 + b I_p$ ), where we assume $P_{IN} \gg P_{RX}$ , and b is a numerical factor. We cannot neglect the pinch term $\Gamma_{pinch}$ if we compare at so many points in the radial direction where $\partial n/\partial r$ becomes small.

When $T_e$ increases by additional heating the plasma resistance decreases and $E_Q$ becomes small with $T_e$. This reduction of $E_Q$ in eq.(3) may weaken the pinch effect in tokamaks and finally we can reach the scaling as $\tau_E \propto P_{IN}^{-0.58}$. An example of calculation is shown in Fig.3,where the empirical data on $T_e$ and $\bar{n}_e,I_p$ are used.

In LHCD the Z pinch characteristics must also exist since the wave transfers momentum to plasmas same as joule heated plasma. In this case $n$ and $E_Q$ in eq.(3) are replaced by resonant electrons density $n_{res}$ and time averaged rf electric field in the toroidal direction $E_{rf}^{\theta}$,respectively. Typical example in the LHCD case is calculated corresponding to Fig.3 as shown in Fig.4. In this calculation,neo-classical prediction is assumed for ions and the anomalous electron diffusion is reduced by the pinch effect. As a anomalous diffusion Kaye-Goldston scaling is used for $D_B$. Some numerical factors are used in the electron thermal conductivity $X_e$ $(=\beta D_{RFCD}/\sqrt{q})$ and in the value of $\tau_E = (\gamma(a_p^2/X_e)^{-1}+\tau_{Ei}^{-1})^{-1}$.

The pinch effect is also expected when we perform the fast wave heating with travelling wave. The rf electric field $E_y$ in the poloidal direction can couple to $B_t$ and $E_y \times B_t$ term can form the inward flux to improve the particle confinement time. The experiment that the density profile is peaking by rf wave of 800 kHz[7] may correspond to this mechanism.

## 4 Discussion and conclusion

It should be noted that even absolute value of $\tau_E$ at $I_p=0$ in Fig.1 coincides with the Bohm time[8]. It is seen that stellerator and tokamks connects smoothly. In eq.(3) $\tau_E$ does not become zero even at $I_p=0$,which corresponds to the stellerator case,whereas the conventional theory of the trapped particle and the drift wave turbulence indictates that $\tau_E=0$ at $I_p=0$ since the variation of loss term itself is only considered in the theory[2]. It is shown using Dawson like treatment that $E_z \times B_\theta$ and $E_y \times B_t$ term do not cancell when the wave is travelling[9]. If the pinch effect discussed here is essential on the plasma confinement,we can say as follows. Flowing current in torus may yield not only the poloidal field to form the rotational transform and magnetic surface but also confine hot plasmas by $\vec{E} \times \vec{B}$ pinch. So,it must be checked whether stellerators and/or other current drive methods such as NBI be able to have such a mechanism to confine hot plasmas.

In conclusion,a simple physical picture teach us that transport is still Bohm type diffusion in additional heating and that flowing plasma current in torus can reduce the cross field diffusion by $\vec{E} \times \vec{B}$ drift. The RF travelling wave putting momentum to plasmas has a potential to improve the deterioration of confnement since the pinch effect is further enhanced against the diffusive bulk plasma.

## References

[1] R.J.Goldston,Plasma Phys.Contr.Fusion 26,87 (1984) and S.M.Kaye,Phys.Fluids 28,2327 (1985)

[2] For example,F.W.Perkins,Proc. 4th Int.Symp.Heating in Toroidal Plasmas,Rome,vol.2,ENAB,p.977 (1984)

[3] O.Naito et al.,Ibid. vol.1 p.159

[4] For example F.Söldner et al., Proc. 12th European Conf. Plasma Phys. Contr. Fusion,Budapest (1985).

[5] JT-60 Team,Ibid. K-I-2,Vol.I,p.563

[6] K.Uehara;J.Phys.Soc.Jpn. 53,2018 (1984)

[7] R.A.Demirkhanov,et al.,JETP Letters 33,28 (1981)

[8] F.F.Chen,Introduction to Plasma physics,Plenum Press,New York,1974,p.170

[9] K.Uehara,J.Phys.Soc.Jpn. 57,4169 (1988)

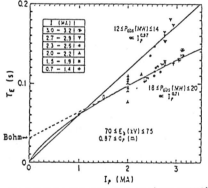

Fig.1 Experimental data of $\tau_E$ vs plasma current $I_p$ in JT-60.

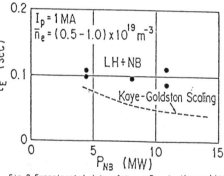

Fig.2 Experimental data of $\tau_E$ vs $P_{NB}$ in the combination of NBI and LHCD case in JT-60.

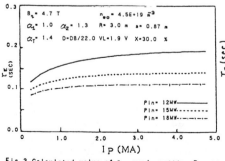

Fig.3 Calculated value of $\tau_E$ vs $I_p$ putting $P_{IN}$ as parameter in JT-60 case,where $\alpha_{n1}=1,\alpha_{n2}=1.3,\alpha_T=1.4$, $\delta=22$,loop voltage $V_L=1.9$ V and $P_{Rx}/P_{IN}=0.3$.

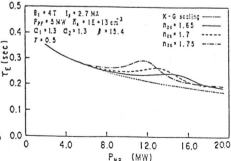

Fig.4 Calculated value of $\tau_E$ vs $P_{NB}$ putting $n_{zc}$ as a parameter in the combination heating of NBI and LHCD in JT-60.

482

# EXCITATION OF WHISTLERS BY CURRENT TO ELECTRODES

J. M. Urrutia and R. L. Stenzel

University of California, Los Angeles, 90024-1547

## ABSTRACT

Current penetration into plasmas is an important topic in fusion studies as well as of general interest. Laboratory studies of currents from(to) pulsed sources(sinks) reveal that the penetration speed is independent of the charge/energy of the emitted(collected) species. Because the current front is a magnetic field perturbation, the propagation velocity is instead determined by the fastest available electromagnetic mode. In our experiments (where $\omega_{ce} \ll \omega_{pe}$), the mode is the whistler branch of the R-wave. Thus, weak beam ($n_b \ll n_e$) particles move ahead of the beam current. In addition, owing to the electron higher mobility, ion currents appear to be initially carried by electrons.

## INTRODUCTION

Discussion of plasma currents often assumes that a steady state has been achieved and, therefore, current is carried by whatever species is collected or emitted.[1] In practical terms, this implies that sufficient time has elapsed for the sheath about the source/sink to become established. On this premise, time-dependent current has been studied in the laboratory[2] and computer simulations.[3] We report here various experiments where the evolution of a plasma current is studied in time and space in a weakly collisional ($\nu_{ei} \ll \omega_{ce}$) magnetoplasma. The currents are induced by switching the bias with respect to the plasma of differently shaped cold electrodes and by employing electron emitters (heated cathodes). Details of these experiments are given elsewhere.[4] Regardless of the method of current drive, we find that current penetration involves the fastest electromagnetic plasma mode excitable in the plasma. Consequently, ion currents appear to be carried by electrons (at least for $t < L/c_s$, where $c_s$ is the ion sound speed and $L$ is a spatial scale lenght) due to electrons coupling with the waves.

## EXPERIMENTAL SETUP

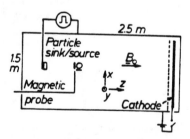

Fig. 1. Schematic of experimental device.

The experimental device (Figure 1) is a large (2.5 m long, 1.5 m diam.) chamber equipped with a 1 m diameter oxide-coated cathode and an external solenoid ($B_0 = 0$-100 G). We choose to perform the experiments in the quiescent afterglow ($t \approx 120\ \mu s$) of the pulsed discharge ($t_{rep} \approx 2$ s, $t_{on} \approx 4$ ms, $n_e \approx 8 \times 10^{11}$ cm$^{-3}$, $kT_e \approx 5kT_i \approx 1.0$ eV ) because it offers a current-free plasma (no inherent sources or sinks). In addition, the diffusion time $(t_d \simeq 1$ ms) is much longer than the time required for the current to propagate across/along the chamber $(t \leq 10~\mu s)$. The cold, non-emitting electrodes consist of discs or spheres made of tantalum or copper. The cathodes range in diameter from 2 to 25 mm. The return electrode for both source or sink is either the chamber wall or one of the cold electrodes. The latter affords a system that "floats" with respect to the plasma potential while the former provides a simpler electrical circuit. Biasing circuits have a typical rise time of 30 ns $(2\pi/\omega_{ci} \simeq 2.6$ ms $> t \simeq 2\pi/\omega_{ce} \simeq 35$ ns at 10 G). The three components of the magnetic field perturbation associated with the currents is measured with magnetic loops capable of motion over all space. The total current density can then be calculated via $\vec{J} = \vec{J}_{cond} + \partial\vec{D}/\partial t = \nabla \times \vec{H}$. The plasma parameters $(n_e, kT_e, \phi_{pl})$ are obtained with suitable Langmuir probes also able of three-dimensional motion. In this way, complete maps of the quantities of interest are available. Such maps, due to their large size, must be handled with computers. In addition, we also monitor plasma fluctuations with rf and microwave antennas as well as employ optical diagnostics to track fast particle motion.

## EXPERIMENTAL RESULTS

We first discuss current collection by a positively biased, non-emitting disc electrode. We observe that an external dc current (Figure 2a, lower trace) does not propagate into the plasma as a single step of average velocity given by the electron temperature $(v_{th,e} \simeq 6 \times 10^7$ cm/s). Measurement of $B_\theta$ (equivalent to obtaining $\mu_o J_z = 1/r~\partial(rB_\theta)/\partial r$ due to symmetry) along a line parallel to the cylindrical axis of symmetry shows, instead, that the front is oscillatory, dispersive, and exhibits various propagation speeds, all lower than $v_{th,e}$. By modulating the current at a fixed frequency (as opposed to launching a single pulse) the oscillatory nature of the current can be established by interferometry. This reveals that the waves posses

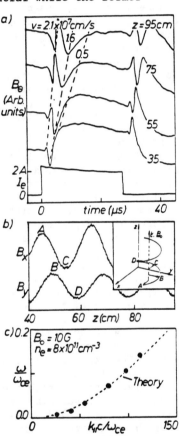

Fig. 2. a) Magnetic field perturbation due to "dc" current. b) Interferometer traces of fundamental oscillation (current pulsed at $\omega/2\pi = 1$ MHz with 50% duty cycle) displaying circular polarization. c) Measured (dots) dispersion relation when current is pulsed at various frequencies compared with plane-wave whistlers.

484

right-hand circular polarization. Variation of the fundamental
frequency yields a dispersion relation in remarkably good agreement
with the theoretical dispersion relation of plane whistler waves,
$c^2 k_{\parallel}^2 / \omega^2 = 1 - (\omega_{pe}^2/\omega^2)(1 - \omega_{ce}/\omega)^{-1}$.[5]

Once the initial front has penetrated the plasma, it is expected that current is constricted to a field-aligned flux tube ($\nu_{ei} \ll$
$\omega_{ce}$). Figure 2 suggests that this may not be correct for our
parameter regime ($\omega_{ce} \ll \omega_{pe}$). It is observed that, as the axial
distance increases, the magnitude of $B_\theta$ (roughly proportional to $J_z$)
decreases, indicating a drop in current density. Thus, axial
current is not conserved along the flux tube. Current must
therefore flow across $\vec{B}_o$ into the flux tube. This may be possible
if a self-consistent electric field
or plasma turbulence is present in
the plasma volume. To date, we have
no convincing evidence that turbulence exists. Electric field measurements, however, are presently under
consideration.

Although not found in the
fusion research field, a most interesting problem is the response of
the plasma to a moving dc-biased
collector, an important problem in
in space-based experiments. Because
of the physical impossibility of
actually moving an electrode in our
chamber, we elect to simulate its
motion by placing the collector at

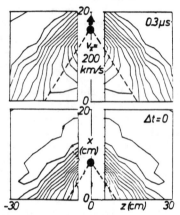

Fig. 3. "Whistler wings" due
to the simulated motion of a
spherical electron collector
across $\vec{B}_o$. Displayed are contours of constant magnetic
field density $B_z$. Similar
results are obtained when ions
are collected.

succesive points along the desired
path ($\Delta x \simeq$ size of the electrode)
and break the dc current into correspondingly delayed pulses ($\Delta t \simeq$
$\simeq \Delta x/v_x$, where $v_x$ is the simulated
speed). The magnetic field pertur-
bation for all space and time is
is then recorded for each step.
The data is then superimposed and maps of the moving perturbation are
available. Figure 3 displays a contour map of $B_z(x,z)$ ($B_z$ is chosen
because it maximizes in the current channel and is associated with
$\vec{E}(t) \times \vec{B}_o$ drifts)[6] for two collector positions. It is observed from
Figure 3 that the current forms a wing-like structure whose angle
with $\vec{B}_o$ is determined by the speed of the collector and the group
velocity of whistler waves. The wings are similar to the predicted
"Alfvén wings" (a consequence of an MHD formulation) caused by the
motion of a metallic structure in space.[7]

The wing structure is also observed if particles are emitted
from the electrode (Figure 4). It then becomes clear that current
in the plasma is not carried by the emitted particles. Instead, the
perturbation appears to be carried by waves coupled to the background electrons. There must be, as suggested by previous experiments[8] involving stationary sources, currents induced in the plasma
to counter the fields associated with the moving particles.

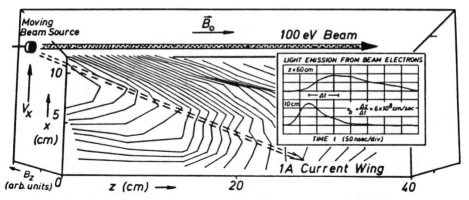

Fig. 4. Three dimensional topographical map of the magnetic field perturbation ("whistler wing") associated with a moving electrode emitting a 100 eV, 1A electron beam parallel to $\vec{B}_0$. The displayed map is taken just before the source (with effective speed $v_x \simeq 200$ km/s, see Figure 3) leaves the measurement volume. Note that the current maximum (dashed arrow) trails the nearly field-aligned trajectory of the fast electron beam (dotted arrow). The presence of the electron beam particles is inferred from time-of-flight techniques employing optical diagnostics (see inset).

## CONCLUSIONS

Our experiments demonstrate that current in plasmas is not necessarily associated with streaming particle motion. In particular, beam energies do not determine current densities. The findings imply that quasi-steady-state theory is incomplete.

## ACKNOWLEDGEMENTS

The research is supported in part by NSF Grants No. PHY87-13829 and No. ATM87-02793, and UC Grant No. UERG W880907.

## REFERENCES

[1] H. M. Mott-Smith and I. Langmuir, Phys. Rev. Lett. 28, 724 (1926).
[2] J. F. Weymouth, J. Appl. Phys. 30, 1404 (1959).
[3] J. E. Borovsky, Phys. Fluids 31, 1704 (1988).
[4] J. M. Urrutia and R. L. Stenzel, Phys. Rev. Lett. 62, 272 (1989); R. L. Stenzel and J. M. Urrutia, Geophys. Res. Lett., May 1989, in press; J. M. Urrutia and R. L. Stenzel, in Proceedings of the Third International Conference on Tethers in Space, 16-19 May 1989, San Francisco, CA, to be published by AAIA.
[5] R. A. Helliwell, Whistlers and Related Phenomena (Stanford Univ. Press, Stanford, 1965).
[6] J. M. Urrutia and R. L. Stenzel, Phys. Rev. Lett. 57, 715 (1986).
[7] S. D. Drell, H. M. Foley, and M. A. Ruderman, J. Geophys. Res. 70, 3131 (1965).
[8] R. L. Stenzel and J. M. Urrutia, Geophys. Res. Lett. 13, 797 (1986).

# BOUNDARY CONDITIONS FOR THE DARWIN MODEL[1]

Harold Weitzner and William S. Lawson

Courant Institute of Mathematical Sciences

New York University, New York, New York 10012

## ABSTRACT

The Darwin model for wave propagation in a plasma is considered in a bounded domain, where boundary conditions affect the wave fields. The consequences of a proposed set of boundary conditions are explored. It is shown that some waves propagating at the speed of light may still be present. The magnitude of these wave fields is shown to be arbitrarily small in an appropriate asymptotic expansion parameter. Such waves could be suppressed by time implicit numerical schemes.

## INTRODUCTION

The Darwin model is an approximate form of Maxwell's equations intended to suppress the effects of waves propagating at the speed of light and to facilitate computation of wave-plasma interactions without the severe restriction of the Courant-Friedrichs-Lewy conditions in explicit numerical schemes. As the Darwin model is usually formulated, no boundary conditions are given since one typically thinks of computations in all space or in periodic domains. In both of these situations the obvious boundary conditions of solutions vanishing at infinity or of periodicity seem perfectly appropriate and physically reasonable and so no problems appear to arise. In bounded domains the boundary conditions are critical and affect the calculated fields. We have proposed elsewhere[1] a reasonable set of boundary conditions and we have given some of their properties. Here we explore further the consequences of these boundary conditions by the solution of an explicit problem for a simple two-dimensional domain. To suppress electromagnetic waves, we conclude that quiet starts in a simulation and time implicit finite differencing schemes in field equations are needed. These conclusions are a consequence of our boundary conditions.

## THE DARWIN MODEL

We start from Maxwell's equations in MKS units with given current and charge sources. We separate the electric field $\mathbf{E}$ into transverse and longitudinal parts $\mathbf{E} = \mathbf{E}_L + \mathbf{E}_T$, where $\nabla \times \mathbf{E}_L = \nabla \cdot \mathbf{E}_T = 0$. In a bounded domain we must give some boundary data on $\mathbf{E}_T$ or $\mathbf{E}_L$, in addition to the usual data on $\mathbf{E}$, in order to define an unique separation. In the displacement current we then replace $\mathbf{E}$ by $\mathbf{E}_L$ to obtain the Darwin model

$$\epsilon_o \nabla \cdot \mathbf{E} = Q = \epsilon_o \nabla \cdot \mathbf{E}_L \tag{1}$$

$$\nabla \cdot \mathbf{B} = 0 \tag{2}$$

$$\nabla \times \mathbf{E} + \frac{\partial \mathbf{B}}{\partial t} = \nabla \times \mathbf{E}_L = 0 \tag{3}$$

---

[1]This work was supported by the U.S. Department of Energy, Grant No. DE-FG02-86ER53223.

and

$$\nabla \times \mathbf{B} = \mu_o J + \frac{1}{c^2} \frac{\partial \mathbf{E}_L}{\partial t} . \tag{4}$$

We assume, of course, charge and current conservation

$$\frac{\partial Q}{\partial t} + \nabla \cdot \mathbf{J} = 0 . \tag{5}$$

We have proposed the boundary condition, that at a point in the boundary with normal vector n̂ at the point

$$\hat{n} \cdot \mathbf{E} = \hat{n} \cdot \mathbf{E}_L . \tag{6}$$

This boundary condition has three desirable consequences. First, the Poynting energy theorem has the natural form in the full domain of interest, $D$,

$$\frac{1}{2} \frac{\partial}{\partial t} \int_D dV (\mathbf{B}^2 + \mathbf{E}_L^2/c^2) + \int_{\partial D} dS \, \hat{n} \cdot \mathbf{E} \times \mathbf{B} + \mu_o \int_D \mathbf{E} \cdot \mathbf{J} \, dV = 0 ,$$

in direct analogy with the original Maxwell system. Second, there is a reasonable form of change conservation on the boundary of $D$

$$\frac{\partial \sigma}{\partial t} + \nabla_S \cdot \mathbf{J}_S + \hat{n} \cdot \mathbf{J} = 0 ,$$

when $\sigma$ is the surface charge density, $\sigma = \hat{n} \cdot \mathbf{E} \equiv \hat{n} \cdot \mathbf{E}_L$, $\nabla_S$ is the surface divergence, $\mathbf{J}_S$ is the surface current, $\mathbf{J}_S = -\mu_o \, \hat{n} \times \mathbf{B}$ and $\hat{n} \cdot \mathbf{J}$ is just the real current flowing into the boundary. Finally, among all curl-free vectors $\mathbf{F}$, the one which minimizes the difference between $\mathbf{F}$ and a specified $\mathbf{E}$, as evaluated in the form

$$I = \int dV (\mathbf{E} - \mathbf{F})^2$$

is exactly $\mathbf{E}_L$ which satisfies the given boundary condition.

We expect that in order to specify a solution of (1)-(4) in a simply connected domain one should give $\mathbf{E}_L$ and $\mathbf{B}$ initially and the tangential component of $\mathbf{E}$ on the boundary as well as the boundary condition (6). In addition, (3) will determine the $\hat{n} \cdot \mathbf{B}$ on the boundary for all time. We introduce a specific electric field $\mathbf{E}_o$ by $\epsilon_o \nabla \cdot \mathbf{E}_o = Q$, $\nabla \times \mathbf{E}_o = 0$ and some boundary condition. We set $\mathbf{E}' = \mathbf{E} - \mathbf{E}_o$, $\mathbf{E}_L' = \mathbf{E}_L - \mathbf{E}_o$ and $\mathbf{J}' = \mathbf{J} + \epsilon_o \frac{\partial \mathbf{E}_o}{\partial t}$. The system reduces to (1)-(6) with $Q = 0$ and with $\mathbf{E}'$, $\mathbf{E}_L'$ and $\mathbf{J}'$ replacing $\mathbf{E}$, $\mathbf{E}_L$ and $\mathbf{J}$. We study this system here and we omit the prime on $\mathbf{E}'$, $\mathbf{E}_L'$, and $\mathbf{J}'$.

## THE TWO-DIMENSIONAL CASE

We consider fields which depend on $x$, $y$, and $t$ only. The Darwin model separates into two uncoupled systems, one for $E_z$, $E_{Lz}$, $B_x$, and $B_y$; the other for $E_x$, $E_y$, $E_{Lx}$, $E_{Ly}$ and $B_z$. Both systems were given in Ref.1 , and we concentrate on the more complex second system. We introduce stream functions by $\mathbf{E} = \hat{z} \times \nabla \psi$ and $\mathbf{E}_L = \hat{z} \times \nabla \psi_L$, and (5) implies there is a function $\chi$ such that $\mathbf{J}' = \hat{z} \times \nabla \chi$ and then $B_z = -\mu_o \chi - \psi_{L,t}/c^2$, while $\psi$ and $\psi_L$ satisfy

$$\Delta \psi_L = 0 = \Delta \psi - \mu_o \chi_{,t} - \psi_{L,tt}/c^2 . \tag{7}$$

The boundary conditions for this system are

$$\psi = \psi_L \quad \text{on the boundary} \tag{8}$$

$$\hat{n} \cdot \nabla \psi \equiv - \hat{z} \cdot \mathbf{n} \times \mathbf{E} \quad \text{given on the boundary} \tag{9}$$

We turn to a very simple explicit case. We consider fields inside a cylinder $r^2 < R^2$. We give $\chi(r, \theta, t)$ inside the cylinder and we specify

$$R \frac{\partial \psi}{\partial r}\Big|_{r=R} = \gamma(\theta, t) . \tag{10}$$

We expand the various functions in Fourier series in $\theta$ in the forms

$$\gamma(\theta, t) = \sum_{n=-\infty}^{\infty} \gamma_n(t) e^{in\theta} \tag{11}$$

$$\psi = \sum_{n=-\infty}^{\infty} b_n(r, t) e^{in\theta} \tag{12}$$

$$\psi_L = \sum_{n=-\infty}^{\infty} a_n(t) r^{|n|} e^{in\theta} \tag{13}$$

$$\mu_o \chi_{,t} = \sum_{n=-\infty}^{\infty} \chi_n(r, t) e^{in\theta} . \tag{14}$$

If we insert the forms (11)-(14) into the differential equation (7) and the boundary conditions (8) and (10), we find that $a_n(t)$ satisfies the equation in time

$$(R/c)^2 a_{n,tt} + 2|n|(|n| + 1) a_n(t)$$

$$= 2(|n| + 1) \left\{ R^{-|n|} \gamma_n(t) - R^{-2|n|} \int_o^R r^{|n|+1} \, dr \, \chi_n(r, t) \right\} \tag{15}$$

and if we define $c_n(t)$ by the relations

$$c_o(t) = a_o(t) - \gamma_o(t)/2 - \int_o^R r dr \, \chi_o(r, t)(\log(R/r) - 1/2) \tag{16}$$

$$2 c_n(t) = (|n|+2) a_n(t) - \gamma_n R^{-|n|} + \{(|n| + 1)/|n|\} \int_o^R r^{|n|+1} \, dr \, \chi_n(r, t)/R^{2|n|} , \quad n \neq 0 \tag{17}$$

then

$$b_o(r, t) = c_o(t) + a_{o,tt} \, r^2/(4c^2) + \int_o^r dr' \, r' \, \chi_o(r', t) \log(r/r') \tag{18}$$

and

$$b_n(r, t) = c_n(t) r^{|n|} - \left\{ r^{|n|} \int_r^R dr' \, \chi_n(r', t) r'/r'^{|n|+1} \right.$$

$$\left. + r^{-|n|} \int_o^r r'^{|n|+1} dr' \, \chi_n(r', t) \right\} /(2|n|) + r^{|n|+2} a_{n,tt}/[4(|n| + 1)c^2] . \quad n \neq 0 \tag{19}$$

The expressions (11)-(19) determine $\mathbf{E}$ and $\mathbf{E}_L$. It is clear from (15) that the principal contributions to $a_n(t)$ for $|n|$ moderately large come from the region near the boundary.

# THE APPEARANCE OF ELECTROMAGNETIC WAVES IN THE DARWIN MODEL WITH BOUNDARY EFFECTS

Throughout we have assumed that the tangential component of **E** on the boundary and the currents are specified. Hence, the right-hand side of (15) is specified, say $A_n(t)$, and $a_n(t)$ is the solution of (15), with $a_n$ and $a_{n,t}$ given initially. The correct time step to advance $a_n(t)$ in time in an explicit scheme would be $R/|n| \gtrsim c\Delta t$, no improvement over the Courant-Friedrichs-Lewy condition $\Delta x \gtrsim c\Delta t$, where $\Delta x$ is the grid spacing. Thus, in explicit calculations we still have a severe time step limitation.

The solution of (15) for $n \neq 0$ is

$$a_n(t) = a_n(0)\cos(t/\tau) + a_n'(0)\tau\sin(t/\tau) + \left[\int_o^t dt' A_n(t')\sin[(t-t')/\tau]\right]/(|n|\tau) \quad (20)$$

where

$$\tau^{-2} = 2(c/R)^2|n|(|n|+1) . \quad (21)$$

The expression (20) shows explicitly the relatively high frequency wave variation associated with electromagnetic waves. If the characteristic time on which $A_n(t)$ varies, say $T$, is much longer than $\tau$, as is typically the case, then we may integrate (20) by parts to obtain the asymptotic expansion in $\tau/T$

$$a_n(t) = a_n(0)\cos(t/\tau) + a_n'(0)\tau\sin(t/\tau) + [A_n(t) - A_n(0)\cos(t/\tau)]/|n|$$

$$- \int_o^t dt' A_n'(t')\cos[(t-t')/\tau]/|n| . \quad (22)$$

We could integrate by parts repeatedly to obtain more terms in this series, but (22) is adequate to make our point. Suppose we consider a problem with a quiet start, specifically $a_n(t)$, $\gamma_n(t)$, and $\chi_n(r,t)$ and many of their time derivatives vanish at $t = 0$. In this case the extension of (22) would show that to some high order in $(\tau/T)$ the solution would vary slowly and on the time scale $T$ and not $\tau$. This asymptotic expansion of the solution can be obtained iteratively from (15) if one assumes the time derivative term to be small. If we compute this solution by an explicit numerical scheme, the time step must be on the order of $\tau$, rather than the larger value of $T$, the time scale of the asymptotic expansion. An implicit scheme may suppress such a short time scale easily, but it is in fact incorrect and omits the small amplitude, high frequency waves which are present.

## REFERENCES

1. H. Weitzner and W.S. Lawson, Boundary Conditions for the Darwin Model, submitted to Phys. Fluids, 1989.

## Author Index

### A

Aikawa, H., 52
Ainsworth, N. R., 14
Airoldi, A., 32
Alcock, M. W., 14
Amagishi, Y., 402
Amin, M. R., 406
Annamraju, R., 130
Annoh, K., 326
Arai, H., 154
Arshad, S., 14
Austin, M. E., 76

### B

Bagdoo, J., 130
Baity, F. W., 214, 250
Balkwill, C., 14
Ballico, M. J., 402
Bartiromo, R., 95
Batchelor, D. B., 218, 282, 378
Becker, H., 346
Becoulet, A., 197
Bernabei, S., 95, 306
Bers, A., 234, 434
Bertschinger, G., 362
Bessenrodt-Weberpals, M., 166
Beuken, J.-M., 362
Bhatnagar, V. P., 205
Blackfield, D. T., 114, 118
Bogen, P., 362
Bonoli, P. T., 68, 114, 118, 126, 134
Bonanos, P., 278
Bornatici, M., 174
Bosia, G., 205
Boyd, D. A., 205
Brambilla, M., 222
Bravenec, R. V., 76
Brower, D. L., 72
Brown, . ` L., 358
Brusati, M., 122
Büchse, R., 95
Bures, M., 205
Burkhart, H.-P., 222
Bush, C., 342
Byers, J. A., 36, 84

### C

Cairns, R. A., 134, 262, 406
Campbell, D. J., 205
Carlson, A., 95, 166
Carter, M. D., 218, 334
Caughman, J. B. O., 226
Cavallo, A., 189, 310
Chan, V. S., 230, 298
Chaudron, G.-A., 130
Chen, J. Y., 76
Chiu, S. C., 230, 298
Cho, S., 474
Chow, C., 234
Christiansen, J. P., 205
Cima, G., 72, 76
Cohen, R. H., 36, 126
Colestock, P., 189, 254, 258, 278, 294, 310, 338, 342, 370
Collins, P. R., 14
Conrads, H., 362
Cook, D. R., 238
Cordey, J. G., 205
Cottrell, G. A., 205
Cox, M., 14
Cross, R. C., 402
Cuthbertson, J. W., 454

### D

DeAngelis, R., 95
Decoste, R., 130
Dellis, A. N., 14
Delvigne, T., 362
Demers, Y., 130
Dendy, R. O., 14, 290
Descamps, P., 362
Devillers, G., 205
Devoto, R. S., 114, 118
Diatchenko, N., 346
Dimonte, G., 418
Dippel, K.-H., 362
Donnelly, I. J., 385, 402, 410
Donohoe, K. G., 466
Dorland, W., 189, 258
Duncan, L. M., 418
Durodie, F., 362

# AIP Conference Proceedings

| | | L.C. Number | ISBN |
|---|---|---|---|
| No. 1 | Feedback and Dynamic Control of Plasmas – 1970 | 70-141596 | 0-88318-100-2 |
| No. 2 | Particles and Fields – 1971 (Rochester) | 71-184662 | 0-88318-101-0 |
| No. 3 | Thermal Expansion – 1971 (Corning) | 72-76970 | 0-88318-102-9 |
| No. 4 | Superconductivity in d- and f-Band Metals (Rochester, 1971) | 74-18879 | 0-88318-103-7 |
| No. 5 | Magnetism and Magnetic Materials – 1971 (2 parts) (Chicago) | 59-2468 | 0-88318-104-5 |
| No. 6 | Particle Physics (Irvine, 1971) | 72-81239 | 0-88318-105-3 |
| No. 7 | Exploring the History of Nuclear Physics – 1972 | 72-81883 | 0-88318-106-1 |
| No. 8 | Experimental Meson Spectroscopy –1972 | 72-88226 | 0-88318-107-X |
| No. 9 | Cyclotrons – 1972 (Vancouver) | 72-92798 | 0-88318-108-8 |
| No. 10 | Magnetism and Magnetic Materials – 1972 | 72-623469 | 0-88318-109-6 |
| No. 11 | Transport Phenomena – 1973 (Brown University Conference) | 73-80682 | 0-88318-110-X |
| No. 12 | Experiments on High Energy Particle Collisions – 1973 (Vanderbilt Conference) | 73-81705 | 0-88318-111–8 |
| No. 13 | $\pi$-$\pi$ Scattering – 1973 (Tallahassee Conference) | 73-81704 | 0-88318-112-6 |
| No. 14 | Particles and Fields – 1973 (APS/DPF Berkeley) | 73-91923 | 0-88318-113-4 |
| No. 15 | High Energy Collisions – 1973 (Stony Brook) | 73-92324 | 0-88318-114-2 |
| No. 16 | Causality and Physical Theories (Wayne State University, 1973) | 73-93420 | 0-88318-115-0 |
| No. 17 | Thermal Expansion – 1973 (Lake of the Ozarks) | 73-94415 | 0-88318-116-9 |
| No. 18 | Magnetism and Magnetic Materials – 1973 (2 parts) (Boston) | 59-2468 | 0-88318-117-7 |
| No. 19 | Physics and the Energy Problem – 1974 (APS Chicago) | 73-94416 | 0-88318-118-5 |
| No. 20 | Tetrahedrally Bonded Amorphous Semiconductors (Yorktown Heights, 1974) | 74-80145 | 0-88318-119-3 |
| No. 21 | Experimental Meson Spectroscopy – 1974 (Boston) | 74-82628 | 0-88318-120-7 |
| No. 22 | Neutrinos – 1974 (Philadelphia) | 74-82413 | 0-88318-121-5 |
| No. 23 | Particles and Fields – 1974 (APS/DPF Williamsburg) | 74-27575 | 0-88318-122-3 |
| No. 24 | Magnetism and Magnetic Materials – 1974 (20th Annual Conference, San Francisco) | 75-2647 | 0-88318-123-1 |
| No. 25 | Efficient Use of Energy (The APS Studies on the Technical Aspects of the More Efficient Use of Energy) | 75-18227 | 0-88318-124-X |